The Tree Experts

A History of Professional Arboriculture in Britain

by Mark Johnston

WINDgather
PRESS

Windgather Press is an imprint of Oxbow Books

Published in the United Kingdom in 2021 by
OXBOW BOOKS
The Old Music Hall, 106–108 Cowley Road, Oxford, OX4 1JE

and in the United States by
OXBOW BOOKS
1950 Lawrence Road, Havertown, PA 19083

Hardback Edition: ISBN 978-1-91118-888-9
Digital Edition: ISBN 978-1-91118-889-6 (epub)

A CIP record for this book is available from the British Library

Printed in Malta by Melita Press
Typeset by Frabjous Books – www.frabjousbooks.com

For a complete list of Windgather titles, please contact:

United Kingdom
OXBOW BOOKS
Telephone (01865) 241249
Email: oxbow@oxbowbooks.com
www.oxbowbooks.com

United States of America
OXBOW BOOKS
Telephone (610) 853-9131, Fax (610) 853-9146
Email: queries@casemateacademic.com
www.casemateacademic.com/oxbow

Oxbow Books is part of the Casemate group

Cover background: Richard Woods's elevated Elysian Walk on the opposite side of the canal at Hare Hall, Essex. Author's image.
Front cover: Sir Henry Steuart's transplanting machine in action in the 1820s moving a large tree in his landscape park at Allanton in Scotland. Author's image.

Contents

List of Figures

The images are from the author's personal library and image collection, unless otherwise credited in the caption, and then the necessary permission has been obtained.

Preface

My interest in the history of arboriculture began in 1982 when working on the tree gang at the Royal Botanic Gardens, Kew. I had recently resigned as Senior Arboricultural Officer at a London Borough and wanted to spend time at Kew while waiting to apply for a suitable lecturing post. I am grateful to John Simmons, the Curator at Kew, for recognising my genuine desire to gain further experience of tree work at the world's most famous botanic garden. It was an inspiring experience, spending every day in an historic landscape shaped by great names in horticulture and arboriculture, and working on trees with their own fascinating history. But I also found something equally inspiring just outside the gardens on Kew Green. In my lunch hour I would often wander over to the Lloyds of Kew bookshop and browse the old and rare books on trees, arboriculture and landscape. I became friendly with Dan Lloyd and he kindly introduced me to some of the great writers, such as Switzer, Loudon, Repton, Robinson, and Elwes and Henry. I made my first purchases of old books with Dan and it was the beginning of my library.

During my career in arboriculture I spent many years as a tree officer, consultant and academic immersed in the challenges of modern urban tree management. By necessity I needed to keep abreast of the latest developments and at times I was fortunate to also be involved in some pioneering initiatives. However, that constant focus on the present and the future gave me little time to indulge in the historical aspects of trees, arboriculture and landscape. Some ten years ago I began to think about my 'official' retirement and how I might continue to contribute in my own time to the arboricultural industry and profession. The weekly commute across the Irish Sea was becoming increasingly unsustainable for many reasons and I was longing to work from home in Ireland on a permanent basis. It was then that I decided to make real use of my library and focus my future research and publications on the historical aspects of arboriculture.

My first history book, *Trees in Towns and Cities: A History of British Urban Arboriculture*, looked at the trees in our urban areas since earliest times. However, this was not just presented as something of 'historical interest'; we must understand how our treed landscapes have developed over the centuries if we are to manage them appropriately today. That theme was also taken up in my next book, *Street Trees in Britain: A History.* This was published when concern about the felling of street trees in cities like Sheffield was attracting national media attention. The response to the book was tremendous, not only from professionals but also from the public. While people are increasingly aware of the environmental value of street trees, they have also become much more conscious of their social and cultural significance as 'living history' in the midst of our modern streetscapes. However, the book which you are now reading is the one I have always wanted to write. *The Tree Experts* is the story of my own industry and profession, from Roman times to the present. Arboriculture has been very good to me for nearly fifty years and I have been fortunate to work with many excellent people along the way. This book is for them.

When this book is published it will mark the end of a four-year journey. From a personal perspective this has been a difficult journey. My research started in June 2017 but within four months I was diagnosed with blood cancer. The hospital consultant said that if I had waited a few more weeks to see my GP the prognosis could have been very different – and this book may never have been written. However, the shock of the illness increased my determination to press ahead with the research. When I was feeling sick from chemotherapy, it gave me a reason to get out of bed and make the most of every day. After months of intensive treatment I went into remission and thankfully I remain that way. Sadly, my elder brother and sister were not so fortunate with this deadly disease. Ian died of cancer in March 2019 and Christine passed away with this in April 2020. Both lived in Dundee and, mercifully, Anne and I were able to be with them near the end. With Christine, we just made it over to Dundee and back before the coronavirus pandemic closed down all travel. For the past year, as a 'clinically extremely vulnerable' person, I have rarely been out of the house. However, that has not been a problem since I have used the time to write up my research and produce this book. And that has been a real 'labour of love' as my extensive research yielded a wealth of fascinating and original material.

Many people have assisted me with this book and those who made a particularly helpful contribution to the research and production aspects are listed in the Acknowledgments. On a personal level there are two people who played a crucial role in enabling me to complete the book. The first is my late sister, Christine Truesdale, a very warm and generous soul who made a substantial contribution to funding my research, particularly in the acquisition of old books, papers and images. The second person is my dear wife, Anne. Her love and support has always meant so much to me but during these past four years it has been amazing in so many ways. I am, indeed, a very fortunate man.

Mark Johnston
Belfast, February 2021

Acknowledgements

Many people have made some contribution, however great or small, to the research and production process for this book and without that assistance it could never have been written. This has included sourcing and supplying information, providing images, provision and assistance with technology, making written submissions and participating in telephone interviews. If there is any incorrect information in the text then I accept full responsibility for this and it is not a reflection on anyone else. The following is a list of individuals who have made a particularly helpful contribution and they are listed in no particular order. I offer my sincere apologies to anyone who believes I have omitted them.

Staff at the Library, National Botanic Gardens, Dublin
Carmel Norris, College Liaison Librarian, James Joyce Library, UCD, Dublin
Eugene Roche, Special Collections, James Joyce Library, UCD, Dublin
Staff at the Information and Business Department, Belfast Central Library
Dianne Heath, photographer, Belfast
Guy De la Bédoyère, historian and author
Linda Farrar, garden historian and author
Dr Sally O'Halloran, Department of Landscape, University of Sheffield
Terence Reeves-Smyth, Chairman, Northern Ireland Heritage Gardens Trust
Tony Kirkham, Head of Arboretum, Gardens and Horticultural Services, Royal Botanic Gardens, Kew
Paul Wood, thestreettree.com
Dave Ford, arboriculturist, instructor and lecturer, York
Peter Bridgeman, former Head of Arboriculture at Merrist Wood Agricultural College and former Chairman of the Arboricultural Association
Alistair Smith, Director at the Association of Tree Officers
Dr Andrew Hirons, Senior Lecturer in Arboriculture, University Centre Myerscough
Caroline Davis, former Arboriculturist at the Department of the Environment and former Chair of The Tree Council
Annette Howard, Information Officer, Royal Institution of Chartered Surveyors
Jenny Milledge, garden historian
Christopher Dingwall, landscape historian and heritage consultant, Perth
Simon Richmond, Senior Technical Officer, Arboricultural Association
The late Bill Matthews, founder of Southern Tree Surgeons and former Chairman of the Arboricultural Association

Prof Paul Elliott, College of Arts, Humanities and Education, University of Derby
Dr Helen Read, Ancient Tree Forum
Jon Stokes, Director of Trees, Science and Research, The Tree Council
Dr Brent Elliott, garden and landscape historian
Dave Dowson, Tree Life Arboricultural Consultancy, Syston
Brynley Andrews, arboricultural consultant, Yeovil
Dr Dealga O'Callaghan, Dealga's Tree Consultancy, Liverpool
Esmond Harris, former Director, Royal Forestry Society of England, Wales and Northern Ireland
Dr Giles Biddle, arboricultural consultant, Wantage, and former Chairman of the Arboricultural Association
The late Henry Girling, former Vice-President of the Arboricultural Association
Dr Glynn Percival, Bartlett Tree Research Laboratory, Reading
Peter Thurman, The Thurman Consultancy, East Sussex
Jago Keen, Keen Consultants and former Chairman of the Arboricultural Association
Lesley Adams, Symbiosis Consulting, Leicestershire
Jo Ryan, arboricultural consultant, York
Debbie Gee, J. Gibbs, agricultural machinery suppliers, Bedfont
Deric Newman, General Manager and Tree Specialist, Glendale Civic Trees
Shane Lanigan, Urban Forestry (Arboricultural Contractors), Bedford
Betina Blake, Fishbourne Roman Palace, Chichester
Molly Wright and Olwen Hawkes, Edwinstowe Historical Society
Reg Harris, Director of Arboriculture, Urban Forestry (Bury St Edmunds) Ltd
Izi Banton, Estates and Conservation Manager, RSPB, Sherwood Forest
Pete Wells, founder, Barcham Trees, Cambridgeshire
Adam Winson, AWA Tree Consultants, Sheffield
Sharon Durdant-Hollamby, Sharon Hosegood Associates, Chelmsford, and Vice-President of the Institute of Chartered Foresters
Nick Beardmore, Principal Consultant, Oakwood Tree Consultants, Findon
Russell Ball, founder and trustee, Fund4Trees
The late Prof Hugh Miller, former Head of the Forestry Department, University of Aberdeen
Roy Finch, Roy Finch Tree Care, Malvern
David Lambert, Director, The Parks Agency
Derek Patch, former Director, Arboricultural Advisory and Information Service
Trevor Preston, former Chief Leisure Services Officer, London Borough of Ealing, and former Chairman of the Arboricultural Association
Dr Felicity Stout, arboriculturist and historian, Sheffield
Kim Gifford, former Chairman of the Arboricultural Association
Zoe Knight, Royal Forestry Society of England, Wales and Northern Ireland
Brian Higginson, arboricultural consultant, Coventry
Bob and Kathleen Hollingsbee, local historians for Dover

Philip Crummy, Director, Colchester Archaeological Trust
Prof Liz Herbert McAvoy, medieval literature specialist, University of Swansea
Dr Jill Francis, early modern garden historian
John Phibbs, Debois Landscape Survey Group, Chalford
Doreen Wilson, garden historian, Northern Ireland
Paul Doyle, Sales Representative, Vermeer UK
Peter Payne, M.D. Payne and Son, tree work contractors, Old Alresford
Maria and Tim McLaughlin, Belfast
Fiona McLaughlin, proofreading services, Belfast

Telephone interviews

The following twenty-three individuals took part in telephone interviews, ranging from thirty minutes to over two hours, as part of the research process. My sincere thanks to all of them as these interviews yielded a wealth of information and opinion, particularly about the development of British arboriculture from 1946 to the present. Their occupations and locations are given in the list above.

Dave Dowson
Peter Thurman
Dr Giles Biddle
Lesley Adams
Tony Kirkham
Caroline Davis
Shane Lanigan
Trevor Preston
Peter Bridgeman
Dave Ford
Dr Glynn Percival
Jo Ryan
Dr Felicity Stout
Russell Ball
Roy Finch
Kim Gifford
Dr Dealga O'Callaghan
Esmond Harris
Henry Girling
Alistair Smith
Pete Wells
Derek Patch
Sharon Durdant-Hollamby

CHAPTER ONE

Introduction

Trees are now in the public eye as never before. The threat of tree diseases, the felling of street trees and the challenge of climate change are just some of the issues that have put trees in the media spotlight. At the same time, the trees in our parks, gardens and streets are a vital resource that can deliver environmental, social and economic benefits that make our towns and cities attractive, green and healthy places. Ever since Roman times when amenity trees were first planted in Britain, caring for those trees has required specialist skills. This is mainly because of the challenges of successfully integrating large trees into the urban environment and the risks involved in working with them, often at height and in close proximity to people, buildings and roads. But who are the people with the specialist expertise to care for our amenity trees? While professionals such as horticulturists, landscape architects, conservationists and foresters have a role to play, it is the arboriculturists who are the 'tree experts'. For centuries arboriculture was often synonymous with forestry or considered an aspect of horticulture, until it emerged in the nineteenth century as a separate discipline. There are now some 22,000 people employed in Britain's arboricultural industry, including practical tree surgeons and arborists, local authority tree officers and arboricultural consultants (Oxford Economics 2018, 19).

This is the first book to chart the history of professional arboriculture in Britain, from the Romans to the present day. Previously, published information on this subject was quite limited and scattered through a small number of books and academic papers. Prior to this book, the most detailed account was the first chapter of my book entitled *Trees in Towns and Cities: A History of British Urban Arboriculture* (Johnston 2015, 1–38). In the United States an excellent book was published some twenty years ago on the development of professional arboriculture in North America (Campana 1999). It was clear that a book devoted to the story of professional arboriculture in Britain was long overdue.

For many people in Britain today the word 'arboriculture' means little or nothing. However, for those who have heard of it, most will probably respond by stating that it has something to do with trees. Others will often say it means tree surgery. This is encouraging, even though it only gives part of the broad scope of the subject. With regard to any precise definition of arboriculture, this book will demonstrate the challenge this presents. The nature and scope of the term has changed over the centuries and a small book could be written on just this topic. Particularly in the last fifty years, this has been constantly shifting and those now working in the arboricultural industry often have different views on this. When I went to Merrist Wood College in 1975 as a naive tree surgeon to begin studying arboriculture, one of the tutors gave our class what was an accepted definition of the subject at that time. This was: arboriculture is the propagation, selection, planting, maintenance and management of woody plants for amenity. Woody plants meant not only trees but also shrubs and many climbing plants. The term amenity was used to embrace the ornamental, leisure and recreational benefits of trees, as opposed to their economic benefits, such as those from forestry plantations and orchards. Many years later, when I was teaching arboriculture and urban forestry at MSc level, I would often give this definition to my students in their introductory lecture just to prompt a debate. As readers of this book will realise, there are now aspects of that definition which are very much open to debate. Another aspect of the traditional scope of arboriculture is the scale of its focus, which has been: the culture of trees singly or in small groups (Wilson 2013, 9). While this was largely true in the past, the appointment of arboriculturists as local authority tree officers in the twentieth century completely changed that. Suddenly, many tree officers found themselves with some degree of responsibility for all the trees and woodlands within their borough, in both public and private ownership. This amounted to many thousands of trees in what would now be called the 'urban forest'. Furthermore, since the 1960s the concept of urban forestry has emerged with its emphasis on a planned, systematic and integrated approach to the management of all the trees and woodlands in and around an urban area. While urban forestry has become an internationally recognised concept and term, the distinction between this and what some people still call urban or municipal arboriculture remains unclear.

Many books on garden and landscape history refer extensively to trees. However, this book differs from most previous studies in some significant respects. Much of the existing literature is concerned with the design of various gardens and landscapes and how trees and shrubs provided the overall framework or many of the prominent features. Nevertheless, there is often little reference to practical arboriculture in terms of how those trees were planted, established, maintained and managed. Where this is mentioned, the account is invariably written by someone with limited practical experience in arboriculture. This book not only focuses on the arboricultural practices necessary to create and sustain those landscapes, it has been written by an arboriculturist with nearly fifty years' experience who has also spent much of the last ten years engaged in historical research. I believe this background was invaluable in evaluating the arboricultural content of countless historical texts on gardening, landscape, forestry and related topics.

Many modern books and academic papers on garden and landscape history have focused on the monarchs and aristocrats who commissioned them or the upper class who were in a privileged position to enjoy them. Fortunately, this is beginning to change, with some excellent recent studies on the role and status of professional gardeners and some accounts of the gardens of the middle and working classes. This book follows those pioneering studies and draws on their findings, particularly in regard to professional gardeners. However, this book focuses on those professionals throughout the centuries who had specialist expertise in arboriculture – the tree experts. The story begins with the Roman *arborator* and concludes with the modern chartered arboriculturist. At this point, some clarification should be given with regard to the meaning of 'professional' as it is used in this book. First, it is used to identify those who are employed to undertake arboricultural work, in contrast to those who are garden owners or skilled amateurs. Secondly, the term 'professional' is used to confer a degree of competence in arboricultural practices, in the sense that the progressive development of professional arboriculture in Britain charts the steady rise in standards of practice. Thirdly, it is used in the context of professional status, for example, in charting the progress of arboriculture from when it was regarded solely as a craft occupation or trade to the point where it is now also regarded as a profession with equal status to architecture, landscape architecture and engineering.

The development of arboriculture in Britain has been influenced greatly by ideas and practices from overseas. Beginning with the Romans who brought arboriculture to Britain, many other countries and cultures, particularly in continental Europe and North America, have introduced their own exciting developments. The extent to which these ideas and practices have been incorporated into British arboriculture is a major theme of the book. There now follows an outline of the contents of this book.

Each of the main chapters follows a similar structure. They begin with a brief outline of the historical period that is covered, which includes reference to some of the major social, economic and political developments. This is to give an immediate historical context to the developments in professional arboriculture that are described in the rest of the chapter. From my experience of teaching arboriculture at various educational levels, it is apparent that many students of whatever age and ability do not have a good knowledge of basic British history. Therefore, this brief reference to general history was considered necessary. As each chapter progresses, more historical context is given which relates to specific events and activities that have had a major influence on British arboriculture. There is a need to place the history of professional arboriculture in its social, political and economic context. Gardens and other amenity landscapes are a fairly reliable barometer of economic activity. These tend to be created and maintained to a satisfactory standard in times of relative prosperity. In times of economic recession there are few resources for amenity landscape creation or renovation and expenditure on maintenance is substantially reduced. This is especially the case with well-treed landscapes that generally require a long-term financial commitment. Furthermore, for a variety of reasons, tree-related work can often be expensive in comparison to other gardening work. This in turn influences the demand throughout history for tree experts and other staff with arboricultural expertise. Just as amenity landscapes reflect the financial climate, they are also an expression of the artistic fashions and social pursuits of the time. Some account of this in the different historical periods is necessary to understand the context for various garden design styles.

The next section of each chapter gives a brief account of the sources of information that were particularly relevant to the research for that historical period. These sources vary considerably as the book progresses. For example, while manuscripts and woodcut images are significant in the early chapters, newspapers and photographs become increasingly useful from the nineteenth century onwards. An account of the specific research methods used in each chapter is also included. Throughout the book primary sources from each era have been used wherever possible, such as contemporary books, documents, manuscripts and images. One difficulty that was found with using some secondary sources on this topic was one of accuracy. For example, accounts of the history of British arboriculture from some American sources are not reliable (Chadwick 1970; Campana 1999; Gerhold and Frank 2002). While these are excellent texts in many respects, their coverage of British arboriculture contains significant errors. Some of these errors have also been repeated in quite recent books on arboriculture and urban forestry that have used these secondary sources. This underlines the value of going directly to primary sources and carefully checking any secondary sources for accuracy and balance.

The arboricultural content of each chapter begins with an account of the era's gardens, parks and other open spaces that contained amenity trees and shrubs. The extent and nature of this varies enormously over the ages, from the very sparse distribution of amenity trees in the Roman era, mainly in the villa gardens of the rich and powerful, through to the great complexity and variety of our urban forests in the twentieth century. While reference is made to the typical garden and landscape styles of different eras, the focus is always on those features that are tree-related. Furthermore, while some mention is made of fruit trees and woodland, especially in the first half of the book, this is mostly where these had an ornamental or amenity role as well as a utilitarian function. The nature and extent of those tree-related features in amenity landscapes has always influenced the number of professionals with arboricultural skills who were required to create, maintain and manage those landscapes. For example, there has been some debate in recent decades about whether the formal Baroque gardens of the late seventeenth century were more labour intensive, and therefore more expensive, to create and maintain than the England Landscape Garden that became fashionable in the mid-eighteenth century. Both these styles of landscape relied heavily on trees for their effect, although the range of arboricultural skills required in each was often quite different.

The historical development of various arboricultural skills is then discussed in some detail. Of course, basic skills such as tree planting, pruning and felling were undertaken in all the historical periods covered by this book. However, the precise techniques used to achieve these tasks have changed over time. For example, tree pruning has evolved from the basic lopping of branches where snags were invariably left, which happened in all ages and still continues, through to flush cutting and the painting of tree wounds, practised widely by tree experts from the sixteenth to the twentieth century. Then, in modern times, we have seen the development of natural target pruning that takes account of research in the 1970s and 1980s into the processes of tree decay and wound healing. Some arboricultural practices have seen significant developments over the centuries, while others have remained largely the same. The extent to which the latest methods of arboricultural practice were employed by tree-related professionals in any particular era was often seen as an indication of their level of professionalism. The 'cowboys' of the day would be rooted firmly in archaic practices, whereas the tree experts would be familiar with the latest innovations and techniques. As well as describing the development of arboricultural practices, a detailed account is given of the various tools, equipment and machinery that have been used in this work through the centuries.

Nurseries that supply trees and shrubs for amenity landscapes have always had a major role in the development of arboriculture in Britain. This includes those nurseries located on private estates that mainly supplied their own gardens, as well as the commercial nurseries that supplied a range of clients from monarchs to local authorities. While this book is not overly concerned with propagation techniques and tree production, there are other reasons for focusing on nurseries. Since the early eighteenth century many commercial tree nurseries have expanded their businesses into other areas such as landscape design and creation, landscape maintenance, large tree transplanting and the supply of tools and equipment.

Since this book is about the tree experts, those professionals who specialise in arboricultural work, it is not surprising that much of the book is taken up with a detailed historical account of their role and status. In the early chapters, most professionals working with amenity trees were employed in private sector establishments, ranging from prestigious royal gardens to the smaller gardens of the gentry. From the late seventeenth century, we see the emergence of commercial companies and consultants that specialised in arboricultural work for wealthy clients, invariably as part of their other horticultural and landscape services. In the early twentieth century the number of these commercial enterprises grew rapidly. From the 1840s, with the emergence of the public park and street tree movements, we see an ever greater number of tree-related professionals employed by local authorities and other public bodies. In the twentieth century this increased dramatically and included the creation of the first specialist tree officer posts in local authorities.

From the early twentieth century, with significant private and public sector developments, we can begin to refer to the arboricultural *industry*. From the late twentieth century, arboriculture became firmly established as a *profession*.

Two closely related initiatives were responsible for that achievement. The first was the development of specialist qualifications and courses in arboriculture. The second was the founding of specialist organisations to provide professional representation for arboriculturists. Considerable space is devoted in the book to charting how this finally came about.

This introduction concludes with a few words concerning what this book is *not* about.

Some subject areas have been omitted because they are not directly relevant to the development of professional arboriculture and others have been omitted because they have been covered adequately elsewhere. There is little coverage of the history of tree introductions or the detailed science behind advances in tree biology, physiology and pathology. Also, the history of tree propagation techniques is not covered in great detail. Where these topics are mentioned, the emphasis is on what has influenced standards in professional arboriculture and its impact on the arboricultural industry and the status of arboriculturists. Furthermore, the book does not dwell much on the lives of individual tree experts as there are some excellent biographical accounts in various books and academic papers. Instead, the focus is on how those individuals contributed to the advancement of professional arboriculture. Finally, I want to stress that this is a history book covering some 2,000 years and due to the limitations of space many recent developments in British arboriculture could not be mentioned. Indeed, an entire book could be written from the research material that was gathered concerning the development of British arboriculture over the past sixty years. In determining what should be included regarding recent times, together with how that should be presented, this was guided by the views of twenty-three well-respected individuals from arboriculture and related fields who were interviewed as part of the research process.

References

Campana, R.J. (1999) *Arboriculture: History and Development in North America.* Michigan State University Press: East Lansing, MI.

Chadwick, L.C. (1970) 3,000 Years of Arboriculture – Past, Present and Future. *Proceedings of the 46th International Shade Tree Conference*, 73–78. International Shade Tree Conference: Urbana, IL.

Gerhold, H.D. and Frank, S.A. (2002) *Our Heritage of Community Trees.* Pennsylvania Urban and Community Forestry Council: Mechanicsburg, PA.

Johnston, M. (2015) *Trees in Towns and Cities: A History of British Urban Arboriculture.* Windgather Press: Oxford.

Oxford Economics (2018) *The Economic Impact of Ornamental Horticulture and Landscaping in the UK. October 2018.* A report for the Ornamental Horticulture Round Table Group. Oxford Economics: London.

Wilson, P. (2013) *A–Z of Tree Terms: A Companion to British Arboriculture.* Ethelburga House: Lyminge.

CHAPTER TWO

The Romans Bring Arboriculture to Britain

It seems likely that tree work in its broadest sense began in Britain with the dawn of the Neolithic era, which in Britain and Ireland ranged from *c.* 4000 to *c.* 2500 BCE. The period was characterised by extensive forest clearance for agriculture and the establishment of permanent settlements which dramatically transformed the landscape. It was forestry skills that emerged first, significantly before any skills associated with tree work in an amenity context.

The invention of stone axes had made it possible to fell timber as well as to graze and burn it down (Noble 2017, 45–68). Along with this ability to fell trees came the development of early forms of woodland management such as coppicing. The term 'coppice' comes from the French word *couper*, meaning to cut, and coppicing is a traditional method of management that exploits the capacity of many tree species to put out new shoots from their stump or roots if cut down. Coppice woodland is normally cut down on a rotation ranging from eight to twenty-five years. New growth then emerges and after a number of years, depending on the type of wood product required, the coppiced trees are harvested and the cycle begins anew. Archaeological evidence from the Somerset Levels and elsewhere has revealed Neolithic trackways across the soft peatland that were made from rods generated by coppicing (Rackham 2006, 62). The earliest of these trackways, at Plumstead in south-east London, has been dated by dendrochronology (tree-ring dating) as being built around 4000 BCE, at the start of the Neolithic period. While some woodland was managed, the large-scale destruction of Britain's forests that began in the Neolithic continued through the Bronze and Iron Ages. When the Romans arrived in Britain much of the original wildwood had already been cleared and vast tracts of the lowland landscape, especially in what is now England, were largely devoid of trees. Great swathes of this had been converted to arable and pasture farmland, which was now inhabited by prosperous, trading Iron Age tribes.

Roman forestry

With the arrival of the Romans in 43 CE forest clearances continued along with some major advances in woodland management and the utilisation of timber and wood products. While there is little direct information available on forestry practice in Roman Britain, inferences can be made from accounts of military campaigns, reports of archaeological excavations, and analogies with other parts of the Roman Empire.

The Roman military campaign to establish their new province of Britannia required considerable volumes of timber, both during the incursions into new territories and then to control these when any local resistance had been overcome (Linnard 2000, 15–16). Their military columns carried some essential timber supplies with them from their base but local timber was used wherever possible. Their network of military roads involved major forest clearances, not only for the actual roadways but also for cleared areas on either side, in order to reduce the possibility of ambush or sniping. Large amounts of wood were used in the construction of forts in newly conquered districts and extensive tree felling in the locality usually supplied most of these requirements.

Substantial amounts of timber would also have been required in the construction industry, to build and repair residential, commercial and industrial properties. On the domestic front, the typical Romano-British villa would have been the centre of a farming enterprise where an expanse of woodland would have been a vital part of the estate. The timber and wood products from the woodland would have been an important resource, providing raw materials for fences and tools, fuel for heating and probably pannage (grazing) for pigs.

One of the reasons for the Roman invasion of Britain was its extensive mineral wealth, and they mined copper, lead, silver, gold, tin and iron (Meiggs 1982, 185). The Roman metal smelting industry was another major consumer of wood as charcoal was used as a fuel in the smelting process. Their iron industry in the Weald has been estimated to have consumed around 6,000 tons of charcoal per year in the period 120–240 CE (Cleere 1981, 299). In practice it is unlikely that the ironworks thrived by destroying much wildwood. Coppice management of nearby woodland made economic sense rather than having to continually go further afield to cut virgin wildwood. The Roman writer Columella (1941, 457–461) describes the coppicing of oak and chestnut and, while he is referring to Italy and Gaul, the same basic practice was transferred to Britain. However, with most Roman writers on agriculture, timber is a minor concern and there is little advice on forest management other than on coppicing (Meiggs 1982, 270).

All this enormous volume of timber production, for military, industrial and domestic use, would have required a veritable army of forestry workers with a variety of practical skills. However, little is known about the composition, training and organisation of that workforce. There is evidence that, to meet its extensive timber requirements, the Roman legions included engineering corps

FIGURE 1. Trajan's Column is a Roman triumphal column completed in 113 CE that survives intact in a public forum in Rome. Some of the reliefs that decorate the column depict scenes of soldiers from the engineering corps engaged in forestry operations.

that were very efficient in forestry operations. Trajan's Column, a triumphal column in Rome completed in 113 CE, depicts scenes of soldiers from the engineering corps engaged in tree felling and harvesting (Figure 1). However, the nature and organisation of the civilian forestry sector in Roman Britain remains something of a mystery (Rackham 2003, 107). The exception to this would have been those villa estates with woodland where there was an obvious need to engage workers with a range of forestry skills in order to ensure the productivity of the woodland. On those estates with substantial tracts of woodland, these forestry workers would probably have formed a permanent part of the estate labour force.

The Roman impact on life and the landscape

The Roman province of Britannia lasted for about four centuries and its enormous impact only began to dissipate with the end of the occupation and the departure of their armies in 410 CE. During this time the Romans radically transformed the social and economic life of much of Britain. Along with new urban centres, new social classes emerged such as merchants and shopkeepers, prompted by greater social mobility (Wacher 1974, 47). As with the rest of the Roman Empire, the economy was dependent on slave labour, which was an integral part of everyday life (De la Bédoyère 2006, 226). Many thousands of Britons were enslaved during the conquest, though we know almost nothing about individual cases. Life for any slave could be hard, but especially for the rural slave who was lumped together with farm animals and tools as just another means of cultivating the land (Varro 1935, 225).

A significant part of the transformation of Britain's landscape under the Romans involved the creation of gardens and other amenity landscapes. While there may have been some gardens in the earlier settlements of the Celtic Britons, these were likely to have focused on vegetable and fruit production and there is no significant evidence of amenity landscapes or ornamental gardens. The Romans created gardens for pleasure and profit in both urban and rural locations, as components of their countryside villas and as part of some residential and public built developments in the towns. The extent of this garden creation in Roman Britain is impossible to gauge but excavations in the provincial town of Pompeii in southern Italy have revealed that 15 per cent of the land area was made up of gardens before it was buried in the volcanic ash of Mount Vesuvius (Floud 2019, 223). There is no reason to believe that the provincial towns of Roman Britain were much different.

The practice of arboriculture presupposes that people care about trees for more than just their timber and wood products. The Romans viewed trees not only for their amenity value but also for their spiritual significance. As a result, trees and shrubs featured to a greater or lesser extent in their gardens, and as well as employing gardeners to create and maintain these gardens, they also needed people with the skills to plant and care for trees. Thus, the arrival of the Romans and their creation of amenity gardens also heralded the beginnings of professional arboriculture in Britain.

Sources of information on Roman gardens and arboriculture

Our knowledge of Roman gardens is based on four sources: literary testimony, archaeological evidence, pictorial records and horticultural traditions (Bowe 2004, 8).

In terms of the literary testimony, we are fortunate that four great Roman manuals on agriculture and horticulture have survived. In chronological order, these works of *De Re Rustica* (translated as *On Agriculture*) are those of Cato (234–149 BCE), Varro (116–27 BCE), Columella (4–70 CE) and Palladius (writing *c.* 380–395 CE). However, these often say little about Roman garden design and gardening and much of our knowledge of that comes from other, scattered, literary sources. The most significant of these are the *Letters* of Pliny the Younger (61–*c.* 113 CE), who gives detailed accounts of the gardens of his own villas. His uncle, Pliny the Elder (23–79 CE), also describes some horticultural practices. In recent years, the works of Cicero (106–43 BCE) in respect to gardens have gained significant attention (Pagán 2016). While those renowned Roman writers on agriculture and horticulture often gave little detail on gardening practices, information on some basic arboricultural practices, such as planting and pruning, can be gauged from their accounts of the maintenance and management of trees and vines in a commercial context. In this respect the works of Columella have been particularly useful to this current research. As well as his *De Re Rustica* being the most comprehensive, systematic and detailed of the Roman agricultural works, he also wrote a separate treatise entitled *Trees* (*De Arboribus*), which focuses on vines, olives and various trees.

Archaeological research has provided much information about Roman garden design and planting. While this began as early as the fifteenth century, it was with the discovery and excavation in the eighteenth century of the ancient cities of Herculaneum and Pompeii that more extensive knowledge of Roman gardens became available. These cities were buried under volcanic ash from the eruption of Mount Vesuvius in 79 CE, resulting in the preservation of a huge amount of archaeological evidence. A considerable amount of information on Roman gardens has also been gained from the excavation of some significant Roman villas in Britain.

Much useful information about Roman gardens can be gained from pictorial representations that have survived on the walls and floors of excavated Roman buildings. These are in the form of frescos, scenes painted on wall surfaces, and mosaics, pictures composed of small pieces of stone and located mainly on floors. However, caution is needed with regard to their accuracy as the artist may have taken artistic licence in the interpretation of the subject (Bowe 2004, 8). Lastly, ancient horticultural traditions, some dating back to Roman times, are still practised in various regions of the world that were part of the Empire. Furthermore, as will be noted below, many of today's gardening tools and sundries are almost the same as those recovered from Roman excavations or depicted in their paintings or mosaics.

While much of our understanding of Roman gardens comes from contemporary authors and excavated sites, this evidence relates mainly to the warmer climes of Rome and surrounding regions and the plants that were cultivated there. How much of that is relevant to the cooler and wetter conditions of the Roman garden in Britain is a matter for some debate among modern garden archaeologists and historians. Nevertheless, on the basis that ideas and practices covering a wide range of disciplines and activities were known to be shared throughout the Roman Empire it seems reasonable to suppose that there were many similarities in terms of basic principles and practices, if not of scale. An analysis of these Roman gardens and the extent and manner in which they used amenity trees and shrubs can give some indication of Roman arboricultural practices.

Roman gardens – design and function

Roman gardens were developed to serve a variety of functions, the most important being the provision of outdoor locations for the enjoyment of leisure and the arts, for the promotion and facilitation of health and exercise, and for the production of fruit and vegetables (Bowe 2004, 13–14). Along with these different functions, two distinct types of Roman gardens can be identified: the enclosed garden, in which the buildings entirely or partially surround the garden, and the open garden, which surrounds a building. It is interesting to note that by 100 BCE a Roman estate and garden was called a *hortus*, with Cato the first known person to use the term. This is the origin of the word 'horticulture', meaning the cultivation of the *hortus*.

Until the first century CE, Roman gardens were simple, informal affairs given over mainly to growing food. Then, the decorative garden became a popular feature of palaces, villas, townhouses and other residential properties. The Romans enjoyed outdoor living and they often furnished their gardens with as much care as they did the rooms of their houses. In addition to private residential gardens, the Romans also created various forms of greenspace in the public sphere (Farrar 1998, 177–186). These included public parks and open spaces, sacred groves, and funerary and temple gardens. With regard to public parks, the archaeological evidence of these is extremely rare. Those that are known are generally called public portico gardens and were enclosed public parks, not unlike the older residential squares now found in London. The first of these was the *Porticus Pompeii* constructed in Rome in 55 BCE by the general Pompey the Great, comprising a *porticus* (colonnade) surrounding a *nemus* (grove) (Gleason 1994, 13). The Roman public considered this open space so delightful and significant in the life of the city that when they referred to many subsequent imitations as *porticus*, they envisioned gardens. By the first century CE, *porticus* were the urban parks of central Rome and similar *porticus* were constructed in other parts of the Empire, almost certainly including Britain.

The Roman writer Propertius describes some of the garden elements of the *Porticus Pompeii*, which included 'the dense avenue of plane trees rising evenly' (Propertius 1990, 229). Other evidence from excavations indicates that the promenade down the central axis was lined on either side with an avenue of plane trees, possibly the same avenue mentioned by Propertius. With the even spacing of the trees this was said to recall the even ranks of troops or the procession of a military triumph (Gleason 1994, 19). There was also a group of trees known as a *nemus*, or sacred grove, with an appropriated dedication to Venus, who protected gardens as well as Pompey in his military conquests. Pompey is known to have taken an interest in trees during his military campaigns and around this time the introduction of new species of tree from across the expanding Empire was the focus of much interest. Pliny the Elder (1968, 81) notes that 'ever since the time of Pompey the Great even trees figured among the captives of our triumphal processions'.

The sacred grove was a feature of many Roman settlements throughout the Empire and from earliest times the public had access to these areas of woodland. The Romans believed that 'the sacred groves of the immortal ones' were the dwelling places of semi-divine beings (Carroll 2018, 13–15). According to inscriptions and literary references, the Romans considered violations of sacred groves as religious offences. For example, a municipal decree from Spoletium in Italy, dating to the period after 241 BCE, prohibited the removal of anything belonging to the sacred grove and the cutting of wood in it, except on the day of an annual festival. In some cases, cutting wood in a sacred grove might even be punishable by death. Garden tombs, called *cepotaphia*, were the most developed form of funerary monument and often included a sizeable plot of land where trees were planted for their shade (Farrar 1998, 177). A number of Roman temples throughout the Empire were sited within a sacred enclosure or portico and evidence increasingly shows that this reserved area may have been used as a sacred grove or garden. Along with sacred groves, the Romans had a close spiritual connection with trees in general. Pliny the Elder (1968, 5) writes that 'different kinds of trees are kept perpetually dedicated to their own divinities' and gives a number of specific examples. He also cites Silvanus as the god of the forests.

Although public open spaces of various types contributed much to the Roman gardening tradition, it has been the residential gardens of the rich and powerful in Roman society that have given us the greatest insights into their approach on garden design and their horticultural and arboricultural practices.

The Roman style of laying out gardens and arranging plantations was usually stiff and formal, regulated by geometrical principles. This applied especially to trees and shrubs but also to flowers, which were displayed in beds and along pathways that were laid out geometrically. Fountains, statues, seats, plant pots and troughs provided ornamentation. In terms of their gardens and estates, the Roman taste was not for natural landscape but for landscapes that had been 'improved' by human hands. Indeed, the artificial enhancement of nature in their gardens was a constant activity demanding much labour. Where trees and shrubs were concerned, this involved arboricultural work. The quincunx was a favourite planting arrangement with the Roman gardeners and is a strictly geometrical form consisting of a central tree surrounded by four trees equidistant from the centre, resembling the five spots on dice. Cicero (1884, 46) speaks of rows of trees arranged in quincunx and states this had previously been done by the Persians when laying out trees in ornamental groups. Most Romans preferred this planting arrangement in their gardens, orchards or vineyards to the straight rows and square arrangement of the Greeks. In the words of Pliny the Elder, 'This was not only advantageous in allowing the passage of air, but also agreeable in appearance, as in whatever direction you look at the plantation, a row of trees stretches out in a straight line' (Pliny the Elder 1950, 55). The quincunx pattern also allowed the maximum density of trees to be planted and thus maximum utilisation of the land. There was also a more esoteric reason for this preference. Virgil (1999, 157) indicates that in the Roman mind the quincunx was closely related to their admiration for military order and discipline.

The most sophisticated Roman gardens boasted an aesthetic unity between a garden's planting and its architecture and sculptural decoration (Bowe 2004, 43). For example, formal garden planting often closely followed the architectural lines of a house or villa. Trees were sometimes planted in lines and trimmed upwards so that their clear trunks would mirror the columnar porticoes.

The selection of plants in the Roman garden was characterised by the use of many evergreen trees and shrubs, such as bay laurel (*Laurus nobilis*), arbutus (*Arbutus unedo*) and box (*Buxus sempervirens*) (Bowe 2004, 43). Conifers of various species were also cultivated as ornamentals and were frequently used as a feature. The yew, however, was not utilised because of its highly poisonous nature. In common with other inhabitants of warm climates, the Romans liked to sit beneath the shade of a tree, but in the case of yew it was thought this could be fatal (Pliny the Elder 1968, 421). Ivy (*Hedera* spp.) was used extensively and was trained on walls and columns and formed 'swags' between trees, columns and architraves in a formal design. A number of deciduous trees, such as Oriental plane (*Platanus orientalis*) and poplar (*Poplus* spp.) and elm (*Ulmus* spp.), were a common choice for gardens. As well as being selected for their horticultural value, trees and other plants were also chosen for their association with one god or another.

The literary evidence of species grown in Roman gardens is complemented by their contemporary depiction in frescos (Farrar 1998, 143–144). Some of these are painted in a rather naive style where the plant details are so simplified that they are barely recognisable. There are, however, several notable frescos of such meticulous draughtsmanship that a positive identification and analysis of species is possible. Of the trees in these frescos, the most common appear to be pine and cypress species. In many fresco panels a large or particularly fine tree was situated in a central position. In the main fresco of the Garden Room at the Villa of Livia, in Primaporta, Rome, painted *c.* 30 BCE, the central tree is a pine, which is highlighted by a recess in the decorative wall (Figure 2). Fruit trees are the main accompaniment, in this case quince and pomegranate. In other parts of this beautiful fresco that decorates the four walls of the dining room, trees such as deciduous oak (*Quercus* spp.), evergreen oak (*Quercus ilex*) and Mediterranean cypress (*Cupressus sempervirens*) are depicted.

The description by Pliny the Younger (1969, 349) of the hippodrome (horse and chariot racing ground) in his Tuscan garden contains one of the most detailed accounts we have of the planting of a Roman garden. The centre of this was quite open so that the full extent of the course could be seen on entering. However, the whole area was 'planted round with ivy-clad plane trees, green with their own leaves above, and below with the ivy which climbs over trunk and branches and links tree to tree as it spreads across them. Box shrubs grow between the plane trees, and outside there is a ring of laurel bushes, which

FIGURE 2. The main fresco of the Garden Room at the Villa of Livia, in Primaporta, Rome, painted *c.* 30 BCE. The central tree is a pine, highlighted by a recess in the decorative wall, accompanied by fruit trees such as quince and pomegranate.

add their shade to that of the planes'. At either end of the long lines of plane trees were shady walks through semicircular groves of cypress (presumably tall, columnar Italian cypresses) and topiaries. The Romans had a particular love of topiary, the clipping of trees and shrubs into ornamental shapes, which will be discussed later in the context of pruning. In terms of garden design, it was a major feature of their gardens and mentioned by many Roman writers, not least Pliny the Younger when describing the garden of his Tuscan villa. His topiary was mainly in the form of 'figures' (representations of human and animal forms) and 'names' (words formed along the ground from clipped box or other plantings). As well as the main formal garden with its geometric arrangement of trees and astonishing array of topiary, there were also lawns with fruit trees.

From Pliny's descriptions of his gardens, especially the one in Tuscany, it is clear they must have taken a lot of specialist tree work to create and maintain. In the Victorian era, the British horticulturist Noel Humphreys (1874, 455) was one of the first to comment on the extent of arboricultural work that must have been involved.

Roman gardens in Britain

While our knowledge of Roman gardening and arboricultural practices in ancient Rome and some other parts of the Empire is now quite extensive, the same cannot be said of Roman Britain. What little we do know of those activities here has been gathered from a meticulous excavation of just a few sites, mainly of prestigious villas which would have been the centre of a farming estate.

The countryside of south-eastern Britain, especially that area to the south and east of the rivers Severn and Trent, is densely scattered with the traces of Romano-British villas (Cunliffe 1981, 99). In excess of 700 have been recorded, of which a reasonable percentage have been excavated to some extent, which has given us some information on the nature and extent of any gardens. No doubt most of them would have been enlivened by kitchen gardens and perhaps small flower gardens and orchards. With those from the earlier period of Roman occupation, only in very rare cases have the physical traces of more formal gardens been recognised. However, in the late third and early fourth centuries there is widespread evidence for the investment of considerable capital in country houses and estates. It is at this time that the 'courtyard villa' becomes popular and it is likely that many of the inner, or residential, courtyards were laid out as formal gardens. Nevertheless, most of our current knowledge relates to just a small number of sites, such as those at Fishbourne, Frocester and Lullingstone.

Fishbourne

The discovery and excavation from 1961 to 1969 of the Roman palace at Fishbourne in West Sussex finally provided proof that the lavish and sophisticated gardens made by the Romans in Italy and many parts of the Empire were also made in Britain (Cunliffe 1971). While it is clearly not typical of Romano-British villas, our knowledge of Roman gardens has been dominated by this one site. Although the earliest buildings at Fishbourne belong to the period immediately following the Roman invasion of 43 CE, extensive additions and alterations were made from the second century (Cunliffe 1998, 25 and 132). It is thought that the architectural, engineering and decorative skills embodied in the Fishbourne palace were developed abroad, quite possibly in eastern Gaul, and introduced into Britain (Cunliffe 1981, 102). This implies that the gardens must also have been created by an immigrant craftsman well versed in the contemporary techniques of garden design. While four different types of garden can be recognised at Fishbourne, in terms of ornamental trees and shrubs our interest here is in the formal garden of the central courtyard. Many features of this garden reflect the formal garden design and gardening techniques associated with the Romans.

Excavations revealed planting trenches on either side of the path in the main courtyard in a pattern of alternate semicircular and rectangular

FIGURE 3. A modern replanting of the low box hedging at the Roman palace at Fishbourne, West Sussex. It is based on the original excavation and is thought to be the first evidence of topiary in Britain.

CREDIT: FISHBOURNE ROMAN PALACE/SUSSEX ARCHAEOLOGICAL SOCIETY.

recesses. These recesses may well have framed garden statuary or ornamental shrubs. The design is strongly reminiscent of the frescos at Herculaneum, except that there the plan is defined by trellis fences and in Fishbourne it is thought that box hedges were used (Bowe 2004, 139). The symmetry of the trench pattern implies symmetrical planting, and it is most likely that it was low box hedging, clipped to a uniform height. It has been speculated that this is the first evidence of topiary in Britain (Thacker 1994, 13) (Figure 3). Evidence of hedging has also been found along subsidiary pathways which run around the edge of the garden in front of the colonnade (Cunliffe 1981, 105). The north-west path was enlivened along both sides with regularly spaced, constricted niches, which are more likely to have been intended to relieve the visual monotony of the straight hedges than to provide settings for shrubs or garden furniture. At the north end of the inner (westernmost) row of hedging, close to the corner of the garden, a large pit filled with topsoil was found against which the bedding trenches stopped. It is likely that the pit was placed here to support a tree, so sited that it formed a pleasing visual stop to the long run of colonnade and hedge, while at the same time masking the somewhat awkward junction between the two wings. It would also have added an attractive emphasis to the end of the west wing colonnade and would presumably have been balanced by another tree at the south-west corner. This possible arrangement would be very early evidence of the use of specimen trees in landscape design in Britain.

Parallel with the east pathway, some 12 ft (3.5 m) to the west of its edge, was a row of stone-packed postholes which, as the central path was approached, became more widely spaced alternating with bedding pits (Cunliffe 1981, 106). Clearly, the eastern side of the garden was more elaborately treated than the north and west, presumably because it formed the backdrop to the principal view from the terraced colonnade in front of the west wing. It seems likely that the row of posts and bedding pits formed an upstanding framework along which ornamental trees were trained, probably cordon fruit trees and similar to an arrangement described by Pliny the Younger (1969, 349) in one of his gardens.

The treatment of the two large areas of garden enclosed by the hedges remains a matter of speculation (Cunliffe 1981, 106). Originally the entire area would have been soil covered, but in spite of reasonably extensive excavation in the northern half only one bedding pit for a tree was discovered. The implication would seem to be that the area was not extensively planted and may well have been grassed.

Frocester

The excavation of the site at Frocester Court in Gloucestershire, which began in 1961 and is still in progress, has provided evidence of a modest sized villa and its garden (Gracie 1970). It dates from the third century CE, quite late in the Roman occupation of Britain, although there were Roman buildings in the immediate vicinity that were built much earlier. The stone-built house had a portico looking out onto a courtyard garden surrounded by walls on two sides and bounded by a fence on its southern edge. The courtyard was nearly an acre in size (0.4 ha) and divided in half by a cobbled roadway that entered through a wide gate in the fence. With regard to the garden and landscaping, five separate planting beds have been discovered lining the roadway that led to a gravel sweep in front of the villa (Cunliffe 1981, 101). Various holes and trenches have been uncovered in the courtyard that were probably planting holes or beds. One long trench ran along the wall to the east, suggesting a hedge. Other, shorter trenches may have been planting holes for trees or shrubs. A line of regularly spaced postholes running across the southern third of the courtyard suggests a fence or trellis for vines or espaliered trees. An orchard may have been planted in the south-western area of the courtyard. The remains of a formally planted grove of trees have also been uncovered at Frocester (Bowe 2004, 139).

Lullingstone

Lullingstone is one of a number of Roman villas that have been discovered in north Kent and is located in the Darent valley about 20 miles from London. Although the excavation of the villa itself has revealed some remarkable finds, our interest here is in its wider estate. As a typical Romano-British villa it was the centre of a farming estate, the extent of which may have changed over time (Wilson 2009, 22–23). English Heritage has commissioned an artist to illustrate a reconstruction of Lullingstone villa in its landscape, as it may have appeared in the late fourth century (Figure 4). While there is no direct evidence for either the gardens or the surrounding fields, the reconstruction used excavations from other sites and the local topography to provide an idea of a prosperous late Roman farming establishment. The resulting image shows a number of features that would require arboricultural or forestry expertise, such as ornamental garden trees, clipped shrubs, an orchard and managed woodland.

FIGURE 4. An artist's impression of Lullingstone Roman villa, West Sussex, in its landscape in the late fourth century. It shows a number of features requiring arboricultural or forestry expertise, such as ornamental garden trees, clipped shrubs, an orchard and managed woodland.

CREDIT: © HISTORIC ENGLAND ARCHIVE.

Other types of Roman gardens in Britain

Apart from the many villa gardens in Roman Britain, other types of gardens were created here that were not associated with residential properties. In truth, our knowledge of these is very limited and derived largely from speculation rather than any direct evidence. While there were areas of public open space in Roman towns in Britain with evidence of hard landscaping, research for this book has found almost no mention of any planting involved and how this might have been arranged. With regard to sites associated with religion, it is known that the Celts used sacred groves called *nemeton* for performing rituals, based on their mythology, which was largely but not exclusively associated with Druidic practice (Koch 2006, 1350). Evidence of such groves has been found in some locations in England, although these are likely to be natural woodland rather than planted. These sites pre-date the Roman conquest but some continued to be active during the occupation. For example, at Kirklinton in Cumbria there is the image of a Romano-British goddess carved into rock that presides over a still pool surrounded by ancient woodland that was once a sacred grove (Edwards 2018, 10).

It is also known that a small number of temples existed in Roman Britain, the largest being the Temple of Claudius at Camulodunum (Colchester).

As already noted, some temples in other parts of the Empire had gardens associated with them. It is tantalising to speculate whether there was a garden or any trees connected with this Temple of Claudius, especially as archaeologists investigating another temple by this name in Rome identified what appear to be L-shaped planting beds that may have held trees or shrubs (Crummy pers. comm., 2020). However, no similar evidence has been found here.

Before concluding this section on Roman gardens in Britain, it should be mentioned that a number of tree and shrub species arrived on British shores at this time. These plant introductions not only enriched the gardens of Britannia but many also remain popular today. These are thought to include box (*Buxus sempervirens*), sweet chestnut (*Castanea sitiva*) and stone pine (*Pinus pinea*), although there is still much debate around this overall topic.

Arboricultural and horticultural practices

Having examined the Roman approach to garden design, with the focus on trees and shrubs, we now consider the various arboricultural and horticultural practices that were necessary to achieve this. Archaeological evidence for gardening practices outside of Rome is still meagre in many parts of Italy and in the provinces. However, the evidence we have to date indicates that these were generally similar throughout the Roman Empire (Jashemski 2017, 454). When Romans went to the provinces they tended to apply gardening practices they had used at home, but, as in all other aspects of their lives, they were also quick to employ useful new techniques encountered through their travels and trade. Some arboricultural practices can be seen in a number of frescos and mosaics. While a similar level of caution that applies to species identification needs to be used with these pictorial sources, where the same practices are depicted frequently and in a range of locations, this gives greater confidence regarding their authenticity.

Most of our evidence in this field has been gleaned from literary sources written by contemporary Roman writers. Despite these being written as long as two millennia ago, many of their practices would be recognised today. John Claudius Loudon, the great nineteenth-century authority on gardening, had read most of the relevant Roman texts and stated, 'The practical directions for garden culture given by the Roman authors are in general excellent' (Loudon 1835, 27). At the same time, he was also aware that their practices were often governed by 'superstitious observance dictated by a religion founded on polytheism. Almost every operation had its god, who was to be invoked or propitiated on all occasions'. In the following account of some Roman arboricultural and horticultural practices, reference will also be made to superstitious beliefs.

Planting

We are fortunate that the conscientious Romans have left us much written information on how they planted trees. Columella suggested that the planting holes should be made a year in advance, or at the very least two months; in the latter case, straw should be burned in the holes to make the soil friable (Columella 1968a, 387). He added, 'The planting holes ought to resemble an oven, being wider at the bottom than the top, in order that the roots may spread more widely and less cold in winter and less heat in summer may enter through the narrow aperture'. He also suggested that, before transplanting, trees should be marked with ochre on one side, so that they could be replanted in the same relation to the sun and prevailing wind (*ibid.*, 385). Virgil also made this point after watching workers take trees from the nursery and plant them in the fields. He observed, 'They print on the bark of trees the quarter of the sky each faced, so as to restore the position in which they stood, the same side bearing the southern heat and the same back turned to the North Pole' (Virgil 1999, 155).

With regard to pruning the tops of young trees immediately before or after planting, there is some conflicting advice. However, Columella (1968a, 383) is clear on this when he states, 'Do not touch it with a knife for three years' after planting. Cato (1935, 45) and others stress the importance of preserving the root ball and suggest placing it in a basket to transfer it to its new site. Where the site is damp, Pliny the Elder (1950, 57) endorses Cato's advice that stones should be placed in the bottom of the planting hole to assist drainage. He also stresses the importance of keeping the roots in good condition, reminding us that if there is any delay in transplanting, the roots should never be left exposed to the elements as this will cause the tree to die soon after planting (*ibid.*, 59). Columella (1968a, 385) is one of a number of writers to recognise the need to give some form of support and protection to newly planted trees. A few frescos have also been discovered that illustrate how gardeners used stakes to protect, support and train various plants (Farrar 1998, 171).

Archaeological excavations have discovered the use of broken amphorae as planters for trees and shrubs (Jashemski *et al.* 1992, 582–583). Amphorae were tall pots used for the transport and storage of various products, both liquid and dry. The discovery of broken amphorae re-used as planting pots at various sites makes it apparent that such containers often had a 'second life' in Roman gardens. The amphorae had been cut in half and turned upside down, with the open neck serving as the drainage hole. There are indications from excavations that these planters were generally used where the existing soil was poor.

For the Romans, decisions about when to plant trees did not just depend on the weather and site conditions, they were also regulated by more esoteric factors such as the phases of the moon. For example, Columella (1968a, 349) states, 'You should make your plantations from the new moon to the tenth day and from the twentieth to the thirtieth day. The latter is the better time for planting; but when you are planting, avoid the cold winds'. Columella also has some other unusual advice regarding planting: 'Before you set a nut-tree, you should soak it in honey-water which should not be too sweet; it will then, when it comes to maturity, produce fruit of a pleasanter flavour, and meanwhile it will grow better and quicker' (Columella 1968a, 393).

There is evidence that the Romans undertook the transplanting of large trees, often moving them considerable distances. Of course, they were not the first ancient civilisation to get involved in this advanced arboricultural practice. Some 3,500 years ago, Hapshedsut, one of the most successful female pharaohs of Egypt, had ordered an expedition to the land of Punt (possibly modern Somalia) to acquire frankincense and myrrh trees (Hepper 1967; Dixon 1969). When her ships returned after a two-year voyage, among the treasure on board were thirty-one healthy frankincense trees. These were then planted in the courts of her Deir el Bahari mortuary temple. Some of the arboricultural techniques involved in this remarkable Egyptian enterprise were almost certainly passed on to later civilisations, including the Romans. No operation of a similar scale is recorded in the Roman literature, but the art of transplanting trees does receive attention and seems to have been carried out with careful regard to physiological principles. For example, the principle involved in moving young trees from the nursery to the same orientation in the field was applied to removing large trees from one site to another. Unlike their treatment of young trees, the Romans generally believed that large trees should have their top growth heavily cut back before being transplanted to another location. On a related matter, Pliny the Elder (1968, 473–474) comments on the practice of re-establishing large trees that have blown over. He notes it as a common occurrence, especially with plane trees. When these trees had blown over their branches were then pruned back drastically before the tree was hauled upright again in its original position.

Pruning

Roman writers and gardeners were much preoccupied with the subject of tree and shrub pruning. There are many references to this in their great works on agriculture and horticulture, not only from a commercial perspective with the cultivation of fruit trees and vines, but also in the context of amenity trees in gardens. The results of pruning can also be seen in some of their frescos and mosaics.

Roman writers generally agreed that trees and shrubs were improved by pruning out unwanted or less productive branches. They also liked to prune off the lower branches of trees for a variety of reasons. As noted above, this was sometimes done for aesthetic effect, where the clear trunks would mirror the columnar porticoes to follow the architectural lines of a building. In other cases this had more practical reasons. Many frescos show trees in a less formal setting that have few branches at a low level (Farrar 1998, 171). This is partly because the leaves of many trees were cut to provide fodder for a variety of animals but also because a browsing animal could reach up to help itself.

With regard to the Romans engaging in major pruning on large mature trees, there are some interesting instances of this from a variety of sources. There is evidence from both pictorial and written sources that the Romans pollarded trees, especially planes, but nothing conclusive from Britain. There is also archaeological evidence from an internal courtyard at Oplontis which indicates that major pruning was carried out on a huge chestnut tree (*Castanea* spp.) (Jashemski 1993, 293–294). The tree had been placed at the centre of the small courtyard and would have completely shaded the garden and portico. Considering the size of the tree's trunk, established through the excavation, the tree's natural crown spread and the constricted nature of its location, it would have required some regular pruning to keep this in check.

Another fascinating example of major tree pruning relates to a tree growing in a sacred grove. The Romans believed that trees in sacred groves had magical or prophetic properties and whenever one of these trees declined rapidly in health or actually died it was thought to be a portent (Carroll 2018, 17). Pliny the Elder (1968, 475) records the growth of an elm in the sacred grove of Juno at Nocera in south-west Italy and an incident that related to the fortunes of the Roman Empire. The crown of the tree had to be severely cut back because the trunk had started to lean drastically towards the altar and it was feared the tree would crash down on it. This intensive and potentially life-limiting tree surgery coincided with an invasion of the Germanic Cimbri into northern Italy in the late second century BCE. However, the elm soon recovered from its heavy pruning and flowered again, by which time the Cimbri were then in retreat.

There is not yet any evidence of major tree pruning being undertaken in Roman gardens in Britain, although this certainly would have happened from time to time. However, for various reasons this was likely to have been relatively unusual. From what little we know about Roman gardens in Britain, it seems likely that there were just a few scattered ornamentals in the larger gardens and nothing particularly large in size or spread. It should be remembered that the creation of shade was the first task of a Roman gardener in the hot Mediterranean climate (Bowe 2004, 44). The most effective and permanent way of doing this was through planting trees, and those trees in close proximity to buildings would have required pruning to maintain their appearance and for safety reasons. However, in the cool and often damp climate of Britain the provision of shade was not such a concern and therefore there was not the same demand for garden trees.

Many writers emphasise the importance of using appropriate and sharp tools for pruning and employing the correct pruning technique. While this often refers to vines, the same basic principles also apply to tree pruning. Columella (1968a, 367) recommends that after pruning a branch 'with a sharp instrument, smooth off the cut'. He adds, 'For a knife that is blunt and dull and soft delays the pruner, and for that reason he accomplishes less work' (Columella 1941, 425). He explains that 'ragged and uneven wounds' will tear the branches and damage the vine or tree. Broken branches should be cut off because if this is not done then the vine or tree 'is rotted with moisture and its wounds do not heal'. Cato (1935, 59) recommends that pruning wounds and grafts should be daubed with a mixture of clay or chalk, a little sand and cattle dung, to protect against harmful agencies. However, many Roman writers do not mention using any form of wound sealant.

Superstitious beliefs also pervade pruning practice. An example of this relates to a novel method of pest control to prevent damage to vines caused by a 'leaf-roller'. Columella declares:

> To prevent this, when the pruning is finished, smear the pruning-knifes with which you have pruned the vineyard with bear's blood; or, if you have a beaver's skin, during the actual pruning, whenever you sharpen your pruning-knife, wipe the edge of it on this skin, and then recommence your pruning.
>
> Columella 1968a, 379

This highlights the Romans' desire to ensure the wellbeing of their plants and brings to mind another interesting practice related to plant health, although not concerned with pruning. Columella (1941, 197) records the widely held belief that olive or wine lees (yeast deposits) were considered a tonic for ailing trees. This was not simply a superstitious belief, as modern science has shown that the minerals these substances contain would in fact be beneficial to many plants. One example of this practice is from a small Roman mosaic in the Vatican that shows a man, who has perhaps just returned from a feast, emptying his wineglass over a newly planted tree or shrub (Farrar 1998, 172).

As with tree planting, the timing of any tree pruning or felling could be governed by the phases of the moon. Columella (1968a, 75–77) states that mid-March is 'a fit time for making vine-props or even stakes, and it is also equally suitable for cutting down trees for buildings; both these operations are better carried out when the moon is waning … since all wood so cut is considered not to be attacked by decay'.

Roman topiary and 'bonsai'

Topiary has already been mentioned in the design of Pliny the Younger's gardens and at Fishbourne, but is discussed here in more detail as a distinct form of pruning. The Romans had a love of topiary and miniaturised forms of trees and shrubs, which were a major feature of their gardens throughout the Empire, including the province of Britannia. According to Pliny the Elder, writing around the time the garden at Fishbourne was created, its origins are well known. He claimed that *nemora tonsilia* (clipped arbours or topiary) was 'invented within the last 80 years [*c.* 5 BCE] by a member of the Equestrian order named Gaius Matius, a friend of his late Majesty Augustus' (Pliny the Elder 1968, 11).

The range of trees and shrubs available to a gardener in ancient Rome was limited, by today's standards (Bowe 2004, 51). Therefore, creative topiaries added to the variety of interesting forms and were especially attractive in a small garden as they allowed the owner to have trees, such as the cypress, normally associated with a larger garden but in a trimmed, miniaturised form.

Box was the main subject for topiary in Roman times, as it still is today. Pliny the Elder (1968, 433) noted that it was 'esteemed for a certain toughness and hardness and for its pale colour, while the tree itself is valued in ornamental gardening (*topiario opere*)'. As well as forming artistic and playful shapes, box was also used as hedging, often trimmed in one low level but sometimes in tiers. It is thought that the Romans introduced box to Britain specifically to be used in topiary and low hedging, as at Fishbourne. Our native yew, much used in topiary today, was excluded from gardens throughout the Empire because it was poisonous. Pliny the Elder (*ibid.*, 421) claims, 'its poison is so active a poison that people who go to sleep or picnic beneath a yew-tree die'.

Cypress was also frequently used specifically for taller hedges that formed wall-like divisions within the garden. In other instances, according to Pliny the Elder (1968, 479), cypress was 'evenly rounded off with trim slenderness' to make those tall, columnar forms that have characterised the Tuscan landscape ever since. In both gardens and in the wider countryside across the Empire it was often closely planted in rows to give shelter from the wind or to create shade in warmer climates.

In addition to their love of topiary, some Romans also had a passion for miniaturising various species of trees. Pliny the Elder (1968, 11) comments on dwarf varieties of plane and other trees where the effect was obtained through a method of branch and root pruning. Although a popular feature in many Roman gardens, Pliny himself was not an advocate of this practice. He complains, 'For there is also the variety called the ground-plane, stunted in height – since we have discovered the art of producing abortions in trees, and consequently even in the tree class we shall have to speak of the unhappy subject of dwarfs'.

Tree propagation and fruit trees

Much of the Romans' focus on trees concerned fruit trees and orchard production. They even had a specific goddess, Pomona, who watched over fruiting trees and orchards. Furthermore, they also had a minor goddess of agriculture, Puta, who presided over the pruning of trees (Janson 1996, 16). While the cultivation of fruit is an aspect of commercial horticulture, the techniques that were applied to fruit trees, such as planting, pruning and grafting, can tell us much about arboricultural practice at that time.

The subject of grafting as a means of propagation and improvement with fruit trees received much attention from the classical Roman authors. It was widely recognised in Roman times, and in some previous civilisations, that a scion and stock from two different trees that were not closely related could not be grafted together successfully. Varro (1935, 269) addressed this issue directly when he stated, 'You cannot, for instance, graft a pear on an oak, even though you can on an apple'. With our modern knowledge of tree physiology we understand the limits of graft compatibility. However, with some writers in Roman times and even much later this was not the case. For example, Columella thought that there was no limit to the types of trees that could be grafted together. In an example of knowledge going backwards, even in ancient times, he famously declared:

> Since the ancients denied that any kind of scion can be grafted on any kind of tree and established that limitation, which we quoted just now as a hard and fast rule, namely, that only those scions can unite which resemble the tree in which they are inserted, in their bark and rind and fruit, we thought it advisable to disprove the error of this opinion and to hand down to posterity a method by which any kind of scion can be grafted upon any kind of tree.
>
> Columella 1968a, 403

Columella then describes a standard method of grafting at that time. However, his statement about compatibility raises questions about the extent of his own practical experience of grafting and it is likely he was just repeating what some others writers had said. Despite errors such as this, the Romans were generally advanced in their knowledge of pomology and propagation. They grafted trees to improve fruiting and they also experimented to produce different flavoured fruits, for example, by grafting plums onto apples and peaches onto plums. Strains of good fruiting trees were grafted on to others as they believed that 'a tree which is engrafted is more fruitful than one which is not' (Columella 1968a, 389). We even hear of two or more fruit branches grafted onto one tree. The Romans employed a wide range of grafting methods, notably approach grafting, cleft grafting, root grafting and budding, and are usually credited with introducing the art to Northern Europe.

This knowledge of tree propagation gained from their study of pomology was transferred over to ornamental trees to the benefit of the garden. Indeed, fruit trees were often included in their ornamental gardens, not just as individuals or in groups as small orchards but also trained along walls as espaliers. Supporting evidence for espalier-trained fruit trees has been found at Fishbourne where the trees were trained to grow against a row of posts on one side of the garden (Farrar 1998, 169).

Tools and equipment

That most basic tree work tool, the axe, dates back to Neolithic times in Britain. The oldest axe-heads were chipped flints wedged into wooden hafts, although under the force of the blows such a tool would tend to split its handle (Edlin 1949, 5). To reduce this likelihood, those early foresters made their hafts from knotty branches, or from timbers such as elm, which do not cleave readily. In the later Neolithic, axe-heads were sometimes laboriously bored with a drill, so that the handle could be set in the pierced hole, as done with metal heads today. Brittle stone axe-heads were later succeeded by bronze as the cutting medium. From the start of the Iron Age in Britain, *c.* 800 BCE, axe-heads were made from iron and usually cast with a hole to take the haft that was secured to the head with a wooden wedge. Serrated flint saws had also been used in Britain since at least the Neolithic period and these were sometimes mounted in grooved wooden handles (Jones and Simon 1961, 12). These were copied in bronze and iron in the Bronze and Iron Ages.

The arrival of the Romans in Britain not only brought more advanced forest management but also quite sophisticated forestry equipment and technology in comparison to that previously used by the native tribes. Major advances credited to the Romans included improved tooth setting for saws and the introduction of the framesaw, although this was not invented by them (Jones and Simon 1961, 18–20). They also used long two-handled saws for the conversion of felled timber, initially developed by the Etruscans. Records also exist of crosscut saws in use during the Roman Empire, although not widely and not specifically in Britain (Miller 2003, 2). It is interesting that no mention could be found in the Roman literature of saws of any description being used for tree felling. It seems this was always done by axe.

Other forestry tools used by the Romans included the *Securis*, the general term for the implement which corresponds to our axe and hatchet (White 1967, 60–61). This had various shapes and uses but was frequently referred to in Roman writing as the woodman's axe. Those used for felling trees may be easily distinguished from the short-handled types used for chopping small timber by differences in the width and convexity of the cutting edge. The woodman's axe was both wider and more convex than that used by the carpenter and the butcher. Trajan's Column in Rome, with its numerous tree-felling scenes, provides six examples of soldiers felling trees with the wide-bladed *securis* and two with the shorter hatchet (Figure 1). The *dolabra* or hatchet was another tool much used in forestry and the long-handled form of this was ideal for grubbing out stumps and surface roots (*ibid.*, 63).

The creation and maintenance of Roman gardens in Britain required a wide range of horticultural tools, many of which had some application in practical arboriculture (Figure 5). Cato (1935, 23) lists a number of tools for the vineyard which could also be used within the *hortus*, and some other authors, such as Columella and Palladius, have similar lists.

With regard to pruning tools, the Romans used a wide range, although the different names often given to these in contemporary and modern literature can be confusing. *Falces silvaticas* were sharp tools like billhooks with a curved blade that could cut down the branches of trees (Farrar 1998, 162). The *falx arborarias* was a smaller version, also with a long curved blade, that was used when pruning small diameter branches on trees and shrubs in the garden. They were also used for light pruning in the orchard and in Cato's system of viticulture they would be in demand for pruning and shaping the supporting trees in the method of vine growing known as *arbustrum* (Cato 1935, 47–49).

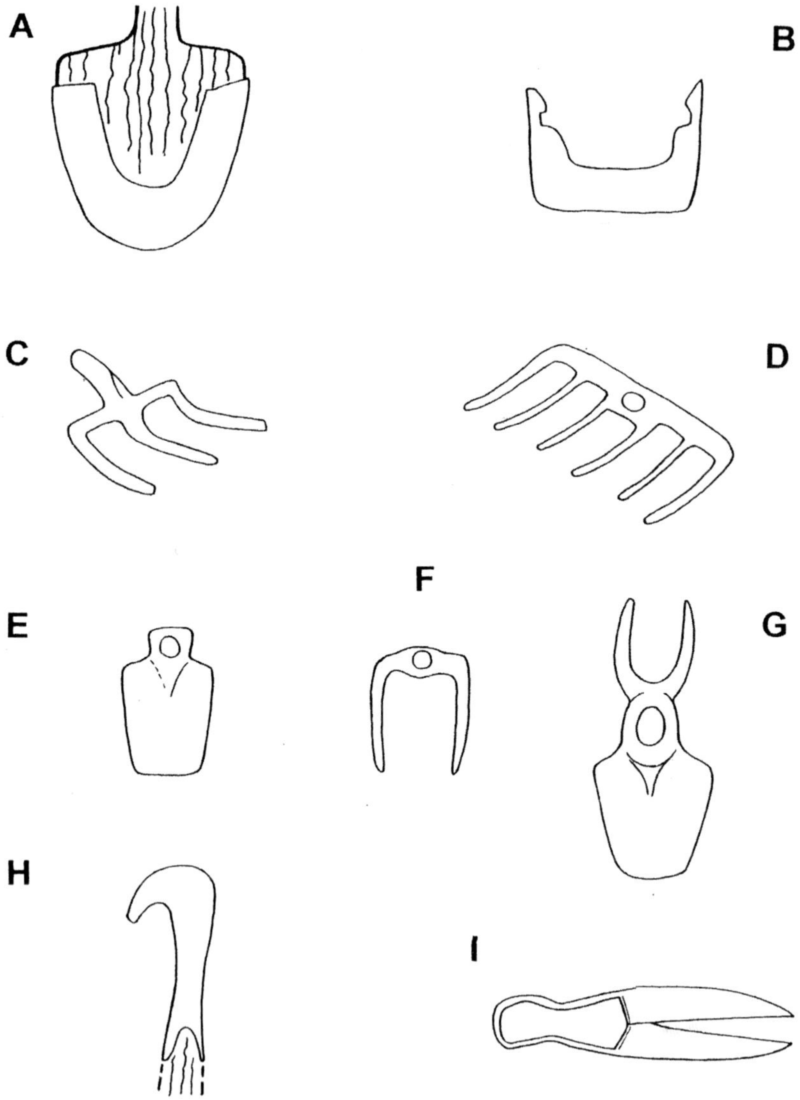

FIGURE 5. A selection of Roman horticultural tools including some with an arboricultural application. A: *Pala lignea*, a wooden spade, shown here with a curved iron shoe. B: *Palas*, iron spade shoe. C: *Ferrea*, iron three-pronged fork. D: *Rastrum*, six-pronged rake. E: *Sarculum*, hoe. F: *Bidens*, two-pronged hoe. G: *Ascia-rastrum*, double-headed hoe. H: *Falx arboraria*, pruning hook. I: *Forfex*, shears.
CREDIT: LINDA FARRAR.

For the more delicate work in the vineyard there was the vine dresser's hook or knife (*falx vinitoria*) which Columella (1941, 425–426) describes in detail. This was derived from the billhook and was designed to deal with several different tasks. Besides the billhook's long sharp edge and recurved 'beak', it had a fine edge for paring and smoothing (*scalprum*), a pointed projection for hollowing and gouging (*mucro*), and a miniature axe blade at the back of the main blade for making straight, defined cuts as in grafting (*securis*). There was a distinct technique required in the use of various forms of *falx*, designed to avoid loss of sap and damage to the growing tree, which is described in detail by Columella (1941, 427). He also gives his own version of the lopping and pruning operations required in order to mould the supporting tree in vineyards to the desired shape and size (Columella 1968b, 51–55). Palladius (1873, 42) mentions the *lupus* as a special type of small pruning saw designed for working in a confined space (White 1967, 6). Various knives were also used for more delicate pruning. A common type was the *cultellus*, a knife of the simple 'machete' or 'panga' type, consisting of a long, slightly curved blade but lacking the typical hook of the *falx arboraria* (*ibid.*, 71).

The Romans developed a range of shears or scissors, known as *forfex*, for use in agricultural and horticultural work (White 1967, 119). Pivoted implements of this type have been known from about 300 BCE. However, the improved form of this implement used by the Romans, the true scissors operated independently by two fingers, date only from the first century BCE. Shears and scissors were used in the vineyard as grape cutters, and there are references to this in the Roman literature. In the garden they were used extensively for clipping topiary and hedges and for other light pruning work.

The Romans had various types of spade, some of which were used for tree planting (White 1967, 17–28). The *pala* was normally used for turning over light or well-worked soil in gardens and orchards, although in hard, stony ground this was done with various types of mattock or *bidens*. The *pala lignea* was usually made of wood and with a curved iron shoe fitted to the lower half of the blade to get maximum wear and a good cutting edge. This spade was often used for planting trees and shrubs. The *bipalium* (foot-rest spade) was essentially a trenching implement. As Cato informs us (1935, 137), the implement was also regularly used for the dual task of digging over the soil for planting, and incorporating into the top spit a layer of manure previously spread over the surface. Varro (1935, 263) recommends this spade for most tree planting.

While there is scant evidence of major tree work being undertaken in Roman Britain, gardeners would have had access to some of the necessary equipment, if only that borrowed from elsewhere. For access into smaller trees they could have used fruit-picking ladders from the orchard; for larger trees ropes were available for climbing and also for lowering heavy branches. For tree felling, axes and similar tools were usually located elsewhere on the villa estate and available from those who managed the woodlands. Where trees needed to be pulled over or large stumps pulled out, this physically demanding work could be done by oxen, the main beasts of burden on the Roman farm.

Having examined the some of the arboricultural and horticultural operations undertaken by the Romans, and some of the tools and equipment used to achieve this, we now consider the people who were engaged to carry out this work.

The *arborator* and professional gardeners

As the Roman Empire expanded, the more affluent property owners in many regions were able to possess or hire servants or slaves to perform various tasks on their estates. This included workers specially trained to work in garden areas. Many had specialist roles in different aspects of horticulture and arboriculture and had job titles named after these roles. In many cases these titles can be gained from the epigraphic evidence on funerary inscriptions found in the *Corpus Inscriptionum Latinarum* (*CIL*) (Farrar 1998, 160). This is a comprehensive collection of ancient Latin inscriptions from the whole territory of the Roman Empire based at the Berlin-Brandenburg Academy of Sciences and Humanities.

There does not appear to have been an all-embracing term for a gardener, although there are some references in the *CIL* to a *hortulanus* (Farrar 1998, 161). However, these are thought to be later inscriptions from the second century CE and may refer to workers in market gardens as opposed to ornamental gardens. A *vinitor* was somebody who cared for vine plants, both in a vineyard and on garden pergolas covered by vines. The *olitorius* took care of the olive trees and undertook various tasks in the olive groves. A *topiaruis* was originally employed to clip hedges, *nemora tonsilia*, and their skilful work in shaping trees, shrubs and hedges was referred to as *opus topiarii*; this led to our word topiary. The name *topiaruis* also appears when referring to a landscape gardener or someone who designed gardens. There are a surprising number of surviving tombstones recording *topiarii* and this could indicate that they were held in high regard and had a certain status as craftworkers. A further indication of the high status of the *topiarius* was that they were sometimes allowed to clip out their name in box topiary in bas-relief alongside that of his master (Pliny the Younger 1969, 349).

Of major significance to the theme of this book, the Romans also had a category of garden worker that was specifically concerned with arboriculture: the *arborator*. The Romans were, as far as we know, the first to give a name to the tree worker or someone who was a specialist in practical arboriculture. Although the title *arborator* was used originally for just a tree pruner, it was later used to embrace a wide range of responsibilities for all types of trees. These included not only pruning and grafting of small ornamental trees and fruit trees but also the care of large trees and the felling and removal of major branches when needed. With regard to major tree work in the villa garden or estate, if an *arborator* was employed by the owner it seems likely that they would have undertaken this work alongside some of the estate's forestry workers. This was potentially dangerous work and the forestry workers' experience of dealing with large trees and their access to the necessary equipment would have been invaluable in carrying this out safely and successfully.

The job title *arborator* is derived from the Latin *arbor*, meaning tree, and it can clearly be seen that this is the origin of our modern words 'arboriculture', 'arboriculturist' and 'arborist'. In modern translations of the original Roman texts in Latin, the term *arborator* has been variously translated from the Latin as tree pruner, forester or arborist. It appears quite often in many of the classic Roman texts on agriculture and horticulture and in the context of a wide range of tree-related activities. However, only one reference could be found to any aerial tree work and this was in the context of the poem by Columella (1968a, 27) where a verse refers to an *arborator* singing while pruning branches perched up in a tree!

Almost all Roman gardeners of any description working throughout the Empire would have been slaves, and this applied equally to the *arborator*. However, some types of gardener could have a higher status as a slave than others. Cicero (1948, 289) described the *topiarii* or landscape gardeners as being among the slaves 'of a higher class', although he added, 'But all the same they are slaves'. An educated *topiarius* could, with good fortune and a sympathetic master, earn money, buy their freedom and even become a Roman citizen.

Overseeing the slaves working on the villa's gardens, farms and wider estate would be a bailiff who would usually report directly to the owner. This bailiff, normally a freedman, would be expected to have a good grasp of all aspects of estate work. Columella (1968a, 57) discusses the difficulty of a bailiff attaining 'knowledge of all the precepts of agriculture from, as it were, a single authority'. He claims, 'You will find scarcely anyone acquainted with the whole art, yet you will find many masters of part of it'. In finding suitable experts to pass on their knowledge he mentions an experienced *arborator* as an example.

Conclusion

When the Romans abandoned the province of Britannia to focus their military resources on difficulties close to home, the Emperor Honorious told the people of Britain that they no longer had a connection to Rome and should now determine their own destiny. In their 400 years of occupation the Romans had not only created many beautiful gardens and landscapes but had also established the foundations of ornamental gardening. As part of this, they had brought arboriculture to Britain by pioneering the planting, maintenance and management of trees and shrubs for their amenity benefits. Furthermore, this can also be described as 'professional' arboriculture: first, in the sense of establishing recognised standards of practice in the planting and care of amenity trees, and secondly, in instituting a category of professional gardener specifically concerned with arboriculture: the *arborator*. As the Romans departed these shores it remained to be seen whether this remarkable legacy would survive in the uncertain times that lay ahead.

References

Bowe, P. (2004) *Gardens of the Roman World.* Getty Publications: Los Angeles, CA.

Carroll, M. (2018) The Sacred Places of the Immortal Ones: Ancient Greek and Roman Sacred Groves. In: Woudstra, J. and Roth, C. (eds) *A History of Groves.* Routledge: Abingdon.

Cato (1935) *Cato and Varro On Agriculture.* Translated by W.D. Hooper and H.B. Ash. Loeb Classical Library, revised edition. Harvard University Press: Cambridge, MA.

Cicero (1884) *Cicero de Senectute (On Old Age)*. Translated with an Introduction and Notes by A.P. Peabody. Little, Brown and Co.: Boston, MA.

Cicero (1948) *Cicero de Oratore in Two Volumes, Book III, together with De Facto, Paradoxa Stoicorum, Di Partitione Oratoria*. Translated by H. Rackham. Loeb Classical Library. Harvard University Press: Cambridge, MA.

Cleere, H.F. (1981) *The Iron Industry of Roman Britain*. PhD thesis, Institute of Archaeology, University College London.

Columella (1941) *On Agriculture, Volume I: Books 1–4*. Translated by H.B. Ash. Loeb Classical Library. Harvard University Press: Cambridge, MA.

Columella (1968a) *On Agriculture, Volume III: Books 10–12. On Trees.* Translated by E.S. Forster and E.H. Heffner. Loeb Classical Library, revised edition. Harvard University Press: Cambridge, MA.

Columella (1968b) *On Agriculture, Volume II: Books 5–9*. Translated by E.S. Forster and E.H. Heffner. Loeb Classical Library, revised edition. Harvard University Press: Cambridge, MA.

Cunliffe, B. (1971) *Fishbourne: A Roman Palace and its Garden*. Thames and Hudson: London.

Cunliffe, B. (1981) Roman Gardens in Britain: A Review of the Evidence. In: MacDougall, E.B. and Jashemski, W.F., *Ancient Roman Gardens*. Dumbarton Oaks Trustees for Harvard University: Washington, DC.

Cunliffe, B. (1998) *Fishbourne Roman Palace.* Tempus Publishing: Stroud.

De la Bédoyère, G. (2006) *Roman Britain: A New History*. Thames and Hudson: London.

Dixon, D.M. (1969) The Transplantation of Punt Incense Trees in Egypt. *The Journal of Egyptian Archaeology* 55(1), 55–65.

Edlin, H.L. (1949) *Woodland Crafts in Britain*. B.T. Batsford: London.

Edwards, A. (2018) *The Story of the English Garden*. National Trust Books: London.

Farrar, L. (1998) *Ancient Roman Gardens*. Sutton Publishing: Stroud.

Floud, R. (2019) *An Economic History of the English Garden*. Allen Lane: London.

Gleason, K.L. (1994) Porticus Pompeiana: A New Perspective on the First Public Park of Ancient Rome. *The Journal of Garden History* 14(1), 13–27.

Gracie, H.S. (1970) *Frocester Court Roman Villa, Gloucestershire: First Report, 1961–67, Building A*. Bristol and Gloucestershire Archaeological Society: Bristol.

Hepper, F.N. (1967) An Ancient Expedition to Transplant Living Trees. *Journal of the Royal Horticultural Society* 92(10), 435–438.

Humphreys, H.N. (1874) Gardening in the Time of Pliny: Ancient Roman Villas. *The Garden* 5, 455–456.

Janson, H.F. (1996) *Pomona's Harvest: An Illustrated Chronicle of Antiquarian Fruit Literature*. Timber Press: Portland, OR.

Jashemski, W.F. (1993) *The Gardens of Pompeii: Herculaneum and the Villas Destroyed by Vesuvius, Volume 2*. Caratzas Brothers: New Rochelle, NY.

Jashemski, W.F. (2017) Gardening Practices and Techniques. In: Jashemski, W.F., Gleason, K.L., Hartswick, K.J. and Malek, A.A. (eds) *Gardens of the Roman Empire*. Cambridge University Press: Cambridge.

Jashemski, W.F., Ricotti, E.S. and Foss, J. (1992) Preliminary Excavations in the Gardens of Hadrian's Villa: The Canopus Area and the Piazza d'Oro. *American Journal of Archaeology* 96(4), 579–597.

Jones, P.d'A. and Simon, E.N. (1961) *Story of the Saw*. Newman Neame (Northern) Ltd for Spear and Jackson: Sheffield.

Koch, J.T. (2006) *Celtic Culture: A Historical Encyclopedia.* ABC-CLIO: Oxford.

Linnard, W. (2000) *Welsh Woods and Forests: A History.* Gomer Press: Llandysul.

Loudon, J.C. (1835) *Encyclopaedia of Gardening.* A new edition. Longman, Rees, Orme, Brown, Green and Longmans: London.

Meiggs, R. (1982) *Trees and Timber in the Ancient Mediterranean World.* Clarendon Press: Oxford.

Miller, W. (2003) *Crosscut Saw Manual.* Revised edition. USDA Technology and Development Center: Missoula, MT.

Noble, G. (2017) *Woodland in the Neolithic of Northern Europe: The Forest as Ancestor.* Cambridge University Press: Cambridge.

Pagán, V.E. (2016) Horticulture and the Roman Shaping of Nature. *Oxford Handbook Online.* Oxford University Press: Oxford.

Palladius (1873) *Palladius on Husbandrie.* The Early English Text Society: London.

Pliny the Elder (1950) *Naturalis Historia V, Books 17–19.* Translated by H. Rackham. Loeb Classical Library. Harvard University Press: Cambridge, MA.

Pliny the Elder (1968) *Naturalis Historia IV, Books 12–16.* Translated by H. Rackham. Loeb Classical Library. Harvard University Press: Cambridge, MA.

Pliny the Younger (1969) *Letters, Books 1–7.* Translated by B. Radice. Loeb Classical Library. Harvard University Press: Cambridge, MA.

Propertius (1990) *Propertius Elegies.* Edited and translated by G.P. Goold. Loeb Classical Library. Harvard University Press: Cambridge, MA.

Rackham, O. (2003) *Ancient Woodland: Its History, Vegetation and Uses in England.* New edition. Castlepoint Press: Dalbeattie.

Rackham, O. (2006) *Woodlands.* The New Naturalist Library. Harper Collins: London.

Thacker, C. (1994) *The Genius of Gardening: The History of Gardens in Britain and Ireland.* Weidenfeld and Nicolson: London.

Varro (1935) *Cato and Varro On Agriculture.* Translated by W.D. Hooper and H.B. Ash. Loeb Classical Library, revised edition. Harvard University Press: Cambridge, MA.

Virgil (1999) *Eclogues. Georgics. Aeneid: Books 1–6.* Translated by H.R. Fairclough. Loeb Classical Library, revised edition. Harvard University Press: Cambridge, MA.

Wacher, J.S. (1974) *The Towns of Roman Britain.* University of California Press: Berkeley and Los Angeles, CA.

White, K.D. (1967) *Agricultural Implements of the Roman World.* Cambridge University Press: Cambridge.

Wilson, P. (2009) *Lullingstone Roman Villa.* English Heritage: London.

CHAPTER THREE

Keeping the Flame Alight in the Dark Ages

For many people in Britain today, what is called the Middle Ages, or the medieval period, is a largely unknown and often misunderstood time in our history. Among historians there is widespread recognition that our knowledge of the nature and context of many of the era's events is limited by a lack of historical evidence, particularly in the early part of it. Even the duration of the Middle Ages in Britain varies according to different authorities. Here, it covers the period from the end of the Roman occupation in 410 CE through to the crowning of Henry Tudor in 1485 that heralded the start of the early modern period and the dawn of the English Renaissance. During most of this time Britain was divided into several kingdoms.

The Middle Ages represent just over a thousand years of British history, a major part of the timescale covered by this book. Therefore, it may seem surprising that so little should be known about it. Indeed, the era has been referred to, rather disparagingly, as the Dark Ages, partly because of our limited knowledge of it but also because it has been regarded as a time of cultural and economic decline after the fall of the Roman Empire and before the advent of the Renaissance. In terms of medieval arboriculture and horticulture, there are certainly major gaps in our knowledge, and this is particularly true of the early part of the era.

Sources of information

As with the Roman era in Britain, there are no surviving gardens from the medieval period (MacDougall 1986, 4). Their sites may still be identifiable and there may still be gardens on those sites, but neither the design, plantings nor the decorations survive at any one site. Thus the study of medieval gardens must rely heavily on visual sources, such as illuminated manuscripts, woodcuts and paintings. In this research Sir Frank Crisp's collection of 539 plates of contemporary illustrations, published in his magnificent work *Medieval Gardens*, was very helpful (Crisp 1924). However, the dating of medieval images can be difficult and only an approximation of their age is often possible. There is also a limited amount of literary evidence, such as the descriptions that appear in manuscripts or in the poetry and prose of the era. However, none of these sources is entirely reliable as many have been inspired partly by imagination with a tendency to exaggerate. While this needs to be taken into account, both visual or written descriptions nearly always represent things that are based in reality (*ibid.*, 1).

It has been said that medieval gardening records in Britain effectively begin with the Normans, as we have no real knowledge of any gardens in the Anglo-Saxon period, from 410 to 1066 (Harvey 1981, 52). Other authorities state that documentary evidence is almost non-existent before the fourteenth century (MacDougall 1986, 4). Apart from some illuminated manuscripts, other written sources of information about gardens and horticultural practices include the financial accounts of monasteries which mention gardens and gardening practices and some early legal records which are often concerned with actions for damages relating to gardens and their plants. Were it not for John Harvey's (1981) classic work *Mediaeval Gardens*, the by-product of over forty years of studying archival sources, we would not have our current understanding of these gardens. Although archaeological surveys could provide some reliable information and help to fill some of the gaps in our knowledge, these have not been undertaken to anything like the extent that has occurred with Roman sites. In the research for this chapter the absence of published books during the period and author's difficulty in accessing original manuscripts meant a greater reliance on secondary sources than elsewhere in this book.

In assessing the surviving evidence of medieval gardens and trying to place this in context it is important to remember that the Middle Ages witnessed some dramatic fluctuations in climate (Meyvaert 1986, 41). For almost 400 years, from the ninth to the thirteenth century, Northern Europe enjoyed a period whose annual mean temperature was several degrees higher than it is today, resulting in a much warmer and drier period (Mann *et al.* 2009). The Domesday Book, that record of the Great Survey of 1086, records nearly forty vineyards established in England at that time. In the context of ornamental gardens this was reflected in the widespread use of vines and the planting of scattered individual trees to create shade. This Medieval Warm Period, as it is now known, was followed by a period of significant cooling called the Little Ice Age. Although not a true ice age, it was another climatic event with huge impact on the landscape, which some experts now date from the late fourteenth century to about 1850. In medieval Britain it is associated with a decline in the acreage of vineyards that even occurred in the warmer southern counties of England.

Design and function of medieval gardens

As a general rule throughout history, the development of horticultural and arboricultural practices is largely dependent on the social and economic conditions of the day. In this respect the medieval period is no different. Most authorities on this era of garden history in Britain agree that there was a general trend where subsistence gardening, for fruit and vegetables, largely replaced the ornamental gardening that was quite common in the Roman era. This was especially true of the time from the departure of the Romans to the arrival of the Normans, a period of over 600 years that was marked by considerable social upheaval, Viking invasions and civil strife. It was only in the latter part of the medieval period that gardens for recreation and leisure became more abundant. The Norman Conquest brought Britain into much closer contact with continental events than it had been in Saxon times, and in the absence of any firm information on the ground we can accept the evidence of what was happening in northern France and Germany as generally applicable to Britain (Harvey 1981, 54).

Even with the arrival of the Normans there is still a scarcity of facts about gardens and it is only in the twelfth century that the picture seems suddenly to expand (Thacker 1994, 22). From early in the reign of Henry I (1100–1135), sites of gardens are named, sometimes with novel and intriguing features, and in the reign of Henry II (1154–1189), we are made aware of a variety of gardens, large and small, spread broadly over England and belonging to different categories of people. By the end of the thirteenth century the fashion in laying out gardens and vineyards seems to have made significant strides. After the death of Edward I in 1307 the art of gardening becomes more obviously luxurious and even extravagant (Harvey 1981, 86). However, this was to cause problems later as the climate took a disastrous plunge into lower temperatures and extremes of weather.

Even if Norman England was still considered a rather backward country, from the reign of Edward I this was certainly not the case. From the crowning of Edward I in 1274, the English king was the arbiter of Western Europe and the Edwards generally were sovereigns of high international standing. Their gardens were to reflect this and the royal accounts of the time show the great outlay on horticulture year after year. The high esteem in which amenity horticulture and arboriculture was now held was confirmed by Richard de Bury, the Bishop of Durham and former tutor of Edward III. Writing in his famous *Philobiblon* of 1344, he considered 'the planting of groves or sowing of seeds' to be surpassed only by the writing of books 'destined to benefit endless generations of posterity' (De Bury 1902, 35).

While England was to experience a slow but steady increase in number and extent of ornamental gardens and the development of horticultural practices throughout the Middle Ages, the situation in Scotland and Wales is regarded differently by many historians. Christopher Thacker (1994, 22) is typical in believing, on the limited evidence available, that both Scotland and Wales lagged behind England in this respect, even though there are individual examples of some impressive gardens in some of the larger noble or ecclesiastical establishments.

Having given a brief outline of the historical context for the development of gardens and gardening in the medieval period, we now examine some of the gardens types that emerged, with reference to the trees and shrubs that were planted and the arboricultural practices that maintained them.

Early monastic gardens

When we think of medieval gardens it is often the monastic garden that comes to mind, with its physic garden of medicinal herbs as well as vegetable plots, vineyards and orchards. There is no doubt that in the years after the Roman occupation and before the arrival of the Normans it was those monastic gardens that played a crucial role in keeping alive the theory and practice of gardening in Britain. The homogeneity in style of many medieval gardens, both religious and secular, is in part the result of the widespread monastic system (Hazlehurst 1966, 9). In the Church's cloistered retreats the rationale which governed the disposition of Roman gardens, whether consciously or not, was in fact preserved and is reflected in the careful ordering of the medieval monastery gardens into squares and rectangular beds, bordered by rigidly laid-out paths. The monastic gardens of the medieval era throughout Europe were first and foremost utilitarian, answering the specific needs of the monastery. A ninth-century German poem called *Hortulus* (The Little Garden) underlines the importance of monastic gardening in Europe at this time. Its author, Abbot Walafrid Strabo of Reichenau Abbey, was a gardener and his focus on the subject of gardening was obviously very popular, ensuring that *Hortulus* was widely read among the nobility and literate population throughout the Middle Ages.

In Britain, a number of ecclesiastical establishments played a vital role in creating gardens and continuing our horticultural tradition. This was especially the case at the larger abbeys and the palatial manor houses of the abbots, where some of the greatest horticultural display was made outside of royal circles (Harvey 1981, 12). An ecclesiastical garden in the priory at West Tarring in Sussex has a particular claim to fame. Here, in the 1140s, Thomas à Becket is thought to have planted the first fig tree to have survived in England, having been brought back from Italy after a pilgrimage to Rome. The half-acre (0.2 ha) fig orchard remains in part within the old flint and brick walls, and the existence of these trees and their connection with Becket is attested from the mid-thirteenth century, when the Bishop of Chichester, Richard de la Wych (*c.* 1198–1253) cared for the garden. John, the prior at Spalding in Lincolnshire in 1253–1274, is said to have made an impressive garden there and planted vineyards and orchards (Harvey 1981, 13). Apart from the monastic orchards, most ecclesiastical gardens do not seem to have had many trees. As these gardens were often small in size, there would hardly have been room for ornamental trees, especially those of substantial size when mature.

It is unfortunate that most of the Scottish monastic records have been destroyed, which has limited our ability to assess the nature and extent of their ecclesiastical gardens and orchards (Cox 1935, 7–9). However, sufficient survive to show us that many of these were extensive and, for the times, well cultivated.

As in England and Wales, the size of monasteries in Scotland varied greatly and consequently the size and importance of their gardens were similarly varied. One of the earliest records relate to gardens at Perth that were given by Alexander II to the Dominican friars on their arrival around 1230. The Tironensian houses, established by Benedictine monks, were particularly famous for their well-kept gardens and orchards. The Cistercians were also great gardeners. Their abbey at Balmerino in Fife, founded in 1229, was especially renowned for its fine gardens that included a number of magnificent trees.

Royal gardens

While monastic and other ecclesiastical gardens and orchards were particularly significant in early Norman times, the royal gardens became predominant from the reign of Henry I (1100–1135). This was especially the case with ornamental gardening, where the royal gardens were seen as the pinnacle of achievement.

In the many turbulent decades after the Norman Conquest, the royal family and the aristocracy lived mainly in fortified castles built on hilltops. This gave little room for a garden within the glacis of the feudal castle and consequently trees there of any description would have been a rarity. As it was often not safe for the residents to venture beyond the castle walls without an armed guard, it was scarcely worth making any garden or orchard outside, merely to see it plundered by some aggressive neighbour. Furthermore, trees were not allowed to grow anywhere near the castle because they might give cover to an attacking force. However, as times became less turbulent the interest in gardens and gardening grew among the powerful and wealthy in the land. As with architecture, it seems likely that the quality and style of gardening was influenced considerably by the personal taste of the king and of a small number of courtiers, nobles and churchmen; and also by England's political prestige at a given time (Harvey 1981, 3). From 1066 onwards, the monarchy was the least insular component of medieval English society. The Norman kings (1066–1154) were not only as much at home in their French duchy as they were in their English kingdom, but maintained close connections with their cousins in Sicily (Colvin 1986, 9). Later kings had strong connections not only with France but also with other parts of Europe. This overseas influence in culture and the arts came largely through various marriages, and it was the royal wives who were to have the major role in creating and shaping many royal gardens.

The medieval English kings would have perhaps a dozen houses at their disposal and could not occupy them all on a permanent or even regular basis (Colvin 1986, 12). In fact, the medieval English court was itinerant, moving around the country from one house or castle to another and, apart from Windsor and the metropolitan palace at Westminster, only a few specially favoured residences were occupied regularly for long periods. These included Clarendon and Woodstock in the twelfth century and Eltham, Sheen and

King's Langley in the fourteenth and fifteenth centuries. Therefore, most of the royal gardens were only for occasional use. Notable among the royal gardens in London were those at the Tower of London and Westminster Palace (McLean 1981, 77). These were also the most innovative gardens in London, as the king could afford to experiment. In terms of tree planting in these gardens, there is only evidence of plenty of fruit trees. While all the thirteenth- and fourteenth-century English kings, especially Henry III, seem to have taken a personal interest in their palace gardens, they always specialised in growing fruit in their London ones.

What little we know about the royal gardens in medieval England is chiefly from their financial accounts (Colvin 1986, 12). Unfortunately, there are no plans, topographical drawings or formal descriptions, either in the royal archives or elsewhere. The earliest drawings we have of royal English gardens were made in the middle of the sixteenth century, and they show gardens profusely adorned with heraldic ornament that was not evident before the Tudor era. With regard to Scotland, the exception to the general dearth of early gardens appears to have been those attached to the royal castles (Cox 1935, 19). During the reign of David I (1124–1153) there was a garden at the base of Castle Rock in Edinburgh and others attached to the royal castles of Roxburgh and Jedburgh. From 1286 to around 1450, little is recorded of even royal gardens. Despite the example of the monastic gardens in Scotland, the nature of the sites and the architectural features of the Scottish castles did not allow much space for a garden or orchard, and certainly not for any ornamental trees. After 1450, the number of gardens in Scotland appears to have increased, at least in those areas under the direct control of the Crown. For example, George Chalmers states in his *Caledonia Depicta* (1807) that gardens were universal in Midlothian during the reign of James III of Scotland (1460–1488). Unfortunately, with all these Scottish gardens there is little or no information on any trees they may have contained, apart from fruit trees.

Non-royal gardens

In the fourteenth and fifteenth centuries large areas of England, Wales and Scotland were enjoying comparative peace and stability. In the countryside, the poorer sections of the community were gradually asserting their rights against their Feudal lords (Amherst 1896, 45). There was an emerging class of farmers who paid some yearly tribute in service, or in kind, to their overlord. Around these small farms and manors, gardens and orchards were planted, but these were concerned with commercial horticulture and it was still only the rich and powerful who indulged in any ornamental gardens.

We will now examine some of the gardens types that emerged in the medieval period, with a focus on those that had some arboricultural elements.

FIGURE 6. Manuscript image dated *c.* 1400 showing an enclosed garden at an English royal castle with a king and queen playing chess. The garden is subdivided by close-boarded fences. In another garden in the background a gardener is pruning a small tree with a billhook.

CREDIT: THE BODLEIAN LIBRARY, UNIVERSITY OF OXFORD.

The **hortus conclusus** ***and the pleasure garden***

A form of garden much associated with the late medieval and Renaissance periods is the *hortus conclusus*. As well as being a genre of actual garden, it was also an emblematic tribute and title of the Virgin Mary in medieval and Renaissance poetry and art. We are not concerned here with this latter aspect of the *hortus conclusus* but with its horticultural and arboricultural significance. *Hortus conclusus* is a Latin term that means a garden enclosed by a fence or a wall. As this would apply to practically all gardens in the Middle Ages, it must be emphasised that this was something more (Crisp 1924, 115). The *hortus conclusus* was essentially an enclosed garden placed within a larger garden, from which it was distinctively separated by hedges or other divisions, and having no relation to the general arrangement of its surroundings. It was typically rectangular or square in shape, divided into quadrants by four paths and with a well or fountain at its centre.

Many authorities on the medieval era make little distinction between the *hortus conclusus* and the enclosed pleasure garden, which was introduced into Britain at the time of the Norman Conquest (Figure 6). The difference seems to focus more on the symbolism associated with the *hortus conclusus* rather than any major physical distinction (McAvoy pers. comm., 2020). The medieval pleasure garden is not to be confused with the pleasure gardens of Georgian and Victorian times that were the forerunners of our modern public parks (Johnston 2015, 144–150). The name 'pleasure garden' is indicative of its purpose, and it was to this end that ornamental trees and shrubs, fruit trees and other plants, accentuated by a grassy lawn, were selected and protected (Stannard 1986, 77–78). Most of the plant species recorded from pleasure gardens, about which we possess good records, also served purposes other than sensual gratification and had alimentary and medicinal purposes. The medieval pleasure garden is sometimes referred to as the pleasance or pleasaunce and in its early days this was often described as an orchard, although not in its modern sense as an area of fruit trees devoted entirely to fruit production (Huxley 1978, 32). The medieval pleasance was much more than a collection of fruit trees and other kinds of tree were often found there. It was usually situated at the back or side of the castle or mansion. In the warmer months of the year the pleasance was used for receiving guests, for playing games, dancing, singing and, on occasion, for large dinner parties. Gradually, the orchards were replaced by more formal designs, 'gardens of pleasure' with elaborate fences and arbours, turf seats, fountains and a main grassy area with wild flowers.

The first written description of the medieval pleasure garden is found in a chapter of *De Vegetabilibus et Plantis* by Albertus Magnus of Cologne, written in about 1260. In his treatise on plants and their cultivation he added a chapter on pleasure gardens, which has been translated by John Harvey. Part of Magnus's hugely influential text refers to the planting and layout of these pleasure gardens and it makes significant mention of trees:

> There are, however, some places of no great utility or fruitfulness … for these are what are called pleasure gardens. They are in fact mainly designed for the delight of two senses, viz. sight and smell. [Having prepared the ground,] then the whole plot is to be covered with rich turf of flourishing grass. … Upon the lawn too, against the heat of the sun, trees should be planted or vines trained, so that the lawn may have a delightful and cooling shade, sheltered by their leaves. For from these trees shade is more sought after than fruit, so that not much trouble should be taken to dig about and manure them, for this might cause great damage to the turf. Care should also be taken that the trees are not too close together or too numerous, for cutting off the breeze may do harm to health. The pleasure garden needs to have a free current of air along with shade. It also needs to be considered that trees should not be bitter ones whose shade gives rise to diseases, such as the walnut and some others; but let them be sweet trees, with perfumed flowers and agreeable shade, like grapevines, pears, apples pomegranates, sweet bay trees, cypresses and such like. …

> If the [midst of the] lawn were to have trees planted on it, spiders' webs stretched from branch to branch would interrupt and entangle the faces of the passers-by.
> Harvey 1981, 6

While Albertus Magnus's thoughts on the pleasure garden had very limited readership and little direct impact in Britain, his whole chapter on this topic was reproduced in the *Liber Ruralium Commodorum* (Book of Rural Benefits) written by Piero de' Crescenzi from Bologna between 1304 and 1309 (Calkins 1986, 157). By the end of the Middle Ages this had become one of the most frequently reproduced works on gardening. Crescenzi added to Magnus's text with two additional sections, which appear to be his own, on larger types of pleasure gardens. Again, these make significant mention of trees.

Crescenzi proposes a garden of medium size for persons of moderate means, occupying a somewhat larger plot of between 1¼ to 4½ acres (0.5–1.8 ha) (Calkins 1986, 165–166). He suggests that it should be surrounded by ditches and hedges of thorns and roses, or plums and quinces, and in which the flowering fruit trees are to be planted in rows interspersed with vines. Some of these trees might be shaped and the grass mowed so that the lawn will remain beautiful and permanent. The largest of the gardens, appropriate for royalty, should be about 12½ acres (5.1 ha) and surrounded by a high wall. A grove of trees should be planted in the northern part as a shelter for all manner of wild creatures, including deer and game fowl. The trees should be arranged in radiating rows away from the palace, so members of the court could watch the wildlife flitting back and forth along the alleys. In addition to flowering fruit trees, more substantial specimens are proposed, such as willows and elms. Crescenzi also mentions elaborate grafts and prunings, perhaps implying espaliered or estrade trees. This garden should also contain trellises entwined with vines in the form of a pavilion where the king and queen might linger without too much sun or rain.

The importance of Crescenzi's book is that it gives a complete picture by a cultured observer of the medieval garden at its most expansive, before the onset of the Renaissance (Harvey 1981, 76). Nevertheless, how much of that picture can be transferred to medieval gardens in Britain is a matter for speculation. While many of the elements from the gardens described by both Magnus and Crescenzi can be identified in the archives on British medieval gardens, our interest here is with the presence and use of any trees and shrubs. One likely difference relates to the frequent use of trees for shade in the warmer gardens of Italy and elsewhere in continental Europe. It would seem that this was not undertaken to anything like the same extent in Britain. The Medieval Warm Period in Britain lasted from the ninth to the twelfth centuries, when the evidence indicates there were few ornamental gardens here. Certainly, after that time, when the British climate became cooler and wetter, as it is today, far fewer trees would have been planted for shade than in the warmer regions of the Continent. There is, however, evidence that vines were a prominent feature of our royal English gardens throughout the medieval period. There are innumerable references in the royal accounts to the framework of rods and timbers on which vines were trained in the form of covered walks, pergolas and arbours (Colvin 1986, 14).

The garden historian, Teresa McLean (1981, 243–248), writes about a type of medieval garden that she calls 'tree pleasure gardens' and states that these were not an uncommon form of the medieval pleasance. Indeed, McLean's description of these appears to be similar to the pleasure gardens of Magnus and Crescenzi where fruit, forest and ornamental trees often grew together in a single garden. As an example of the tree pleasure garden, McLean cites the gardens of the Cistercians at Balmerino Abbey in central Scotland and believes these must have been similar to those described by Geoffrey Chaucer in his fourteenth-century poem *The Parliament of Fowls*. As mentioned above, it is known that the gardens at Balmerino had some fine trees and it is interesting to note that at the time of the Dissolution some of these gardens had names that made reference to trees. The layout of these tree pleasure gardens varied from the artistically formal to the chaotically informal. If they were attached to institutions with specialist gardeners, such as castles, palaces, big monasteries and manors, they probably had some sort of formal layout, such as rows. This would facilitate the harvesting of their fruit trees in the autumn and make delightfully colourful and fragrant walks for their owners and visitors in the spring. In addition, these gardens often had a single central tree, reminiscent of the central tree in Eden, with its ambivalent symbolism as the tree of knowledge and the forbidden tree. It is likely that this prominent tree would have been given special care and attention.

Orchards

Although orchards are now regarded as an aspect of commercial horticulture, in medieval times they often had an additional ornamental or recreational function. Furthermore, when medieval records refer to an orchard, they mean a fruit tree garden as opposed to a garden in which there were some fruit trees growing, such as the *hortus conclusus* or pleasure garden (McLean 1981, 225). While fruit trees and orchards featured prominently in the Middle Ages, it should be noted that the eating of raw fruit was not particularly popular. However, fruit such as apples and pears were valued greatly as a source of drink, particularly cider, and as material for sweet, spicy desserts, preserves and comfits (*ibid.*, 224).

Medieval documents reveal that orchards were invariably surrounded by a variety of boundaries, such as the traditional ditch and quickset hedge, wattle, palisade or walls, and also moat, which were reasonably animal-proof (Landsberg 1995, 17). As well as keeping out animals and unwanted people, the additional purpose of a boundary was legal demarcation of the territory, in order to obtain maximum compensation against theft. Some medieval orchards were of considerable size. The largest recorded is the 12 acre (4.9 ha) orchard at Llanthony Priory in Wales in 1199. Depending on spacing this would have contained between 400 and 1,000 trees. Orchards were a monastic speciality and many of these were around 7 acres (2.8 ha). In Scotland, monastic orchards were known before 1202. By 1245 those belonging to the monastery of Haddington in Lothian were called old, a fame which endured for several centuries and spread far beyond Scotland (Cox 1935, 11). Throughout the medieval era monastic gardeners were famous for their skill in designing and managing orchards and tree gardens, some of them for pleasure rather than fruit production, many for both. The cemetery orchard was also a feature of some monasteries, symbolising Paradise and providing a delightful resting place for the monks after a lifetime of service to God. Fruit trees would blossom between the graves, along with vines, roses and flowers.

The most common size of manorial orchards was around 4 acres (1.6 ha) and these provided the nobility and their friends with a source of culinary as well as recreational pleasure. An early fifteenth-century account described the orchards at Raglan Castle in Wales as 'full of apple trees and plums, and figs, and cherries, and grapes, and French plums, and pears, and nuts, and every fruit that is sweet and delicious' (Whittle 1989, 83). The king was the leading fruit grower in medieval England and the planting of orchards at royal castles and manors was a regular function of the king's works (Harvey 1974, 16). He could afford to experiment with new varieties of orchard fruit and, like monastic fruit growers, had the international contacts that made this possible. Most other orchard owners were far more conservative.

Other amenity landscapes

Apart from those trees in the different types of medieval garden described above, trees were also planted for ornamental purposes in other locations, both urban and rural. However, we do not include here any planting in forests and parks that were managed primarily for timber or hunting, even though this was a preoccupation of the ruling class in the medieval era (Manwood 1615; Young 1979). This is the realm of traditional forestry, not arboriculture, and beyond the scope of this book.

In terms of tree planting in our medieval towns and cities, this was generally very limited (Johnston 2015). However, thanks to a few early chroniclers we know of some exceptions to this. With regard to London, Holborn was the leafiest of the capital's suburbs. In William FitzStephen's *Description of the City of London*, written in Latin in 1180, he describes Holborn as a place where 'the Citizens have Gardens and Orchards planted with trees, large, beautiful, and one joining to another' (FitzStephen 1772, 25). In the City of London itself the newly founded livery companies had built magnificent halls and it was only natural that these should have beautiful gardens attached to them (McLean 1981, 71). Only a few of the City companies had gardens as big as those of the Grocers and Drapers; most were quite small. The tiny Cutlers' garden had a vine, pruned each year by a gardener who pruned their fruit trees at the same time. These were probably mulberry trees, the favourites of the livery companies. Sadly, there is not a single one of these original livery gardens surviving today.

In some of Britain's medieval cities there were urban manor houses, with courtyards, outbuildings and gardens that were often bigger and more ambitious than many attached to manors in the country (McLean 1981, 72–75). The Earl of Lincoln's fourteenth-century house and garden in London is an example. He grew apples, pears, large nuts and cherries in it, and took a keen personal interest in pomology. There are records of him sending to the Continent for slips of new apple and pear trees, employing an expert gardener to graft and look after them. The present Lincoln's Inn Gardens are on part of the Earl's famous garden.

Apart from the monastic gardens and orchards that contained trees, there is evidence of ornamental tree planting around some ecclesiastical buildings. In 1091 at Crowland Abbey in Lincolnshire the archives record that there was a great fire 'which caught the green trees, ashes, oaks and willows growing about the building' (Harvey 1981, 58). These had undoubtedly been planted deliberately and formed part of an amenity landscape that surrounded the building. In 1222 there are records of a tempest in the neighbourhood of Worcester 'that brought down great trees in gardens as well as woods' (*ibid.*, 73). This mention of 'great' trees in gardens implies ornamental gardens rather than fruit orchards and may refer to the recreational grounds of the monks in the cathedral priory. In 1243, elms were planted outside the west front of Wells Cathedral, then still unfinished (*ibid.*, 16). They were near the south side of the front and extended towards the west, implying one or more rows of trees parallel to the axis of the cathedral church. The gardens at Durham belonging to the Cathedral Priory seem to have been mostly utilitarian but there were parks around the city that had a recreational and sporting function. The archives show that payments were made in 1331 for cutting and pruning branches in Shincliffe Park, possibly as part of general woodland management and possibly for landscaping reasons (*ibid.*, 92). At Lambeth Palace in 1428 on the west side of the walled garden twenty-nine elms were felled in 'le Rowe Garden', which suggests that the palace had elms planted in lines like the earlier ones at Wells Cathedral (*ibid.*, 98). Later at Wells in 1487 the Cathedral Treasurer was involved in a dispute concerning the pollarding and lopping of the trees in the churchyard (*ibid.*, 140). This is a very early example of where major tree maintenance was considered necessary by the owners but the resulting work did not meet with the universal approval.

Another category of medieval urban gardens is that of university colleges. By the fourteenth century there were small college gardens at both Oxford and Cambridge and these probably contained a few trees, although as the gardens were mainly utilitarian in purpose these were likely to have been fruit trees, valued both for their fruit and attractive blossom (Amherst 1896, 42). At King's Hall, Cambridge, there was a 'serjeant of the garden' who was paid for nailing up and pruning the vine in 1339 (Harvey 1981, 107). At Oxford many of the present college gardens are on the sites of old monastic establishments dating from medieval times and have inherited their gardens (Batey 1982, 6).

Arboricultural and horticultural practices

When the Romans abandoned their province of Britannia, their ornamental gardens fell derelict and much of the specialist knowledge of horticulture and arboriculture that sustained them was slowly forgotten. While they left behind many people of Romano-British descent who had enjoyed the material benefits of Roman occupation, including their ornamental gardens, those lives were to change radically without the support of the Empire that had sustained Roman rule, ideas and values in much of Britain over some 400 years.

During the so-called Dark Ages in Britain and even in the early days of the Normans it was in the monasteries with their gardens, orchards and vineyards where the flame of horticultural and arboricultural knowledge was kept alight. A good knowledge of plants and their cultivation was encouraged in the monasteries through well-stocked libraries, often containing laboriously hand-copied herbals. The monastic orders played a vital role in the dissemination of this horticultural, botanical and medical knowledge through the collecting, copying, illustration and distribution of manuscripts. These included the works of those notable Roman authors on agriculture and horticulture, ensuring they continued to be read in monasteries and aristocratic circles throughout the Middle Ages. Monks were natural importers and exporters, constantly improving their knowledge and resources by exchanging plants, seeds, cuttings and horticultural gossip with members of other monastic families not just in Britain but internationally (McLean 1981, 25). The Cistercian Order played a particularly prominent role in this exchange network of plant material and ideas that flourished among the monasteries. Throughout the medieval period the monasteries and their gardens made an enormous contribution to practical horticulture and land management by developing an understanding of practices such as soil enrichment, marling, land reclamation and drainage (Landsberg 1995, 34). They were also at the forefront of water and woodland management, and of the cultivation of vineyards, orchards and pleasure gardens.

Medieval manuscripts giving practical instruction

One of the earliest of the medieval European texts containing practical advice on horticultural and arboricultural practices was *De Vegetabilibus et Plantis* by Albertus Magnus, mentioned above in the context of pleasure gardens. Magnus was born about 1206 at Lauingen in Swabia (Germany), studied at Padua and became a Dominican friar and bishop, remaining in the order until his death in 1280. His collected works, eventually published together in 1899, went to thirty-eight volumes and covered an enormous range of subjects. To what extent

FIGURE 7. A tree pruner at work in late medieval times. This woodcut is from an edition of Crescenzi's *Liber Ruralium Commodorum* (*c.* 1480), although it probably dates from a much earlier edition.

Magnus's *De Vegetabilibus* was read in Britain during the Middle Ages is impossible to say, although it is likely to have been quite limited.

Magnus's *De Vegetabilibus* relies heavily on the classical Roman authors and repeats much of their good advice on the planting, maintenance and management of trees and shrubs. He also included some of the ancient doctrines, such as transplanting in the first quarter of the moon. However, he was also able to express his own views, which were not necessarily accepted practice at the time. For example, in the general management of fruit trees during medieval times, it was usual for the soil surrounding the trees to be scrapped back from the roots and new soil added in the spring. However, Magnus recommends not doing this with trees in the pleasure garden to ensure the grassed area retains its green and aesthetically pleasing appearance. As given in the quotation above, he states, 'For from these trees shade is more sought after than fruit'.

The *Liber Ruralium Commodorum*, written by the retired Bolognese lawyer Piero de' Crescenzi between 1304 and 1309, seems to have been far more widely read than Magnus's text and by the end of the Middle Ages was one of the most frequently republished works on gardening (Calkins 1986, 157). A major reason for its popularity may have been the tone of its presentation, which is that of a 'How to' manual. Also, the text was occasionally illustrated or illuminated, and these images depict the agricultural and horticultural practices of the day. A number of these illustrations relate directly to arboricultural practice and show people planting, pruning and grafting trees (Figure 7).

The twelve books of Crescenzi's treatise are devoted to the effective management of the farming estate and describe various practical aspects of farming and horticulture. Of particular interest here is Book V, 'On the Care of Trees, Including Fruit Trees and Shade Trees' which covers arboriculture, and Book VIII, 'On Pleasure Gardens', already mentioned above, which includes information on the planting and arrangement of amenity trees. Crescenzi's work also follows in the tradition of the classical Roman authors, especially Columella, where he replicates closely the contents and structure of his *De Re Rustica*. In addition, he copied extensive passages from Magnus's *De Vegetabilibus* and not just his entire section on the pleasure garden. Some modern writers have regarded Crescenzi as a plagiariser and 'armchair compiler' who had no effect on the agricultural and horticultural practices of his day. While it is true that Crescenzi often just repeated what others had said, he was also a keen observer and recorder of practices at that time and his text is invaluable for that reason alone.

Crescenzi's *Liber Ruralium Commodorum* was extensively copied throughout the fourteenth and fifteenth centuries (Calkins 1986, 159–162). We know of at least 141 manuscripts that have survived, of which thirty-five copies date from the fourteenth century. There may have been many more copies which have now been lost. With the translation of the work from Latin into the vernacular its accessibility increased, with three copies in Italian in the late fourteenth century and fifteen in French in the fifteenth century. It was reputed that there was an edition in English in 1616, but no copy is known to survive, and there is no modern English edition. However, Edward IV of England (1442–1483) owned an illustrated copy in French written in Flanders between 1473 and 1483, which is now in the British Library. On the evidence available it seems likely that from the fifteenth century copies of Crescenzi were to be found in the most important royal and aristocratic households in England.

After circulating in numerous manuscript copies, the invention of the printing press by Johannes Gutenberg in Germany around 1440 greatly increased the circulation of Crescenzi's treatise. Indeed, it became the first printed modern text on agriculture when it was published in Augsburg by Johann Schüssler in 1471 (Anderson 1997, 66). Some fifty-seven editions in Latin, Italian, French and German appeared during the following century.

Having discussed two of the most influential continental European contributions to the medieval literature on horticulture and arboriculture, we now consider some works emanating from Britain. Walter of Henley (Walter de Henley) was an English agricultural writer of the thirteenth century, writing in French. His known work is called *Husbandry*, written about 1280, which deals with the agricultural management of a manor and is one of the earliest English manuscripts to address this subject (Oschinsky 1971, 145–147). It was eventually published in an edited version in the late Victorian era (Lamond 1890). Walter of Henley was a Dominican friar who had served in the office of bailiff where he gained practical experience of managing a manorial estate, including its woodlands. His book is packed with practical information on different aspects of husbandry, including forestry, although there is no significant mention of amenity trees. Nevertheless, he is notable for being possibly the first English author to refer to sustainable woodland management when he urged landowners to survey their woods and establish the volume of standing timber to determine 'what you can sell without loss or destruction, and what it is worth yearly beyond the return' (*ibid.*, 7).

The earliest known original work on gardening written in English is a treatise in verse by Mayster Ion (Master John) Gardener entitled *The Feate of Gardening* (Amherst 1896, 71). This exists as a unique manuscript in the Library of Trinity College, Cambridge, and although this has been assigned an approximate date of 1440, it is thought to be a copy of an original made at least fifty years earlier (Harvey 1985, 83). This short document was first published in 1894 by Alicia Amherst, the Victorian garden historian, who added a glossary (Gardener and Amherst 1894). The verse has 196 lines in total, consisting of a prologue and eight divisions where Master John gives recommendations on sowing, planting, pruning and grafting for a wide range of plants. The first two divisions focus specifically on trees. The first is entitled 'Off Setting and Reryng of Treys' and deals with the planting and raising of trees. This is followed by a section entitled 'Of Graffyng of Treys', which deals with grafting various types of trees. While the text contains little on mature tree care, Master John does state that where branches have been pruned, a bandage of clay is recommended as a wound treatment, to exclude the air and prevent haemorrhage. As well as giving sound advice on basic horticultural and arboricultural practice, the text does include instructions on what work should be done under various signs of the zodiac. This appears to reflect Master John's own beliefs rather than being just a repetition from other authorities.

Master John's pioneering treatise is invaluable for a number of reasons. Although turned into doggerel verse, its content is entirely practical and shows no sign of borrowing from other sources. In this it differs markedly from almost everything else of a comparable kind in its field. It also gives incontrovertible evidence of the plants that could then be found in the medieval English garden, and this includes some trees and shrubs.

There are a number of delightful medieval manuscripts still in existence in which the occupations of the seasons are depicted in a variety of ways (Landsberg 1995, 94). Some of these date from as early as the eleventh century and they become more common in the thirteenth and fourteenth centuries. These calendars are usually in the form of a set of twelve miniatures which show contented workers in an ordered society, with each month depicting a different activity. Of interest here are those activities frequently depicted that relate to tree work. February was often shown as the month for pruning, whether this was the pruning of fruit trees or the cutting of coppice for fuel or straight rods to be utilised in various ways. June was often depicted as the month for felling large trees for timber (Figure 8).

While there is evidence in medieval archives of major tree work being undertaken, such as felling, pollarding and crown reduction, there does not seem to be any written accounts or images describing precisely how this was carried out. Furthermore, in contrast to some Roman texts, there does not seem to be any mention of transplanting large trees, although it is likely there were instances in medieval times when mature tree moving was attempted.

In concluding this section on medieval manuscripts giving practical instruction, mention should be made of the unique surviving account book of Beaulieu Abbey, a Cistercian abbey in Hampshire. This includes the Forester's Account for the forest year 1269 to 1270 and the *Tabula Forestarii*, a guidebook or set of rules governing forest yield, woodland management and specifications

FIGURE 8. Images of tree work from the eleventh-century English manuscript known as the Julius Calendar (Cotton Julius A. IV). The top image represents February and shows workers pruning trees. The image below represents June and shows workers felling trees with axes and loading a cart.
CREDIT: BRITISH LIBRARY.

and prices for forest products (Hockey 1976). This is Britain's earliest known forester's *vade mecum*, a handbook or guide that is kept handy for frequent consultation. As well as being skilful gardeners the Cistercians were also accomplished foresters. Surveys of some Cistercian monastery woods during the Dissolution (1536–1539) reveal age structures that indicate careful silvicultural management for over a century (Linnard 2000, 62).

Topiary in medieval gardens

The art of topiary that had been practised widely in Roman times remerged to some extent in the late medieval gardens in Britain (Crisp 1924, 78–80). Evidence of this comes from illustrated manuscripts, rarely before 1400 but

FIGURE 9. The most popular medieval topiary came in the form of *estrade* trees. The example shown on the left has two tiers encircled by iron bands. The image is dated 1461 but its origin is unknown.

more common in the fifteenth century, and in the financial accounts of some royal houses and gardens. Medieval topiary work was far less extravagant than the earlier Roman examples from the great villas of Pliny's time. It usually featured yew and box cut into simple shapes without any attempt at imitating people, animals or other imaginative representations.

The most popular medieval topiary came in the form of *estrade* trees (after the French word meaning platform), which would be clipped and trained in Burgundian fashion, the leaders in European fashion at that time. This involved tiers of two or three discs of foliage, sometimes encircled by a metal ring to accentuate the form (Figure 9). Those with three tiers, narrowing towards the top, were said to represent the Holy Trinity. The height of these *estrades* could vary between 3 ft and 12 ft (1–4 m) and they would make a dramatic feature in the garden. Most of the larger examples are shown growing in garden beds but some smaller ones also appear in pots.

Tree nurseries and propagation

The propagation of trees and other woody plants through grafting, joining together a scion and stock from two separate plants, has been practised since ancient times. It was first mentioned in the sacred books of the Chinese about 6000 BCE and was widely undertaken in the Roman era (Janson 1996, 29). It is likely that after the fall of the Roman Empire the art of grafting survived in the medieval Christian monasteries of Europe until it regained popular appeal during the Renaissance (Mudge *et al.* 2009, 459). In the Middle Ages grafting was used mainly in the propagation of fruit trees, usually to improve the quality and quantity of fruit. However, it can also be applied to ornamental trees, most frequently to improve their aesthetic features.

Throughout its history the Christian Church has always had an ambivalent attitude towards the practice of grafting. It has often been condemned as an interference with the work of the Creator and contrary to 'the natural order of things'. This might explain why during the Middle Ages grafting was considered a secret by the initiated, who were frequently based in the monastic orders, and a miracle by the rest of the population. In contrast to this perception of grafting as something of a 'dark art', other elements of the Church had a more enlightened view. In 400 CE, Saint Jerome, the founder of monasticism, told his monks to spend more time grafting and budding fruit trees 'to escape sloth and the devil' (Janson 1996, 21). This positive view of grafting slowly gained acceptance throughout the Middle Ages to the point where a good understanding of the practice was considered the most essential part of the husbandman's education (Amherst 1896, 53).

Master John Gardener's *Feate of Gardening* was one of the first English texts to state the vital role that grafting should play in the work of the gardener (Gardener and Amherst 1894). He also gave excellent instructions on how to undertake the successful grafting of fruit trees that have stood the test of time. Gardeners with good soil and limited time could use one of the two standard methods that he described, either apples with apples or pears with hawthorns. An apple or pear stock was to be planted and, when it had grown into a healthy young tree, grafted either by shortening the branches and inserting scions of the desired variety or cutting off the head and inserting the scions into the stem, just below where the head had formed. To ensure a successful union, clay had to be laid on the stock, to keep the rain out, and moss bound over the clay with 'a wyth of haseltree rynde'.

Some gardens had a 'resting tree', one of the principal trees in the garden, and what would now be termed a stock tree from which vegetative material for grafting would be taken. Ranulph III, sixth Earl of Chester (1181–1231), in a charter no later than 1191, granted William, the keeper of his Castle garden and orchard, his 'restingtre' as well as the residue of the Earl's apples (Harvey 1974, 16). A long chain of later evidence in the Castle's account rolls confirms the importance of this 'resting tree'. In 1353, it was stated that Philip Raby, gardener at Chester Castle, was entitled to the tree and the residue of the apples by ancient custom. It would seem that this apple tree must then have been well over 150 years old, or else it had been replaced with a new stock tree from time to time.

The work of grafting and other methods of propagation appear regularly in many monastic and royal accounts (McLean 1981, 240–241). In 1265, at Rimpton near Taunton, an expert gardener was paid the high wage of 1*s* for eight days' work grafting new trees in the apple and pear garden. A few gardeners, notably those in charge of monastic and big manorial orchards, became grafting specialists. A few others applied their skills to develop commercial tree nurseries to provide gardeners and farmers with ornamental and forest trees. However, the history of medieval nurseries for ornamental and forest trees is a difficult subject to research. This is because fruitless garden trees are not mentioned in garden accounts, and these nursery trees were often regarded as agricultural rather than horticultural, being grown in coppices rather than gardens.

From his extensive research, John Harvey (1981, 17) discovered that in London by the reign of Edward I (1272 to 1307) there was a nursery trade able to supply a wide range of trees and other plants. Even in the countryside the acorns and nuts sown were sometimes bought, not gathered. In Teresa McLean's (1981, 241) detailed research she found that nurseries grew young forest trees for sale to gardeners, farmers and estate managers, mainly for fencing and sometimes for park and woodland planting. Grafts, scions, shoots and saplings were all called 'imps', and 'impgarths' or 'impyards' were enclosed plantations of imps, used in the sense of a nursery-plot for grafts in the first instance, and then for transplanted seedlings generally, usually of trees. From the reign of Henry I (1100–1135) onwards we find the field names 'impgarth' and 'impyard' appearing in accounts and on maps (Harvey 1981, 58). At Pontefract in Yorkshire, famous five centuries later for its trade nurseries and liquorice beds, there was an 'impecroft' by 1215. It is hard to distinguish an impgarth from a field plantation or a coppice. An impgarth was a garden wood as much as it was a wood garden, full of young forest and fencing trees that would be taken out of their garth as soon as they were big enough to be planted out permanently. Willows and osiers, alders, poplars and aspens were the most common impgarth trees. Ash trees were also grown, because they provided good poles for making hurdles, and hazel because their young stems were pliant enough to be made into wattle screens, hurdles, thatch frames and fences.

From the number, frequency and size of medieval nursery transactions, it is clear these were commercial ventures operating on a large scale to meet a significant demand, certainly for forest-type trees. Thorold Rogers, the Victorian economist and historian, uncovered evidence of this in his research into early agriculture. For example, he found 2,000 hazelnut plants bought for 4*s* 1*d* in 1335, the transaction taking place in Oxford (Rogers 1866, 594). That there was a commercial outlet for transplanted hazel shows an advanced concept of woodland management which would not otherwise be suspected.

Arboricultural standards in medieval times

FIGURE 10. Is this an example of poor tree-planting practice in medieval times? The bare-rooted tree has few roots and is being planted in full leaf. This woodcut is from an edition of Crescenzi's *Liber Ruralium Commodorum* (*c.* 1480), although it probably dates from a much earlier edition.

This section on arboricultural and horticultural practices concludes with some comment on the standard of tree work that prevailed during the medieval period. Many garden historians would agree with Miles Hadfield (1979, 25) when he stated, 'The evidence of progress in gardening in early medieval times is mostly indirect and devoid of detail'. Certainly, our lack of knowledge of the early medieval period and, to a lesser extent, of that whole era, tends to encourage the view that nothing particularly innovative occurred and that gardening largely stood still. This is also reinforced by our awareness of the heavy reliance throughout the era on the works of the classical Roman authors on land-based subjects, which were accepted largely without question. Roman theories and practice regarding horticulture and arboriculture were widely endorsed by medieval writers and authorities within monasteries and aristocratic circles even when these were often unrelated to the simple facts demonstrated by the professional gardeners of the time.

Some garden historians believe that the overall standard of arboricultural work during medieval times was quite poor. Anthony Huxley (1978, 142) describes the approach to tree planting in the Middles Ages as 'distinctly casual' and bases this judgement on some of the woodcuts and other images that 'show trees with bare roots left exposed to the air, waiting to be planted'. Other images show bare-rooted trees with few roots being planted while they are still in full leaf (Figure 10). Peering back through the mists of time with often questionable evidence, it is almost impossible to take an objective view of this. We know that medieval images were frequently less than accurate representations of what was being depicted. Nevertheless, with many images demonstrating poor arboricultural practice, the suspicion remains that this was often the case.

Tools and equipment

While it is unlikely that any Roman garden in Britain survived intact into the early Middle Ages, we have noted that many of their gardening practices continued. With the evidence available from written, pictorial and archaeological sources it also appears that many of the gardening tools and equipment that were commonly used in Roman Britain often continued to be used in much the same form in the Middle Ages. Indeed, this would hardly be surprising since the tools and utensils used in medieval gardens do not differ greatly from those that were still in regular use in the early years of the twentieth century (Harvey 1981, 114). After all, the fundamental techniques of horticulture do not change much.

Early manuscript texts describing horticultural and arboricultural tools and equipment are not common and these descriptions can be vague and confusing. The precise shape of the tool, its dimensions and uses may not be clearly stated. By contrast, images of tools and equipment found in illustrated medieval manuscripts, woodcuts or paintings can be far more enlightening. While these images often depict religious or symbolic scenes, they nevertheless include people undertaking various horticultural and arboricultural activities. They show gardeners at work using spades, rakes, forks and buckets that have changed little over the centuries, along with a formidable range of axes, pruning knives and billhooks, the latter being the multipurpose Swiss army knife of the medieval gardener. Many medieval pictures depict the use of ladders in gardens and orchards and there are even images of rudimentary wheelbarrows.

Alexander Neckham, the medieval English scientist and theology tutor at Oxford, wrote his treatise *De Utensilibus* (Concerning Tools) around 1190 and this contains one of the first lists of tools and instruments then used in a wide range of activities (Wright 1857, 96–119). As part of this work Neckham give a list of what the *rusticus* (labourer) would normally use and it covers tools for both farm and garden work, including some of arboricultural use, such as axes, saws and grafting knives.

In exploring the evidence for tools and equipment known to have been used in medieval gardening, very little could be found that relates directly to major tree work. Therefore, we can only speculate on this matter. It seems likely that ropes and ladders enabled access into the tree for aerial work, as they have always done, and some images from late fifteenth-century manuscripts indicate this. In terms of the tools and equipment used to fell mature trees or remove large branches, this is likely to have involved the use of equipment more commonly associated with forestry work rather than gardening, such as axes, saws and mattocks.

Various types and sizes of handsaw were used in forestry work at this time. During the Middle Ages, and for many centuries later, pit-sawing of large timber was the norm (Foot 2010, 132). In the conversion of timber following forestry felling operations, saw-pits were usually temporary and filled in immediately after use (Rackham 1976, 118). These were often located in the woodland close to where the trees had been felled. Crosscut saws had been known in Roman times and continued to be used in the Middle Ages. A significant advance in handsaw technology in this period was the gradual change from saws that cut on the pull to those that cut on the push. The earliest handsaw known in England to cut on the push stroke was excavated from an archaeological dig at a thirteenth-century midden on the Isle of Wight (Dunning 1939). The teeth were crosscut and very large, there being only four teeth to the inch. This thirteenth-century saw embodies most of the advantages required, such as a tapered blade, teeth raked and set, and cutting on the push. Several paintings from fourteenth- and fifteenth-century British churches depict this type of saw, but there is no earlier pictorial evidence.

With handsaws commonly used for timber conversion in woodland, it seems odd that there is no record of any attempt to use them for felling trees. Axes were the tools always used to fell all sizes of trees, a practice that continued right up to the nineteenth century.

Status of gardeners and tree-related professionals

The social and economic relations between the ruling class and the working population of Britain throughout much of the Middle Ages were dominated by the feudal system. This had developed around 700 CE and flourished up to the early fourteenth century, thereafter declining steadily until its legal abolition in England in 1660. The essence of the feudal system involved the aristocracy granting land and protection to their tenants, who then became villeins, workers with more rights than slaves but under various legal restrictions which differentiated them from the freeman. In return, these villeins swore an oath of allegiance to their lord and agreed to undertake a specified amount of work annually for their lord and also engage in military service on his behalf, when required. While this social and economic system operated without major challenge for hundreds of years, for various reasons it eventually became necessary to devise a more flexible arrangement (Robinson 1988, 16). This involved converting fixed labour services into money payments or rents and to hire people to work the land for wages. This changeover occurred by degrees from the twelfth century onwards, leading to a gradual approximation to modern conditions. The manorial demesne along with its farm and gardens came to be cultivated by hired workers who received wages in return, while the villeins (now tenants) bought out their service obligations by paying quit-rents. The commutation of these obligations was beneficial to both sides. It gave lords a freer hand in the management of their estates, it improved the standard of farm work since forced labour was grudging and inefficient, and it was obviously to the tenants' advantage to pay a fixed rent.

One of the main contributing factors to the demise of the feudal system in Britain was the Black Death, a bubonic plague pandemic that swept through Europe in the middle of the fourteenth century. When it reached England in 1348 it had a devastating impact, resulting in the death of 40–60 per cent of the population. In Scotland as much as a third of the population may have died and in Wales this was around a quarter. This decimation of the population created a severe shortage of workers of all kinds, and skilled craftsmen, such as gardeners, were particularly in demand. For workers generally and especially skilled craftsmen, the labour shortage encouraged them to break the longstanding ties with their feudal lord and seek work wherever they were wanted, often with much better terms and conditions than before. By the end of the fifteenth century the long-term impact of all this had led to a significant increase in the cost of labour, which for much of the medieval period had ranged from cheap to extremely cheap (Harvey 1981, 115).

In comparison to the Roman era, less is known about the work and status of gardeners in the Middle Ages, certainly in terms of many specialised roles such as arboricultural work. However, an examination of early manuscripts reveal words such as gardener, *ortolanus*, *hortolanus* or *gardinarius*, which all distinguish the gardener from William Langland's common labourer, 'Davy, the ditcher and delver', as described in his late fourteenth-century poem *Piers Ploughman*. The extent to which a gardener might be involved in the many additional crafts exhibited in a garden would depend on the wealth of its owner (Landsberg 1995, 86). In a simple garden, one gardener would normally be expected to be self-sufficient and undertake, to the best of their ability, all the required tasks. In addition to working with plants, this could involve constructing pole trellises or building a turf seat. At the other extreme, in a royal garden many craftsmen with their own specialist tools would be employed, supervised by a head keeper or gardener.

The ecclesiastical institutions with their often extensive gardens and orchards required skilled gardeners to maintain them. In some monasteries each obedientiary looked after his own garden, but in others there was a monk gardener responsible for the gardens (McLean 1981, 36–37). In either case, casual labour from outside was hired to help if the monastery possessed extensive gardens. Helpers did the digging, weeding, planting, fruit picking and so on, and all the unskilled jobs that the abbot considered should not take up too much of the monks' time. As well as casual help, some monasteries had one or two resident lay gardeners, who lived and ate with the other monastic servants. It was customary for monasteries, like all employers, to give their most important lay employees a robe or livery of office, and gardeners were often given boots, gloves and leggings as their livery, to honour their status and for basic practical reasons.

In some ecclesiastical institutions the head of the religious community was renowned for their skill and knowledge in a particular aspect of horticulture or arboriculture. In the late tenth century, Brithnod, the first Abbot of Ely, was famed for his skill in tree planting and grafting (Amherst 1896, 7). Together with Leo, his assistant gardener, to whom he passed on these skills, he is credited with making major landscape improvements around the Abbey. According to a contemporary chronicler, Leo is said to have laid out 'gardens and orchards elegantly, and planted choice fruit trees there in regular and beautiful order', as well as shrubs and shady trees (McLean 1981, 247). The garden historian Sylvia Landsberg (1995, 37) says this is the earliest English description of an aesthetically pleasing garden and it is notable that it refers to a monastery.

The royal gardens were another major employer of professional gardening expertise in the medieval era. In the late thirteenth century the principal royal gardens at Westminster, Windsor, Eltham and Sheen had salaried gardeners in charge, and so did some others (Colvin 1986, 12). They were usually paid considerably less than master masons or master carpenters, and although occasionally designated as 'Master', seem normally to have enjoyed an inferior status. Each held an independent post and there was no head gardener with general authority over all of the royal gardens. Any major expenditure on gardens was accounted for by clerks who supervised the royal works. As with the monasteries, distinctive robes were issued to head or master gardeners to indicate their status.

While the status of many gardeners, such as those in the more prestigious gardens, was steadily improving, it would be some time before any of these individuals came together in an organised body. In London, the Worshipful Company of Gardeners was first mentioned in the City of London Corporation records in 1345 (Barnes 1994, 5). After existing for centuries as a 'mystery' or 'fellowship', the Guild was finally incorporated by Royal Charter under James I in 1605. With their status rising steadily, several men who described themselves as gardeners were doing well enough to be subject to the 1381 poll tax taken in York (McLean 1981, 80). As gardener and herberer were listed as separate occupations, the latter probably meant a specialist in pleasure gardens and arbours, indicating they might have some additional arboricultural expertise. Gardening in general was regarded as one of the 'miscellaneous' crafts, like malting, bookbinding and scrivening, that could be pursued by a freelance operator who did not belong to a guild or company. By the mid-fourteenth century gardeners of this kind were making a good living in towns like London, York, Oxford and Cambridge.

With the emphasis on fruit trees in the medieval garden and the importance attached to this by the monarchy, the position of royal fruiterer was created. It was the fruiterer's responsibility to supply the royal household with sufficient quantities of fruit for the kitchens and fruit trees for their gardens, and to ensure that all this was of the highest quality. Although the fruiterers had, like the City gardeners, effectively existed as a company since 1463, they remained as a 'mystery' until receiving their Royal Charter, along with the gardeners, in 1605 (Gould 1912, xii–xx).

In the royal gardens and larger manorial gardens, it was sometimes the practice to appoint gardeners from overseas who had particular skills to offer. Their surnames were often suggestive of their origin, such as William Gascoigne at Eltham Palace, or William Florentyn who directed the Queen's garden works at Guildford Castle. They were usually given permanent appointments and relatively generous terms and conditions of employment. Other specialists were employed temporarily, as at the Manor of Cuxham, where the recording of an isolated expense of 1*d* for a gardener in connection with vine-pruning indicated only half a day's work (Landsberg 1995, 86). Other specialists engaged temporarily for arboricultural work would include those with expertise in grafting or pruning. In the more prestigious gardens when large trees needed to be removed or heavy branches pruned, and care was needed to avoid damage to nearby garden features, it is possible that specialist tree workers were engaged. As in Roman times, these are likely to have been forestry workers who used their own equipment. They would have been familiar with aspects of heavy tree work in regular forestry operations and in land clearance of mature woodland to convert this to agriculture. They may have been assisted in this work by the garden's regular full-time workers.

Those medieval workers engaged in forest-based activities, such as timber production or facilitating the hunting of game, are beyond the scope of this book. This includes those with the title 'Woodward', who were concerned with all matters relating to trees, timber and underwood as well as the protection of the deer (James 1991, 66). Also, the title 'Forester' in medieval times applied to an individual who had more to do with the hunting of game than the trees which sheltered them. However, we are concerned with the acquisition of skills by medieval forestry workers that might at some point have been applied to amenity trees. A fascinating example of this on a grand scale has come out of research by the historian Bill Linnard (2000, 27–35). We know that during the latter part of the thirteenth century in parts of Wales and England there was an abundance of men with tree-felling skills. This came about as the result of military campaigns by Edward I to subdue the rebellious Welsh. As part of these campaigns Edward ordered his armies to cut through often dense forest with invasion roadways deep into the heart of hostile territory. To reduce the risk of ambush the roadways were also very wide, between 150–250 metres. The tactic resulted in the felling of millions of trees and the permanent or temporary clearance of vast areas of forest. These felling operations were carried out by a large numbers of specialist workers, recruited from the English border and Midland counties. Each man was given a strong axe or hatchet suitable for felling large or small trees. While there is no record of the total number of men recruited, there must have been many thousands. During the 1277 campaign on one section of roadway from Flint to Rhuddlan, between 1,500 and 1,800 woodmen were engaged, and between 700 and 1,000 on the section from Rhuddlan to Degannwy. The men were carefully guarded, not only for protection but also to prevent desertions, although they were always paid. We can only speculate on what training these men may have received, and if they tried to use these tree work skills again after their military service, perhaps by working in the gardens and grounds of the aristocracy or monastic orders.

Conclusion

In the historical period covered by this book, the medieval era embraces over 1,000 years. However, professional horticulture and arboriculture in Britain during this era did not witness the same degree of development and innovation that occurred during the four centuries of Roman occupation. Among the medieval gardening cognoscenti there was a heavy reliance on the theories and practical advice of the classical Roman authors, with little appetite to question this. The priority, certainly before the Norman Conquest, was to keep alive existing ideas and practices rather than develop new ones.

Several factors may have contributed to the limited progress in arboricultural practice and the status of those engaged in this during the medieval era. First, there were relatively few ornamental gardens, with most gardens focusing on herbs, fruit and vegetables. Even where gardens had some ornamental function there were usually few trees, not least because many gardens were small and located in enclosed and confined spaces adjacent to buildings. This was especially the case in the early part of the medieval era. Furthermore, when trees were present in gardens, they were often fruit trees that do not reach large dimensions when mature. Any ornamental trees also tended to be small-growing types, although there were some notable exceptions to this. Consequently, these generally smaller trees required less demanding maintenance in contrast to large forest-type trees. Moreover, when the time came for their removal this did not constitute a major arboricultural operation requiring specialist skills. However, while the medieval period witnessed quite slow progress in the development of professional arboriculture there were signs toward the end of the era that more rapid advances might be made in the future.

References

Amherst, A. (1896) *A History of Gardening in England.* Second edition. Bernard Quaritch: London.

Anderson, F.J. (1997) *An Illustrated History of the Herbals.* Columbia University Press: New York, NY.

Barnes, M. (1994) *Root and Branch: A History of the Worshipful Company of Gardeners of London.* The Worshipful Company of Gardeners: London.

Batey, M. (1982) *Oxford Gardens: The University's Influence on Garden History.* Avebury Publishing: Amersham.

Calkins, R.G. (1986) Piero de' Crescenzi and the Medieval Garden. In: MacDougall, E. (ed.) *Medieval Gardens.* Dumbarton Oaks Research Library and Collection, Trustees for Harvard University: Washington, DC.

Colvin, H.M. (1986) Royal Gardens in Medieval England. In: MacDougall, E. (ed.) *Medieval Gardens.* Dumbarton Oaks Research Library and Collection, Trustees for Harvard University: Washington, DC.

Cox, E.H.M. (1935) *A History of Gardening in Scotland.* Chatto and Windus for Flora and Sylva: London.

Crisp, F. (1924) *Medieval Gardens.* John Lane Bodley Head: London.

De Bury, R. (1902) *The Love of Books: The Philobiblon of Richard de Bury.* Newly translated into English by E.C. Thomas. De La More Press: London.

Dunning, G.C. (1939) A Thirteenth-Century Midden at Windcliff, Near Niton. *Proceedings, Isle of Wight Natural History and Archaeological Society* 3(2), 128–137.

FitzStephen, W. (1772) *Description of the City of London.* Newly translated from the Latin original. B. White: London.

Foot, D. (2010) *Woods and People: Putting Forests on the Map.* The History Press: Stroud.

Gardener, J. and Amherst, A. (1894) A Fifteenth Century Treatise on Gardening by 'Mayster Ion Gardener'. With remarks by the Honourable Alicia M. Tyssen Amherst. *Archaeologia* 54(1), 157–172.

Gould, A.W. (1912) *A History of the Worshipful Company of Fruiterers of the City of London.* Printed privately: London.

Hadfield, M. (1979) *A History of British Gardening.* Third edition. John Murray: London.

Harvey, J. (1974) *Early Nurserymen.* Phillimore and Co.: London and Chichester.

Harvey, J. (1981) *Mediaeval Gardens.* B.T. Batsford: London.

Harvey, J. (1985) The First English Garden Book: Mayster Jon Gardener's Treatise and Its Background. *Garden History* 13(2), 83–101.

Hazlehurst, F.H. (1966) *Jacques Boyceau and the Formal French Garden.* University of Georgia Press: Athens, GA.

Hockey, S.F. (ed.) (1976) *Account Book of Beaulieu Abbey.* Camden Fourth Series, Vol. 16. Royal Historical Society: London.

Huxley, A. (1978) *An Illustrated History of Gardening.* Paddington Press: London.

James, N.D.G. (1991) *An Historical Dictionary of Forestry and Woodland Terms.* Basil Blackwell: Oxford.

Janson, H.F. (1996) *Pomona's Harvest: An Illustrated Chronicle of Antiquarian Fruit Literature.* Timber Press: Portland, OR.

Johnston, M. (2015) *Trees in Towns and Cities: A History of British Urban Arboriculture.* Windgather Press: Oxford.

Lamond, E. (1890) *Walter of Henley's Husbandry, together with an anonymous Husbandry, Seneschaucie, and Robert Grosseteste's Rules.* Longmans, Green and Co.: London.

Landsberg, S. (1995) *The Medieval Garden.* British Museum Press: London.

Linnard, W. (2000) *Welsh Woods and Forests: A History.* Gomer Press: Llandysul.

MacDougall, E. (1986) Introduction. In: MacDougall, E. (ed.) *Medieval Gardens.* Dumbarton Oaks Research Library and Collection, Trustees for Harvard University: Washington, DC.

Mann, M.E., Zhang, Z., Rutherford, S. *et al.* (2009) Global Signatures and Dynamic Origins of the Little Ice Age and Medieval Climate Anomaly. *Science* 326, 1256.

Manwood, J. (1615) *A Treatise of the Lawes of the Forest.* The Society of Stationers: London.

McAvoy, L.H. (2014) The Medieval *Hortus conclusus*: Revisiting the Pleasure Garden. *Medieval Feminist Forum* 50(1), 5–10.

McLean, T. (1981) *Medieval English Gardens.* William Collins: London.

Meyvaert, P. (1986) The Medieval Monastic Garden. In: MacDougall, E. (ed.) *Medieval Gardens.* Dumbarton Oaks Research Library and Collection, Trustees for Harvard University: Washington, DC.

Mudge, K., Janick, J., Scofield, S. and Goldschmidt, E. (2009) A History of Grafting. *Horticultural Reviews* 35, 437–493. Wiley-Blackwell: Hoboken, NJ.

Oschinsky, D. (1971) *Walter of Henley and Other Treatises on Estate Management and Accounting.* Clarendon Press: Oxford.

Rackham, O. (1976) *Trees and Woodland in the British Landscape.* J.M. Dent: London.

Robinson, J.M. (1988) *The English Country Estate.* Century in association with The National Trust: London.

Rogers, J.E.T. (1866) *A History of Agriculture and Prices in England, Vol. II.* Clarendon Press: Oxford.
Stannard, J. (1986) Alimentary and Medicinal Uses of Plants. In: MacDougall, E. (ed.) *Medieval Gardens.* Dumbarton Oaks Research Library and Collection, Trustees for Harvard University: Washington, DC.
Thacker, C. (1994) *The Genius of Gardening: The History of Gardens in Britain and Ireland.* Weidenfeld and Nicolson: London.
Whittle, E.H. (1989) The Renaissance Gardens of Raglan Castle. *Garden History* 17(1), 83–94.
Wright, T. (ed.) (1857) *A Volume of Vocabularies.* Printed privately: London.
Young, C.R. (1979) *The Royal Forests of Medieval England.* Leicester University Press: Leicester.

CHAPTER FOUR

Green Shoots in Tudor and Early Stuart Times

This chapter covers the Tudor and early Stuart period, beginning in 1485 and ending just before the outbreak of the English Civil War in 1642. The Tudor period began with the end of the War of the Roses in 1485, which led to the crowning of Henry VII of England and the founding of the House of Tudor, an English royal house of Welsh origin. After nearly 120 years the Tudor dynasty in England and Wales ended with the death of Elizabeth I when she died without issue in 1603. She was succeeded by James VI of Scotland, of the House of Stuart, who then became James I of England, thus bringing together England and Scotland in a unified kingdom covering both Britain and Ireland. This period embraced the English Renaissance, a cultural and artistic movement dating from the late fifteenth century to the early seventeenth century. This was associated with the pan-European Renaissance, although developments in England only began to blossom about a century later.

The relative peace that descended on England after the War of the Roses largely continued, with some notable relapses, through to the unification of the kingdoms of England and Scotland. England's preoccupation with civil war during the War of the Roses was probably a factor in Scotland's relative peace at that time which continued into the early part of the sixteenth century. Thereafter, the Anglo-Scottish Wars resumed periodically until the latter part of the sixteenth century, causing disruption to life on both sides of the border, only ending with the Union of the Crowns in 1603. This heralded the start of the Jacobean era that witnessed times of comparative stability and prosperity throughout Britain. When James's son, Charles I, became king in 1625 Britain was once again in a period of unrest that escalated during his reign, culminating in the outbreak of civil war in 1642.

Sources of information

While evidence of gardens and horticultural and arboricultural practice during the Tudor and early Stuart era is greater than that for the medieval period, most historians agree that this is still quite scarce in comparison to the latter part of the seventeenth century and thereafter. In the first instance, almost no garden has survived from this period; the vast majority have either disappeared or been subject to such change and renovation over the centuries that little or nothing remains of their original state. There are rare exceptions, most notably the recent rediscovery and renovation of the haunting landscape at Lyveden New Bield in Northamptonshire (Bradshaw 2004).

The lack of evidence on the ground is compensated to some extent by the documentary evidence available for this period, although even this is scarce in comparison to that existing a century later. From the early sixteenth century we see the emergence of the first printed works in English that contain information about gardens and gardening. Some contain fascinating information about tree planting and tree care and represent significant landmarks in the development of practical arboriculture. Thanks to the research of a few historians in recent years, we also have information on gardens and practical gardening taken from the account books of some of the landed estates of this era. Some useful information on arboricultural practices has been extracted from this research. Another source of information has been the garden books of some of the aristocracy and landed gentry that have been published in recent times, which give personal accounts of the work undertaken in their gardens. While some relate to a slightly later period than this chapter, given the limited progress in practical gardening around this time they do provide some useful insights into the early Stuart period. One further source of written information is the diaries of contemporary travellers, some from overseas, who often give enlightening descriptions of the gardens they visited. In terms of the research for this book, one major drawback of the various account books, garden books and diaries of this era is that they invariably focus on *what* was done rather than on *how* it was done.

Visual sources have also played a role in the research for this chapter. Images located in some of the early horticultural manuals have been especially informative and have provided some useful insights into the arboricultural practices of the time, along with the tools and equipment used to undertake these tasks. In contrast to the medieval era, images outside of books, such as separate paintings and drawings, were not used to any great extent in the research for this period.

Most studies by garden historians that deal with this period have concentrated on the few well-documented gardens of the royalty and aristocracy rather than on gardens in general or the work of the average gardener. However, there have been some notable exceptions to this, such as the research of Jill Francis (2018).

Garden design in Tudor and early Stuart times

During the medieval period Britain's royalty and nobility had lived in fortified castles or manor houses that allowed little space for gardens other than small utilitarian plots. Trees were largely excluded and were also kept away from the fortified walls because they might give cover to attacking enemies. However, the invention of gunpowder, together with more powerful and stable central government, made these defences and precautions redundant. In their place, the ruling class were able to build large unfortified red brick mansions with wide windows. These were no longer built on the tops of hills, but usually on lower-lying land, often surrounded by a moat to offer some degree of protection. As peaceful times continued this diminished the necessity of keeping all property within the protecting lines of the moat and it became customary to have gardens beyond it. With this additional space there was much scope for the development of gardens in terms of both design and size, although this took many decades to be realised. Although these developments took place across Britain, the pace of change was slower in Scotland than in England (Cox 1935, 3).

In the early days of the Tudor era gardens still had an essentially utilitarian purpose and were much the same as they had been for centuries, as mainly enclosed spaces providing food, medicines and scent for the household. They generally stayed that way for the remainder of the Tudor period, with some famous exceptions such as the gardens at Hampton Court Palace, which Henry VIII acquired when he stripped Cardinal Wolsey of his power in 1529, and those created at Henry's new palace of Nonsuch after construction started in 1538. While some authorities believe that garden history in Britain only begins as a coherent subject in the reign of Elizabeth I (1558–1603) (Thacker 1994, 54), it would be some years into her reign before the more prominent members of her privy council started making gardens (Jacques 2013a, 13). They were led by Sir William Cecil where the formal gardens at his magnificent new mansion, Theobalds, begun in 1564, were modelled after the Château de Fontainebleau in France.

When Robert Dudley, Earl of Leicester, constructed a glorious new pleasure garden at Kenilworth, he had only a few recent English examples of any magnificence to refer to (Jacques 2013a, 13). His garden was to be in the vanguard of a new wave of gardens in the middle years of Elizabeth's reign. It was created as a key element in Dudley's efforts to impress Elizabeth I, along with some extravagant hospitality and festivities, when she eventually visited Kenilworth in 1575. We know much about the design of this garden from a long letter written by Robert Langham, a member of Dudley's household. Langham devotes several pages to what has been called the 'earliest and longest description of an Elizabethan pleasure garden' (Strong 1979, 50). However, while he gives a detailed account of the garden's architecture, sculpture and overall ambience, his description of the planting scheme is brief and rather vague (Goldring 2013, 62). Although he praises the garden's plants, he provides few specific details. With regard to its trees, the reader must make do with references to 'holly treez', and to 'frute Trees bedecked with their Applz, Peares and rype Cherryez'. Langham's letter formed the basis of a major reconstruction of Dudley's Kenilworth garden in 2009.

The extravagant showpiece gardens of the Tudor elite were not representative of gardens at this time since only a few major landowners had the wealth or inclination to indulge in an extensive garden specially set aside for flowers and pleasure. Nevertheless, the second half of the sixteenth century was a time of great prosperity when yeomen and country gentlemen invested both capital and social aspirations in improving and updating houses or building new ones (Roberts 1999, 89). This included a greater focus on their gardens, although it is likely that many 'new' gardens were redevelopments or extensions of existing ones. This was less expensive than creating a new garden from scratch, which could cost a considerable amount of money. However, while there has been much research on the buildings of the late Tudor nobility and gentry, very little is known about what might be called their associated 'vernacular gardens' or landscapes (*ibid.*, 90). What form did these take and how important were aesthetic or ornamental considerations? Access to an understanding of these gardens cannot, on the whole, be gained through literature or detailed plans or through paintings. Nevertheless, a useful starting point is the evidence from contemporary practical literature produced by and for the country gentry engaged in managing their estates and gardens.

What comes through clearly from most of these contemporary authors is their preference for a geometrical layout for these gardens in what has been called the *compartiment* system in Tudor and early Stuart gardens (Jacques 1999). In *The English Husbandman* by Gervase Markham (1613, 112–127) he gives a diagram of his 'great square', a garden divided into four equal quarters by alleys, and suggests ways of ornamenting them. In William Lawson's (1631, 12) *A New Orchard and Garden* of 1618, he says, 'The forme that men like in generall is a square'. In *Paradisi in Sole* by John Parkinson (1629, 3) he also recognises the usual preference for a geometric layout and states, 'The foure square forme is the most usually accepted with all, and doth best agree to any mans dwelling'.

The inspiration for this formal geometric arrangement came from the Renaissance gardens of Italy and France. From the late fifteenth century Italian architects became increasingly concerned with the organisation of the garden area (Jacques 2013b, 95–96). They developed a system in which square *compartimenti* could be appreciated together, usually in groups of four or eight, to make a symmetrical and harmonious whole. This Renaissance tradition in geometric garden layout spread to England by the middle of the sixteenth century. However, the word 'compartiment' was unknown in England where the terms used were 'plot' (sometimes 'plat'), 'square' and 'quarter', although none of these are direct equivalents.

While there is little visual evidence of the planting arrangement in gardens in Elizabethan times, an indication of this can be found in northern continental sources (Watkins 2013, 166–167). Of particular interest are the numerous images of geometric gardens produced by Hans Vredeman de Vries (1527–1606), a Dutch Renaissance architect, painter and engineer. Not only are his paintings and engravings of imagined gardens wonderfully detailed,

FIGURE 11. An engraving of a geometric garden produced in 1583 by Hans Vredeman de Vries for his *Viridariorumque Elegantes et Multiplicis Formae*. It shows elaborately clipped trees in prominent positions within a labyrinth or maze-like design.

but a positive connection exists between him and the Earl of Leicester as the design of one of the tapestries produced for Leicester in 1585 was taken directly from a Vredeman image. Vredeman published a series of ornamental pattern books that were highly influential abroad. *Hortorum Viridariorumque Elegantes et Multiplicis Formae* is the part of this series devoted to gardens, published in Antwerp in 1583. In a total of twenty-nine engravings, Vredeman showed gardens subdivided into compartments by fences, hedges and pergolas. The area within the compartments was patterned with maze-like designs in squares and curves. In the beds a distinct planting hierarchy can be seen, in which the largest plants or trees are given the most prominent position. The tallest, often an elaborately clipped tree or large shrub, is placed in the centre or at the corners of the compartment (Figure 11). There is no axial relationship between garden and house in most of the plans, so the garden is seen as sufficient unto itself, as it had been for centuries. Vredeman de Vries' book was hugely influential throughout Europe and must have been invaluable to the next generation of garden designers, such as Salomon de Caus and Jacques Boyceau.

On the limited evidence available it seems that the garden in Britain was quite similar to the geometric European model. Early *compartimenti* or quarters were surrounded by waist-high rails with latticing down to the ground, which helped to keep the unwanted off the beds (Jacques 1999, 36). Rails and lattice were also used to support vines and other climbing plants. On the Continent, a progression from rails to hedges seems to have been underway by the mid-sixteenth century and it was not long before there was a gradual trend in England where privet or whitethorn came to be used without the support of rails. In the early Stuart era at Nunwell on the Isle of Wight, Sir John Oglander (1936, 203) advised that a hedge should replace the timber fencing that had become rotten, stating, 'A good white thorne hedge will doo as well as ever the pales did'. For the embellishment of the quarters, planting fruit trees at the corners became common as well as planting them just within the rails. When hedges appeared, fruit trees sometimes continued to be planted just within the hedge or even in the hedgeline. Quarters did not always mean there were four of these and often there were six or eight. While the total of these was usually an even number this was not always the case. When the Great Garden at Theobalds was created sometime between 1575–1585 this was set out in nine squares (Andrews 1993, 135). When described in 1650, they were bordered by hedges of whitethorn and privet and 'cutt into a handsome fashion with cherry trees planted at the angles'.

A fascinating insight into a typical garden for the gentry during this period can be obtained from William Lawson. In his book *A New Orchard and Garden*, first published in 1618, he gives a plan for what he regards as his ideal garden and this may have been based on land held by Lawson at the mouth of the Tees (Harvey 1982). This shows a simple geometric arrangement of square plots, divided by alleys and bounded by fences and wooded walks (Lawson 1631, 13). There are six squares in total and only one, the knot garden, has a purely ornamental purpose. The others include two squares devoted to the kitchen garden, two set with trees (likely to include fruit trees) that are interspersed with other garden features and ornaments, and another area that is set with trees (probably fruit trees) in a formal triangular arrangement. Apart from the six squares, other features are identified in the plan, such as the various walks or *allées*, the mounts set at each of the four corners of the garden, and the stairs connecting the squares that must have been at slightly different elevations. What may initially be confusing to the modern reader is Lawson's tendency to use the terms orchard and garden synonymously. However, anyone reading through his book and studying his plan would be left in no doubt that, for him, the orchard *is* the garden. What is also apparent from Lawson's plan, from an arboricultural perspective, is that many of the squares and features involve the use of trees (Figure 12).

FIGURE 12. William Lawson's plan of an ideal garden from 1618. The parts where he refers to trees include A: 'All these squares must be set with trees'; B: 'Trees 20 yards asunder'; H: 'Walks set with great wood thick'; I: 'Walks set with great wood round about your orchard'; L: 'The out fence set with stone fruit'.

While Lawson's plan was not typical of aristocratic gardens of this period, it does include many features that were to be found in the larger gardens across Britain. We now consider those garden features from Tudor and early

South.
N
N
R
M
A
E
A
M
K
I
P
C
G
G
B
F
C
L
H
East.
West.
G
G
G
D
D
O
O
M
North.
M
R

Stuart times that involve the use of trees and whose creation and maintenance necessitated a degree of practical arboricultural work.

Tree-related features in Tudor and early Stuart gardens

In discussing various tree-related garden features, we will try to avoid placing them into particular styles of garden often used by garden historians. Instead, we will rely more on identifying the time period when they may have become apparent or when they ceased to be widespread. As much as garden historians like to talk about Elizabethan, Renaissance, Jacobean or Caroline gardens, the contemporary gardeners of the day were clearly oblivious to these retrospective labels and simply acted according to their own inclinations (Francis 2018, 328). Furthermore, many of the garden features described below were popular throughout this period and do not fit conveniently into any particular category.

Knots and topiary

Of all the features associated with gardens at this time perhaps knots are the most archetypal and well known. These were square or rectangular beds laid out in complicated geometric patterns and planted with interlacing ribbons of low, evergreen plants such as germander, santolina, hyssop or rosemary. Sometimes the pattern worked in living green was an end in itself, the interstices being filled with coloured earth or gravel. Otherwise, the design was more open, and the spaces filled with flowering plants. The more complicated the pattern of the knots, the more they were admired, and knot-masters were regarded as highly skilled craftsmen. In Cardinal Wolsey's garden at Hampton Court it was said that the knots were 'so enknotted, it cannot be exprest' (Cavendish 1825, 11). While knot gardens were common in Tudor and early Stuart gardens, it was not until the 1620s that the tree species box (*Buxus sempervirens*) appeared in them. When describing the methods employed for setting out 'trayles, knots and other compartiments', Parkinson (1629, 5–6) recommends the dwarf French or Dutch box. Although box is now commonly used as an edging plant in British gardens, this is thought to be the first time it was recommended for that use in a published book.

Writing in 1613, Markham (1613, 120) regarded knots as traditional ornaments for gentry gardens, although he stated that they were 'ancient and at this day of most use amongst the vulgar though least respected with great ones, who for the most part are wholly given over to novelties'. This suggests that by this time knot gardens were becoming unfashionable, at least in the grandest gardens. Indeed, some garden historians believe that in the Jacobean period the knot garden was replaced by the French *parterre de broderie* (embroidered parterre) (Strong 1979, 40). However, contemporary evidence indicates that knots continued to be laid out in the gardens of the gentry long into the seventeenth century (Francis 2018, 183).

Another novelty to emerge as a feature of gardens in this period was topiary work. This had been prominent in Roman gardens but became much less common and less extravagant during the medieval period. It had been brought back into Britain from Italy via France and from early Tudor times no garden of any size was complete without topiary or 'antike worke'. By the 1530s, John Leland, the English poet and antiquarian, would refer to several gardens with topiary work (Leland 1907). Some modern historians believe that the gardens of Hampton Court had lots of fantastic topiary figures in Henry VIII's time and this was certainly the case during the reign of Elizabeth I (Strong 1979, 33). The account of its Privy Garden in 1599 by Thomas Platter, a German visitor, describes the most elaborate topiary comprising human figures, animals and monsters, all presented in massed displays (Williams 1937, 200). In contrast to the situation in England, exaggerated topiary work did not become fashionable in Scotland until after the Restoration in 1660 (Cox 1935, 36).

Only a very limited number of tree species were used in topiary at this time. While juniper and privet were used, yew was often preferred as it grew slowly and was long-lived. In the 1620s, Parkinson (1629, 6) noted that box was still not widely used and was only growing in 'the gardens of those who are curious'. Nevertheless, he sees its advantages in being 'ever greene, it being reasonable thicke set, will easily be cut and formed into any fashion one will'. He also noted that it was slow-growing, an advantage in terms of the maintenance of mature topiary. While recognising the versatility of box and yew for topiary, Parkinson also encourages the more imaginative use of privet 'to make hedges or arbours in Gardens, whereunto it is so apt, that no other can be like unto it, to bee cut, lead, and drawne into what forme one will, either of beasts, birds or men armed or otherwise' (*ibid.*, 445).

William Lawson was full of admiration for the range of fantastic topiary images that were being created by highly skilled gardeners. In 1618 he wrote, 'Your Gardner can frame your lesser wood to the shape of men armed in the field, ready to give battle: of swift-running Greyhounds, or of well-sented and true running Hounds, to chase the Deere, or hunt the Hare. This kind of hunting shall not waste your corne, nor much of your coyn' (Lawson 1631, 71). However, not everyone was enthusiastic about the fashion for topiary. Francis Bacon famously disliked both knots and topiary images, although he approved of topiary in architectural forms as pyramids and columns. Writing in 1625 in his famous essay *On Gardens*, he declared, 'I, for my part, do not like images cut out of juniper or other garden stuff; they be for children' (Bacon 1902, 15).

Mazes and labyrinths

The labyrinth and the maze were similar garden features in the Tudor and early Stuart period and both involved the use of tree species. In the literature on gardens from those early times 'labyrinth' and 'maze' were frequently used as synonymous terms, although some authorities maintain there are distinct differences. Perhaps the most significant difference is in how they are actually experienced. Frank Crisp (1924, 70) put this very simply when he stated, 'The labyrinth is more properly applicable to arrangements which are simply a question of length. If you go on (and on) you are bound to get to the centre. The term "maze" is more properly applicable to those arrangements which have at various points a "block"' which obliges the visitor to retrace his steps'. However, we will consider them together.

The concept of the labyrinth, important in some ancient civilisations, was seized on by medieval Christianity in the twelfth century as a symbol of man's tortuous journey towards salvation. Representation of labyrinths and mazes appeared on cathedral and church floors and in gardens where they were constructed from turf (Matthews 1922, 54–99). The Tudors, with their passion for puzzles and conceits, delighted in mazes and developed them further as a garden feature. Early hedge mazes were generally only knee-height, composed of hyssop, lavender or thyme, and were popular as attractive, fragrant and entertaining garden features.

Sir William Cecil's magnificent gardens at Theobalds had a maze by the late 1560s, described by a contemporary visitor as being 'large and square and a beautiful *jet d'eau* in the centre' (Matthews 1922, 114). Another famous Elizabethan maze was at Nonsuch, where Platter noted in 1599 that the hedges were so high that one could not see through them (Williams 1937, 197). That would have been atypical of early mazes, in which the hedges were only a little above ground level, so that the whole maze could be seen at one view and the divisions stepped over. In the Jacobean era the taller hedge maze started to become more popular and tree species were commonly used. William Lawson described this in 1618 when he wrote, 'Mazes well framed a mans height may perhaps make your friend wander in gathering of berries till he cannot recover himselfe without your helpe' (Lawson 1631, 71).

Mounts, banqueting trees and arbours

The larger formal gardens of this period were best viewed from a high vantage point to show off their intricate patterns and overall design. To help facilitate this, the 'mount', a raised pile of earth above the general level of the garden, became an important garden accessory. While they were often covered in trees or sometimes with topiary, they also had views out across the broad expanse of the garden to the landscape beyond. To accommodate a relaxed contemplation of the scene, a seat or arbour was usually provided. These mounts were frequently accessed by a spiral path winding its way to the top, as with the mounts at Lyveden New Bield (Bradshaw 2004, 24) and at Wressle Castle in Yorkshire. The latter was described in the 1530s by John Leland (1907, 53), who wrote, 'And yn the orchards were mountes *opera topiario* [with topiary work], writhen about with degrees [steps] like the turnings of cokilshelles, to cum to the top without payn'. The Copperplate Map of London, surveyed between 1553 and 1559, has a wonderful stepped example of a mount topped by a tree in a garden at Bishopsgate, just outside the walls of the City of London.

Another garden feature designed to view the garden from a height was the banqueting tree. As the name suggests, its primary function was to allow a large number of people to dine not only outdoors but in a large tree where they had views over the garden. At this time the ultimate status symbol in a gentleman's garden was the banqueting house. Some of these were lavishly decorated buildings with two or more storeys. The banqueting tree was just a less expensive but perhaps more exhilarating version of this. In contemporary literature the dining platform in a banqueting tree was usually referred to as an arbour and this may have had some similarity with those arbours on the ground in the garden. Lime trees were commonly trained to make these elaborate arbours, two and even three storeys high. Parkinson (1629, 610) says of the lime, 'It is planted both to make goodly Arbours and Summer banqueting houses … a second above it and a third also'. He then describes a three-storey arbour of this kind he had seen at Cobham Hall, in Kent, which was so roomy that there was space for at least fifty men in each of the first two storeys:

> And I have seene at Cobham in Kent, a tall or great bodied Lime tree, bare without boughs for eight foote high, and then the branches were spread round about so orderly, as if it were done by art, and brought to compasse that middle Arbour: And from those boughes the body was bear againe for eight or nine foote (wherein might be placed halfe an hundred men at the least, as there might be likewise in that underneath this) and another rowe of branches in that encompassing a third Arbour, with stayres made for the purpose to this and that underneath it: upon the branches were boards to tread upon, which was the goodliest spectacle mine eyes ever beheld one tree to carry.
>
> Parkinson 1629, 610

The building of this banqueting tree was a remarkable feat of construction, especially for those early times, and would have involved a range of skilled craftsmen. Looking back on this today, we can still only marvel at the expertise in tree surgery and carpentry this would have required.

Arbours were a common feature at ground level in Tudor and early Stuart gardens. The term arbour normally describes a shady garden alcove, often with a curving arch shape, with the sides and roof formed by trees or fragrant climbing plants such as roses and honeysuckle trained over a framework. A tunnel-arbour is an extended version of this, a form of covered walkway where people can shelter from the sun or rain. In some respects an arbour is similar to a pergola, although that word originally meant a projecting eave. In Thomas Hill's (1577, 1) book *The Gardener's Labyrinth* there is an illustration of an arbour being built as a shade against the sun (Figure 13).

FIGURE 13. Elizabethan gardeners building an arbour as a shade against the sun. The image is from Thomas Hill's book *The Gardener's Labyrinth* (1577).

Wildernesses and avenues

One garden feature in this period that was all about trees was the wilderness. This was essentially a designed grove or woodland with paths cut through it. Despite its name, the wilderness was in no sense a wild place but a piece of landscape gardening in the grand manner, in which any apparent wildness of nature had been carefully contrived by the gardener's art. In some wildernesses the hand of the gardener seems to have been so evident in the finished result that the whole area was no more than a formal garden, with trees instead of flowers and shrubs. Whatever the extent of its formal design, the essence of the wilderness is that it should be a beautiful and stimulating area in which to wander or pause for contemplation. In formal gardens of the time the wilderness was laid out at some distance from the knot garden or parterre, features that were usually close to the house.

One of the first wildernesses to be created in Britain was at Nonsuch. After the death of Henry VIII the gardens at Nonsuch were transformed in the Elizabethan period by Lord Lumley (Strong 1979, 63). During his visit to Italy in 1566 Lumley had seen Italian gardens with their extensive use of trees and shrubs planted in clumps or *boschetti* focusing on sculptural and architectural groups, each with its symbolic or allegoric meaning. The French equivalent to this was the *bosquet.* Lumley was resolved to reproduce a similarly powerful effect in his redesign of Nonsuch. However, within a few years the wilderness was being used mainly for recreation and had lost it original meaning as a place of quiet contemplation.

One of the earliest descriptions we have of a wilderness at Nonsuch is from Anthony Watson, the rector of Cheam. Although his written description of Nonsuch is not dated, this is likely to have been around 1590, and it was the first time that the term wilderness was used to describe part of a garden. He comments on the trimmed trees at the end of the south walk in the wilderness that were known as the 'canopies'. Watson wrote that some of these trees 'now by wonderful skill are trimmed into a circular shape and offer perpetual moist shade where the heat of the sun is not felt' (Watsoni 1590, 9–10). He added, 'The number, the arrangement, and the beauty of the trees, which have not just come about by chance, flourish to such a point of praise and admiration through the care of skilful hands'. Clearly, a considerable amount of expert pruning work was undertaken regularly to maintain the appearance of these trees. In another reference to trees, Watson describes the terraces at Nonsuch as 'set out with lofty and magnificent tree-lined walks'.

Thomas Platter, the German traveller, was another contemporary visitor to Nonsuch, arriving in 1599 when Queen Elizabeth was in residence. He gives the following description of the wilderness:

> In the very densest part of the wood about here a great many trees are uprooted and cleared, within a breadth of some eighteen to twenty feet, along a straight course, so that there is a vista from one end to another ... while the delicious song of the birds in the tall trees densely planted along the sides in ordered array afford one great delight.
>
> Williams 1937, 197

Where the trees had been uprooted in the wilderness, this was not only to provide alleys and vistas but also the space for a bowling green and a tennis court. Just to the west of the wilderness was another area of densely planted trees known as the Grove of Diana (*ibid.*, 195). This was ornamental woodland with topiary arches, shady walks and a fountain.

Knole is a house and estate at Sevenoaks in Kent belonging to the Sackville family. When Sir Thomas Sackville began creating the garden in 1603 he included a large area known as the Wilderness. This is especially interesting because it was a working, practical landscape that was very different from other wildernesses during this period, which were viewed as designed *bosquets* within the pleasure garden (Taylor 2008, 242). In June 1612, Lady Anne Clifford, the wife of Sir Thomas's grandson, mentions in her diary that wood was being harvested in the wilderness. Furthermore, research suggests that its use as a place to grow and harvest valuable wood crops continued throughout the seventeenth century.

By 1634 what a wilderness should look like and how it was used were defined by that developed at Wilton House in Wiltshire (Taylor 2008, 237). The magnificent garden at Wilton was created for the fourth Earl of Pembroke by Isaac de Caus, the French landscape gardener and architect. The wilderness was a major part of this and was in the style of an Italian *boscetto* designed with symmetry and neatness. The rigidly balanced groves on each side were the backdrop for statues of Flora and Bacchus, as beautifully illustrated in de Caus (1645) book *Le Jardin de Wilton* (Figure 14). The location of the wilderness

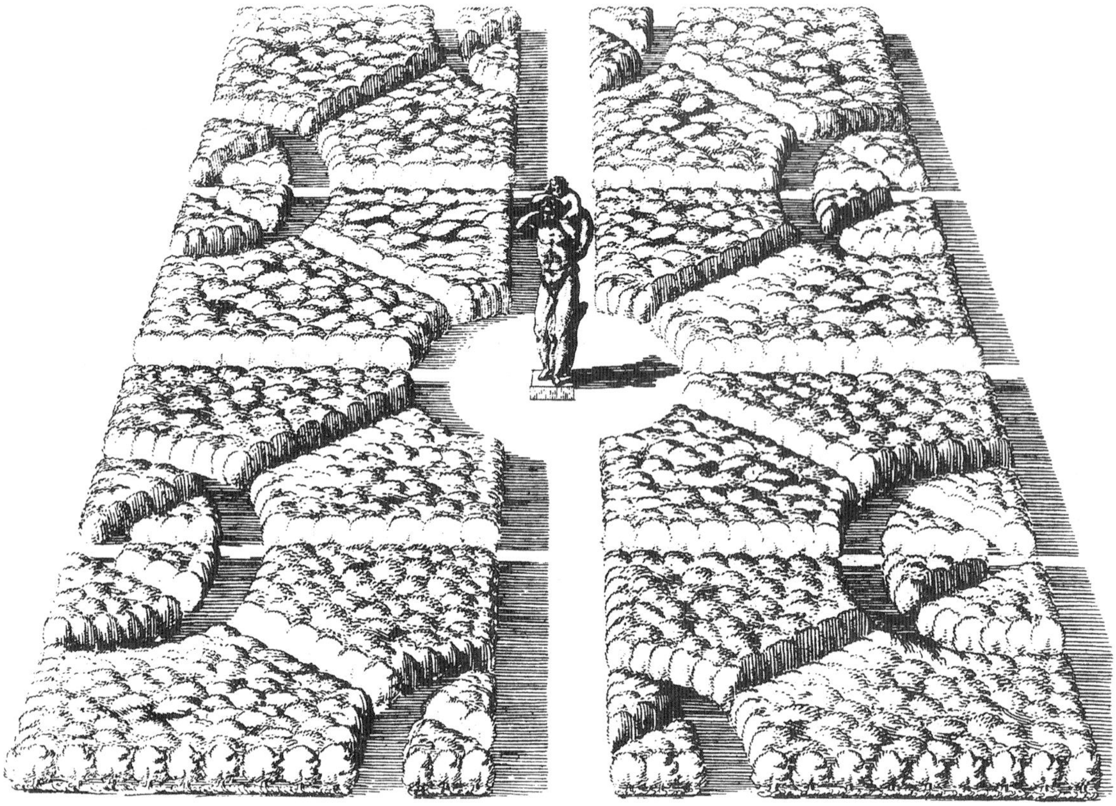

FIGURE 14. Part of the Wilderness at Wilton with its rigidly balanced groves on each side of the statue of Bacchus. This was a celebrated feature of the magnificent garden created for the fourth Earl of Pembroke around 1634 by Isaac de Caus.

in the garden was probably dictated by an effort to conceal the river running through it with a mass of trees and shrubs. The very fame of Wilton meant that whatever the other earlier wildernesses may have looked like, or were used for, they were either gradually forgotten or overshadowed by this.

By the time of the Civil War the wilderness was a well-established feature within aristocratic gardens. In 1639, the preacher Michael Jermin (1639, 36) remarked on the popularity and benefits of 'a widernesse which often is adjoyned to great gardens belonging to great houses, and by a multitude of thick bushes and trees, affecting an ostentation of solitariness in the midst of worldly pleasures'. However, the creation of a wilderness was now no longer just the preserve of the aristocracy, having reached the landed gentry in many parts of England and Wales, although less so in Scotland.

Another feature of gardens that became increasingly popular among the monarchy and the aristocracy in this era was the planting of long and impressive avenues leading up to their great houses. Although there are a few Tudor examples, by the early seventeenth century the avenue of lime, elm or horse chestnut had become a recognised aristocratic symbol. While it has not been possible to find any use of the term 'avenue' at this time in Britain, these distinct landscape features were often covered by the term 'walk'.

In 1600, a Czech nobleman called Baron Waldstein went to see Burghley House in Lincolnshire and noted 'the drive which leads up to the main entrance is an usually long one and great care has been taken to choose the best sites for planting the trees on either side of it' (Groos 1981, 111). Audley End, built by the Earl of Suffolk in 1605–1614, had an avenue of double rows of lime trees which led to the great entrance gateway. Extending the formality of the garden into the landscape beyond was not only becoming fashionable with the nobility, it was also about making a statement regarding their own grandeur and influence. This trend towards long and impressive avenues confirmed that the ruling class were now ready to embrace and control the whole landscape, rather than seeking to protect themselves against it. The trend was much influenced by continental European ideas on garden and landscape design, which will be discussed later.

Developments with other gardens and greenspaces

Having examined some tree-related features of gardens in the mainly rural residences of the monarchy, aristocracy and gentry, we now look at tree-related developments with other gardens and open spaces. These include monastic gardens, orchards, urban public walks and gardens, and plant collections.

Monastic gardens and their demise

As noted in the previous chapter, medieval monastic gardens played a crucial role in preserving the theory and practice of horticulture and arboriculture inherited from Roman times. The dissolution of these monasteries in England, Wales and Ireland between 1536 and 1541 on the orders of Henry VIII sparked a revolution in land tenure as enormous tracts of land passed into new hands. The separate kingdom of Scotland under James V was not part of this. In the 1520s, church lands covered about one-sixth of Henry's kingdom, with much of this devoted to agricultural and horticultural production. The confiscations of land were quickly followed by the sales and grants to new owners, which gave Henry VIII money to pursue his policies of centralisation and altered the social balance of England as a whole. Three-quarters of the Church's lands were sold off by 1558, mostly to the gentry class which expanded rapidly and thus became the strongest supporters of the Tudor monarchy, its land reforms and the new Protestant Church.

When the monasteries were destroyed by Henry VIII's commissioners, many of their gardens were destroyed with them. This represented a tremendous blow to the extent of gardens and gardening expertise in England and Wales. When the new owners took over, many of those gardens and orchards that remained were not maintained to anything like their previous standard or extent. There was one significant exception to this general pattern of destruction and neglect.

Westminster Abbey was dissolved in 1540 but, most unusually, the Benedictines returned to it in Mary's reign under Abbot John Feckenham (McLean 1981, 55). However, after Mary's death they lost it for good when Elizabeth I confiscated the abbey in 1560. When she sent for Feckenham to inform him of her decision he was busy planting trees in the Abbey grounds. In the words of a contemporary historian: 'Queen Elizabeth, coming to the throne, sent for Abbot Feckenham to come to her, whom the messenger found setting of elms in the orchard of Westminster Abbey. But he would not follow the messenger, till first he had finished his plantation'. This was Feckenham's last activity as a free man as he was subsequently imprisoned, where he remained until his death. His abbey was made into a college, although his elms survived as a row of trees until 1779, when they were cut down.

Orchards

Great attention was paid to fruit trees in Tudor days with Henry VIII setting the lead in the 1530s and 1540s when he sent some of his gardeners on several trips to France to obtain apple trees and grafts for the royal orchards. The royal orchard at Teynham in Kent was established around 1533 and soon became famous for the quantity and quality of its fruit. The history of the orchard is recorded in a rare little pamphlet entitled *The Husbandmans Fruitfull Orchard* written by N.F. (unknown author) and published in 1608. This recounts how Richard Harris, Fruiterer to King Henry VIII, 'fetched out of Fraunce great store of graftes, especially pippins, before which time there were no pippins in England' (N.F. 1608, Epistle to the Reader). Harris then used the fruit trees he propagated to plant an orchard of 140 acres (57 ha), which was a vast area to dedicate to fruit production and must have caused quite a sensation at that time.

While England was trying to develop its own fruit industry, there was general recognition that continental Europe had for many years been leaders in pomology. At the start of *The Husbandmans Fruitfull Orchard* the author looks back over recent history, noting the reliance on continental horticulture. Several writers in the late sixteenth century comment on how proficient in pomology were members of French religious orders and bemoan the loss of such skills which followed the Dissolution of the Monasteries in England. Consequently, those English gardeners who were experts in the care and management of fruit trees were highly regarded and much in demand.

The common practice in medieval times where orchards had both a utilitarian and ornamental or recreational function continued into the Tudor period. However, as time progressed there appears to have been a steady transformation in the gardens of the monarchy and nobility where the focus was increasingly on their ornamental and recreational role. At the same time the orchard steadily became a distinct and mainly utilitarian feature that was quite separate from any pleasure garden. This trend was epitomised by John

Parkinson when he discusses the choice of trees for the pleasure garden in his *Paradisi in Sole*. Where previously there may have been many different fruit trees he was now far more selective. He states, 'The beautifull shew of these three sorts of flowers [cherry, apple and peach] hath made me insert them into this garden … although the rest of their kindes I have transferred into the Orchard, where among other fruit trees, they shall be remembered' (Parkinson 1629, 402).

This trend, however, was less marked with the gardens and orchards of the gentry. As noted, for William Lawson the garden and orchard were synonymous, and this also comes through very clearly in Gervase Markham's *The English Husbandman*. In the organisation of their ideal gardens the concept of both 'profit and pleasure' is fundamental and the one which defines them as being the distinctive landscape of the 'yeomanly gentry' (Roberts 1999, 101). That there was a contemporary understanding of a recognisable 'gentry' landscape is suggested most clearly by Markham. Although he refers to the practices and designs seen in the great gardens to support some of his ideas, he makes a clear distinction between the gardens of the nobility and those that are appropriate for his own class. As noted above in Markham's (1613, 120) comments on knots, he was dismissive of the gardens of the 'great ones' since these are 'given over to novelties' and instead he prefers simple traditional designs. Furthermore, in his characterisation of the 'great ones' what strikes him most is their ability to take land out of production and devote this entirely to the creation of an ornamental garden governed solely by the dictates of fashion. By contrast, in the gentry garden of the 'playne and honest husbandman' Markham does not see any contradiction between pleasure and profit, and he confidently states, 'I will therefore joyne them both together, and make them onely but one Orchard' (*ibid.*, 33).

With regard to the planting design for trees in the orchard, both Markham and Lawson maintain their preference for geometric arrangements, which also has benefits for 'profit and pleasure'. They both recommend that trees are planted at equal distances in rows, in quincunx pattern. This is not only for efficient fruit picking and management of the orchard, but also for ornamental effect, so that 'which way soever a man shall cast his eyes yet hee shall see the trees every way stand in rowes, making squares, alleyes and devisions, according to a mans imagination' (Markham 1613, 37).

The growth in the number of orchards in Tudor and early Stuart times seems to have rendered their legal protection increasingly necessary and various laws were enacted to this end. In England, a curious act was passed in the time of Henry VIII stipulating that his subjects must not, after the first day in May, 'maliciously, willingly or unlawfully, bark any apple trees, pear trees or other fruit trees of any other person or persons' (Amherst 1896, 103). The penalty for contravening this law was to give a payment to the owner for treble the cost of the damage caused and also to pay a hefty fine to the authorities.

Other amenity landscapes with trees

Not all tree planting in the vicinity of houses during this period took place in formal or ornamental gardens. In Scotland, where there were fewer of these gardens than in England, tree planting around houses was often primarily for shelter from adverse weather conditions, although this could also have some aesthetic role. An ordinance of 1535 laid down that everyone who possessed land worth 100 pounds (Scots) had to plant woods and orchards round the house (Brown 1904, 9). The landowner also had to compel his tenants to plant one tree for every merkland that they held for him. Merkland was an old Scottish unit of land measurement based on land rent, not actual area. This insistence on planting by tenants indicates it was difficult to persuade them to plant trees of their own volition. In a royal rental book of 1541 for what must have been a model estate, strict conditions were laid down for tenants that required the creation of gardens as well as tree planting (Cox 1935, 28). Each tenant had to have near his house a large well-furnished garden, hedged with thorn, willow, alder or aspen. For every acre of land for which he paid a silver merk he had to plant three trees, ash, plane (sycamore) and elm. Such conditions were generally obeyed by the richer tenants, but it is very doubtful whether the poorer ones paid any attention to them, so imbued were they with the widely held belief that trees around a field were more of a curse than a blessing. Similar legal inducements to plant and retain trees on residential properties were quite common in England, and this included many urban properties and their gardens. Archival sources reveal an awareness of the responsibility of tenants and landowners to regularly plant and take care of trees on their properties (Stout 2015, 161). The importance of planting trees was written into many people's tenancy agreements and such tenancies could also be revoked if tenants cut down trees 'contrary to custom'.

With the increased population in the Tudor era the spaciousness in the urban fabric following the Black Death began to disappear as gardens and yards were built over to provide new accommodation. This random infilling of open space not only removed any existing trees in the way, it also meant that the smaller houses in London and other major settlements often had to make do with urban gardens of equally small size and often awkward shapes that were not conducive to trees, particularly those of any size (Johnston 2015, 105). However, many of the wealthy Tudor and Stuart citizens of London and other cities still retained their large gardens which gave scope for quite extensive designs and various garden features, often with some trees. While less wealthy citizens with smaller urban gardens could not compete with these, much of the basic style of those gardens and many of their more popular features were still reproduced on a limited scale. 'Alyes' of arched trees and shrubs together with rose- and jasmine-covered wooden arbours were very popular, as were small-scale knot gardens. Hedges of various species were also planted around the knot garden and because of their convenient height they were often used for spreading out and drying laundry, a use of hedges also found in the much smaller gardens of poorer people. As the urban population increased steadily during this period and towns and cities expanded outwards, more garden space was created, offering at least some opportunity for more trees, both for ornament and utility. In London, in the Bishopsgate and Moorfields areas immediately north of the City walls, gardens were leased separately, without houses, rather in the manner of modern allotments. The steady increase in the number of urban trees in these areas would offer future opportunities for practical arboriculture.

A new form of British urban landscape that arose in the seventeenth century was 'the walk' or tree-lined promenade (Borsay 1986; Johnston 2017, 9–25). These were essentially garden *allées* outside the garden, many of which were used for promenading rather than for games. Although privately owned, these walks could be made available to members of the public, usually on request. Seeing the fashionable world let loose in these places, visitors from abroad sometimes mistook them for public parks. In some respects these early seventeenth-century walks marked the transition from the introspective and essentially non-public enclosed garden of medieval times to the relatively public one which was conceived as integral to later town planning.

As early as 1597, a walk or promenade was laid out by the Benchers of Gray's Inn, then as now a centre for the legal profession (MacLeod 1972, 105). The Benchers engaged Francis Bacon, later Lord Chancellor and an authority on gardens, to organise the laying out of the walk and the planting of the trees that comprised mainly elms. Known as the 'Walkes', this remained a fashionable London promenade for over 200 years. One of the most influential public walks was on Moorfields in London, a connected series of fields just beyond the old city walls that the Corporation of the City of London leased from St Paul's Cathedral (Harding 1990, 49). In the sixteenth century this undeveloped piece of land was subject to common rights and was the nearest open space to the city centre. It was used for a variety of activities but was in danger of encroachment and abuse as a dumping ground. In 1593, the Corporation ordered that it be inspected and subsequently kept clean and tidy. In 1605–1607, two gardeners were contracted to drain and level the ground and plant elm trees along new paths. The result was a transformation of this marshy and neglected area into what the contemporary writer Richard Johnson (1607, 3) described as: 'Those sweet and delightfull walkes of More fields as it seems a garden in this Citty, and a pleasurable place of sweet ayres for Cittizens to walke in'. The success of the new walks at Moorfields was replicated elsewhere in London. In June 1619, there were reports that 'Lincoln's Inn Fields are to be laid out in walks, like Moorfields' (Stout 2015, 162). Provincial towns then followed London's lead by creating their own walks. In 1638 in Tunbridge Wells, the Walks (later called the Pantiles) were laid out with a double row of trees (Johnston 2017, 17–18).

Privately owned living plant collections and the first botanic or physic gardens were other locations were trees could be found in urban areas. Before the development of the first botanic gardens, there were a small number of individuals and institutions that built up plant collections in their own gardens for both study and enjoyment. One of the earliest of note in Britain was the private endeavour of a dedicated individual. Friar Henry Daniel's fourteenth-century garden at Bow in London, in which he grew 252 different kinds of plants, has been described as possibly the first early botanic garden in Europe (Keiser 1996, 70). While Daniel's principal interest was herbal medicine, there is written evidence from his herbal that his garden contained a range of trees and shrubs. Before the monasteries were dissolved some built up collections of plants, including trees and shrubs, mainly for medicinal purposes but also for the enjoyment of their residents. One plant collection of note attached to a private residence was created by John Tradescant (1570–1638) from 1629 in his garden in Lambeth, south London (Leith-Ross 2006, 98). In his various overseas travels he collected a huge number of plants, including many trees and shrubs, and cultivated these in his garden. Along with his house, which contained his collection of rarities and curiosities known as The Ark, this was open to members of the public.

During the reign of Henry VIII (1509–1547) great advances were made on the Continent in the science of Botany. The first true botanic garden was established in Italy in 1543 at the University of Pisa and other Italian universities quickly followed. These were not botanic gardens as we understand them today as places for the collection, cultivation, preservation and display of a wide range of plants. Instead they were known as physic gardens, created specifically for the cultivation and study of medicinal herbs. The first in Britain was at the University of Oxford, founded in 1621 as the Oxford Physic Garden, with a donation from Henry Danvers, first Earl of Dandy. Initially it was able to generate an income by selling fruit grown within the garden walls. Always a very small garden close to the heart of the city, it eventually managed to accommodate a remarkable number of trees, some of great age and rarity. However, the establishment of further botanic gardens in Britain would have to wait until after the Restoration, and their reputation as centres of excellence for practical arboriculture was something for the future.

One particular development relates to a national tree-planting effort involving many thousands of ordinary citizens. This was the initiative of James I to establish a silk industry in Britain by encouraging the planting of mulberry trees. The King had been impressed by Henri IV of France's work to promote the silk industry there and he decided to do the same in Britain. In 1609 a short treatise and circular letter entitled *Instructions for the Increasing of Mulberrie Trees* (Stallenge 1609) was sent to the Lord Lieutenant of every county. This ordered them to announce that the following March 1,000 mulberry trees would be sent to each county town to be sold at 3 farthings a plant or 6*s* a hundred. A vast amount of seed was also distributed around Britain for those who wished to propagate mulberry trees themselves. The living trees were imported from continental nurseries and all who were able were required to buy them. However, the organisers ordered the black mulberry (*Morus nigra*) instead of the white version (*Morus alba*). The latter is the natural food of silkworms but grows less well in England. At the time, it was said that James was deliberately wrongly advised by the 'perfidious French', although others believed his own advisers chose the black mulberry as it was more readily available and would thrive better.

This episode could be the earliest incidence of a national campaign to plant trees on a grand scale, other than for forestry purposes, which was promoted by the monarchy and government. The project undoubtedly introduced many people to the propagation, planting and care of trees who previously had no experience of practical arboriculture. While some basic instructions on propagation and planting were included in the royal directive sent to the Lord Lieutenants (Stallenge 1609), many individuals participating in the project would have sought arboricultural advice from knowledgeable individuals within their own communities. People naturally took this tree project seriously because it was a directive from the King and they would have wanted to ensure their trees flourished, or at least did not die.

European influences on extensive tree planting in early Stuart times

With the advent of the Jacobean era there was a noticeable increase in the use of trees in gardens and their role in creating ornamental and recreational landscapes, although this trend was limited to the gardens of the very rich and powerful. These were often more densely planted than had been the case in Tudor gardens and the planting often extended out into the landscape at the edge or even beyond the formal garden. Much of the influence for this came from continental Europe.

During the Elizabethan era England was largely isolated from developments in mainland Europe (Strong 1979, 73). This was due to the ongoing political tensions and open warfare that had embroiled England, and this lack of contact with the Continent was reflected in many spheres of English life, including trade, commerce, the arts and the sciences. The accession of James VI of Scotland as King of England coincided with the re-establishment of some degree of peace in Europe and this led to greater European contacts and influence. There are many examples of the Jacobean aristocracy travelling to France, Italy and the Low Countries to establish economic links and also to visit the homes and gardens of the European ruling class. In terms of the influence of continental European developments on British gardens and gardening, the French and Italian traditions would be particularly significant. A number of individuals played a leading role in this.

Salomon de Caus (1576–1626), the older brother of Isaac de Caus, came from a Huguenot family in France and established a formidable reputation across Europe as an engineer with a particular interest in gardens. De Caus was in England from 1607 to 1613, initially in the service of James I's queen, Anne of Denmark, to lay out gardens at Somerset House and at Greenwich

(Strong 1979, 74). These were influenced by what has been called the Mannerist style, which was associated with the arts and gardens of the late Renaissance, especially in Italy. The creation of the garden at Somerset House was perhaps the most elaborate manifestation of this new style epitomised by de Caus. The garden was symmetrical with a quatrefoil laid onto a square, the patterns accentuated by trees and with a central walk down the middle. In these gardens the main role of trees, apart from as topiary and clipped hedges, was to line walks and boundaries, which were in geometrical patterns. Another famous garden influenced by de Caus was Hatfield House, Hertfordshire, where Robert Cecil, Earl of Salisbury, had inherited his father's passion for gardening. The layout of the gardens took place between 1607 and 1612 and was one of the most important gardens of the Jacobean era. The extent of the tree planting at Hatfield was enormous and went on for over three years (*ibid.*, 106). This task was entrusted to John Tradescant, then at the start of his remarkable career, who had been employed by Cecil as his head gardener in 1610 (Leith-Ross 2006, 28). From those who sought Cecil's favour came mighty gifts of trees and other nursery stock, for example, 500 fruit trees from Marie de Medici, wife of Henry IV of France, who also sent over two of her gardeners to supervise the planting.

The marriage of Charles I to Henrietta Maria, daughter of Henry IV of France, in 1625 continued the French influence on gardening in Britain and with it came French gardeners to provide advice on garden design. The first was André Mollet (*c.* 1600–1665) who arrived at the invitation of Henrietta Maria to lay out the gardens of St James's Palace at some point in the late 1620s (Pattacini 1998, 5). We know little of what the gardens looked like at that time but a description by a French visitor mentions many walks 'bordered on both sides with an infinity of fruit trees, which rendered walking so agreeable that one could never get tired' (Strong 1979, 188). A leading figure in French garden design at that time was Jacques Boyceau (1560–1633), who was in charge of Louis XIII's gardens. He designed the gardens of the Luxembourg Palace and laid out the original plan for the gardens at Versailles. Boyceau's *Traité du Jardinage*, published five years after his death, is the first French gardening book to emphasise the aesthetics rather than the practicalities of gardening (Boyceau 1638).

Some elements of landscape design, while not unknown in Tudor times, were given a substantial boost through continental European influences. One of these was a taste for planting long, straight avenues or lines of large forest trees across the deer parks and into the surrounding countryside. The Italian architect Vincenzo Scamozzi (1548–1616) advised in his major treatise *L'Idea Della Architettura Universale* that approaches to palaces should be lined with trees to focus attention on the principal facade and end in a semicircular parade large enough for coaches to turn (Scamozzi 1615, 356–360). The Duke of Buckingham was one of the first in England to take up Scamozzi's advice and in 1621 had planted a mile-long great walk at New Hall with a half-round at the house-front. He also planted a double walk, though without a parade termination, at Burley-on-the-Hill. The work was supervised by John Tradescant, who had been appointed as Buckingham's gardener to manage and possibly design the grounds of both of his country residences. The traditional tree for the few avenues planted in Tudor times had been the English elm (*Ulmus procera*), grown from hedgerow suckers. The European lime (*Tilia x europaea*) now came into favour, largely through Dutch influence, and John Tradescant was paid in 1624 to travel to the Low Countries to purchase them for the New Hall avenue.

The creation of the gardens at New Hall was a wonderful example of the emerging fashion for extensive tree planting and tree-dominated garden features among the monarchy and leading aristocracy. When John Evelyn visited New Hall in 1656 he wrote admiringly of the grounds: 'The hall is noble; the garden a fair plot, and the whole seat well accommodated with water; but, above all, I admired the fair avenue planted with stately lime trees, in four rows, for near a mile in length. It has three descents, which is the only fault, and may be reformed' (Evelyn 1901, 312).

As with the owners of some other notable gardens of the early Stuart era, Buckingham received generous donations of trees from the monarch and fellow aristocrats. James I had given directions for 'one thousand timber trees of Oake' to be transplanted to New Hall from his various woods in Kent (Leith-Ross 2006, 80–81). However, because these were 'not sufficient by reason of the smallness of them in bulke', when Charles I came to the throne he added another 500 to the gift. The Earl of Northumberland sent 2,000 walnut trees and elms were also planted. The supervision of all these tree plantings was almost certainly the responsibility of John Tradescant.

The magnificent garden at Wilton House has been mentioned in the discussion on wildernesses and is a fine example of the continental European influence. Wilton is widely regarded as the greatest of the English Renaissance gardens and has had a huge impact on the overall development of gardens in Britain (Strong 1979, 151–165). It was created in the 1630s by the fourth Earl of Pembroke who engaged Isaac de Caus to lay out an immense Italianate garden, divided into three parts with a broad, straight walk up the centre. The first section was laid out in 'embroidered' parterres of box in the latest French manner, each adorned with a marble fountain. Next came the densely planted wilderness and then a third section beyond this that featured an oval circus marked out with cherry trees. The end of the central path terminated at a splendid grotto, full of surprising waterworks. This magnificent garden became famous all over Europe and was a great favourite with Charles I. As with other gardens influenced by continental European fashions, the creation of the garden at Wilton involved a huge amount of tree planting and thereafter a considerable amount of routine tree maintenance was required to ensure it looked its best.

Arboricultural practices and early authors

The printed book was an important instrument of the Renaissance and played a crucial role in spreading ideas and information throughout Europe on all aspects of human endeavour, particularly the arts, science and philosophy. In Britain, the late Tudor and early Stuart period saw the publication and dissemination of some significant practical texts that gave guidance on a range of arboricultural and horticultural practices. Although this only began in earnest quite slowly in the 1550s it steadily gathered pace in the Jacobean era so that by 1640 there were many relevant publications.

In the early seventeenth century the previously allied subjects of horticulture and botany, both particularly relevant to trees, began to split into separate fields of study. During this period and for some time after, much of the British literature on trees reflected economic preoccupations and tended to focus on forest (timber) trees and fruit trees. Nevertheless, these often contained information about basic arboricultural practices such as propagation, planting and pruning. Also, from the mid-sixteenth century came the publication of the first horticultural manuals to be written in English and devoted solely to the 'arte' of gardening. While these generally covered commercial horticulture, such as pomology and the kitchen garden, some also included useful practical information about the propagation, planting and care of trees. However, it was only at the start of the seventeenth century that we see published works with any significant focus on ornamental gardening or amenity trees. At this time there was also a change in the way gardening writers approached their work, reflecting a tendency to question the place of traditional beliefs and philosophies and instead to pursue knowledge through practical experiment. An early advocate of this approach was Sir Francis Bacon (1561–1626), now regarded as the father of the scientific method, who proposed a new framework for the study of natural history in his book *Novum Organum*, published in 1620. In Bacon's view, knowledge was born of experience, not authority, and he rejected knowledge based on the secrecy of alchemy, magic and the occult. However, while this new scientific approach was gaining credibility, many writers on horticultural and arboricultural topics were still largely repeating without question the ideas and practices set out by the classical Roman authors. Also, many early English works on these topics freely plagiarised from continental European sources, often without acknowledgement.

The increasing volume of relevant literature at this time indicates that these subjects were now perceived as worthy of consideration and publication. For the historian these books provide an excellent body of accessible research material that gives invaluable information on arboricultural and horticultural practices. However, it cannot be assumed that techniques described in books were commonly practised at this time as this is almost impossible to assess. Nevertheless, these books provide some remarkable insights into the world of at least some professional arborists and gardeners during this period.

While these early texts played a crucial role in promoting arboricultural and horticultural practices to members of the landed aristocracy, the extent to which they may have been read beyond those privileged circles should not be overestimated. It seems unlikely that these books would have been read by most professional gardeners or labourers, with the possible exception of those working in prestigious gardens where they may have been allowed access to their employers' copies. In the first instance, even the relatively cheap manuals were too expensive for most people to purchase. Secondly, there was the question of literacy and, while literacy rates at this time are difficult to establish, the ability to read was not thought to be widespread among the working class.

Arnold, Fitzherbert and Hill

Perhaps the earliest published text of relevance is a curious little book entitled *The Customs of London* which contains a chapter on 'The Crafe of graffynge and plantynge and alterynge of Frutys' (Anon 1811, 164–170). It was first printed in Antwerp around 1502 and later published in London in 1521. Its compiler, Richard Arnold, an entrepreneur haberdasher, thought it 'good and necessary for every citizen to understand and know' how to graft and plant fruit trees (Janson 1996, 40). However, the brief information it contains seems to be largely a summary of practices from the classical authors.

The first work published in the English language of purely gardening interest is also one that focuses on trees and is the anonymous treatise entitled *The Crafte of Graffynge and Plantynge of Trees* (Anon *c.* 1520). It came from the press of Wynkyn de Worde and was published initially about 1520 (Henrey 1975, 55). There is a single copy in the British Museum (Bloomsbury), and although this is incomplete the entire work is thought to run to little more than ten pages. The woodcut on the title page depicts two men with axes, one in the process of cutting down a tree growing in a small plantation (Figure 15). Two later editions of the work appeared from the press of William Copland around 1563 and 1565 and have the same image. The contents of this work appear to be a clever combination of Arnold's chapter and sections from another book concerned with husbandry published about 1510 (Janson 1996, 40). It contains some rather spurious advice and fantastic notions, such as how to graft to obtain coreless fruit. However, it also includes some practical advice that would endure for centuries, such as that for wound treatment: 'Also it is good for all

FIGURE 15. The woodcut on the title page of *The Crafte of Graffynge and Plantynge of Trees* (*c.* 1520), the first work published in the English language of purely gardening interest. It depicts two men with axes, one in the process of cutting down a tree growing in a small plantation.

manner of trees when a bough is cut awaye that ye place that is cut be covered with cleye or some other defensable plaster for the defence of the rayne that it make not the tree to rot'. Despite its ancient misconceptions, this very rare book represents a milestone in arboricultural literature. In 2018, a copy of the later 1565 edition sold at auction in London with a hammer price of £16,000.

One of the earliest British works on agriculture is *The Boke of Husbandry* by Anthony Fitzherbert, published in 1523, which refers to trees and woodland, focusing mainly on planting and grafting (Fitzherbert 1882). It includes advances in some of the common practices of the time, for example, with the cropping or shredding of trees. Shredding is a form of pollarding involving the regular removal of the side-branches from a single-stemmed tree. Instead of starting at the top of the tree and working downwards when removing branches, which was then common practice, Fitzherbert advocates starting at the bottom and working up (*ibid.*, 85). This approach is often used in modern tree surgery for the same reason that Fitzherbert gives: it avoids the risk of the falling branches tearing out the branches below and damaging the bark, which encourages decay. Fitzherbert was also probably the first British author to give advice on transplanting large trees, advocating heavy prior pruning: 'And if the tree be very longe, cut of the top, two or three yards' (*ibid.*, 82). He also reiterated classical Roman advice to mark the side of the tree before moving to ensure the same side faces the sun when transplanted.

The earliest book on general gardening is a small treatise by Thomas Hill (or Hyll), a Londoner and author of many popular works on science. This appeared initially under the title *A Most Briefe and Pleasaunte Treatise, Teachying how to Dress, Sowe, and Set a Garden* and, although undated, was probably published in 1558 (Hill *c.* 1558; Henrey 1975, 57). Although Hill died in 1574, his second and most important gardening book entitled *The Gardener's Labyrinth* was published in 1577 under the pseudonym Didymus Mountain (Hill 1577). While the publication of this book is rightly regarded as a milestone in the development of British horticulture, it suffers from two weaknesses that were common to many early gardening books. It relied heavily on examples and precedents from classical literature and it also asserted that all gardening activities should be determined by the moon and its phases. The book is very much a general gardening manual with information about flowers, vegetable, soils, planting, cultivation and threats to plants, which explains its popularity at the time and the fact that it went to several editions. However, from an arboricultural perspective Hill does not have much to say about trees, although there are some references to fruit trees.

Mascall, Church and Markham

In 1569, the publication of a small book by Leonard Mascall was a major landmark in the development of British arboriculture. Entitled *A Booke of the Arte and Maner Howe to Plant and Graffe all sortes of Trees*, this was the first readily available text on the propagation, planting and care of trees to be published in the English language (Mascall 1569; references here are from the 1575 edition). Leonard Mascall (d. 1589) was the owner of a mansion called Plumpton Place, near Lewes in Sussex, who became Clerk of the Kitchen in the household of Matthew Parker, Archbishop of Canterbury (Henrey 1975, 64). As well as this classic work on arboriculture he also wrote books on a number of other rural topics. There is no doubt that much of the technical content of Mascall's book comes from continental European sources, as he freely acknowledges at the start: 'Gentle Reader thou shalt understand I have taken out of divers authours this simple work, into our Englishe tongue' (Mascall 1675, To the Reader). He notes that England compares poorly in planting and grafting to some other counties, including Italy and France. A large part of the book is a translation of a work by Davy Brossard, the French Benedictine monk, entitled *L'art et Maniere de Semer, et Faire Pepinieres des Sauvageaux.* At the end of the book is a section entitled 'Here followeth certain ways of planting and graffing' that Mascall translated from a Dutch text. However, to dismiss this book as largely plagiarism would be most unfair. Anyone with knowledge of practical arboriculture would notice there is much in this remarkable book that most likely has come from Mascall himself. The introductory sections that are entirely Mascall's own are especially poignant. He addresses his book to everyone, which includes the 'poore man', the 'riche man' and the 'noble man'. He is an undoubted champion of the professional gardener and arborist and has no time for those who do not respect their valuable contribution to society. He criticises the many people 'which be of the base and abject sort of the commonwealth' and the 'faire personages' who do not want to get their hands dirty with manual work but look down on their fellow men 'labouring of the earth which we must all live by'. These were radical sentiments to express in print at this time and were later echoed by other writers on agrarian topics several decades later, around the time of the Commonwealth.

In another original section, entitled 'An Exhortation to the Planter and Graffer', Mascall stresses the value of practical experience and knowledge of local conditions. He states, 'Always before ye do intende to plant or graffe, it shall be meete to have good experience in thinges meete for this Art, as in knowing the Natures of all trees and fruits, and the differences of climates, which be contrarie in everie lande' (Mascall 1675, An Exhortation). To that end, Mascall's book gives a considerable amount of practical advice on various aspects of arboriculture, and, although much of the information may have been drawn from overseas sources, he gives the impression he can support this from his own experience. The following passage on branch removal typifies his approach to practical arboriculture:

> And if there be in your trees certaine branches of superfluous wood that ye wyll cut of, tary untyll the tyme of the entering in of the sappe, that is, when they begin to bud, as in Marche and Aprill: then cut of as ye shall see cause, all such superfluous branches harde by the tree, that thereby the other branches may prosper the better,

> for then they shall soner close their sappe upon the cutte places than in the Winter, which should not do so well to cut them, as certain do teach which have not good experience. …. Take heede therefore in cutting then of your great branches hastily, that through their great weight, they do not cleave or separate the barke from the tree in any part thereof. And for the better remedie; fyrst you shall cut the same great branches half a foote from the tree, and after to sawe the rest clean of harde by the body of the tree, then with a broade Chisell cut all cleane and smooth upon that place, then cover it with Dre dung.
>
> Mascall 1575, 45

FIGURE 16. This illustration from Leonard Mascall's pioneering book on practical arboriculture, first published in 1569, shows contemporary arboricultural tools, many of which were used for pruning and grafting.

Regardless of how much of Mascall's advice is still relevant today, he displays an awareness of basic practical arboriculture that is far in advance of other British texts from the sixteenth century and much later. This is underlined by the inclusion of a full-page illustration of contemporary arboricultural tools (Figure 16). Mascall's approach to the planting and care of trees is essentially utilitarian rather than ornamental. However, regardless of the purpose of the garden or orchard, he stresses the need to 'make a sure defence' for this 'to the ende, that not only rude persons and children may be kept out, but all kinde of hurtfull cattell [and other animals] in damaging your plantes or trees' (Mascall 1575, An Exhortation). Mascall's remarkable work on practical arboriculture deserves to be far more widely appreciated today. A truly pioneering book, it has remained obscure and underestimated for too long. Always a very rare collector's item, a copy of the 1569 first edition sold at auction in London in 2018 with a hammer price of £15,000. Fortunately, the 1575 edition can be accessed on the internet.

An account has already been given of the project ordered by James I in 1609 to plant many thousands of mulberry trees throughout Britain. Given that this was an initiative of the monarch and the government, the basic practical instructions on propagation and planting that were included in the royal directive sent to the Lord Lieutenants were almost certainly what were considered 'best practice' at that time. The advice on tree planting is as follows:

> In the moneths aforesaide [September–December, March and April] (according as your plants are waxen strong) you may remoove them either into the hedges of your fields, or into any other grounds. If in hedges, you must set them sixteen feete the one from the other; if in other ground, intending to make a wood of them, eighteen feete at the least. But, a moneth before you doe remoove them, you must make the holes, wherein you propose to set them, about foure feete, in bredth, and so deepe as that their rootes may be well covered, and some halfe a foote of loose earth left under them, having always a speciall care so to place them, that they may receive the benefit of the sunne, and not be shadowed or overspread by any neighbouring trees.
>
> Stallenge 1609

Forestry texts often have material that is relevant to arboriculture, especially for basic practical operations such as tree planting and pruning. An important work on forestry entitled *An Olde Thrift Newly Revived* was published in 1612. No author's name appears on this work, but the title page states that it is 'by

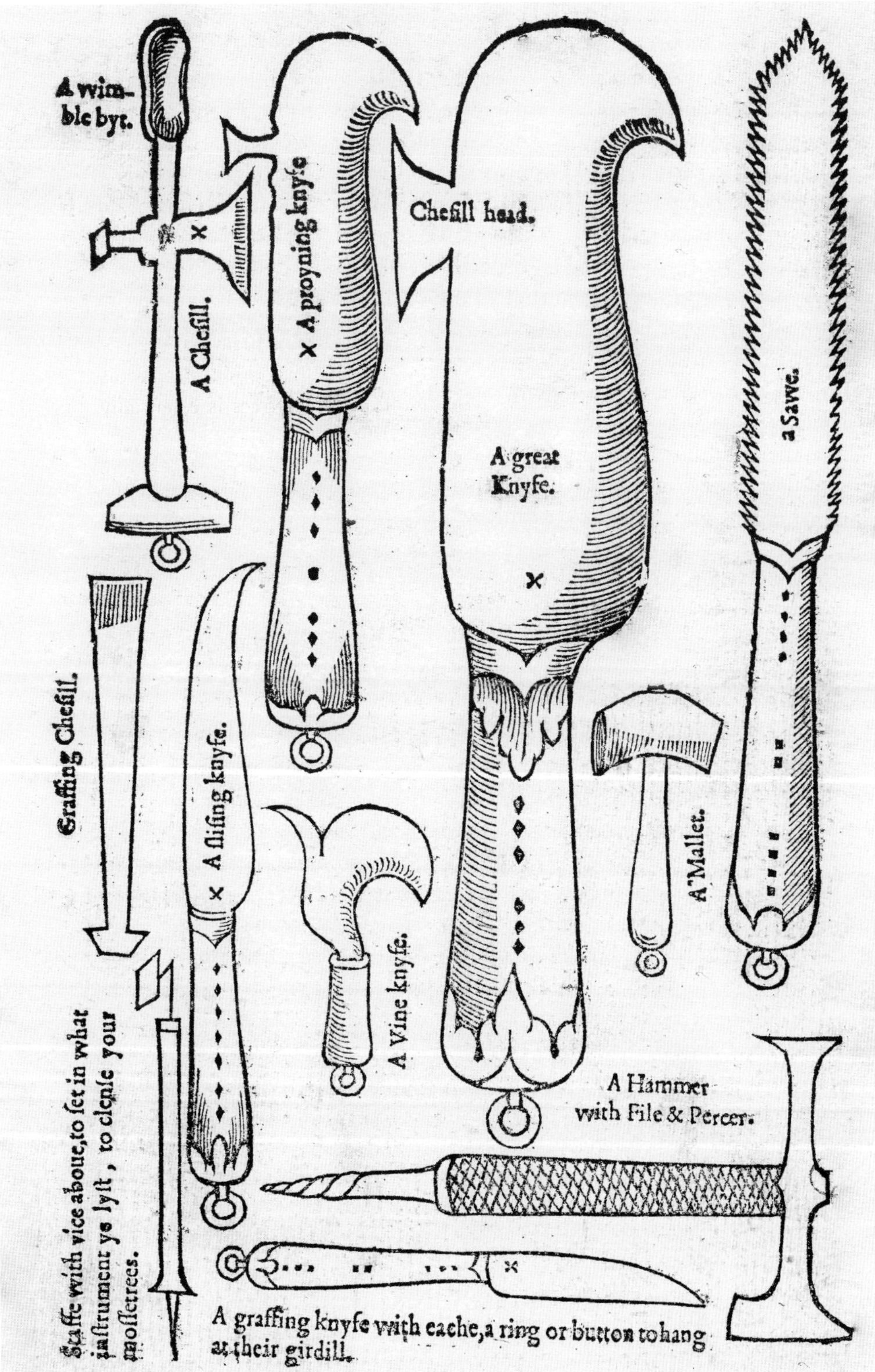
A wimble byt.
A Chesill.
A proyning knyfe
Chesill head.
A great Knyfe.
a Sawe.
Graffing Chesill.
A slising knyfe.
A Vine knyfe.
A Mallet.
A Hammer with File & Percer.
Staffe with vice aboue, to set in what instrument ye lyst, to clense your mossetrees.
A graffing knyfe with eache, a ring or button to hang at their girdill.

R. C.' and the preface is signed 'R. Ch.'. As the author was a surveyor, and it is known that about 1608 the King's Surveyor to Lord Salisbury was called Rocke Church, it is reasonable to assume that Church wrote this book (Henrey 1975, 115). The contents are a combination of a passionate call for national reforestation and a manual on how to raise and manage woodlands. For those who believe that British works on tree management and calls for extensive forestry planting only began with John Evelyn's *Sylva* in 1664, this book written over fifty years before will be a revelation. The title page states that the text is written in the form of a dialogue 'between a surveyour, woodward [forester], gentleman, and a farmer'. Given that Church was a surveyor, it is amusing to note that the surveyor has all the good practical advice. In terms of the book's relevance to arboriculture, Church gives some helpful recommendations for tree planting and post-planting maintenance. He begins by highlighting the need to carefully select trees that are appropriate for the soil conditions and to plant them at the right time (Church 1612, 6). He emphasises the importance of good ground preparation and then recommends a method of forestry planting that is similar to the modern method of furrow planting (*ibid.*, 10). He is emphatic about the need to maintain newly planted trees and states, 'And the chiefest labour about young trees that are newly planted, is to keep the ground cleane from weeds, as it will much hinder them from growing to that handsome and full forme, as otherwise they would be' (*ibid.*, 14). Unfortunately, this basic truth about post-planting maintenance is still being ignored in so many tree-planting schemes some 400 years later.

The classic work by Gervase Markham (1568–1637) entitled *The English Husbandman* has been mentioned in the context of garden design and is rightly regarded by horticulturists as an important early text (Markham 1613 and 1614). It also contains valuable information on arboriculture and tree management, which tends to be overlooked. The first part of the book was published in 1613 and the second part in 1614; it proved to be a relatively popular work and the revised and enlarged edition of 1635 was reprinted three times before 1654. Markham, who was born into a wealthy Nottinghamshire family, was a prolific author who published numerous books, many of which are concerned with practical subjects such as farriery and herbal medicine. Along with Arthur Standish's *The Commons Complaint* (Standish 1611) and Rocke Church in 1612, he recognised the poor state of the nation's depleted forests, that its economic strength depended on the availability of timber and that a major afforestation effort was required (Markham 1614, 50–51).

The arboricultural relevance of Markham's *The English Husbandman* is considerable and the title page declares that the book focuses on the 'Art of Planting, Grafting, and Gardening after our latest and rarest fashion' (Markham 1613, title page). Markham states as his justification for writing the book that all similar works are just translations from Latin or French writers, who are 'all forrainers and utterly unacquainted with our climbes' (Markham 1613, Epistle to the Reader). While Markham does not acknowledge the earlier contributions of

British writers such as Mascall, he is equally keen to point out the importance not only of planting well, but also of preserving, protecting and maintaining trees once they are established. He is probably the first author to advocate a regular and systematic approach to tree surveying as an essential tool for effective woodland management, something that would later become a cornerstone of arboricultural management. For the woodland owner to ensure his trees remain healthy and vigorous, 'he shall daily walke into his Woods, and with a searching eye survay every tree which is of any account, and see if he can finde any fault or annoyance about the same' (Markham 1614, 63). In his advice on transplanting trees, some of this reiterates the classical Roman authors. Nevertheless, when it comes to drastically pruning young trees before transplanting, he differs from the usual Roman advice and what he calls 'the fashion of France'. According to Markham, you should not 'dismember, or disrobe the tree of its beauties, that is to say, cut off all his upper branches and arms, but you shall diligently preserve them: for I have seene a tree thus replanted after the fall of the leafe to bring forth fruit in the Summer following' (Markham 1613, 59). However, 'if the tree you replant be olde then it is good to cut off the maine branches within a foote of the stocke, least the sappe running upward, and so forsaking the roote too suddainely doe kill the whole tree'.

William Lawson and John Parkinson

William Lawson's classic work, *A New Orchard and Garden*, was first published in 1618 and is one of the most important early works on horticulture and arboriculture (Lawson 1618). Bound together in the same volume was his work entitled *The Country Housewifes Garden*, dated 1617, the first British gardening book written specifically for women. This goes into some detail on the design and layout of the garden and orchard, 'together with divers new knots for gardens' (Lawson 1617, title page). *A New Orchard and Garden* and *The Country Housewifes Garden* were then republished many times together and also included in some of the works of Gervase Markham. References here are taken from the 1631 edition.

William Lawson (*c.* 1554–1635) was the vicar of Ormesby, a country parish in Yorkshire. As well as undertaking his clerical duties, Lawson was a keen gardener with extensive land. His *New Orchard and Garden* was addressed to the garden owner rather than the gardener, which is clear from his introductory advice on how to employ the right sort of gardener. While Lawson wanted his text to be widely read by his fellow gentry, there are still Latin quotations, proverbs and classical allusions scattered through the work. Nevertheless, this was still a readable and practical manual, based on 'the experience of 48 yeares labour', which focuses on the best way to establish and maintain an orchard, 'particularly in the North' (Lawson 1631, title page). In the text Lawson frequently refers to his own particular knowledge of gardening in the north of England. He gives advice, for instance, as to which fruit trees are 'most common, and meetest for our Northern countries', suggesting the planting of apples, pears, plums and damsons rather than the more tender fruits such as apricots and peaches which 'will not like in our cold parts' (Lawson 1631, 3).

Lawson's thoughts on the design and management of orchards and gardens have already been mentioned and there can be no doubt about his pioneering role in British horticulture. However, within his *New Orchard and Garden* there is also a wealth of information relating to practical arboriculture. While the book is concerned mainly with the orchard, Lawson does not limit his remarks to fruit trees and to some extent includes trees in woodlands. He seems aware that some readers might question his experience in other areas when he asks, 'But why do I wander out of the compasse of mine Orchard, into the Forests and Woods?' (Lawson 1631, 44). He responds by pointing out that both need careful 'dressing' or pruning to be profitable and many of the same basic principles of successful cultivation apply. As well as embracing the wider aspects of arboriculture, Lawson is also one of the first writers to use the term 'arborist'. A few of his more significant contributions to practical arboriculture are given here.

The frontispiece of *A New Orchard and Garden* immediately signals Lawson's focus on trees. It is a woodcut showing three arborists planting and pruning fruit trees (Figure 17). Next to the frontispiece Lawson has inscribed the wise words: 'Skill and paines bring fruitfull gaines'. At the beginning of the text he

FIGURE 17. The frontispiece of William Lawson's *A New Orchard and Garden* (1618) immediately signals his focus on trees. This attractive woodcut shows three arborists planting and pruning fruit trees.

is quick to stress the need for skilled arborists and the huge damage that can be done by those who are not. He declares, 'And what a hinderance shall it be, not onely to the owner, but to the common good, that the unspeakable benefit of many hundreds of yeeres shall be lost, by the audacious attempt of an unskilled Arborist' (Lawson 1631, 2). This wonderful quotation deserves to be widely publicised by our modern arboricultural organisations in their efforts to promote high standards of work among professional arboriculturists. However, Lawson was not convinced of the need to always employ a professional to undertake work in the garden or orchard. He highlights the pros and cons of employing gardeners and arborists, thus losing some of the profits, and gardening for oneself, with the consequent hard work: 'If you bee not able, nor willing to hyre a gardner, keep your profites to your selfe, but then you must take all the paines' (Lawson 1631, 3).

In Lawson's insistence on high standards of work with all trees he refers to poor standards in woodland management: 'How many forests and woods wherein you shall have for one thriving tree, foure (nay sometimes twenty foure) evill thriving, rotten and dying trees, even while they live?' (Lawson 1631, 41). As well as asserting his own views on good practice, Lawson is not shy about informing his readers that these often run contrary to received wisdom. In his discussion on tree planting he declares, 'And herein I am of a contrary opinion to all them which practice or teach the planting of trees, that ever yet I knew, read or heard of' (*ibid.*, 29). He then gives his own thoughts on the correct planting and spacing of trees, based on his own observations over forty years.

For Lawson, pruning is 'a point of knowledge most requisite in an Arborist as well as for other woods as for an Orchard' (Lawson 1631, 40). In a further reference to the need for competent arborists, the importance of skilful pruning and the previous lack of guidance on this, he declares:

> So that a skillfull and painfull Arborist need never want matter to effect a plentifull and well drest Orchard: for it is an easier matter to take away superfluous boughs (if your Gardner have skill to know them) whereof your plants will yield abundance, and skill will leave sufficient well ordered. All ages both by rule and experience doe consent to a pruning and lopping of trees: yet have not any that I know described unto us (except in darke and generall words) what or which are those superfluous boughs, which we must take away, and that is the chiefe and most needful point to be knowne in lopping. And we may well assure our selves (as in all other Arts, so in this) there is a vantage and dexterity, by skill, and an habite by practice out of experience, in the performance thereof for the profit of mankind; yet doe I not know (let me speake it with the patience of our cunning Arborists) any thing within the compasse of humane affaires so necessary, and so little regarded, not onely in Orchards, but also in all other timber trees, where or whatsoever.
>
> Lawson 1631, 40–41

As well as highlighting the damage caused by poor standards of tree pruning, Lawson gives guidance on what he regards as correct pruning technique:

> Deadly wounds are when a mans Arborist wanting skill, cut off armes, boughs or branches an inch, or (as I see sometimes) an handfull, or halfe a foot or more from the body: These so cut cannot cover in any time with sap, and therefore they die, and dying they perish the heart, and so the tree becomes hollow, and with such a deadly wound cannot live long.
>
> Lawson 1631, 56

Lawson also gives advice on wound treatment after pruning:

> Cover the middle of the wound, the next summer when he is dry, with a salve made of tallow, tarre and a very little pitch, good for the covering of any such wound of a great tree: unless it be barke pild, and then a sear-cloth of fresh Butter, Hony and Waxe presently (while the wound is green) applied, is a soveraigne remedy in Summer especially. Some bind such wounds with a thumbe rope of Hay, moist, and rub it with dung.
>
> Lawson 1631, 49

In the treatment of hollow trees, Lawson recommends: 'If he be hoald, cut him close, fill his wound, tho never so deepe, with morter, well tempered and so close at the top his wound with a Seare-cloth doubled and nailed on, that no aire nor raine approach his wound' (Lawson 1631, 56).

One of the earliest methods of tree training was to grow fruit trees in a single vertical plane, often against a wall. This was done partly for a tidy effect but also to promote growth and ripening. If the vertical surface is a south-facing brick wall, this provides shelter and absorbs warmth, some of which is given out at night, so that tender fruit has a much better chance of ripening than if grown on trees in the open. Lawson (1631, 8), working in the north of England, was very conscious of this and wrote that the cold climate 'provoketh most of our great Arborists, to plant Apricockes, Cherries and Peaches, by a wall, and with tackes and other means to spread them upon, and fasten them to a wall, to have the benefit of the immoderate reflexe of the Sunne'.

Although Lawson gives plenty of useful practical advice, he does occasionally overreach himself. For example, he maintains there are no limits to what form can be achieved through pruning: 'Never let any man ever so much as thinke, that it is unprofitable, much less unpossible to reforme any tree of what kind soever. For (believe me) I have tried it, I can bring any tree (beginning by time) to any forme. The peare and holly may be made to spread, and the Oke to close' (Lawson 1631, 44). This claim is accompanied by a woodcut illustration of the 'perfect forme of a Fruit tree', which is made to conform to the rectangular shape of the woodblock itself (Figure 18). In the text Lawson apologises for what he calls the deformity of the drawing, but only 'because I am nothing skillfull either in painting or carving', and not because it might be a deformed tree (*ibid.*, 46).

It is clear from the above extracts from Lawson's *New Orchard and Garden* that he made a huge contribution to our knowledge of practical arboriculture and the development of high standards in professional tree care. He certainly left his readers in no doubt about the need to employ people for this work who

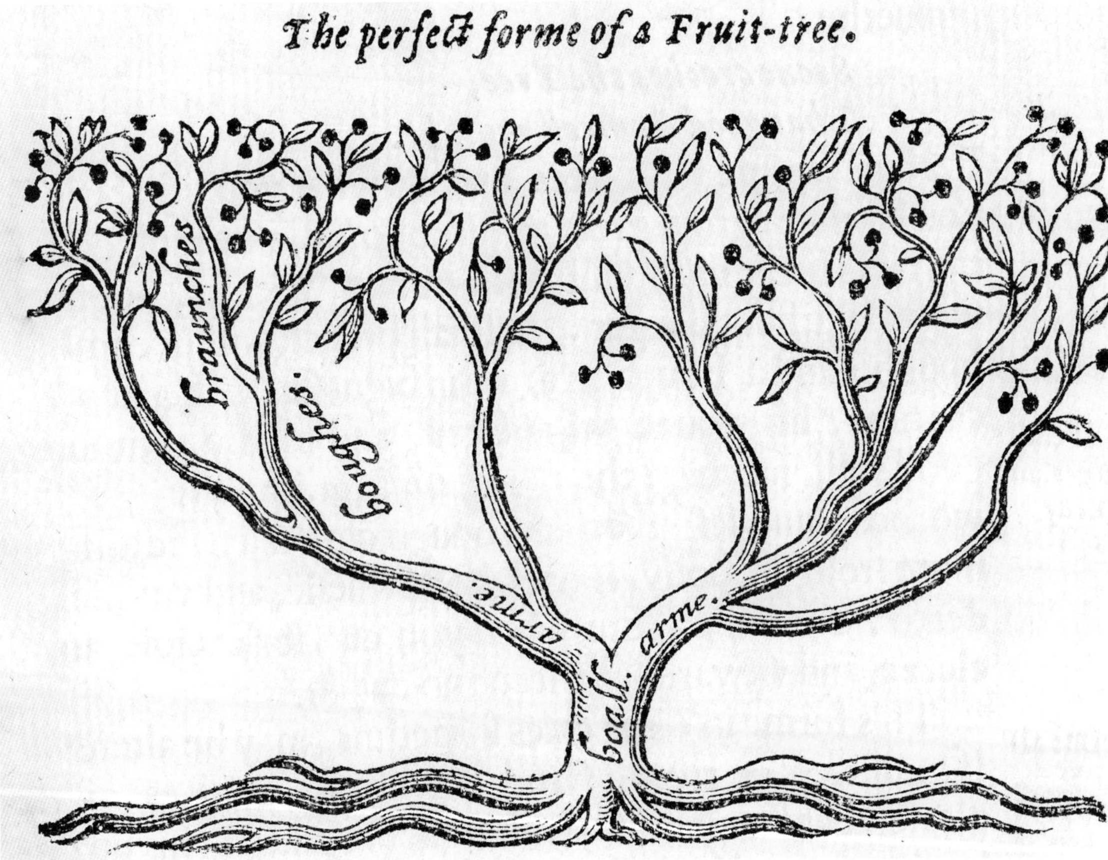

FIGURE 18. 'The perfect forme of a Fruit-tree' from William Lawson's *A New Orchard and Garden* (1618). Lawson believed there are no limits to what form can be achieved through pruning.

had the necessary specialist knowledge and experience. However, his written work is also imbued with a strong sense of spirituality and altruism that is so appropriate for a truly religious man. This brief account of Lawson's work ends with the following quotation taken from the last page of his book in which he refers to his own legacy:

> To conclude, what joy may you have, that you living to such an age, shall see the blessings of God on your labours while you live, and leave behind you to heires or successors (for God will make heires) such a worke, that many ages after your death, shall record your love of their Countrey? And the rather, when you consider to what length of time your worke is like to last.
>
> Lawson 1631, 74

One further author should be mentioned here as his name has already arisen several times and will do so again before the end of the chapter. John Parkinson (1567–1650) spent his early life in Yorkshire before moving to London to become an apprentice apothecary, eventually being appointed as apothecary to James I. He also took a keen interest in gardening and his beautiful book *Paradisi in Sole* is regarded by many authorities as the most important seventeenth-century work on gardens and horticulture (Parkinson 1629). In contrast to the books that preceded it, *Paradisi in Sole* represents a revolution in horticultural writing as Parkinson champions the merits of the ornamental pleasure garden as an end in itself, devoting the first and longest chapter to the 'Ordering of the Garden of Pleasure'. Nevertheless, from an arboricultural perspective this monumental work gives surprisingly little practical guidance on the planting and care of ornamental trees, although Parkinson has more to say on tree care in the context of fruit trees.

Tree health and tree protection

Concerns about tree health and methods to protect trees against pests and diseases are raised in some of the early literature on horticulture, arboriculture and forestry. Some of these weird and wonderful 'remedies' are clearly taken from the classical Roman texts, while others seem to be more in the nature of indigenous folk remedies handed down through generations of people working on the land. Many of these treatments for ailing trees would remain popular for decades to come and are discussed in the next chapter.

Of interest here was the increased awareness in the Tudor and early Stuart era of the need to protect trees against the ravages of an adverse climate or occasional weather extremes. As more exotic and tender trees were brought into Britain from the Continent, many of them fruit trees, gardeners realised the importance of protecting these plants from the cold and trying to replicate the conditions in their natural habitat. Markham in particular offered a number of ingenious solutions. These generally involved some form of construction, either freestanding or against a wall, made of bricks, wood and glass, that went over the trees and gave them protection from storm-force winds and frosts (Markham 1613, 129). Some of these constructions sound like forerunners to the glasshouse, although historical evidence suggests that very few were actually built. In Markham's other great work on gardening and husbandry, *Maison Rustique, or The Countrey Farme*, his edited translation in 1616 of Charles Estienne's great French text, he addresses at length the problems of growing tender fruit trees, particularly orange trees (Markham 1616, 300–301). He eventually admits that it might be better if they were left to grow in their native countries. However, he acknowledges the French practice of growing them in containers and removing them in winter to a purpose-built house for protection. Nevertheless, he believes that if attempts are made to plant them in Britain, they should be planted in the ground and completely covered with branches and mats over the winter period. He also recommends the lighting of fires around tender trees when the weather is exceptionally cold. Parkinson was also aware that tender trees and other plants can be damaged by cold winds and frost and recommends the use of shelterbelts. He states that the pleasure garden can be protected, 'by the helpe of high growne and well spread trees, planted on the North side thereof, to keep it the warmer' (Parkinson 1629, 1).

While these early gardening manuals recognise most of the basic necessities for tender trees and other plants to grow successfully, such as good soil and warmth, they omit to mention that vital ingredient of adequate light. This was not really appreciated until the late eighteenth century when the mysteries of photosynthesis were revealed. When these tender plants arrived in Britain from counties with relatively long, sunny summer days and mild winters, it was not surprising that they invariably failed to thrive. Not only was the British summer, especially in the north, often dull and wet, but in the winter the plants were routinely wrapped up in protective cladding or brought inside to overwinter in outbuildings that usually allowed very little natural light. It was only in the late seventeenth century that the purpose-built orangery became a regular feature of fashionable residencies.

Tree nurseries and propagation

The steady increase in new gardens and the extension of existing ones during this era created a rising demand for nursery stock of various types and sizes. As tree-related features became popular elements of these gardens, especially with the trend towards densely planted wildernesses and groves, tree nurseries were needed to meet this demand.

With regard to the rarer and choicer varieties of fruit trees, it was common practice to purchase young trees from continental sources. We have already recounted how Henry VIII sent his fruiterer, Richard Harris, to France. It was said that France and the Low Countries were chosen as the source of supply because of the need to acquire 'both rare fruite and lasting fine fruite'. Two hundred pear trees also came to Henry VIII from France, sent by M. de Sens (Dent 1962, 114). They arrived via Rye, with three horses making three journeys to bring them to London, with most of them going on to Nonsuch Palace. There is an interesting account of the purchase of fruit trees for the Llantrithyd estate of Sir Thomas Aubrey, although the origin of the trees is unclear. On 6 March 1623, a payment of 3*s* 10*d* is recorded 'For carriage of the trees from London', followed by a further payment of 5*s*, 'Geven unto Richard Herberts servant whome he sent hether to plant the trees that was sent from London' (Bowen 2006, 55–60). These trees must have been rather special and quite large for a gardener to be engaged specifically to plant them, implying that the resident gardener did not have the necessary skill or experience to undertake this work (Francis 2018, 282). While the accounts do not state how the trees were transported from London, because of the distance and locations this was probably by sea rather than by road.

Many of the books on husbandry and gardening from this period devote at least some space to the practice of propagating trees by vegetative means. Almost all of this is in the context of fruit trees. In their account of this it would appear that many of the basic techniques and the materials used had not changed significantly since those of the late medieval period. However, some early books

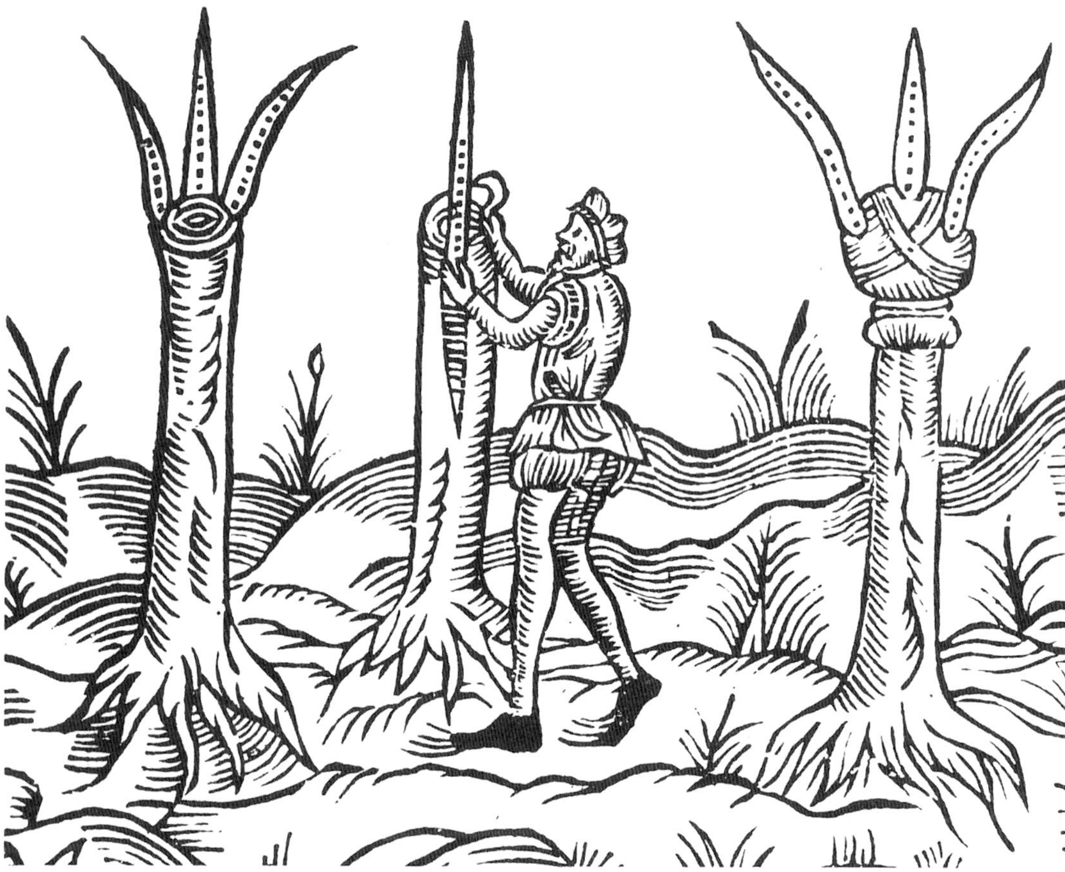

FIGURE 19. This woodcut from Leonard Mascall (1569) shows three stages in the practice of cleft grafting. This is where a piece of sion wood is inserted into a slit made in the top of the stock. Note the thick coating of wax or grafting clay around the graft on the right.

included excellent illustrations of different methods of grafting and budding, which can be very informative in our modern understanding of the techniques that were used. For example, Mascall's book gives an illustration of the practice of cleft grafting (Figure 19). Another book by an anonymous author in 1594, entitled *The Orchard, and the Garden*, illustrates the practice of patch budding (Anon 1594) (Figure 20). Some books also contained illustrations of various tools and equipment, including many used in grafting, and these are discussed later.

With regard to the supply of forest trees for the large gardens and estates of the early Stuart monarchy and aristocracy, most tried to be self-sufficient. It was then common practice to broadcast or plant seed over the planting site in the hope that there would be adequate germination and plenty of young seedlings. However, this method was challenged by both Rocke Church and Gervase Markham, who recommended the use of tree nurseries to raise young trees, which having reached a sufficient height could then be planted out. This approach would overcome the frequent problem of tree seed being eaten by mice and other little creatures. In advocating the use of tree nurseries, Church states, 'And this Garden or Nurcerie thus made, you may, when they

FIGURE 20. This illustration from the anonymous work *The Orchard, and the Garden* (1594) shows an arborist pruning and patch budding a fruit tree.

are growne to three foot high, remove of them how many you please, and to what other convenient place about your ground you thinke fit … and then being planted, they will prosper much the better' (Church 1612, 26) .

In Markham's promotion of tree nurseries he assured his readers that each native species, except oak, 'may as well be planted [from transplants] as well as sowne', detailing the conditions required (included the right phase of the moon) and the method of transplanting for each species (Markham 1614, 60). He considered it particularly difficult 'for the Oake to be taken up and replanted'. While the use of transplants was being recommended for woodland-type planting, Markham made a leap forward in practical arboriculture by recognising that this method of tree establishment could also be applied to urban planting schemes. In support of this he cites the walks at Moorfield as an example of how trees in popular urban open spaces can be grown successfully from transplants (*ibid.*, 60). Markham was also conscious of the need to create many more nurseries to meet the growing demand for young forest trees (*ibid.*, 63).

Another type of woody plant much in demand in the early Stuart era were quicksets, used mainly for hedging. These were also often raised in nurseries on the estates where they were required but, when purchased, they were generally supplied in large quantities, usually priced by the hundred. At Trentham Hall over the winter of 1635, there are frequent references in the estate accounts to payments for 'setts' at the rate of 6*d* per hundred (Francis 2018, 281). Thousands of quicksets and osiers were also purchased that winter and planted around the new garden and grounds at Trentham. It appears that large numbers of these plants were readily and cheaply available since gardeners or nurserymen could gather cuttings of easily rooting plants such as willow and hawthorn without much difficulty. A full-size stock tree or an existing hedge would have provided hundreds of cuttings or sets. There is also evidence in some account books of the sale of surplus stocks to neighbouring estates.

While there were inevitably some commercial tree nurseries operating in different parts of Britain in the late fifteenth and early sixteenth century, the evidence for this relates mostly to the London region. It is only in the reign of Henry VIII and the period of the Dissolution of the religious houses that we begin to find nurserymen who are identifiable personalities, and the first links in a chain of continuous histories of individual firms or of particular plots of garden ground (Harvey 1974, 38).

Reading through the contemporary gardening literature it appears that nurserymen and seedsmen were not always held in high regard. One common complaint was that seeds and plants were often sold incorrectly labelled. One of the causes of confusion was the great variety of names for the same plant, a situation that could not be remedied until Linnaeus's binomial classification in the mid-eighteenth century. Parkinson (1629, 571) gives us an example when he discusses the 'many varieties and difference of Cherries'. He complains that 'our Nursery men do so change the names of most fruits they sell, that they deliver but very few true names to any'. He considers them innately untrustworthy and declares: 'Scarce one in twentie of our Nurserie men doe sell the right, but give one for another, for it is an inherent qualitie almost hereditary with most of them, to sell any man an ordinary fruit for whatever rare fruit he shall ask for: so little they are to be trusted' (Parkinson 1629, 571).

Tools and equipment

Thanks to the publication of the first printed books and the increasing number of these books during this period, our understanding of contemporary arboriculture, horticulture and forestry is relatively good. Fortunately, many of these texts also took the form of manuals that not only gave details of practical techniques but some also included illustrations of the relevant tools and equipment.

Fitzherbert's *Boke of Husbandry* of 1523 is one of the first books in the English language to give a list of tools for grafting (Fitzherbert 1882, 87–88). Leonard Mascall's pioneering text on practical arboriculture, first published in 1569, has a superb full-page illustration of various tools used for tree work, and especially those used for grafting (Figure 16). This seems to be the first illustration of a range of arboricultural tools and equipment to appear in a British publication and it shows that the Elizabethan arborist had quite an armoury of knives, saws, billhooks and chisels at their disposal. The modern arboriculturist will recognise many of these tools as versions of them are still in use today, over 450 years later. The first detailed account of how arboricultural equipment was actually used in practice, especially in pruning operations, comes from William Lawson in his classic work *A New Orchard and Garden* in 1618. Referring to general pruning work, he states:

> The Instruments fittest for all these purposes, are most commonly: For the great trees an handsome long, light Ladder of Firpoles, a little, nimble, and strong armed Saw, and sharpe. For lesse Trees, a little and sharpe Hatcher, a broad mouthed Chesell, strong and sharpe, with a hand-beetle, your strong and sharpe Cleever, with a knock, and (which is a most necessary Instrument amongst little trees) a great halfted and sharpe Knife or Whittle. And as needful is a Stoole on top of a Ladder of eight or moe rungs, with two backe-feet, whereon you may safely and easefully stand to graffe, to dress, and to gather fruit thus formed: The feet may be fast wedged in: but the Ladder must hung loose, with two bands of iron.
>
> Lawson 1631, 50–51

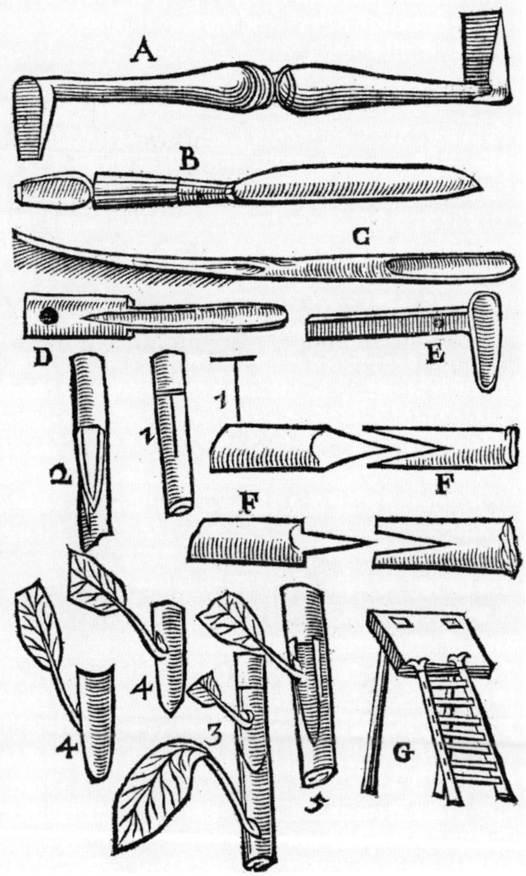

A. The Iron Inſtrument with cheſſels at each end, the one bigger and the other leſſer, to keepe the cleft of the Tree open vntill the graft bee placed in the ſtocke, which with a knock vpwards will be eaſily taken away.
B. The ſmall Penne-knife with a broad and thinne ended hafte, to raiſe the ſides both of the bud and the down-right ſlit in the body or arme of a Tree to be grafted in the bud.
C. A pen or quil cut halfe round to take off a bud from the branch.
D. An Iuory Inſtrument made to the ſame faſhion.
E. A ſhielde of braſſe made hollow before to be put into the ſlit, to keepe it open vntill the bud be put into its place.
F. The manner of grafting called incifing or ſplicing.
G. A Ladder made with a ſtoole at the toppe, to ſerue both to graft higher or lower, and alſo to gather fruit without ſpoyling or hurting any buddes or branches of Trees.

FIGURE 21. An illustration from John Parkinson's *Paradisi in Sole* (1629) showing a range of tools and equipment for grafting and indicates some techniques. Note 'G' which is the ladder and platform used for working at height in a tree.

In the last part of this quotation Lawson is describing a type of ladder and platform used for working at height in a tree, which enables pruning operations to be undertaken safely and efficiently. He also includes a small illustration of the platform in the margins of the text. Such a platform is illustrated in Parkinson's (1629, 543) *Paradisi in Sole* (Figure 21), and as this is freestanding he says that work can be done 'without spoyling or hurting any buddes or branches of Trees'. A similar illustration appears again in John Evelyn's *Elysium Britannicum* at least forty years after Lawson's work. While this platform was occasionally used in the Stuart era, particularly for pruning fruit trees, it seems that an ordinary ladder was still the common means of access into the tree for most work (Figure 20). In the Tudor era, before this platform appeared, ladders seem to have been the standard equipment for all types of aerial tree work. It was important that ladders were as lightweight as possible and of sufficient length. As Thomas Tusser wrote in 1573 in his book *Five Hundred Pointes of Good Husbandrie*, 'Light ladder and long, doth tree least wrong' (Tusser 1812, 3). While ropes were undoubtedly used in various aspects of tree work, such as the removal and lowering of large branches to avoid damaging plants or property

FIGURE 22. This woodcut from 1602 shows an arborist undertaking pruning in a large tree located in a walled garden. The origin of the image is unknown.

underneath, an extensive search of the relevant literature could find no mention of this. Furthermore, research has discovered that images from this period of arborists or gardeners working in trees are very rare (Figure 22).

In Parkinson's *Paradisi in Sole* he devotes several pages to the grafting of trees and shrubs and he includes an illustration of a range of grafting tools (Parkinson 1629, 543) (Figure 21). There are noticeably fewer tools than in Mascall's illustration but those that are shown are quite similar. While Parkinson goes into considerable detail about how these tools are used in the task of grafting, from an arboricultural perspective it is disappointing that he does not have a similar amount of text or relevant illustration relating to tree pruning. In terms of any equipment that might have been used for surveying trees, little is known of this during Tudor and early Stuart times. However, Rocke Church (1612, 74) does describe and illustrate a handy little portable instrument for measuring the height of trees from which the volume of timber can then be calculated.

How was the professional gardener or arborist supplied with the necessary tools and equipment for their work? A local blacksmith is likely to have played a crucial role in this, especially with basic metal tools, and a nearby carpenter might have contributed his skills to provide some of the wooden components of tools and equipment. There is also evidence that some of this material, such as the more complex items, was purchased from further afield from manufacturers who specialised in supplying those working in agriculture, horticulture and forestry. The purchase of tools and equipment is an expense noted in most estate or household account books, although such purchases are often few and far between (Francis 2018, 114). This may be because the tools lasted a long time as the result of meticulous maintenance and repair, or perhaps in some cases they belonged to the gardener or arborist himself and therefore would not appear in the household accounts.

Status of gardeners and tree-related professionals

With the emergence of printed books on relevant subjects, and the greater availability of archival sources such as estate and household account books, our knowledge of gardeners and tree-related professionals during this period is considerably better than that during the medieval era. There are, however, some aspects of this that can lack clarity. During this time the term 'gardener' had a very wide meaning and was often used to embrace such occupations as botanist, florist, arborist, forester, landscape gardener, nurseryman, seedsman, fruiterer, fruit-grower, garden implement dealer, herbalist and others. When reading the accounts of what was being done in the various types of garden, orchard or other greenspace, there is often no indication which of the specialised categories of 'gardener' might have been involved. Furthermore, in many of these descriptions of practical activities it is often difficult to distinguish whether the writer, when referring to the 'gardener', is talking about the garden owner or somebody employed by them. Nevertheless, when William Lawson talks about the 'arborist', we can be sure he is referring to somebody employed by the garden owner since at one point he states, 'And what a hinderance shall it be, not onely to the owner, [for the damage done] by the audacious attempt of an unskilled Arborist' (Lawson 1631, 2).

The conditions under which gardeners and related professionals were employed could vary enormously. At the top end of the labour market were the sought-after positions of head gardener at a royal garden or the prestigious garden of an aristocrat. Then there were positions as head gardener in the gardens of the lower ranked nobility. All these positions came with an annual salary and usually a tied house. Those fortunate to be appointed to important positions were expected to ensure they remained at the 'cutting edge' of horticulture by keeping up with the latest developments and learning new skills such as how to deal with unfamiliar plants arriving from overseas. The head gardener would be responsible for all the gardening staff, the buying and selling of plants and all the planting and maintenance in the gardens. The manual cultivation was usually done by labourers hired by the day, rather than by permanent staff. In many of the gentry gardens there might be only one gardener who did his best to cope with most of the work, occasionally calling on casual labourers to assist with heavy work or freelance specialists for skilled work such as advanced pruning or grafting. Labourers were often paid by the day, although sometimes they may have got a fixed rate for a particular job, such as grubbing out a tree stump, and this was known as payment by the 'greate' (Francis 2018, 129). This appears to be an arrangement more favourable to the employer than the labourer. Sir Thomas Temple, who after his retirement in 1624 created a new garden at Burton Dassett in Warwickshire, often urges his steward to find a good labourer whom he can pay 'by the greate', in one instance, to cut down an elm tree (*ibid.*, 130). Outside specialists undertaking skilled work such as pruning or grafting might also be paid on a piecework basis.

Some contemporary writers, such as Lawson (1631, 3), discuss the pros and cons of owners getting involved themselves in practical gardening on their own property. There were obvious limits to what might be expected of the nobility and gentry, and hard manual work was not included. Leonard Mascall observed, 'What thing is now counted more filthy in these days (among faire personages) than labouring on the earth which we must all live by' (Mascall 1575, To the Reader). However, some of the more skilled and delicate manual tasks were quite acceptable and even to be recommended. John Parkinson is quite specific in what constitutes a suitable activity for gentlemen gardeners. He states that 'Gentlemen and others are much delighted to bestowe their paines in grafting themselves, and esteeme their own labours and handie work farre above other mens' (Parkinson 1629, 538).

Women had some role in the garden during this period but the precise nature of this is often difficult to determine because they are rarely mentioned. However, this certainly did not include any tree work. Contemporary literature sometimes states that the flower garden and kitchen garden were the responsibility of women, although the tending of fruit trees was definitely considered a man's job. While women might have been involved in the pleasure garden or flower garden, the extent of this seems quite limited. When women are mentioned in this connection, as in Lawson's *The Country Housewifes Garden*, the impression is usually given that they are the owners, or, more precisely, the owners' wives. However, we do know that some women were hired as labourers to work in orchards and pleasure gardens on menial tasks. Skilled work such as planting and pruning would have been out of the question. Weeding of every description was certainly women's work, and weeder women were a significant element of the gardening workforce during this period. Those who did the 'wedynge' often remained anonymous as payment to them, along with other casual labourers, was usually made within the reimbursement of a lump sum of expenses to the head gardener. However, the accounts of Hampton Court give a list of weeders in the orchard dated 16 July 1516, and most of them are women (Sands 1950, 16).

The employment of overseas gardeners, mostly from continental Europe, was common during this period. It started in the reign of Elizabeth I, when British horticulture received a boost from an influx of Protestant or Huguenot refugees from France and the Low Countries. A number of these refugees brought with them specialist skills in tree production and care, especially in relation to fruit trees, and many settled and found work in the coastal towns of Kent, Essex and Suffolk. The Dutch gardeners in particular were highly esteemed for their expertise in the planting and care of trees and were regularly invited by local landowners to advise them on their gardens. In John Gage's *The History and Antiquities of Hengrave, in Suffolk* there is reference from that time to a Dutch gardener who was paid 3*s* 4*d* 'for his travayle from Norwich to Hengrave to view ye orchards, gardens, and walkes', and again, 'Paid to the Dutchman for clypping the Knotts, altering the alleys, setting the grounds, finding herbs and bordering the same, 40*s*' (Gage 1822, 16). French gardeners were also commonly found in English gardens at this time. Not only did their knowledge of continental garden fashions and exotic plants give them an advantage over most of their English counterparts, but their presence would have conferred a degree of status on their employers (Francis 2018, 322). They were frequently imported in order to carry out major changes to a garden, which would then be maintained by the resident English gardener.

Some continental gardeners established quite a reputation for themselves and were much in demand. One such individual was Guillaume Lubias from Dieppe, a remarkable gardener who was especially skilled in both planting and grafting (Grant and Leslie 1798, 352). In 1540 he was employed in Scotland at the gardens of the Priory of Beauly and the Abbey of Kinloss where his skills were said to not only have transformed these gardens but also much of the surrounding countryside. His name appears again in connection with Henry VIII's order of 200 pear trees for Nonsuch Palace which were sourced from France (Dent 1962, 114). In an acknowledgement of his exceptional abilities the king commanded Guillaume de Dieppe to be at Nonsuch when they were planted.

There were some major fluctuations in the labour market for gardeners and tree-related professionals during this period, although it is hard to find much detailed evidence. In the late fifteenth and early sixteenth centuries many of the best gardeners had been working in monastic gardens. With the Dissolution of the Monasteries these gardeners would have been out of work and looking for alternative employment, unless they were lucky enough to be retained by the new owners (Harvey 1972, 24). In Scotland it is thought that one of the reasons for the less rapid growth of gardens before the seventeenth century, in comparison to England, was the lack of experienced professional gardeners (Cox 1935, 30). Professional gardeners who were paid a salary were few in Scotland at this time.

During this period we begin to get more frequent references in the literature and archival sources to people who specialised in tree-related activities in the garden and other amenity landscapes. This was undoubtedly due to the increasing popularity of tree-related features during the Tudor and early Stuart era, such as topiary, mazes, wildernesses, avenues and walks, which required quite skilled and often extensive use of trees. As we have noted, William Lawson was one of the first to use the tree-specific term 'arborist' in this connection. However, due to the widespread use of the term 'gardener' during this time, this would also have been used to embrace many who specialised in tree-related work.

Examples of named gardeners undertaking tree-related work can be drawn from the household or estate account books associated with some of the famous gardens of the time. In this respect we are fortunate to be able to draw on the research of garden historians, especially John Harvey, David Jacques and Jill Francis. Examples have already been given in this chapter, such as John Tradescant and Guillaume Lubias, but a few more are given here. At Hatfield House, the master gardener, Thomas Tudor, commanded a salary of £25 per year and was apparently the highest paid member of the household (Francis 2018, 127). The maintenance of the garden with its extensive tree planting required plenty of tree work, especially skilful and intricate pruning, and not just with its knots and topiary. The vast gardens at Hatfield House were renowned throughout the country at this time, so perhaps we should expect their gardener to be exceptionally highly skilled and therefore suitably rewarded. In 1620, at Halland House in Sussex, 'Grove the gardner' was paid 1*s* a day for specific tasks such as 'pruning and dressing of the trees in the garden' (*ibid.*, 130). However, in 1640 at Holm Lacy in Herefordshire, Brake the gardener was paid just 7*d* a day for the same task. Other examples of named gardeners engaged on tree work refer to urban locations such as the first tree-lined walks or promenades. At Moorfields in 1599, the gardener Richard Brooks was paid for 'beating up' the young trees (replacing those that had died) and from March 1600 he began to receive additional wages of 50*s* every quarter 'for weedinge and keepinge the walks' (Jacques 1989, 45). Between August 1608 and January 1609, among numerous payments to the gardener John Mortimer, there was one for 'stettinge sicamore, elme and birche trees' (*ibid.*, 48–49). There is also a record that Mortimer was paid for 150 tree stakes that he used in these plantings.

The number of skilled tree workers at any time or place is always limited by the extent of any recognition of the need for these tree experts. If there is no recognition then there is no demand and other less skilled and experienced workers will be employed, probably undertaking a basic and possibly unsafe approach to the work. It is interesting that Sir Thomas Temple, noted above, regarded felling trees as a labouring job (Francis 2018, 130). In this case it should be remembered that he was creating a garden from scratch and the felling was part of general land clearance. However, tree felling and major branch removal in an established garden can be another matter entirely, with the need to avoid damage to nearby plants, garden features and even buildings. This sort of tree work was undoubtedly undertaken, but little evidence of it has been found in the archives. As in medieval times, it is likely that skilled personnel and specialist equipment from the estate's forestry section were often employed in the garden for these purposes.

The first steps to bring gardeners together in a formally recognised organisation were made at the start of the seventeenth century with the foundation of the Worshipful Company of Gardeners. In 1605, James I granted royal charters to gardeners and fruiterers so that they could form their own companies. A 'mystery' or fellowship of gardeners had existed in London from the mid-fourteenth century, ensuring that training took place through an organised apprenticeship scheme and acting as a benevolent fraternity looking after members and their families. However, they had not previously enjoyed the freedom of the City and other privileges accorded to livery companies. In the earliest days of the Company it was composed almost exclusively of working gardeners, although some of the leading members had in their employment several apprentices and journeymen (Barnes 1994, 45).

The preamble to James I's charter sets out the operations controlled by the Worshipful Company, namely, 'the trade, crafte, or misterie of Gardening, planting, grafting, setting, sowing, cutting, arboring, rocking, mounting, covering, fencing and removing of plantes, herbes, seedes, fruites, trees, stocks, setts, and of contriving the conveyances to the same belonging' (Barnes 1994, 31). While the activity of 'arboring' is listed, this refers to the making of arbours rather than anything to do with mature trees. This catch-all list applied to gardeners within a 6 mile radius around the City. The Charter also stated that 'certain ignorant and unskilful persons' who had taken upon themselves to practice their trade 'not having been experienced thereto' had sold dead and diseased plants, seeds, stocks and trees 'to the great hindrance and deceit of many of our loving subjects'. One of the main reasons for incorporating the Company was to prevent these abuses and ensure that 'cowboy' nurserymen and gardeners did not give the entire gardening profession a bad name.

Conclusions

While the thousand years of the medieval era saw little progress in arboriculture, this Tudor and early Stuart period of around 150 years witnessed the emergence of some 'green shoots' in the early development of professional arboriculture in Britain. The main drivers were the comparative political stability and absence of conflict across much of Britain, leading to greater prosperity among the ruling class and growing numbers of gentry. This in turn gave a boost to the creation of many new gardens and the expansion of existing gardens. In contrast to the small enclosed gardens of medieval times, many of these much larger gardens were focused on the ornamental, particularly in the early Stuart era, and had important tree-related features, some of which involved extensive tree planting. There were also more urban gardens and the first tree-lined walks and promenades. All this required an increasing number of gardeners, especially those with specialist arboricultural skills.

New developments in this period also included the first books to focus on practical arboriculture, with guidance on how to undertake a range of practical tasks such as tree planting, pruning and propagation. Some books also included detailed descriptions and illustrations of a range of tools and equipment used in practical arboriculture. Furthermore, we see the first use of the term 'arborist' in the English language to describe someone with specialist expertise in the planting, care and management of trees.

In the Jacobean era the growing fashion for extensive tree planting among the monarchy and leading nobility was a foretaste of the enormous quantity of tree planting that would come with the Baroque garden in Britain in a few decades time. However, progress towards this stalled in the reign of Charles I amid increasing political instability and grave concerns about what the future might hold. Those concerns were justified when tensions between the King and Parliament deteriorated to the point of civil war in 1642. There would be much death and destruction in Britain before some degree of normality returned and the rich and powerful could once again indulge their love of ornamental landscapes and trees.

References

Amherst, A. (1896) *A History of Gardening in England.* Second edition. Bernard Quaritch: London.

Andrews, M. (1993) Theobalds Palace: The Gardens and Park. *Garden History* 21(2), 129–149.

Anon (*c.* 1520) *The Crafte of Graffynge and Plantynge of Trees.* Wynkyn de Worde: London.

Anon (1594) *The Orchard, and the Garden Containing Certaine Necessarie, Secret, and Ordinarie Knowledges in Grafting and Gardening.* Adam Islip: London.

Anon (1811) *The Customs of London, otherwise called Arnolde's Chronicle.* F.C. and J. Rivington, T. Payne and others: London.

Bacon, F. (1902) *Of Gardens: An Essay by Francis Bacon.* Hacon and Ricketts: London.

Barnes, M. (1994) *Root and Branch: A History of the Worshipful Company of Gardeners of London.* The Worshipful Company of Gardeners of London: London.

Borsay, P. (1986) The Rise of the Promenade: The Social and Cultural Use of Space in the English Provincial Town *c.* 1660–1800. *Journal for Eighteenth-Century Studies* 9(2), 125–140.

Bowen, L. (ed.) (2006) *Family and Society in Early Stuart Glamorgan: The Household Accounts of Sir Thomas Aubrey of Llantrithyd, c. 1565–1641.* South Wales Record Society: Cardiff.

Boyceau, J. (1638) *Traité du Jardinage, selon les raisons de la nature et de l'art.* Michel Vanlochom: Paris.

Bradshaw, M. (2004) *Lyveden New Bield.* The National Trust: Swindon.

Brown, P.H. (1904) *Scotland in the Time of Queen Mary.* Methuen: London.

Cavendish, G. (1825) *The Life of Cardinal Wolsey, Volume II.* Harding, Triphook and Lepard: London.

Church, R. (1612) *An Olde Thrift newly Revived: Wherein is Declared the Manner of*

Planting, Preserving and Husbanding Young Trees of Divers Kindes for Timber and Fuell. W.S. for Richard Moore: London.

Cox, E.H.M. (1935) *A History of Gardening in Scotland.* Chatto and Windus for Flora and Sylva: London.

Crisp, F. (1924) *Medieval Gardens.* John Lane Bodley Head: London.

De Caus, I. (1645) *Wilton Garden (Hortus Penbrochianus: Le jardin de Vuilton).* Thomas Rowlett: London.

Dent, J. (1962) *The Quest for Nonsuch.* Hutchinson and Co.: London.

Evelyn, J. (1901) *The Diary of John Evelyn.* Edited in 1901 from the original manuscript by W. Bray in two volumes. Volume 1: 1620–1664. M. Walter Dunne: New York and London.

Fitzherbert, M. (1882) *The Book of Husbandry by Master Fitzherbert.* Reprinted from the edition of 1534 and edited by W.W. Skeat. The English Dialect Society: London.

Francis, J. (2018) *Gardens and Gardening in Early Modern England and Wales, 1560–1660.* Yale University Press: New Haven, CT.

Gage, J. (1822) *The History and Antiquities of Hengrave in Suffolk.* James Carter and Joseph Booker: London, and John Deck: Bury St Edmunds.

Goldring, E. (2013) The Langham *Letter* as a Source for Garden History. In: Keay, A. and Watkins, J. (eds) *The Elizabethan Garden at Kenilworth Castle.* English Heritage: Swindon.

Grant, J. and Leslie, W. (1798) *A Survey of the Province of Moray: Historical, Geographical, and Political.* Isaac Forsyth: Elgin.

Groos, G.W. (1981) *Diary of Baron Waldstein, a Traveller in Elizabethan England.* Thames and Hudson: London.

Harding, V. (1990) Gardens and Open Space in Tudor and Early Stuart London. In: Galinou, M. (ed.) *London's Pride: The Glorious History of the Capital's Gardens.* Anaya: London.

Harvey, J. (1972) *Early Gardening Catalogues.* Phillimore and Co.: London and Chichester.

Harvey, J. (1974) *Early Nurserymen.* Phillimore and Co.: London and Chichester.

Harvey, J. (1982) William Lawson and his Orchard. *Country Life*, 28 October, 1338–1340.

Henrey, B. (1975) *British Botanical and Horticultural Literature before 1800. Volume 1: The Sixteenth and Seventeenth Centuries.* Oxford University Press: Oxford.

Hill, T. (*c.* 1558) *A Most Briefe and Pleasaunte Treatise, Teachying how to Dress, Sowe, and Set a Garden.* John Day: London.

Hill, T. (1577) *The Gardeners Labyrinth: Containing a discourse of the Gardeners life.* Henry Bynneman: London.

Jacques, D. (1989) 'The Chief Ornament' of Gray's Inn: The Walks from Bacon to Brown. *Garden History* 17(1), 41–67.

Jacques, D. (1999) The Compartiment System in Tudor England. *Garden History* 27(1), 32–53.

Jacques, D. (2013a) The Place of Kenilworth in Garden History. In: Keay, A. and Watkins, J. (eds) *The Elizabethan Garden at Kenilworth Castle.* English Heritage: Swindon.

Jacques, D. (2013b) The Form and Structure of the Garden. In: Keay, A. and Watkins, J. (eds) *The Elizabethan Garden at Kenilworth Castle.* English Heritage: Swindon.

Janson, H.F. (1996) *Pomona's Harvest: An Illustrated Chronicle of Antiquarian Fruit Literature.* Timber Press: Portland, OR.

Jermin, M. (1639) *A Commentary, Upon the Whole Booke of Ecclesiastes or the Preacher.* John Clark: London.
Johnson, R. (1607) *The Pleasant Walkes of Moore-field.* Henry Goffon: London.
Johnston, M. (2015) *Trees in Towns and Cities: A History of British Urban Arboriculture.* Windgather Press: Oxford.
Johnston, M. (2017) *Street Trees in Britain: A History.* Windgather Press: Oxford.
Keiser, G.R. (1996) Through a Fourteenth-Century Gardener's Eyes: Henry Daniel's Herbal. *The Chaucer Review* 31(1), 58–75.
Lawson, W. (1618) *New Orchard and Garden.* Alsop for Roger Jackson: London.
Lawson, W. (1631) *New Orchard and Garden.* John Harrison: London.
Leith-Ross, P. (2006) *The John Tradescants: Gardeners to the Rose and Lily Queen.* Revised edition. Peter Owen: London.
Leland, J. (1907) *The Itinerary of John Leland In or About the Years 1535–1543. Parts I to III.* Edited by L. Toulmin Smith. G. Bell and Sons: London.
MacLeod, D. (1972) *The Gardener's London.* Gerald Duckworth: London.
Markham, G. (1613) *The English Husbandman. The first Part: Containing the Knowledge of True Nature, etc.* Printed by T.S. for John Browne: London.
Markham, G. (1614) *The Second Booke of the English Husbandman, etc.* Printed by T.S. for John Browne: London.
Markham, G. (1616) *Maison Rustique, or The Countrey Farme.* John Bill: London.
Mascall, L. (1569) *A Booke of the Arte and Maner Howe to Plant and Graffe all sortes of Trees.* Henry Bynneman for John Wight: London.
Matthews, W.H. (1922) *Mazes and Labyrinths: A General Account of their History and Developments.* Longmans, Green and Co.: London.
McLean, T. (1981) *Medieval English Gardens.* William Collins: London.
N.F. (1608) *The Husbandmans Fruitfull Orchard.* Roger Jackson: London.
Oglander, J. (1936) *A Royalist's Notebook: The Commonplace Book of Sir John Oglander.* Constable: London.
Parkinson, J. (1629) *Paradisus in Sole Paradisus Terrestris, or a Garden of All Sorts of Pleasant Flowers.* Humfrey Lownes and Robert Young: London.
Pattacini, L. (1998) André Mollet, Royal Gardener in St. James's Park, London. *Garden History* 26(1), 3–18.
Roberts, J. (1999) The Gardens of the Gentry in the Late Tudor Period. *Garden History* 27(1), 89–108.
Sands, M. (1950) *The Gardens of Hampton Court: Four Centuries of English History and Gardening.* Evans Brothers: London.
Scamozzi, V. (1615) *L'Idea Della Architettura Universale.* Volume 2. Published by the author: Venice.
Stallenge, W. (1609) *Instructions for the Increasing of Mulberrie Trees, and the Breeding of Silk-Wormes, for the making of silke in this Kingdom.* Printed for Eleazar Edgar: London.
Standish, A. (1611) *The Commons Complaint Wherein is contained two Special Grievances.* William Stansby: London.
Stout, F. (2015) Before Evelyn: Trees, Tree Planting and Tree Management in Sixteenth- and Early Seventeenth-Century England. *Arboricultural Journal* 37(3), 150–165.
Strong, R. (1979) *The Renaissance Garden in England.* Thames and Hudson: London.
Taylor, K. (2008) The Earliest Wildernesses: Their Meanings and Developments. *Studies in the History of Gardens and Designed Landscapes* 28(2), 237–251.

Thacker, C. (1994) *The Genius of Gardening: The History of Gardens in Britain and Ireland*. Weidenfeld and Nicolson: London.

Tusser, T. (1812) *Five Hundred Points of Good Husbandry.* A new edition by W. Mavor. Lackington, Allen and Co.: London.

Watkins, J. (2013) Patterns of Planting. In: Keay, A. and Watkins, J. (eds) *The Elizabethan Garden at Kenilworth Castle.* English Heritage: Swindon.

Watsoni, A. (1590) *Magnificae, et Plane Regiae Domus, quae vulgo vocatur Nonesuch brevis, et vera descripto 1590*. Unpublished manuscript at Trinity College, Cambridge, MS R.7.22, edited by M. Biddle and J. Dent.

Williams, C. (1937) *Thomas Platter's Travels in England 1599*. Jonathan Cape: London.

CHAPTER FIVE

Arboriculture in the Age of the Formal Garden

The English Civil War (1642–1651) is a dark period in history that had a catastrophic impact across the whole of Britain and Ireland. Following the execution of Charles I, the monarchy was abolished and the Commonwealth was established whereby England, Wales, Scotland and Ireland were governed as a republic. The Commonwealth lasted from 1649 to 1660 when the monarchy was restored and Charles II became king. The Restoration signalled the continuation of the Stuart dynasty with the reign of Charles II (1660–1685), followed by James II (1685–1688). James II was deposed in 1688 by his daughter Mary and her husband William of Orange with his Dutch army, in what is called the Glorious Revolution. The couple then ruled as Mary II and William III from 1689 until Mary's death in 1694 when William continued as sole monarch until his death in 1702. He was then succeeded by Anne, the sister of Mary II, who reigned from 1702 to 1714. In 1707, under the Acts of Union, the kingdoms of England (that included Wales) and Scotland united as a single sovereign state known as Great Britain. After Anne's death in 1714, she was succeeded by her German second cousin who became George I (1714–1727), and this heralded the start of the Georgian era.

This chapter begins with the approach of the Civil War in 1640 and ends in 1730, a span of ninety years. Garden historians covering this approximate period often chose slightly different start and end dates. Some begin with the civil war, others with the start of the Commonwealth and for others it is the Restoration of 1660. This author has chosen the start of the Civil War since many of the subsequent developments that are discussed are hugely influenced by this traumatic episode. In terms of garden history in Britain, this period has been characterised as the Age of the Formal Garden, where French and Dutch styles are said to have predominated. Historians generally consider that this influence was waning in the 1720s with the emergence of new ideas that would become known as the English Landscape Garden. Therefore, our coverage of this period ends around the year 1730.

Sources of information

Unlike previous periods of garden history, a number of authentic gardens from this era have survived, many of them particularly notable. Others have been recreated using extensive archive evidence, both written and pictorial. Nevertheless, a huge number of these formal gardens were swept away in the latter part of the eighteenth century amid the fashion for the English Landscape Garden, which will be covered in the next chapter.

The various archival sources of written and pictorial evidence for this period are similar to those of the previous chapter, although far more extensive. For example, the second half of the seventeenth century in England witnessed a huge increase in the number of gardening books published over that from the first half of the century (Henrey 1975, 169). This period also saw the publication of several important books focused specifically on arboriculture and tree-related topics. These primary sources have been invaluable in providing a wealth of information about arboricultural and tree-related practices during this period. In conducting this research it has been important to actually read these authors and not just accept what modern historians have said about their work.

Thanks are due to a few recent historians who have provided information on gardens and practical gardening taken from the account books of some of the landed gentry of this era. These include John Harvey, David Jacques, David Marsh and Sally O'Halloran. Some useful information on practical arboriculture has been extracted from their research. Another source of information has been the garden books of some of the contemporary aristocracy and landed gentry, such as those by John Evelyn and Sir Thomas Hanmer, which give detailed accounts of the work undertaken in their gardens. Information has also been obtained from diaries of contemporary travellers, such as Celia Fiennes and Daniel Defoe, who often give enlightening descriptions of the gardens they visited.

Visual sources have played a vital role in the research for this chapter. Images located in some of the early books have given some very useful insights into arboricultural practices of the time, along with the tools and equipment used. With regard to the gardens and garden features from this period, one especially valuable source of images is the work of Leonard Knyff and Johannes Kip, who first published their *Britannia Illustrata* in 1707. This contains eighty exquisitely produced plates of bird's-eye views, each one drawn by Knyff and engraved by Kip. For this research, the work of Knyff and Kip gives a remarkable insight into the extent of tree planting, maintenance and management in the tree-dominated formal landscapes of this era.

The Civil War, the Commonwealth and fruit trees

The politics of this era, particularly the English Civil War, the Commonwealth and the Restoration, has always greatly influenced our view of horticulture and arboriculture during these turbulent times. During the Civil War itself, it is understandable that not much of note occurred in the fields of arboriculture and amenity horticulture. The nation's preoccupation with the conflict ensured that even among those who owned large estates and gardens there was little time or inclination to focus on the ornamental aspects of trees and landscapes. However, as amenity horticulture faded into the background, forestry came to the fore since woodlands and their reserves of timber were a vital resource in times of war. Vast numbers of trees were felled to fuel the war effort and this was undertaken by both sides in the conflict. These deprivations would often be visited upon estate owners by their enemies, but some landowners also felled many of their own trees in support of their side in the civil war. Some modern historians would argue that this destruction took place in the context of an already alarming shortage of timber for shipbuilding, iron foundries and other uses. As noted in the previous chapter, these concerns about a national timber shortage, along with calls for an extensive afforestation effort, had already been voiced a few decades earlier by writers such as Arthur Standish, Rocke Church and Gervase Markham. These concerns continued to be expressed by some commentators during the years of the Commonwealth and they became more evident with the restoration of the monarchy. However, the true extent of any timber shortage in Britain in the seventeenth century remains the subject of debate among modern historians (Sharp 1975, 52). Nevertheless, many contemporary commentators believed that shortages existed and this helped to create a general fear of timber starvation.

The call for a national afforestation effort and measures to protect Britain's existing timber reserves was voiced by one writer while the civil war was still raging. Walter Blith (1605–1654) was a captain in Oliver Cromwell's New Model Army who in 1649 wrote *The English Improver or a New Survey of Husbandry* (Blith 1649). His allegiance to the parliamentary cause is immediately obvious from the book's dedication to Cromwell and the frontispiece headed 'Vive La Republick'. While Blith's book was a general text on husbandry, a significant portion focused on trees and forestry. It was well received and a new edition was released in 1652/53 in a much-expanded form as *The English Improver Improv'd* (Blith 1653). The book includes a chapter on how to plant trees in a 'Plantation of divers sorts of Wood, and Timber, in such a way as shall raise as much in twenty years growth, as usually and naturally groweth in forty or fiftie years' (Blith 1653, 153). Blith also recognised that extensive tree planting could have an ornamental dimension, anticipating John Evelyn and Moses Cook. He thought that small blocks of woodland set out as squares, rectangles, triangles or ovals would be appropriate if 'thou aim at thy delight and pleasure' (*ibid.*, 155). Inside, there would be a circle at which the walks and vistas through the woodland would intersect. The various blocks would be separated with grass-covered walks 'which will serve as a Cart-way, or Passage, to fetch out thy Wood at every fall, as well as for walks for thy recreation' (*ibid.*, 156). Blith also advised that 'several Groves or Plumps of Trees may be Erected about any Manour, House or Place', both for the production of timber and 'for delight and pleasure' (*ibid.*, 153). Although Blith contributed to the national debate on forestry and highlighted the ornamental use of woodland, his political allegiances would ensure he was largely forgotten after the Commonwealth. Nevertheless, while Blith was calling for the protection of the nation's trees, Cromwell's agents were requiring the felling of many trees in the gardens and woodlands of the former Crown estates and those of the defeated Royalist nobility. These landscapes were often viewed as symbols of their self-glorification and the removal of the trees for timber, without recompense, was seen as a way for them to pay compensation to the nation.

During the Commonwealth there was a national movement to promote the planting and improvement of orchards, driven by the political and religious ideals of the Puritans and their allies. Before the Commonwealth, fruit tree cultivation was seen as a rewarding pastime for gentleman, but it now became something of a vocation for the faithful. The movement was driven by a small circle of like-minded reformers led by Samuel Hartlib (*c.* 1600–1662), a polymath of German-Polish origin who settled in England. Hartlib's (1652) own contribution was set out in his booklet entitled *A Designe for Plentie* which recommended that private land, commons and waste ground should be planted with fruit trees 'for the relief of the poor, the benefit of the rich and the delighte of all'. Landowners would plant a specified number of trees, a requirement that would be enforced by public authorities, and all England would become the 'Garden of God'. The fruit could then be given to the poor and the wood from the trees used for fuel or furniture. He concluded his pamphlet with an appeal for communal benefit to be put before individual gain. John Beale (*c.*1608–1683), a clergyman from Herefordshire, was a member of Hartlib's circle and was persuaded by him to publish a small book entitled *Herefordshire Orchards: A pattern for all England* (Beale 1657). As well as practical advice on fruit tree cultivation, Beale combines this with spiritual thoughts. He cites a number of authors, but none is examined in such depth as William Lawson. Furthermore, he states that although he was initially sceptical of many of Lawson's statements and general approach, he later found from his experience that much of what Lawson had said was true (*ibid.*, 8–9). Although a confirmed republican during the Commonwealth, Beale later became a Royalist.

The Puritan and republican thinkers of Hartlib's circle were happy to cooperate with any former Royalist supporter with knowledge and expertise. Their considerations were essentially practical and respected individuals in horticulture, such as John Tradescant, fell into this category. He was valuable to them because they were particularly concerned with the cultivation and spread of useful plants and were anxious to improve the hitherto often poor standards of horticulture and agriculture in the country. While these radicals may not have done much to advance the cause of ornamental horticulture, they gave a major boost to commercial horticulture and to pomology in particular.

Ralph Austen

The most influential writer on fruit trees among Hartlib's circle was Ralph Austen (*c.* 1612–1676), whose remarkable work would also make a major contribution to the development of arboriculture in Britain. Austen came from a yeoman family in Leek, Staffordshire, and his mother was a cousin of Henry Ireton the parliamentary leader (Turner 1978, 39). He had no university education and no official status, though he referred to himself as 'gent' in later years. He was in Oxford by about 1646 and remained there until his death in 1676. He may have fought in the parliamentary army and was certainly involved in the military-backed purge of Oxford University which followed the siege of 1646. His closest friends were the military leaders and puritan academics who replaced ejected Royalists. Austen built up a small nursery in the 1650s near his house in south-west Oxford, which was expanded in later years.

As a member of the revolutionary Parliament, Austen voted for the first excise duties on cider and perry and then lobbied Cromwell to invest the proceeds in a national fruit tree planting scheme (Janson 1996, 125). Nothing came of this project, but Austen's dedication to the cause of fruit is evident in his book *A Treatise of Fruit-Trees*, published in 1653. This charming and original work was written with a passion not only for fruit trees but also for his political and religious convictions. Austen (1653, Epistle) declares that he wrote *A Treatise of Fruit-Trees* to highlight the 'great deficiency in England' in the quantity and quality of fruit trees grown. On the title page Austen identifies himself as a 'Practiser in ye art of Planting' and proudly states that his treatise is based 'on experience gathered in ye space of twenty years'. Austen viewed the creation of many new orchards as a project that would benefit the whole of society, especially the rural poor. To provide land for this planting he wanted to see an organised system of enclosure to benefit the poor: 'For in divisions and inclosures of Wast and Common Grounds, (by Persons appoynted for that purpose), why should not the Poore have their share, and proportion, as well as their rich Neighbours, and that to them, and their heires for ever; yea, let the Poore be first provided for' (Austen 1653, Epistle).

In later years such talk would be viewed as treasonable. Furthermore, Austen recommended that the poor be hired to assist in the preparatory work for these orchards 'whereby they might maintaine themselves, and profit others, instead of burthening of them'. The poor would be involved with 'inclosing and preparing grounds for planting', while the task of planting and the supervision of the project would be undertaken by a gardener. The book was dedicated to his 'much Honoured Friend' Samuel Hartlib and on the same page he praised his friend Walter Blith for his work in promoting tree planting.

A constant theme of Austen's book is the 'Profits and Pleasures' of the orchard. Indeed, the title page has an image of an enclosed garden of fruit trees in an ordered arrangement, above which are clasped hands illustrating Profits shaking hands with Pleasures (Figure 23). If Austen's book was purely

FIGURE 23. Ralph Austen's *A Treatise of Fruit-Trees* in 1653 was not just a work on the cultivation of fruit trees but a major contribution to practical arboriculture. Its theme of 'profits and pleasure' was indicated at the top of the title page.

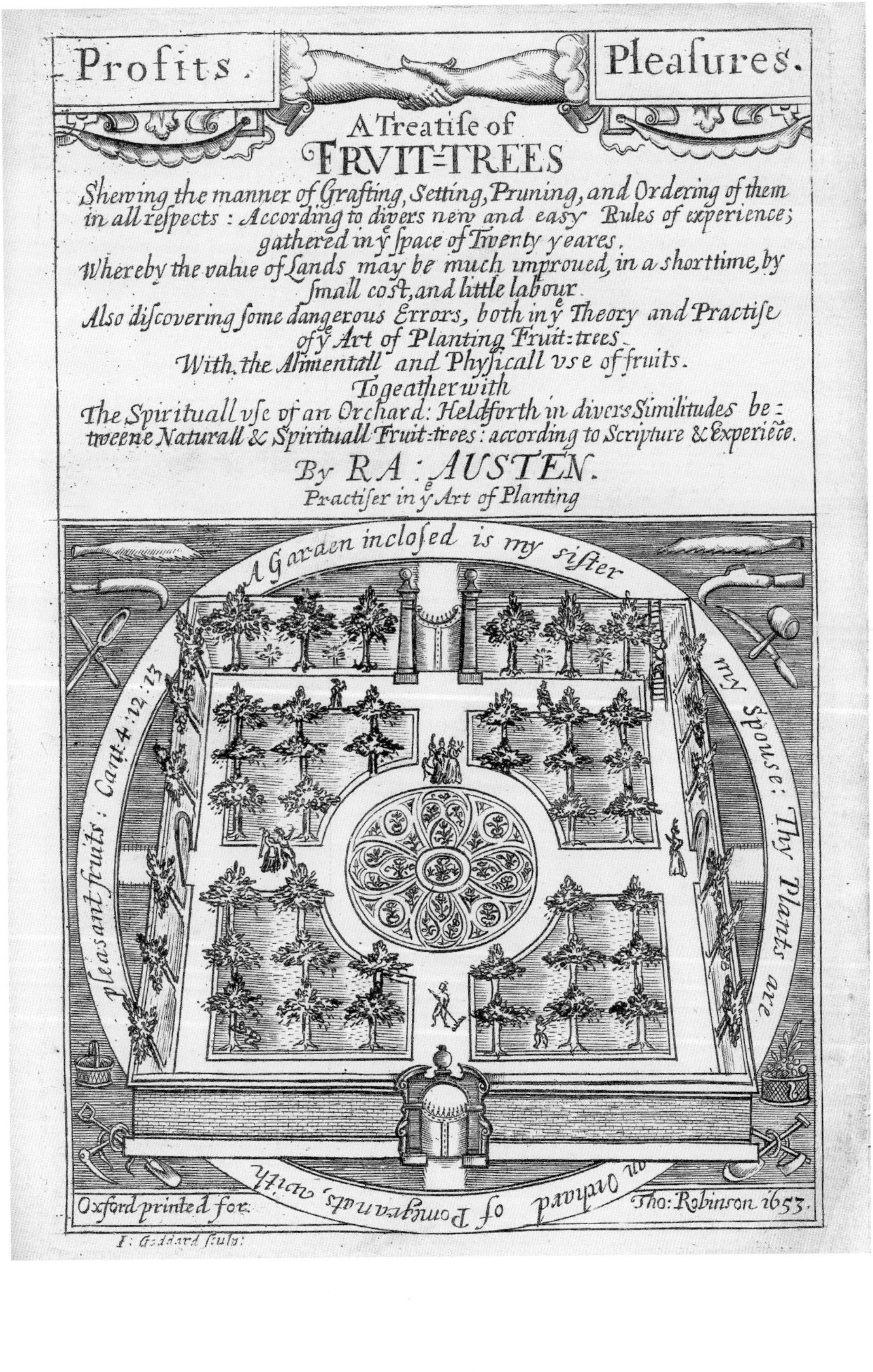
Profits. Pleasures.
A Treatise of
FRVIT-TREES
Shewing the manner of Grafting, Setting, Pruning, and Ordering of them in all respects: According to divers new and easy Rules of experience; gathered in ye space of Twenty yeares.
Whereby the value of Lands may be much improued, in a shorttime, by small cost, and little labour.
Also discovering some dangerous Errors, both in ye Theory and Practise of ye Art of Planting Fruit-trees.
With the Alimentall and Physicall vse of fruits.
Togeather with
The Spirituall vse of an Orchard: Held forth in divers Similitudes betweene Naturall & Spirituall Fruit-trees: according to Scripture & Experiẽce.
By RA: AUSTEN.
Practiser in ye Art of Planting
A Garden inclosed is my sister my spouse: Thy Plants are an Orchard of Pomegranats, with pleasant fruits: Cant. 4: 12: 13
Oxford printed for Tho: Robinson 1653.
I: Goddard sculp:

about the cultivation of fruit trees for profit, it would be of limited interest to this present volume. However, he gives equal value to the pleasures of the orchard and introduces this section of his book with the words, 'Now, we find pleasure in an Orchard or Garden of Fruit-trees suitable to all Senses of the body' (Austen 1653, 35). He then goes through the five senses and gives examples: for the sense of hearing (bird song), for the sense of touch (cool shade in summer), for the sense of sight (planting design with *allées* and arbours), for the sense of smell (blossom) and for the sense of taste (fruit). While Austen is not concerned with the design of ornamental landscapes, he does talk of fruit trees contributing to the 'beauty and comlinesse' of the garden (Austen 1657, 69). He also advocates that it would be of great benefit to all 'if men would Plant Fruit-trees, not only in Gardens, but also in many of their Fields and Hedges' (Austen 1653, Epistle). This interest in tree planting in the wider landscape is extended to forests and woods in the second edition of his book where he echoes his friends Hartlib and Blith in complaining about the lack of woodland and the scarcity and high cost of timber. He declares, 'It is much desired (by many who wish the welfare of the Common-wealth) that men may by Authority be put on, and encouraged to plant, not only store of Fruit-trees, but also all sorts of Wood' (Austen 1657, Epistle). He wants children to be encouraged to join this national tree planting effort: 'If Schoole-boyes and others while they are young, set upon planting Okes, Ashes, Elmes and other trees, they may probably (as others have done) make use of them for building, sale, etc., even themselves in their owne-persons, unto exceeding great profit' (*ibid.*, 133).

Austen's *Treatise* was published together with his text entitled *The Sprituall Use of an Orchard*, which deals with the religious and philosophical aspects of his work. He demonstrates how God is evident in the natural world, declaring: 'The World is a great Library, and Fruit-trees are some of the books wherein we may read and see plainly the Attributes of God' (Austen 1653, To the Reader). In contrast to his *Treatise* this text contained little of practical use and was omitted from the third edition in 1665.

In Austen's extensive account of the propagation, planting and pruning of fruit trees there is much that is relevant to practical arboriculture. The importance of his work can be compared with that of William Lawson a few decades earlier. His recommendations were invariably based on his own experience and some of his more significant contributions to arboricultural practice are covered elsewhere in this chapter. He was always keen to contradict old-fashioned notions and he has a chapter at the end of his book entitled 'Errors Discovered' (Austen 1653, 78–97). As well as highlighting some poor arboricultural practices of the time he also exposes some fantastic beliefs. These included writing an inscription on a fruit stone and planting it, expecting the same writing to appear in the ripe fruit of the tree; soaking the stone in liquor to have the taste in the fruit; and placing apple grafts into pike's blood to give red apples.

In common with many of his contemporary writers, Austen frequently quotes the bible as well as the classics, even in the sections of his book that deal with practical matters. However, he is aware that revered authorities can be wrong: 'Some great and Learned Authors have asserted very weak things, and altogether untrue' (Austen 1653, 78). For example, he is keen to dispel the common belief that the phases of the moon should dictate when to plant and graft trees (*ibid.*, 90).

Austen's work is often overlooked by modern historians and, when mentioned, is regularly dismissed as having little value. For example, Miles Hadfield states, 'It is much concerned with moral matters and of little practical use' (Hadfield *et al.* 1980, 18). Nan Fairbrother (1956, 99–100), the landscape architect, cannot understand how Ralph Austen could have been both a Puritan and a gardener and writes of his 'divided mind'. She refers to the 'strangely twisted consciences' of Austen and his friends and asserts that 'the gentle light of reason could not reach the dark and deformed spirits of the Puritan zealots'. Perhaps these criticisms have less to do with what they said about gardening and much more about their radical ideas on overturning the established order.

Other writers on fruit trees

Following the restoration of the monarchy in 1660, the interest in fruit trees that had flourished during the Commonwealth continued, now championed by individuals loyal to the new king, Charles II. John Evelyn, author of *Sylva*, the classic work on trees, is often credited with writing extensively on fruit trees. Aside from his translations of continental works on fruit, Evelyn had little to say on the subject due 'to my poor knowledge and research' (Evelyn 1664, 50). He acknowledges there were already many good books on the topic and sees no value in repetition. Nevertheless, *Sylva* was published with a twenty-page addition entitled *Pomona; or an Appendix Concerning Fruit-Trees.* However, as Evelyn states, the credit for *Pomona* should go to his friends and various members of the Royal Society, especially John Beale, although Beale's name is not on the title page. It is interesting that in both *Sylva* and *Pomona* the name of Ralph Austen is never mentioned. Evelyn must have known Austen's work, which anticipates his own *Sylva* in many respects, and only political bias can explain the omission (Turner 1978, 39).

In 1672 a useful little book appeared entitled *A Short and Sure Guid[e] in the Practice of Raising and Ordering of Fruit-Trees* (Drope 1672). Written by Francis Drope, it was published posthumously by his brother, Edward. Drope was a prebendary at Lincoln Cathedral who had studied divinity at Magdalen College, Oxford. The book gives a concise and knowledgeable account of the propagation of fruit trees but there is less on planting and pruning. As an amateur connoisseur Drope urged his readers to bestow tender love and gentle nurture on young fruit trees. His brother had practised the opposite when he ordered all the trees to be chopped down that were planted by Cromwell's 'fanaticks' at Magdalen College (Janson 1996, 132).

Charles Cotton, the English poet and writer, was concerned at the number of orange and lemon trees being imported from France and other countries when he believed English growers could be producing them. In *The Planters Manual* he attacks the style of the French, wanting to avoid being 'debauch'd by their effeminate Manners … and fantastick Fashions, by which we are already sufficiently Frenchified, and more than … is consistent either with the Constitution, or indeed the Honour of the English nation' (Cotton 1675, To the Reader). His anti-French sentiment is ironic because, although Cotton's name appears on the title page, the book is his translation of the French *Instruction Pour Les Arbres Fruitiers*, published in 1653 by Antoine Bertier of Paris (Henrey 1975, 178–179).

In 1681 Thomas Langford published his book entitled *Plain and Full Instructions to raise all sorts of Fruit-Trees that prosper in England* (Henrey 1975, 179). This includes a commendatory letter addressed to the publisher by John Evelyn, who writes: 'I have read the Treatise of Fruit-trees, etc. which you lately put into my hand, and … I know of nothing extant which exceeds it, so nor do I of anything which needs to be added to it' (Langford 1681, Letter). This is praise indeed from the great man, but ignores the work of Austen and his friends. The book has plenty of information on the propagation and planting of fruit trees, but much less on the maintenance and management of the orchard. Langford's treatise sold well and in 1696 was published in a new and enlarged edition.

While several British authors had now published authoritative works on fruit tree cultivation, there was a tendency to regard the French as leaders in this field. Some important French works of this time include *Le Jardinier François* by Nicolas de Bonnefons, later translated by John Evelyn and published in London (Bonnefons 1658). Also, a book by Antoine Le Gendre, who had worked in Louis XIII's gardens, which was translated into English as *The Manner of Ordering Fruit Trees*, possibly by John Evelyn (Le Gendre 1660). In 1685, an anonymous English translation of Nicolas Venette's *The Art of Pruning Fruit-Trees* was published (Venette 1685). It is difficult to assess the extent to which these French texts advanced British knowledge of practical arboriculture. While they were widely read by the nobility and gentry, it is impossible to gauge the degree to which their recommendations on the propagation, planting and maintenance of fruit trees were transferred to ornamental trees.

Trees in the aftermath of the Civil War

Just as the Commonwealth had overturned the established order of royalist Britain, so the Restoration of 1660 prompted another dramatic upheaval, this time back towards the old order. It was also a time for the settling of scores and the reversal of fortunes. Those who had stayed loyal to the monarchy were often generously rewarded by Charles II, while those who had not could expect some form of retribution, ranging from social exclusion to execution, depending on the extent of their perceived treachery. This revision of the social order also applied to the world of horticulture and gardens. Some Royalists with a reputation in gardening had decided to 'sit out' the Commonwealth, for example, Sir Thomas Hanmer on his Welsh estate and John Evelyn with his travels on the Continent. With the Restoration their self-imposed exile ended and so began a reversal of fortunes among the leadership of the horticultural world. Evelyn, the Royalist and favourite of the new King, became famous; Ralph Austen, the republican with revolutionary ideas, was ignored. Others, such as John Beale, simply changed sides.

As with conflicts throughout history, those who triumphed set about promoting a narrative of those tragic times that would support their own agenda. This applied particularly to the wholesale felling of trees and woodlands that had been undertaken by both sides in the conflict. The depredations wrought by the Parliamentarians upon the estates of the Crown, Church and supporters of Charles I were exaggerated by Royalist propagandists after the Restoration in such a way as to create an association between the wanton felling of trees and republican politics (Thomas 1983, 209; Stout 2015, 152). Evelyn and other contemporaries also cleverly represented tree planting as a way of affirming a gentleman's loyalty to the restored monarchy, as well as emphasising its supposed 'heroic' nature (Evelyn 1664, To the Reader). Meanwhile, those republicans such as Blith and Austen who had earlier highlighted the need to protect Britain's trees and engage in extensive afforestation were simply ignored.

Regardless of the motivation behind the Royalist calls for tree protection and planting, this period in our history can be viewed as partly responsible for our modern sympathetic attitude towards trees (Thomas 1983, 197–198). Although there was no sudden or dramatic shift from tree destruction to tree preservation, the rise of a more sympathetic attitude, especially among the nobility and gentry, is unmistakable. One outcome of this was to put a focus on the value of ancient trees. While these had been venerated in British cultural and religious life for centuries, individual trees of great age were now being identified and described by some of the writers of the day. The forestry expert John Smith (1670, 77) maintained that some living oaks dated from the first summer after the Flood and that a few went back to the Creation. Other authors were keen to record and publicise these trees to ensure they got the attention and protection they deserved. Both John Evelyn (1664) and Moses Cook (1676) mention some ancient specimens in their respective books on trees. However, the real pioneer was Robert Plot, the naturalist and first Professor of Chemistry at the University of Oxford, who recorded many ancient trees in his books on the natural history of some English counties. In the third edition of *Sylva* Evelyn gives recognition to Plot's work after mentioning several ancient trees: 'To these we add what we find taken notice of by the learned and industriously curious Dr Plot in his Natural History', and he goes on to describe them (Evelyn 1679, 167). The work of Plot and its collation by Evelyn could be said to constitute the first list of ancient and champion trees in Britain.

Early development of the Baroque garden in Britain

The Restoration of 1660 gave a great impetus to gardening and the creation of new gardens. The early part of this was a continuation of gardening in the time of James I, with its increasing use of densely planted trees extending out into the landscape beyond the formal garden. While this was to continue in the new era, it would soon be undertaken on an even grander scale.

When Charles II returned to Britain from exile on the Continent, he was a grateful admirer of the French king, Louis XIV. It was no coincidence that as Louis turned to the creation of vast formal gardens as an expression of his power and cultural supremacy, Charles should emulate his benefactor's enthusiasm, and endeavour to employ French garden designers. The exiled Charles and members of his court had witnessed at first hand some of the glorious early creations of the master of landscape design, André Le Nôtre, such as the spectacular gardens of Vaux-le-Vicomte. Once back in London, they followed closely the remarkable embellishment and expansion of the gardens at Versailles, the most famous example of Le Nôtre's work. His use of palisades, tall green walls made of clipped trees, gave the impression of green architecture stretching out into the distant landscape. Le Nôtre did not publish his thoughts or designs, but such was his towering influence that others did this for him. It fell to Antoine-Joseph Dezallier d'Argenville (1709), nine years after Le Nôtre's death, to put his ideas on paper, and in 1709 he published the hugely influential *La Théorie et la Pratique du Jardinage*. A translation of this by the London architect John James was published in 1712 under the title *The Theory and Practice of Gardening*, without giving credit to the original author (James 1712, title page). Although this English edition appeared over fifty years after the Restoration, Le Nôtre's magnificent Baroque style of garden was already well established in Britain with many fine examples.

When Charles II pondered where his new London court should be based, he decided against any major renovation of his ramshackle palace of Whitehall and instead turned his attention to St James's Palace. He commissioned the French landscape gardener, André Mollet, who had worked there in the late 1620s, to make St James's Park look impressively French without enormous expense. The result of Mollet's commission was a grand *patte d'oie* (goose's foot) radiating out from a quadruple semicircle of trees around the Horse Guards Parade ground (Mowl 2000, 48). One long double avenue flanked a canal basin pointing to the empty site of what is now Buckingham Palace. Two shorter arms of avenues stubbed off left and right, the left arm to ponds for water fowl, the right to a quadruple avenue on the line of the present Mall (Figure 24). Pleased with the effect, Charles ordered an even grander *patte d'oie* with another central canal flanked by double rows of lime trees to be laid out at his Tudor palace

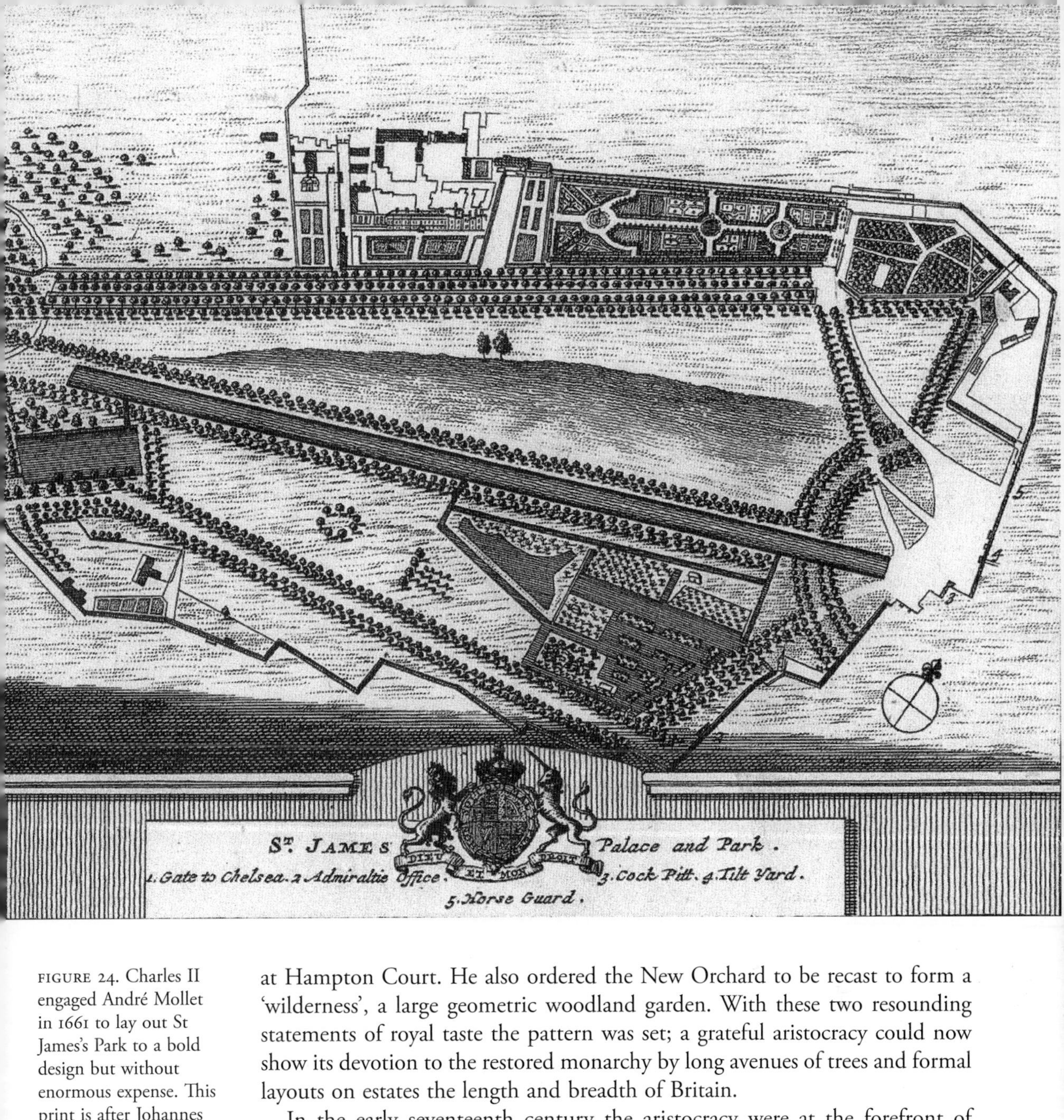

FIGURE 24. Charles II engaged André Mollet in 1661 to lay out St James's Park to a bold design but without enormous expense. This print is after Johannes Kip's engraving and was published in 1707.

at Hampton Court. He also ordered the New Orchard to be recast to form a 'wilderness', a large geometric woodland garden. With these two resounding statements of royal taste the pattern was set; a grateful aristocracy could now show its devotion to the restored monarchy by long avenues of trees and formal layouts on estates the length and breadth of Britain.

In the early seventeenth century the aristocracy were at the forefront of innovative garden creation involving extensive tree planting. Now the pace of planting was being set by the Crown itself. From the start of his reign Charles II had ambitions to create a new palace and had focused on Greenwich on the south bank of the Thames as an ideal location, acting as a grand *entrée* to London for visiting royalty and ambassadors (Jacques 2017, 87). In 1662, Charles wrote to Louis XIV asking if he could borrow Le Nôtre to design and supervise the construction of the gardens. Louis was reluctant to release Le Nôtre and the great man was only able to send a sketch plan based on dimensions provided to him. This involved a large area for a parterre and terraces cut into the hillside

FIGURE 25. This hand-coloured print of Cassiobury Park is from Knyff and Kip's *Britannia Illustrata* and dated 1708. It shows the spectacular scale and nature of what Moses Cook achieved at Cassiobury through his remarkable tree planting and landscaping.

to be planted with walks or avenues of trees, with other avenues branching off to the boundaries of the park, much in the style of what was being planned for Versailles. To meet the requirements of the design, the agent at Greenwich purchased 700 large elm trees, 500 of which were for the terraces, and 7,000 lime trees for the park walks and the nursery. Planting started in the winter of 1663/64. Greenwich Park is unique in being the only British park designed by Le Nôtre.

Towards the end of Charles II's reign a period of intensive garden-making commenced, lasting twenty years, fuelled by an ambitious peerage (Jacques 2017, 117). Whether they were supporters of James II, or Whig lords who opposed his succession, there was a consensus that the French garden style was appropriate to the seats of leading men. To create these landscapes in the style of Le Nôtre and his associates, a considerable amount of land was required. The trees were planted in longer and bolder avenues and there were wide paths and terraces, all set out in a formal axial arrangement that stretched out towards the horizon. Many of the greatest houses of the period were within or close to medieval deer parks and with the rising popularity of the Le Nôtre style they increasingly became the object of embellishment (*ibid.*, 20). At first this was mostly confined to avenue planting, then to woodlands and belts but increasingly the gates and lodges at the forecourt moved to the park entrance, and the parterres, groves and ornamental ponds of the garden became grander and looser as they migrated out into the park as lawns, plantations and lakes.

Most ancient parks had plenty of gnarled old trees, and sometimes enclosed ancient woodland, and more recently enclosed parks might incorporate substantial areas of coppice (Jacques 2017, 111). Some owners, familiar with the more wooded hunting preserves of France, imitated the French treatment of park woods and cut them through with rides intersecting at a *rond-point*. Although the parks suitable for this form of treatment were few, it was not unusual to see walks planted through parks. While these were in effect rides, the term 'ride' was not yet in currency, and increasingly 'walk' seemed appropriate for planting on such a scale. Generally, parkland planting was becoming more varied and complex, with walks issuing from gates and grates, and systems of walks branching off the axial vista to explore the park and its views.

Of all the parks and gardens that were embellished in the new style the most pioneering and one of the most influential was Cassiobury, the seat of the Earl of Essex. This remarkable forest garden was created for Essex by his gardener, Moses Cook, an innovator in both landscaping and arboricultural practice. Cook had begun working for Essex around 1660 and soon after was asked to lead the renovation and redesign of the gardens at Cassiobury. The Knyff and Kip illustration of Cassiobury, published in their *Britannia Illustrata* in 1707, shows the spectacular scale and nature of the tree planting and landscape features (Figure 25). However, the image is not a survey of the gardens and it is necessary to consult additional sources to support this. Cook's seminal book *The Manner of Raising and Ordering, and Improving Forest-Trees* gives valuable

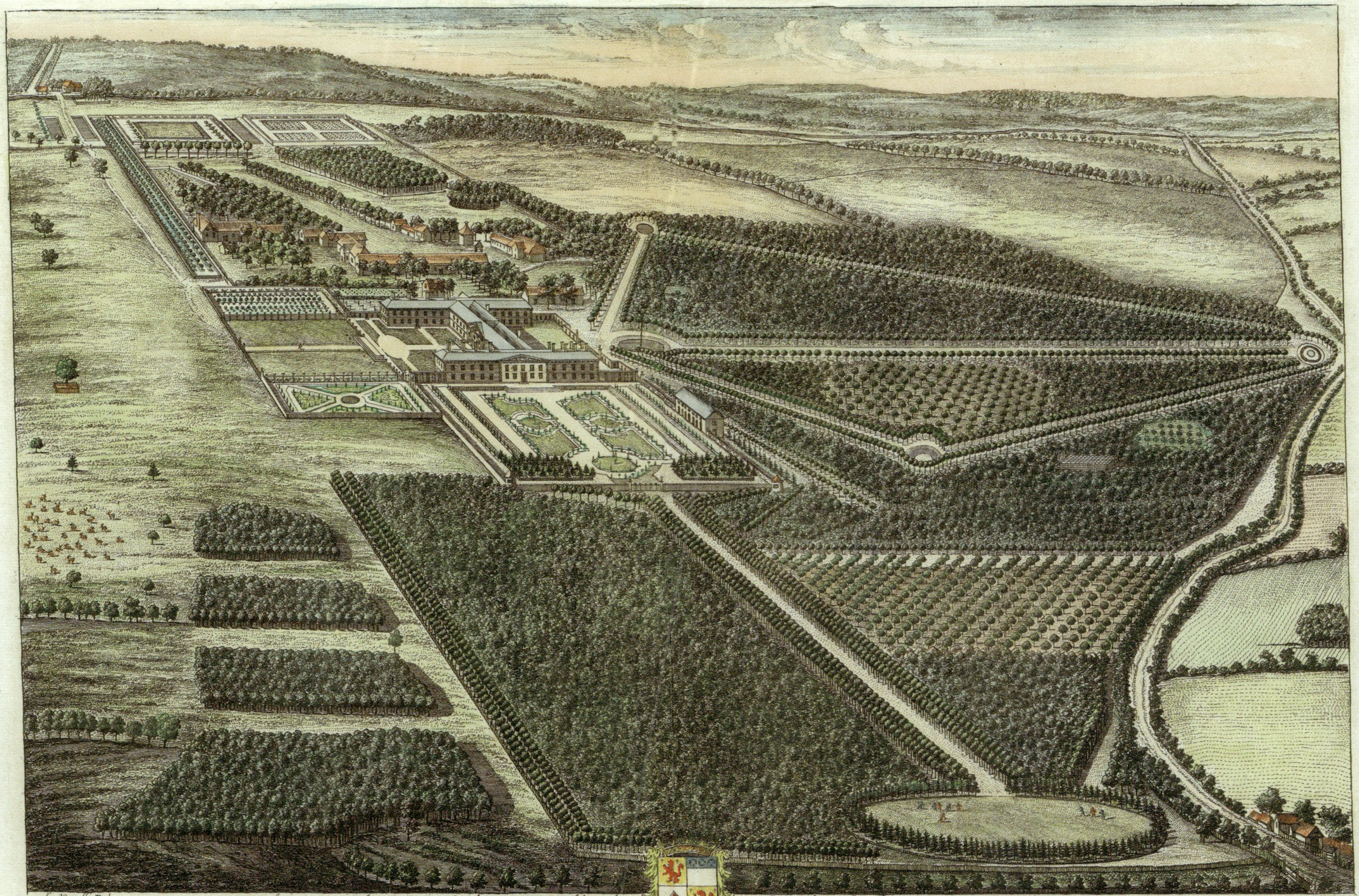

L. Knyff Del. I. Kip Sculp

Cashiobury the Seat of the R.t Hon.ble the Earle of Essex in Hartfordshire.

28

information about his work in the gardens (Cook 1676). This, together with an analysis of cash books of the late 1670s, Evelyn's description of 1680, a valuation of timber in about 1685 and a survey of the estate in 1798 all provide evidence to support the Knyff and Kip illustration (Milledge 2009, 6).

Garden writers through the ages have given credit for the design and construction of the gardens at Cassiobury to various people. For example, a posthumous edition of Daniel Defoe's *A Tour of Great Britain* gives the credit to Le Nôtre (Defoe 1742, 2, 366). Many modern garden historians give the credit to the Earl of Essex and regard Cook as merely his assistant, or even maintain that Essex created the gardens in consultation with John Evelyn (Horwood 2010, 11). This is unfortunate since anyone who has read Cook's book and examined the historical evidence should realise that the ideas and expertise behind this wonderful landscape at Cassiobury came from Moses Cook, albeit ably assisted by his employer (Cook 1676, Epistle). What was achieved there and how this was done is spelt out in detail in his book, giving various designs together with their dimensions and plenty of sound practical advice on the planting, maintenance and management of the landscape. In terms of the scale of the project, Cook went beyond the still-enclosed parterre gardens near the house and extended the pleasure grounds to the Hemel Hempstead road that bounded the estate on the north-west, giving a total area for this of 130 acres (53 ha). This was done with a series of radiating 'wood walks' where avenues of new trees were planted along *allées* cut through the woodland, behind which was the pre-existing woodland. These wood walks terminated in the east with a crow's-foot and roundel and in the south with a large circular bowling green. This was a radical departure from previous landscape practice in Britain both in scale and materials, although it had been anticipated by William Lawson (1631, 13) in his 1618 plan for an ideal garden that included 'walkes set with great wood thicke'. Cook planted different species of trees in different areas, including oak, ash, beech, lime and his particular favourite, black cherry. The walks were also edged with hedges. All Cook's extensive planting schemes seem to have been established successfully with very few losses. When planting 296 trees in four rows adjacent to the bowling green, he reported that all the trees survived their first year and were flourishing. However, in the second year three of the trees were vandalised by 'some base men or boys' (Cook 1676, 22). By 1680 the gardens at Cassiobury were almost complete and a year later Cook left to join the commercial partnership that founded the Brompton Park Nursery in west London. The enormous contribution of Moses Cook to professional arboriculture in Britain will be discussed in more detail later.

As word of the spectacular landscapes at locations such as St James's Park, Greenwich and Cassiobury circulated among the landed nobility, so planting within parks became more adventurous. An increasing number of rides were being established every decade until the 1720s (Jacques 2017, 218). By the 1700s it had become common for full circles to be cut out of woodland, and new ones were planted to form the end of walks, or where two walks intersected. In addition, the 1700s saw plumps or clumps being employed for park decoration. Some of the outstanding examples of gardens and parks that were developed during this period should be mentioned. They include the huge plantation of trees in a regular manner made for Sir Joseph Child, the successful financier, established at Wanstead Abbey in Essex from 1706. The 100 acres (40 ha) of planting with its criss-crossing avenues extended into Epping Forest. The gardens at Melbourne Hall in Derbyshire were designed in 1704 for Thomas Coke, Vice-Chamberlain to Queen Anne, 'to suit with Versailles'. With its extensive and formal arrangement of trees Melbourne Hall now promotes itself as the best surviving gardens in the manner of Le Nôtre. While many of the gardens in the new French style were in the south and midlands of England, this great age of tree planting eventually reached the north. One of the grandest was at Studley Royal in Yorkshire, started around 1710 by John Aislabie and developed in earnest after his disgrace in the South Sea Bubble financial scandal in 1720.

While the formal Baroque style of garden was dominant from the Restoration in 1660 to around 1730, it was not universally adopted throughout Britain. Even in England, West Country gardens may have had some characteristics of the style but were often more restrained and simple (Chambers 2008). In Wales, the enthusiasm for formal gardens in the French style was slower to take hold and less obvious than in England (Linnard 2000, 79). In Scotland, it was thought initially that the formal garden style began early with the formation of the magnificent gardens at Drummond Castle near Crieff around 1630. However, evidence suggests the original garden was enlarged and laid out in its distinctive design of the St Andrew's Cross about 1670 to 1680 (Cox 1935, 37–39). While the French style of garden was embraced by many Scottish nobles, this was often done without the same degree of formality of their English counterparts. In the first Scottish book on gardening, John Reid's influential *The Scots Gard'ner* of 1683, there is an enthusiastic endorsement of the formal style and a lack of symmetry is regarded as unsightly. Reid declares:

> As the Sun is the Centre of the world: As the Heart of the man is the Centre of the man: as the nose the Centre of the face: and as it is unseemly to see a Man wanting a leg, an arme, or his nose standing at one side of his face, or not straight, or wanting a cheek, an eye, an ear, or with one (or all of them) great at one side or small on the other; Just so with the House-courts, Avenues, Gardens, Orchards, etc., where regularity or uniformity is not observed. Therefore, whatever you have on the one hand, make as much, and of the same forme and in the same place, on the other.
>
> Reid 1683, 2

No doubt prompted by the words of Evelyn and Reid, one great advance in Scotland during this period was the more general use of trees in the policies of Scottish estates, those enclosed grounds and gardens surrounding a country mansion. By the 1720s the travellers Daniel Defoe and John Macky both expressed surprise and admiration at the numbers of trees growing on large Scottish estates, many of which, especially conifers, had been planted shortly after the Restoration. In 1724 Defoe wrote, 'You hardly see a House of Note, especially in the South Parts of this Country, but is adorned with Groves and Walks of Firr-trees about it' (Defoe 1742, 4, 52). From the start of the eighteenth century there was a marked increase in forestry planting in Scotland, for example, by the Earl of Haddington at Tyninghame, and this also included some ornamental features. John Macky wrote that Haddington 'hath planted many millions of trees in a sandy down or links and they thrive mightily. He hath also laid out several avenues through his park, which, when full grown, will be as noble as any in Britain' (Macky 1732, 26).

Having outlined the characteristics of the French formal garden in the Baroque style that became fashionable after the Restoration, we now explore some of the elements that relied on the extensive use of trees.

Avenues and tree-lined walks

These are two closely related landscape features and the terms are often used by different authors during this period to mean essentially the same thing. Nevertheless, avenues are generally regarded as parallel lines of trees, often forming an approach way to the front of a large house or mansion. John Evelyn was possibly the first person in Britain to publicise the term 'avenue' when he wrote in his *Dairy* for 1645 the description of a scene in Rome 'with a beautiful avenue of trees' (Evelyn 1901, 1, 166). In his third edition of *Sylva* he defines an avenue as 'the principal Walk to the Front of the House or Seat', showing how the two terms were interchangeable (Evelyn 1679, Advertisement). Walks or *allées* are composed of lines of trees or clipped hedges either side of a walkway and generally located in the garden or radiating out from the rear of the house. In the formal French garden the *allées* constitute the framework and are one of its most important features.

Avenues or walks appeared before the Civil War, but following the Restoration they were much more widely planted and stretched out into the surrounding countryside to embrace the distant landscape. A gentry house would have just one avenue, running to the main facade of the house. This was planted as much to provide an impressive frame for a view of the house as to improve the prospect from the house itself. A common trick with avenues was to widen them as they moved away from the house so that, viewed from the far end, this produced a false perspective of converging lines, making the avenue seem longer than it really was and the house larger. Not only did a long avenue glorify the house on which it focused, it also symbolised the landowner's power and control over his land. An examination of illustrations by Knyff and Kip and others indicates that the majority of avenues were single rows of trees, with most of the remainder being double and only rarely more. A large landowner would plant a network of avenues, comprising several individual avenues and cross avenues stretching out across the park, resembling the rides in French forests. One of the most extensive was planted by the first Duke and Duchess

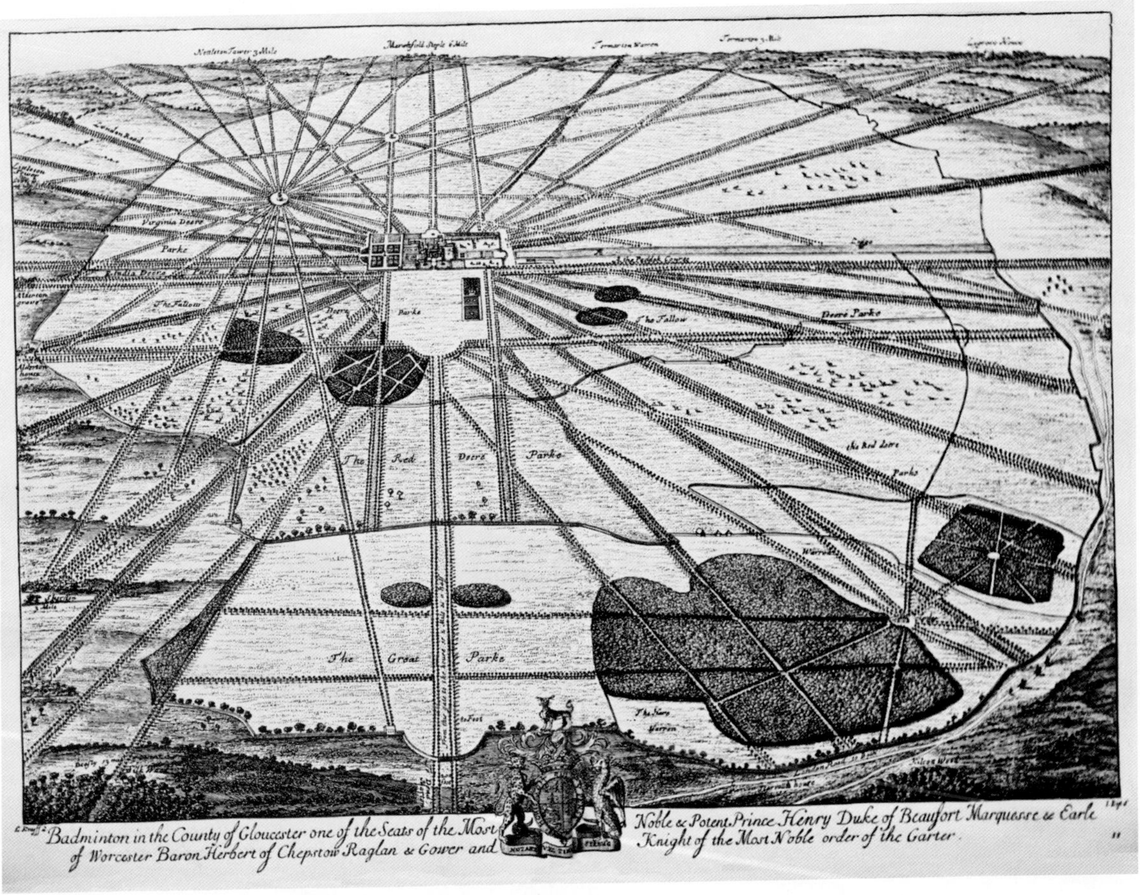

FIGURE 26. One of the most extensive networks of avenues in the Baroque style was the layout planted by the first Duke and Duchess of Beaufort at Badminton. This print is from Knyff and Kip's *Britannia Illustrata* of 1707.

of Beaufort at Badminton. While it was never fully planted, an idea of its sheer scale and complexity can be gained from the Knyff and Kip illustration (Figure 26). Planting avenues, especially on a grand scale, was also considered good economic practice as it was a long-term investment in timber.

What were considered the proper dimensions for tree-lined walks were set out by the Frenchman, Jacques Boyceau, in his *Traité du Jardinage* in 1638. However, his ideas of proportion were little known in Britain until the publication of Mollet's *Le Jardin de Plaisir* of 1651, translated into English in 1670. Mollet was an influential advocate of extensive axial avenues and walks laid out in exact proportions, and he rated the provision of an avenue leading to the house second only in importance to its situation. Mollet's ideas were put into practice at St James's Palace in 1660 and Hampton Court in 1661, although in both these cases the arrangement of avenues was on the garden rather than on the approach side (Couch 1992, 175).

The most influential British author in the second half of the seventeenth century to write about the layout of walks and avenues was Moses Cook. While he was keen to stress that his remarks were only concerned with walks outside the enclosed garden walls, there is much on this subject in his book *Forest-Trees* (Cook 1676). Always at the forefront of Cook's mind were basic aesthetic considerations as well as the practicalities of implementing the scheme. With regard to the width of walks and avenues, his advice is: 'Do not mask a fine Front, nor veil a pleasant Prospect (as too many do) by making Walks too narrow' (*ibid.*, 137). He advises that all walks should end in a circle, semicircle, triangle or other figure, either at the house end of an avenue, or where a walk intersected with another (*ibid.*, 137–139). He draws an analogy with the trunk of a tree and its rounded crown: 'For a walk ending bluntly without any Figure or entring into another, may be compared to a Tree with the Head off'. Another suggestion was that circles should be three times the breadth of the walks. The various ways of starting and terminating planted vistas are also illustrated in Cook's book, with diagrams showing half-circles at the house end and circles where vistas intersect in the park (*ibid.*, figs 16–25). Cook's recommendations for walks did not just apply to new plantings; he also goes into great detail on how walks can be cut through existing woodlands (*ibid.*, 137–148).

The species of trees selected for avenues were usually elm or lime but sometimes beech, horse chestnut or plane. Oak was seldom used as it was considered a timber tree, best grown in dense plantations to produce tall, straight timber. While many favoured the elm for 'its aspiring and tapering growth' (Evelyn 1664, 19), according to Cook (1676, 51) this was only 'if you take care in pruning it up'. Cook favoured limes as they grew fast and tall, thrived on different sorts of ground, had beautiful flowers and 'no Tree keeps such a constant Pyramid-shape as this' (*ibid.*, 70–71). Evelyn advocated the sweet chestnut (*Castanea sativa*), stating that 'for avenues to our Country-houses, they are a magnificent and royal Ornament' (Evelyn 1664, 25). Conifers were also planted in avenues, although the range of species was limited. Evelyn (*ibid.*, 53) recommended the maritime pine (*Pinus pinaster*) for walks and Cook (1676, 140) described planting three rows of spruce firs (*Picea abies*) around the bowling green at Cassiobury. With regard to the spacing of trees in rows along avenues, where the authors differentiated between long or main avenues and other walks, a wide spacing was given for the major avenues.

As grand avenues became increasingly popular for prestigious properties, a new form of avenue planting appeared consisting of blocks of trees. The most famous example was included in the design for the Great Avenue at Blenheim, the Duke of Marlborough's magnificent palace at Woodstock, Oxfordshire. In 1706–1707 blocks of planting were added at regular intervals either side of the main avenue of a double row of elms. These blocks became known as 'platoons', reflecting the military symbolism of the landscaping at Blenheim.

Groves and wildernesses

Groves and wildernesses were forms of ornamental woodland that were an essential element of the formal French style of garden, although the French themselves used the term *bosquet*. They were usually dissected by paths and clearings and contained a range of features such as buildings, sculptures and fountains. They had appeared in British gardens as early as the late sixteenth century and became increasingly common after the Restoration. They took a bewildering variety of forms and changed significantly over time. The French conception of ornamental woodland was neat and orderly and often placed within walls (Jacques 2017, 72). Their *bosquets* were illustrated in several plates published by the French gardener André Mollet (1651, pls 25–28). They were set out as formalised groves of dense trees and shrubs, good for shade and for attracting the birds. On the perimeter high palisades gave crispness to the arrangement and concealed the stems of trees, which were considered to be unsightly and to detract from the effect. Tunnel arbours provided the means of communication and the pathways or alleys were gravelled. When working for Queen Henrietta Maria at Wimbledon Manor in the early 1640s, Mollet had made a *bosquet* with palisades and open walks, which surveyors described as a 'wilderness' (Amherst 1896, 322). Among those writers after the Restoration, Evelyn did not use the term wilderness, preferring to use grove. Cook only occasionally used the terms grove and wilderness, mostly in the context of evergreens and within the enclosed garden, and although the woodlands at Cassiobury were generally ornamented, he talked about all this in the context of walks and specifically those outside the enclosed garden walls. One of the most impressive wildernesses was at Hampton Court, originally created for Charles II. This was much improved by William III around 1701 with a complete refurbishment that included the addition of an arbour on the north side and a large yew tree, nearly 30 ft (9.1 m) high, planted at the centre of the plantation.

In their translation of the classic French work *Le Jardinier Solitaire* (*The Retir'd Gard'ner*), London and Wise (1706, 744) defined the grove as a 'Plot of Ground, more or less … enclosed by Palisades of Hornbeam, the Middle of it fill'd with tall trees, as Elms or the like, the Tops of which make a Tuft of Plume'. London and Wise noted there were other types of groves which consisted 'only of Trees with high Stems'. This type of planting 'forms a sort of little Forest', with the ground beneath the trees 'kept very smooth, and well roll'd'. Such regular groves were particularly suited near a palace, while irregular groves of this kind were more appropriate in 'great Parks' (*ibid.*, 745). Palisades, in the context of groves in the formal French garden, were usually lines of densely planted trees clipped to give the appearance of tall, impenetrable green walls. Palisades could also be 'trimmed after divers Fashions, into Fans, Curtains, low Hedges, etc., according to the Nature of the Place' (James 1712, 47). In these ornate forms they resembled topiary (Figure 27).

Dezallier d'Argenville's classic work *La Theorie et la Pratique du Jardinage* promoted the ornamentation of woods and declared that a country seat without this was 'defective in one of its principal Particulars' (James 1712, 48). D'Argenville provided patterns for six types of *des bois et bosquets* (woods

FIGURE 27. This illustration of palisades is from the second edition of Dezallier d'Argenville's classic work *La Theorie et la Pratique du Jardinage* published in 1713. The two lower examples show arcades, palisades pieced with arches that resemble an architectural arcade. They were previously unknown in Britain but over twenty were made after 1713.

and groves), which were distinguished by their overall design. The categories include: 'Forests, or great Woods of high Trees; Coppice-Woods; Groves of a Middle Height, with tall Pallisades; Groves opened in Compartments; Groves planted in Quincunx, or in Squares; and Woods of Ever-Greens' (*ibid.*, 48–60; Woudstra 2018, 68–71). The first two types, 'Forests' and 'Coppice-Woods', consisting of tall trees or coppice, were 'proper only in an open Country, or in a Park of six or eight Miles over' (James 1712, 49). These did not contain palisades, just rides cut through for hunting, and were similar in appearance to the walks created at Cassiobury. 'Groves of a Middle Height' did have palisades, maintained with careful pruning up to 40 ft (16 m) in height, and were then common in fine French gardens (Figure 28). These groves were 'adorned with Halls, Cabinets, Galleries, Fountains, etc.' and their squares or quarters were 'edged with Palisades and Lattice-work' with neatly finished gravel walks (*ibid.*, 50). After these features had been laid out, the middle of the wood was planted or sown. 'Groves opened in Compartments' had trees planted along the walks and alleys but there were no trees in the squares. Instead, the interior of these were left open and included compartments and grass cutwork 'adorned with Yews and Flowering-Shrubs regularly planted' and shaped. 'Groves planted in Quincunx' had tall trees planted in 'several Alleys or Rows in quincunx or square formation'. There were no hedges or palisades and the ground under the trees was kept 'rolled or laid with Turf, leaving only certain Alleys in the Middle, and some small Cabinets and Vistas'. The sixth kind of grove, 'Woods of Ever-Greens', were 'the finest of all, on account of their continual Verdure' but were the least common because they took so long to establish (*ibid.*, 51). D'Argenville's book focused mainly on the design of these groves and woods and little practical detail was given on their construction.

FIGURE 28. Dezallier d'Argenville's 'Groves of a Middle Height, with tall Pallisades' were maintained with precise and regular pruning. This example in the garden at Chiswick House is shown in a hand-coloured print *c.* 1753 entitled 'A View of Three Walks'.

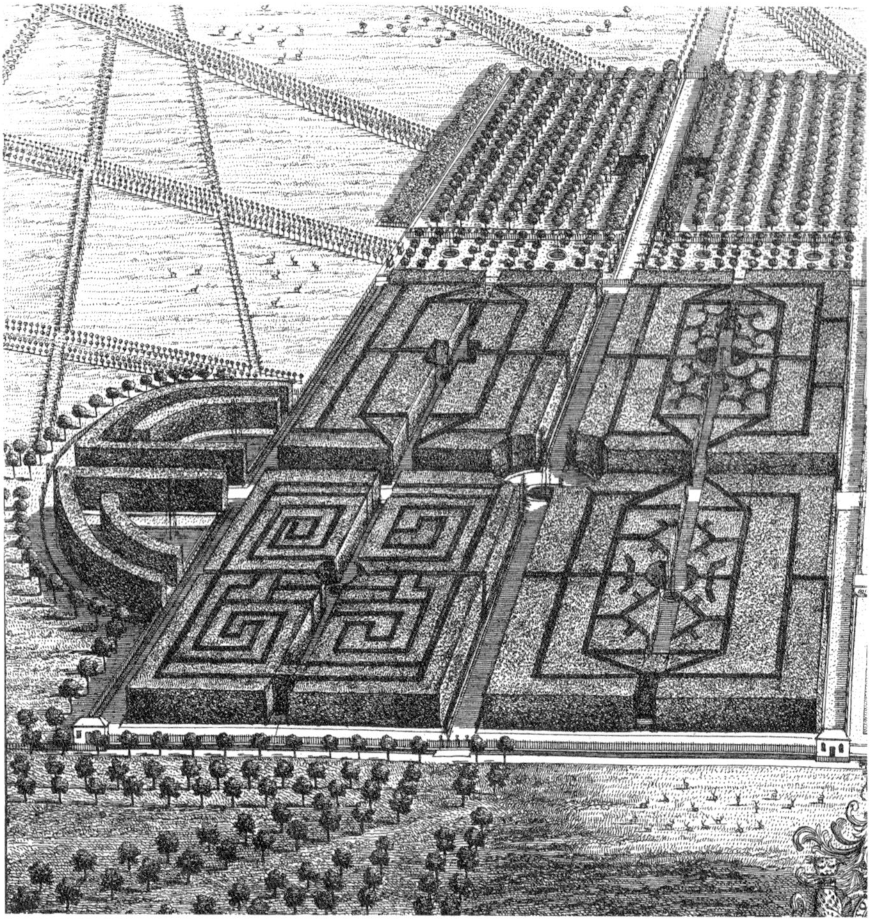

FIGURE 29. The wilderness at Badminton that was located at the side of the house. This detail from the 1707 Knyff and Kip print shows the substantial amount of arboricultural work involved in establishing and maintaining such a complex garden feature.

The publication of James's translation of d'Argenville's classic text had a profound effect on British groves, wildernesses and outward plantations. Beyond the parterre, but within the walls of the garden enclosure, all 'inward' wildernesses continued to be composed with geometric precision (Jacques 2017, 210). The advice was to set them out precisely with high palisades, and ideally the trees within the wilderness quarters would be pruned to a uniform height. Alternative ways of filling palisaded wilderness quarters continued to be practised, with orchards, kitchen-garden plots and, especially, exotic trees and shrubs all being possible. A Knyff and Kip (1707) illustration of the gardens at Badminton shows the wilderness area, initially planted in the late 1660s. This is located just beyond the parterres at the side of the house and gives an excellent impression of how such a garden feature would have looked (Figure 29). Quite apart from the substantial amount of arboricultural work involved in establishing such a complex feature, to keep this looking its best would have required the regular trimming of the trees in a time-consuming task involving a small team of skilled gardeners.

The species of trees used in wildernesses within the enclosed garden varied considerably. Evergreen trees were often used for the body of the planting, giving some colour and shelter in the winter. A few of the larger wildernesses were planted with infills of deciduous forest trees. In the 1690s, Celia Fiennes visited Tixall Hall in Staffordshire and found

> just by the Bowling-green is a very fine wilderness with many Large walkes of a great Length, full of all sorts of trees, scycamores, willows, Hazel, Chesnutts, walnuts, set very thicke and so shorn smooth to the top which is left as a tuft or Crown – they are very Lofty in growth which makes the Length of a walke Look Nobly.
>
> Fiennes 1888, 145

While groves and wildernesses were planted primarily for ornamentation, they could serve a utilitarian purpose in providing timber and fuelwood for the house and the estate. Although this applied mainly to extensive areas of ornamental woodland located beyond the walls of the formal garden, some within the enclosed garden were also exploited economically. Timothy Nourse, whose book *Campania Fœlix* was a popular guide on estate management, promoted both the ornamental and economic benefits of woodland. He stated, 'A Plantation of Coppice-Wood, as it is most profitable to a private Undertaker, so it is most pleasant and ornamental, if near a Gentleman's House' (Nourse 1700, 111).

In the early part of the eighteenth century wildernesses came to occupy a larger proportion of the formal garden and were often planted closer to the mansion. This was one of the last remaining places in the garden that allowed opportunities for a private and meditative experience. Most of the garden could be seen from the house and the house could be viewed from most parts of the garden, and only in the wilderness were any secluded places to be found.

John Evelyn and Moses Cook

The names of John Evelyn and Moses Cook have already been mentioned several times in this chapter. This is not surprising since these men were widely regarded by contemporaries as the outstanding tree experts of their generation. Their subsequent reputations, however, have taken quite different trajectories, with Evelyn having now acquired iconic status and Cook being little known. This seems an appropriate point to give an assessment of their respective contributions to the development of professional arboriculture in Britain. Some of their remarks on specific aspects of practical arboriculture are discussed at various points later in this chapter.

John Evelyn

John Evelyn (1620–1706) was born in Wotton, Surrey, into a family whose wealth was founded largely on manufacturing gunpowder. He grew up in

Lewes, Sussex, and although he declined to attend Eton College, he had a good education at Lewes Old Grammar School followed by Oxford University. Evelyn and his family were firm Royalists and as the Civil War was looming he left England to tour France, Italy and the Netherlands, immersing himself in the arts and sciences and visiting some of the famous gardens and nurseries. In France Evelyn made connections with the exiled court of Charles II, which led to his marriage in 1647 to Mary Browne, daughter of Sir Richard Browne, the English ambassador in Paris. In 1651 he became convinced that the Royalist cause was hopeless and decided to return permanently to England, living at Sayes Court in Deptford, keeping a low profile and transforming his gardens.

Following the Restoration Evelyn's fortunes rose rapidly, beginning in 1662 with his appointment as a member of the first Council of the Royal Society. In September 1662, Sir Robert Moray, President of the Royal Society, initiated a discussion regarding the shortage of standing timber and the lack of forward planning to rectify the situation (Campbell-Culver 2006, 59). A week later the Commissioners of the Navy sent him a paper detailing their concerns on this issue. Moray immediately invited four of his Fellows, including John Evelyn, to consider the problems and find an acceptable solution. Within a month all four had responded and Evelyn was asked to make a digest of the contributions and then deliver this as a lecture. The text of the lecture made such a powerful impression on those present that the Society decided that its important message should reach a wider audience and Evelyn was commissioned to produce a book based on this. The resulting work was *Sylva; or, A Discourse of Forest-Trees, and the Propagation of Timber in His Majesties Dominions*, published eighteen months later in February 1664 (Evelyn 1664).

In the preface to *Sylva*, Evelyn declares that his work was a response to the destruction of woods and trees during the Civil War, and he places the blame for this firmly with the Parliamentarians. He declares, 'But what shall I then say of our late prodigious Spoilers, whose furious devastation of so many goodly Woods and forests have left an Infamy on their Names and Memories not quickly to be forgotten! I mean our unhappy Usurpers, and injurious Sequestrators' (Evelyn 1664, To the Reader). He then goes on to state that the destruction to trees and woods wrought by the 'Gallant and Loyal Gentry' during this time was largely caused 'by an inhumane and unparallel'd Tyranny over them' and their necessity 'to preserve the poor remainder of their Fortunes, and to find them Bread'. To restore the woods and forests he proposes a massive programme of tree planting, declaring: 'Truly, the waste, and destruction of our Woods, has been so universal, that I conceive nothing less then an universal Plantation of all the sorts of Trees will supply, and well encounter the defect' (*ibid.*, 2). In contrast to the republican writers who had urged this, there is no talk of 'trees for the people' or any mention of the poor; Evelyn's focus is on ensuring the 'Power … Wealth and Safety' of the king (*ibid.*, Epistle).

There is no doubt that Evelyn's *Sylva* gave a tremendous boost to tree planting throughout Britain and its impact endured for many decades. Its early editions also coincided with the rising economic fortunes of the aristocracy when money was again available for planting and garden creation. At various times Evelyn claimed his book had prompted the planting of many millions of trees. However, it is impossible to distinguish Evelyn's influence from that of other contemporary writers on the topic. In the 1679 edition of *Sylva* he states, 'I need not Acquaint Your Majesty how many Millions of Timber-Trees (besides infinite others) have been Propagated, and Planted throughout Your vast Dominions, at the Instigation, and by the sole Direction of this Work' (Evelyn 1679, To the King). Obviously, this assertion about numbers of trees planted cannot be verified, but Evelyn's claim that this is due solely to his own work says much about his recognition of the contribution of others.

The content of *Sylva* is mainly about tree propagation, forestry, timber and wood products, and includes descriptions of different trees and an account of their uses. He also deals with the 'infirmities' of trees with brief references to various diseases and other harmful agencies. There are also sections on pruning and on the 'Age, Stature and Felling of Trees' and another 'Of the Laws and Statutes for the Preservation and Improvement of Woods'. Throughout the work there are references to some notable examples of different tree species to be found in parks and gardens, mostly in England. Although *Sylva* is often described nowadays as one of the world's great works on arboriculture, there is only a limited amount that relates directly to amenity trees and designed landscapes. The work is mainly on forestry and timber production and it is likely that he intended to give more coverage to arboriculture and amenity trees in *Elysium Britannicum*, his unpublished work on gardening. There are two appendices to *Sylva*, *Pomona* on fruit trees and *Aphorisms Concerning Cider*, where the credit for writing goes to various named and unnamed authors. The last part of *Sylva* is a work of some thirty pages by Evelyn entitled *Kalendarium Hortense*, a practical guide describing gardening activities in each of the twelve months of the year. There is, however, very little mention of amenity trees.

It has been stated that the extensive tree planting shown in the views of English country seats by Knyff and Kip is John Evelyn's best testimonial (Mowl 2000, 44). As already indicated, such a claim cannot be substantiated as it is impossible to separate the impact of Evelyn from that of other authors of this time. Furthermore, *Sylva* contained only vague guidance on the aesthetics of park planting and would have provided little assistance in designing the type of landscapes depicted by Knyff and Kip. However, once Moses Cook had published his classic work on this subject in 1676, Evelyn offered a little more detail. Nevertheless, it is not until the 1706 edition of *Sylva*, published just after Evelyn's death, that he finally gives a plan for tree planting and this is based on Cook's work. Indeed, Evelyn concedes, 'I should here presume to furnish some farther directions, were it not already done to my hand by the often cited Mr Cooke, in that useful work of his; where, in chapter the 38th he has laid down all that I can conceive necessary' (Evelyn 1706, 2, 175). Unlike Cook, Evelyn never constructed a true forest garden and this might explain the lack of practical detail on this topic in his writing. However, Evelyn's advice on tree planting was sought by several of his friends and acquaintances, such as the Earl of Clarendon at Cornbury in Oxfordshire (Evelyn 1901, 1, 376).

As noted, the text of *Sylva* was based on Evelyn's amalgamation of notes by him and three other individuals. Thanks to a forensic examination of the text by historian Guy de la Bédoyère (1995, 173–332) and the work of other researchers, it has emerged that a substantial amount of the first edition is made up of contributions from many authors, some acknowledged but many not. Much of *Sylva* is essentially an anthology of associated classical quotations concerning wood and trees in general, connected by Evelyn's prose with all its anecdotes, advice and digressions (*ibid.*, 179). The quotations were utilised as they were considered appropriate whether they originated in myth in verse form, or in practical information contained in a treatise. In the latter case much was copied from Pliny the Elder, with some of Evelyn's 'text' being no more than an unattributed translation (De la Bédoyère pers. comm., 2018). Furthermore, a number of his classical references are wrong, misquoted or wrongly attributed, indicating that he made little or no attempt to check his manuscripts. Various passages of *Sylva* were also derived from contemporary authors. For example, much of the chapter on pruning is from William Lawson's *New Orchard and Garden* (Evelyn 1664, 74–76). To Evelyn's credit he does acknowledge 'our Country-man honest Lawson' for the section that is directly copied and this is in a different type. Other examples include Markham's work on hedges and Beale on fruit trees. The nearest Evelyn gets to acknowledging that *Sylva* is a collaborative work is his remark that he was 'Assisted by divers Worthy Persons' (Evelyn 1664, To the Reader). As *Sylva* went into further additions, Evelyn would add additional material that he sourced from other authors. Apart from material taken from Moses Cook, there is Evelyn's rapid and thorough incorporation of material derived from John Smith's *England's Improvement Reviv'd* (Smith 1670). Evelyn wanted to demonstrate through clear calculations that it was as advantageous financially as it was aesthetically for landowners to plant trees. He therefore quoted Smith's own extended calculations on this issue in the second edition of *Sylva* in 1670, almost before Smith had had a chance to publish his own work (Sharp 1975, 66). It appears that Evelyn's real genius lay in his ability to collect and synthesise the work of others and promote this very effectively using his influential connections, rather than as the originator of new ideas and practice.

Evelyn's stilted style of writing, including his use of Latin and Greek, would have restricted the readership of his books to the educated few. However, it would not have appeared especially obscure to people with a similar background to his, and it was these people he aimed the book at: educated, wealthy landed men (De la Bédoyère 1995, 179). The use of classical quotations or references was a kind of *lingua franca* at the time. However, Evelyn discovered that the market for his *Sylva* went beyond his intended audience of 'Gentlemen and Persons of Quality' (Evelyn 1679, Advertisement). Finding that 'Ordinary Rustics' (foresters and arborists) also thought the book of use, he was obliged in later editions to supply a glossary of terms so that there 'may yet be no prejudice to the meaner capacities'.

Brief mention should be made of some of Evelyn's other works. Apart from his translations of French works on fruit trees, he has also been credited with translating Jean de la Quintinie's *The Compleat Gard'ner* (De la Quintinie 1693). Although Evelyn is named on the title page as translator, much of the translation had been made by the King's gardener, George London, with Evelyn responsible for the remainder. Evelyn stated that his name had been placed on the title page for commercial reasons, something that went 'against my intentions' (De la Bédoyère 1995, 18). Evelyn's own great work on gardening, *Elysium Britannicum*, which was only in draft form when he died, was eventually published in 2001, edited by John Ingram (Evelyn 2001). Ingram (1998, 36) observes that much of the content was compiled from a wide variety of sources, both ancient and modern, and included contributions from Evelyn's wide network of colleagues. Some historians consider that Evelyn's *Fumifugium* is his most enduring work (De la Bédoyère 1995, 13). This was a proposal to solve London's obnoxious smoke problem and is discussed later.

Moses Cook

Little is known of the life of Moses Cook (*c.* 1640–1714), particularly his early years, although recent research by garden historian Jenny Milledge (2009, 10) and others has shed more light on this. The date and place of Cook's birth are unknown but it is thought this was around 1640 and somewhere in Lincolnshire, where he was the son of a farmer. Cook's education is unknown, although he had an excellent grasp of mathematics, including logarithms and geometry, which proved invaluable in his landscape design work. Cook commenced working as a gardener for the Earl of Essex around 1660 (Desmond 1977, 145), although this was probably at the Earl's property at Hadham Hall before he got involved with the gardens at Cassiobury. In 1661, after the Restoration, the Earl of Essex, whose Royalist father had been executed by the Parliamentarians, embarked on a major renovation of the family seat at Cassiobury. Cook's leading role in the design and creation of that pioneering landscape has already been outlined. Here, we give an appraisal of his book *The Manner of Raising, Ordering, and Improving Forest-Trees* (Cook 1676) and his other contributions to the development of arboriculture in Britain.

Just as many historians have failed to give credit to Cook for his work at Cassiobury, his remarkable book has also been largely overlooked. However, influential writers who were contemporary with Cook seem very aware of the importance of his book and his outstanding abilities, among them John Evelyn. After spending a few days with the Earl of Essex in 1680, Evelyn wrote, 'The gardens at Cassiobury are very rare, and cannot be otherwise, having so skilful an artist to govern them as Mr Cooke, who is, as the mechanical part, not ignorant in mathematics, and pretends to astrology' (Evelyn 1901, 2, 143). In the 1679 edition of *Sylva*, Evelyn praises him as 'that industrious Planter Mr Cooke, from whose ingenuity and experience (as well as out of gratitude for his frequent mentioning of me in his elaborate, and useful work) I acknowledge to have benefited myself, and this Edition' (Evelyn 1679, 59).

It is unfortunate that as the centuries have passed by the importance of Cook's book has been slowly forgotten. That great nineteenth-century doyen of garden history, Alicia Amherst (1896, 189), describes his book as 'a work on fruit trees'. More recently, it has been stated that Cook extensively plagiarised Evelyn's *Sylva* (Williamson 2018, 46). While Cook quotes Evelyn several times, always with acknowledgement, any plagiarism was the other way. Because of Evelyn's elevated social position, Cook, as the humble gardener, would have chosen his words quite carefully when referring to him in his book. There is no doubt that Cook respected Evelyn as a fellow tree expert, although this did not stop him occasionally correcting or qualifying what Evelyn wrote on technical matters (Cook 1676, 34). There is, however, one sentence that is most enlightening regarding the relationship between the two men. When discussing the ash tree, Cook (*ibid.*, 53) states: 'And as for Raising the Ash, I shall give you the same Rules as I did to the aforesaid Honourable Person the same time, before the *Discourse of Forest-trees* was written'. This tells us that Cook was supplying Evelyn with information about trees before *Sylva* was written. As Essex would have known if this early contact between the two men took place, it seems implausible that Cook would have stated this in his book if it was not true.

Cook's book is a very practical text giving explicit instructions on all aspects of the cultivation of forest trees for both profit and amenity. Like Evelyn, he was keen to highlight the needless waste and destruction of trees and woodlands that had occurred in the past and he appealed to all lovers of trees to demonstrate this by planting and protecting them (Cook 1676, To the Reader). He hoped that his book would provide useful guidance on this 'for the good of this Nation'. While most of it is concerned with practical arboriculture and woodland management, the last third is devoted to landscape design with trees, especially the making of walks, avenues and figures. To accompany his text he included almost fifty illustrated examples of how to measure 'Superficial Figures' and was the first British author to provide detailed illustrated instructions on how to implement various shapes in a garden. In covering matters of design and aesthetics, Cook seems keen to downplay his contribution for fear of upsetting those with a reputation in this field. In one revealing sentence he states, 'But I am too far got into the Orchard or Garden; I must retreat to my Forrest-trees, to shelter me from the Gardiners Anger' (*ibid.*, 98). However, a major part of his book *is* concerned with the ornamental and this was to have a substantial influence on many gardeners who came after him.

Cook's personality shines through in his writing and the impression is given of someone who is passionate about their work, although probably not a patient man. He writes with the quiet confidence of extensive first-hand practical experience and of one who has a sound technical knowledge of their subject, something that is immediately obvious to the modern arboriculturist. It is also clear that the book is his own work, apart from where he quotes other authors, and then acknowledgement is given. While the occasional addition of lines of his own poetry may have seemed quaint to his contemporary readers, there is no attempt to impress them with endless quotations from the classics. Where he disagrees with the findings of both classical and contemporary authors he is not afraid to say so, and it is always done on the basis of his own practical experience.

Around 1680, the political career of the Earl of Essex started to unravel, leading to his death in the Tower of London in 1683. In 1681, possibly prompted by Essex's increasing difficulties, Cook left Cassiobury and joined with George London and two others to establish the Brompton Park Nursery in London. This hugely successful nursery and landscape business will be discussed later. Cook retired around 1689 and died in 1714, his will indicating he was then a wealthy man possessing a substantial amount of property in Essex and Hertfordshire (Milledge 2009, 13). A second edition of his book was published after his death in 1717 and a third edition in 1724.

It is hard to explain the different reputations of Evelyn and Cook in terms of their contributions to arboriculture and treed landscapes in Britain. Unlike those earlier republican authors who suffered from political bias, this did not apply to Cook. Although only a child in the Civil War, his sympathies as a man were with the Royalist side (Cook 1676, To the Reader). It seems likely that Cook suffered from social discrimination. The aristocracy were always liable to cite John Evelyn or the Earl of Essex as the inspiration for their tree planting rather than the gardener Moses Cook, even if they had heard of him. Nevertheless, among his fellow professionals his reputation was immense. Stephen Switzer wrote:

> Evelyn and Cook on Forest Trees … bid the fairest for a complete Account in their respective ways. But the first is full of Erudition and Learning, that there is scarce Room left for Practice; besides the Works of Mr Evelyn, how much forever we revere him, are something immethodical and confused. … Cooke's is indeed an excellent Book, and one that has done more Good in the Nursery and Forest Way than any other that I know of.
>
> Stephen Switzer 1733, liv

Looking back on this from the twenty-first century it is hard to argue with Switzer's assessment. In terms of Cook's recognition in modern times, while garden historians are becoming more aware of him, he is unknown to most arboriculturists. Perhaps the only twentieth-century arboriculturist to really appreciate Cook and contrast his reputation with that of Evelyn was Denis Le Sueur (1954). One factor influencing this today is the availability of their respective books. While the first edition of *Sylva* is freely available in facsimile or on the internet and original copies can usually be purchased, the original first edition of Cook's book is very rare, there are no facsimiles and only the second edition of 1717 is available on the internet, fuelling the belief that he got his ideas from Evelyn and he published much later. Hopefully, this present volume will help to give Cook his due recognition as one of Britain's greatest professional arboriculturists.

Later Baroque gardens and the Dutch influence

Towards the end of Charles II's reign the pace of garden-making had picked up significantly and this had continued under James II (Jacques 2017, 18). The political upheaval of the Glorious Revolution (1688) and the drain on the public purse in funding the Nine Years' War (1688–1697) hardly dampened the pace, which reached a peak in the reign of William III (1689–1702). The War of the Spanish Succession (1701–1714), however, led to further taxes and high interest rates, and as disposable income was thereby eroded garden-making dipped in much of Queen Anne's reign. Once the drain of wartime expenses had ended, the British economy, now among the strongest in Europe, flourished in the reign of George I (1714–1727). A period of stability under Sir Robert Walpole's government created the conditions for the resurgence in building and garden creation in the 1720s.

With William and Mary co-reigning as British monarchs, the Whig-dominated Parliament voted vast sums to promote their international standing through the royal palaces and gardens (Jacques 2017, 117). While William was an admirer of formal gardens in the French style, in his homeland of the Netherlands these had acquired a distinct Dutch influence. The manner of this had just been expressed so eloquently in the magnificent gardens of his new palace at Het Loo, designed by Daniel Marot, a pupil of Le Nôtre. Now that William could embark on some new royal garden projects in England he was keen to see these reflect a similar style of formal garden design. Among garden historians there has always been considerable debate regarding what constitutes distinct national styles of garden design. Much has been made by some authors of supposed Italian, French and Dutch influences on English garden style in the formal gardens period (*ibid.*, 22). There are always difficulties in pinpointing the characteristics that especially denote national styles, but even if some are identified, they rarely transfer in an unambiguous manner. There is no doubt that many landscape features and the manner of their appearance were common to many of the European garden design traditions around this time. Some historians now regard the narrative of a significant Dutch influence on British gardens at this time to be flawed and have questioned the true extent of this in some surviving examples of supposed Dutch gardens (Jacques 2002). Nevertheless, we will outline briefly what are often regarded as some of the characteristics of the Dutch formal style and how these are said to differ from the French style.

Dutch gardens were generally smaller than French gardens and made more use of ornamental canals. In some respects they were a regional accommodation

FIGURE 30. Many trees were displayed in containers, especially those that were tender, which could be removed on trolleys and placed inside over winter. This image is from *Vera et Accurata Delineatio* by Salomon Kleiner, published in 1724.

of the grand French garden, adapted to the constraints of the landscape of the Netherlands with its enduring limitations of too much water and too little space. The Dutch made more use of tidily trimmed topiary and had a passion for 'greens' or 'ever-greens', the terms they used to embrace all evergreen deciduous trees and conifers. These were planted around their gardens in formal arrangements and kept as closely pruned forms. Many were also displayed in containers, especially those that were tender, and these could be removed on trolleys and placed inside over winter (Figure 30). They also planted a greater variety of plants within their parterres, as opposed to French parterres that usually relied upon gravel and grass. The Dutch had a great enthusiasm for flowers that was not especially evident in French gardens. One particular characteristic that set Dutch gardens apart from the French tradition was the custom of dividing the garden into separate compartments with high hedges, which were very effective at keeping out the cold winds that blasted across the polders. Whereas in France the main vista was wide and long and edged with high hedges or palisades, in the Netherlands the entire area was subdivided into separate sections, in which either a basin, fountain or flower parterre would dominate. The reason for this compartment principle was probably to disguise the fact that most Dutch gardens were limited in size by the cost of land in this small, populous country.

A major element in Parliament's strategy to promote the international standing of William and Mary included an immensely expensive project to create a major European palace for them at Hampton Court. William was much involved in the decisions concerning Hampton Court, as he had been at his palaces and gardens in the Netherlands. George London, an English garden designer, and Daniel Marot, previously at Het Loo, were employed at various times to design and deliver the spectacular gardens and its individual features. London developed the existing Wilderness and Tiltyard with formal layouts from 1689 onwards, and the Maze planted sometime before 1695 in the north-west part of the Wilderness. The canal was retained and two great parterres were created, one of which, the Great Fountain Garden on the east front of the palace, provided a semicircle for the *patte d'oie* of radiating avenues of yew trees. Planted in 1707, the yews were kept as 8 ft (2.4 m) high conical-shaped topiary for their first few decades. However, they were then allowed to grow away and after 300 years are now magnificent topiary giants. The other parterre, the splendid Privy Garden on the south side of the palace, was gloriously embroidered with box and gravel arabesques, dotted with shapely yew pyramids, globe hollies, urns and statuary. A reconstruction of this to the original design was completed in 1995. William and Mary's other gardening project in the early years of their reign was at Kensington Palace. Previously called Nottingham House, this had been purchased in 1689 to provide the asthmatic William with a relatively smoke-free atmosphere that was still close to Westminster. London redesigned and extended the gardens in the formal Dutch style.

There were now many fine gardens outside of London, mostly in the formal style and combining a mixture of features that could be described as Dutch, French or both. The diary of Celia Fiennes (1888) of her tour through England, written mostly in the 1690s, is a rich source for descriptions of the gardens of that time. The gardens she describes tend to be quite similar with gravel and grass walks, shady alleys of clipped trees, 'some walks like Arbours Close, others shady, others open, some gravel, some grass' (Fiennes 1888, 22). Standard cypress or yews 'cut in severall forms were dotted about' and trimmed hedges of holly, laurel or box divided the different parts of the garden. Occasionally, mention is made of 'fine greens' and 'dwarfs' (small, closely pruned fruit trees), or tender fruit such as oranges and lemons given shelter or a greenhouse. With regard to trees in the wider landscape, Fiennes has many references to wildernesses and groves with alleys cut through them. Her constant and admiring references to clipped trees, cut topiary and stilted hedges with 'platted' branches wherever she went testifies to the popularity of this formal style of gardening.

The extent of arboricultural work required in the typical Dutch formal garden, with its extensive use of topiary, clipped high hedges, pruned trees in containers, and other tree-related features, would have been similar to that of the French formal garden. Although the typical Dutch garden may have been smaller in scale than the vast expanses of the French garden with its avenues and rides disappearing over the horizon, the incredible amount of pruning required to keep everything looking its best would have made these gardens equally labour-intensive. We will now examine some of these tree-related elements in more detail.

'Greens', topiary, mazes and parterres

The interest in using deciduous evergreens and conifers in the garden was evident quite early in this era. An entire section of Thomas Hanmer's *Garden Book*, written in 1659, is devoted to greens which 'we esteem much' for beautifying walks and gardens as they remain green all winter, 'not casting their leaves totally at any season of the year' (Hanmer 1933, 126). While yew was sometimes chosen for hedging and topiary, as mentioned by John Parkinson in 1629, John Evelyn (1706, I, 300) claimed to 'have been the first which brought it into fashion'. Another tree sometimes used for clipping was the Norway spruce (*Picea abies*), which naturally grew into a regular shape but could be further trained into a cake-stand form and was often 'planted at the Corners of your Squares, or in direct lines at a little distance from your Walks' (Worlidge 1683, 74). John Worlidge (*ibid.*, 77) thought that the addition of 'gilded' or variegated evergreens, such as holly, laurel and box, made groves and walks 'more pleasant'. The quick-growing Mediterranean cypress was also coming rapidly into fashion, no doubt because it was now being raised from British seed. As Worlidge (*ibid.*, 69) observed, 'Being the most uniform, straightest, and most slender of any other' it already had an attractive shape before any pruning. The cypress was reputed to be tender but Evelyn (1664, 60) casts doubt on this, although he states they should not be 'late clipp'd'. In the severe winter of 1683–1684 practically every cypress in England was killed, which must have been a painful lesson for contemporary gardeners and arborists.

The enthusiasm of William III for his Dutch style of garden gave a boost to British gardening generally and to the use of 'greens' in particular. According to Daniel Defoe, writing in the mid-1720s, William introduced to Britain 'the love of gardening [and] his majesty was particularly delighted with the decoration of ever greens, as the greatest addition to the beauty of a garden, preserving the figure of the place, even in the roughest part of an inclement and tempestuous winter' (Defoe 1962, 166).

While topiary had been popular in Britain since Tudor times, it was now one of the most prominent features of the garden. While gardens in the French style had tended to use clipped trees mainly as elements in a geometrical composition, the Dutch style appreciated the more individual qualities of specimen topiary. The bulk of the new clipping was of the traditional pyramid yews and globe hollies, but 'shaped greens' were more evident in the 1700s. Such pieces were those cut into tiered and other complex forms instead of simple shapes. The botanist Richard Bradley (1731, 163), in explaining 'Technical words', gave three shapes for clipped trees and shrubs: 'pyramid', 'headed' and 'pyramid, cut and divided', that is conical, globe-headed and tiered (Figure 31).

Along with topiary, mazes also became popular in the larger gardens. While Lawson tells us that some head-high mazes were to be seen in the 1610s, they really took off in the late seventeenth century. William and Mary, missing their maze or *doolhof* at Het Loo, commissioned a new one at Hampton Court. Originally planted in hornbeam in 1690 and later replanted in yew, it is the oldest surviving example of a puzzle maze. In practically all mazes it was fashionable to relieve the monotony of the walks by placing statues, vases, seats, fountains and other ornaments at focal points. However, among some people mazes appear to have been regarded disdainfully. Switzer observed that mazes 'seem to be calculated for an inferior Class of People' (Switzer 1718, 2, 189). Scotland, meanwhile, did not generally embrace mazes or topiary, although the massive ornamental yew hedges planted in 1702 at Crathes Castle should be mentioned.

Another tree-related garden feature that required regular clipping was the parterre, which had now replaced the older knot garden. These had been pioneered by the Mollet family in France and become a striking feature in the formal French garden. The elaborate *parterre de broderie* (embroidered parterre) could be seen at its grandest in the Baroque gardens of André Le Nôtre. The parterre is essentially a small garden laid out on a level surface consisting of plant beds, typically in symmetrical patterns, which are separated and connected by paths of gravel or grass. Parterres in the time of William and Mary were often embellished with tightly clipped small trees and shrubs, *jets d'eau* and figures. Where the beds of the parterres were bordered with low hedging, this was generally with Dutch box. In the grandest gardens with the largest and most elaborate parterres the total length of the box hedging was vast. The amount of pruning required to keep all this looking neat and tidy would have been a major task for the gardeners.

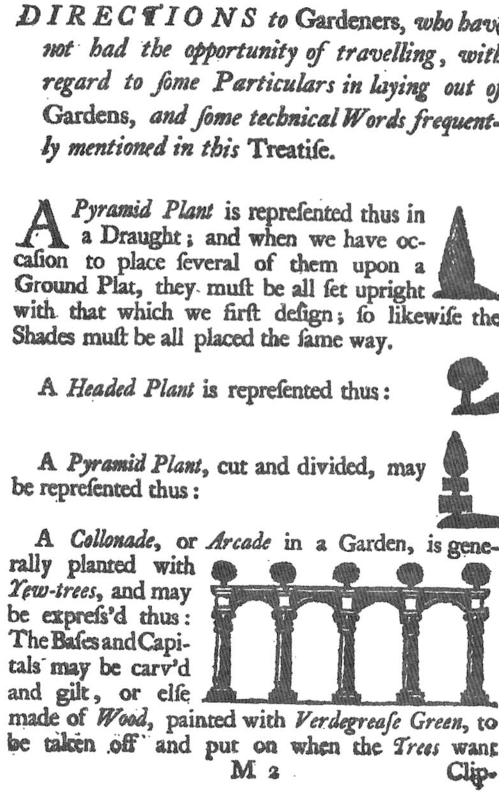
DIRECTIONS to Gardeners, *who have not had the opportunity of travelling, with regard to ſome Particulars in laying out of* Gardens, *and ſome technical Words frequently mentioned in this* Treatiſe.

A *Pyramid Plant* is repreſented thus in a Draught; and when we have occaſion to place ſeveral of them upon a Ground Plat, they muſt be all ſet upright with that which we firſt deſign; ſo likewiſe the Shades muſt be all placed the ſame way.

A *Headed Plant* is repreſented thus:

A *Pyramid Plant*, cut and divided, may be repreſented thus:

A *Collonade*, or *Arcade* in a Garden, is generally planted with *Yew-trees*, and may be expreſs'd thus: The Baſes and Capitals may be carv'd and gilt, or elſe made of *Wood*, painted with *Verdegreaſe Green*, to be taken off and put on when the *Trees* want Clip-

M 2

FIGURE 31. Terms for clipped trees and shrubs from Richard Bradley's *New Improvements of Planting and Gardening* (1731). He also includes the arcade of yews grown against a timber framework, popularised by Dezallier d'Argenville.

While contemporary writers focused on the role of men in the development of these great gardens and parks, some women were also involved, occasionally with the arboricultural aspects. Switzer (1718, 1, 72) speaks highly of Mary Somerset, Duchess of Beaufort, not only for her knowledge and promotion of exotic trees but also for her major contribution to the landscaping at Badminton. He also mentions the Countess of Lindsay, who laid out extensive plantations. According to Switzer, 'This Lady was reputed to be a continual Attendant and Supervisor of her Works, without any regard to the rigid Inclemency of the Winter-season; and not only so, but also in the Measuring and Laying out the distances of her Rows of Trees, she was actually employed with the Rule, Line, etc.' (*ibid.*, 73).

When William III died and Anne became Queen in 1702 Britain had just become embroiled in the War of the Spanish Succession. This heralded a period of constraint in royal and public expenditure and Anne was keen to set an example in thrift (Green 1956, 65). Anne found that the 74 acres (30 ha) of royal gardens had been costing her predecessor £5,000 a year, of which £1,623 was accounted for by Hampton Court. Advised by Sarah Churchill, Duchess of Marlborough, who was obsessional about extravagance, she resolved that the gardens should cost less. In 1707 Anne ordered her royal gardener, Henry Wise, to simplify the Great Fountain Garden at Hampton Court, the largest *broderie* design in England, to a *parterre à l'Angloise* arrangement (Jacques 2017, 218). The box at Kensington was also rooted out because, according to Switzer (1718, 1, 83), Anne disliked the smell of the plant. However, a greater motivation was likely to have been the expense of maintaining the box, and also her preference for a more English style of parterre. This event seems to have signalled the start of a trend away from very elaborate and very expensive close-pruning work to a fashion for a simpler style of topiary and parterre. During the 1710s the propaganda against 'clipped greens' gathered momentum, led by such arbiters of gardening fashion as Alexander Pope and Joseph Addison. In September 1719, Princess Caroline invited Pope, along with several others with an interest in gardening, to a meeting to discuss alterations to her garden at Richmond Lodge. Pope's account of the consultation in a letter to Lord Bathurst reveals some divergence of opinion, not only about suitable trees but also on the subject of pruning. He wrote:

> There were some who cou'd not bear Ever-greens, and call'd them Never-greens; some, who were angry at them only when cut into shapes, and gave the modern Gard'ners the name of Ever-green Taylors; some who had no dislike to Cones and Cubes, but wou'd have em cut in Forest-trees; and some who were in a passion against any thing in shape, even against clipt hedges, which they call'd green walls.
> Pope 1956, 2, 14

It seems that Pope and Addison themselves did not object to the pruning and shaping of trees and shrubs *per se*, just to the highly contrived and 'tortured' topiary shapes that were still popular with many.

Gardens of the gentry

Throughout history the gentry and middle class have invariably made substantially different gardens from those of the landed aristocracy. Until quite recently the rich and powerful were usually the pioneers of fashion and others would follow if they had the money and opportunity to do so. With regard to the formal gardens in this era, the gentry could not afford to follow the same fashions, at least not on the same scale, and their gardens had to fit with the limitations of their situation. Therefore, we find these gardens with scaled-down versions of some of the features in the grander gardens, such as much less topiary, smaller parterres and fewer trees in containers.

Due to some illustrations in gardening books and other images of the time, we are able to gain an insight into how these gentry gardens looked. While the aristocracy could consult people with considerable experience of gardens, gentlemen would often just consult appropriate books. John Worlidge's *Systema Horti-culturae; Or, The Art of Gardening*, published in 1677 and again in 1683, has two excellent illustrations of gentry gardens. The round garden has a path around the perimeter that is described as good for walking without having to turn (Worlidge 1683, 15). There are several beds in the round central area planted with fruit trees and conifers, possibly including cypresses, spaced regularly in a geometric arrangement (Figure 32). The illustration of the square garden shows a more conventional design with a series of rectangular beds filling the square (*ibid.*, 17). The only trees shown are regularly spaced fruit trees in the orchard beyond the fence of the square garden. The frontispiece of *The Gentleman's Recreation* by John Laurence, published in 1716, shows apple trees planted in regular rows in square, grass-covered plots and pruned to tiny rounded heads (Laurence 1716, frontispiece). The presence of plenty of fruit trees in these gentry gardens indicates their utilitarian as well as recreational functions.

FIGURE 32. The Round Garden from John Worlidge's *Systema Horti-culturae* (1683). The beds in the round central area are planted with fruit trees and conifers, possibly including cypresses, spaced regularly in a geometric arrangement.

Trees in the urban realm

The increase in urban greenspace that became noticeable in Tudor times was even more evident by the mid-seventeenth century. Surveys of properties in London at this time give a good indication of this (Harding 1990, 49). On the Bishop of London's estate of more than 700 properties in Shadwell, the houses were small but some 40 per cent still had gardens; a group of 100 houses in High Holborn were on average much larger and over 70 per cent had gardens; in Piccadilly, the houses were of medium size and almost all had gardens. While there is no indication of the extent of trees in these gardens, only the larger ones were likely to have accommodated any and these would have been mostly fruit trees. Following the Restoration the extent of gardens in London increased still further. Writing in 1691, John Aubrey (1982, 86) declared that in the time of Charles II 'gardening was much improved and became common: I doe believe … that there is now ten times as much gardening about London as there was in Anno 1660'.

In the late seventeenth and early eighteenth century the creation of urban public walks became more common in London and in other British towns and cities (Johnston 2017, 13–23). The trees in the early walks were now maturing, as indicated in 'The Prospect of Grays Inn', a print made around 1705 (Jacques 1989, 54). This shows the extensive lines of elm trees which now required regular arboricultural work to keep them in a safe and attractive condition. As a result of research by David Jacques (*ibid.*, 55) we have a good account of the pruning and replacement planting work that was undertaken. Although the development of public walks in Britain would be a significant factor in the eventual planting of street trees in our towns and cities, this would not occur until the early nineteenth century (Johnston 2017). However, the planting of street trees in the Netherlands was attracting the interest of both John Evelyn and Moses Cook. Evelyn, who had visited Amsterdam in 1641, remarked on

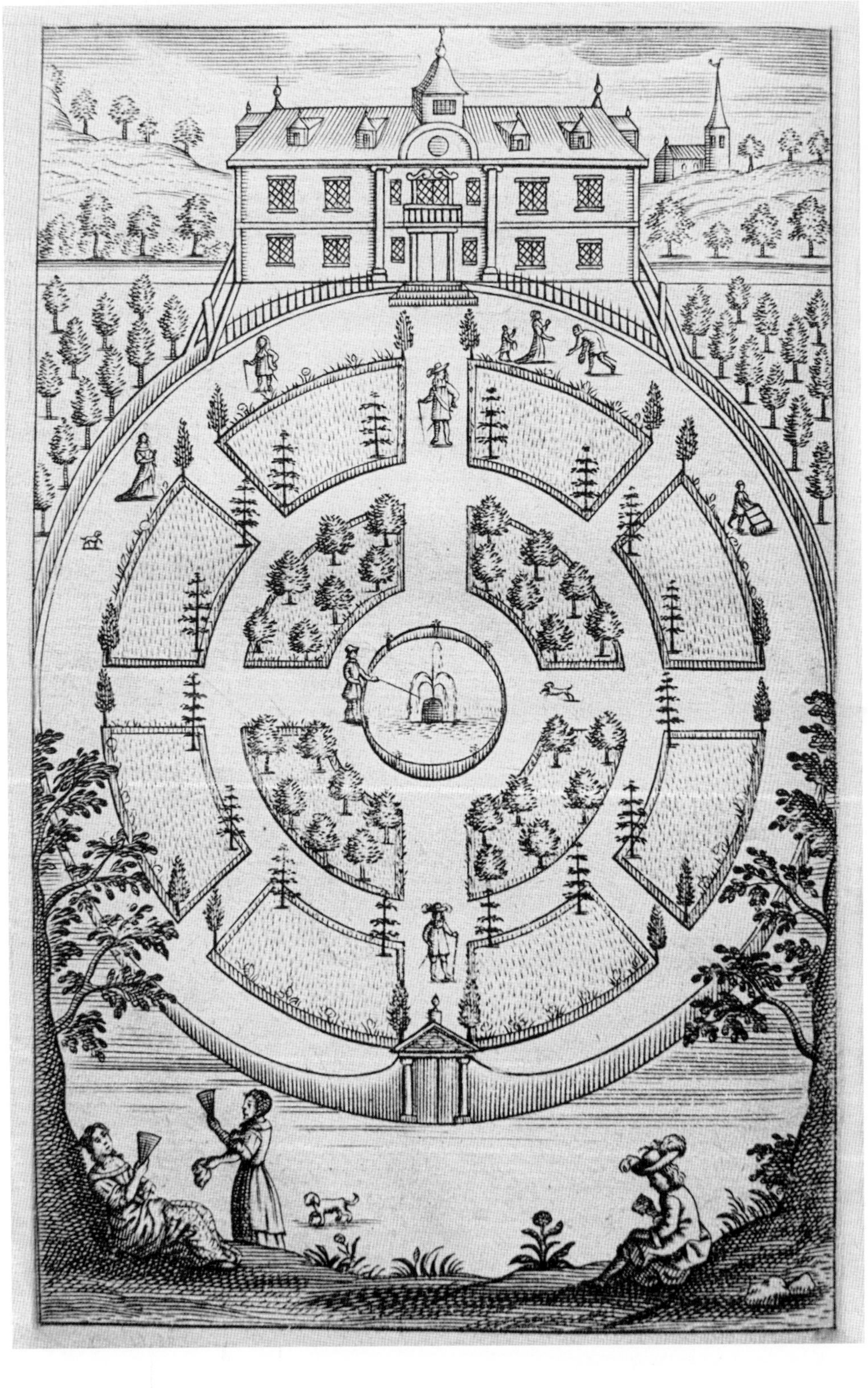

the straight and well-paved streets and 'the houses so uniform and planted with lime trees, as nothing can be more beautiful' (Evelyn 1901, 1, 22). Cook not only admired the beauty of the Dutch street trees but also went further with what might be the first published comment on the health benefits of urban trees. He remarked, 'And (as I have heard) a wise Mans Opinion was that the Lime-trees in the Cities in Holland adde much to the Health of the Inhabitants; and it is my belief' (Cook 1676, To the Reader).

Both Cook and Evelyn were conscious of the potential conflicts between trees and built development. Cook (1676, 73) cautions against planting sycamore and maples near houses because of problems with honey dew, particularly how it attracts insects, which can be a nuisance to those living in the houses. Evelyn (1706, 1, 22) warns against having large trees too close to buildings and states, 'But for the oak, elm, wall-nut, firs, and the taller timber-trees, let the dismal effects of the late hurricane (never to be forgotten) caution you never to plant them too near the mansion, (or indeed any other house) that so if such an accident happen, their fall and ruin may not reach them'. Vandalism of urban trees, including their unauthorised removal, was also being recognised as a problem at this time. Timothy Nourse (1700, 59) declared it was almost impossible to raise a quickset hedge near a city, 'by reason of the great Numbers of Poor who inhabit the Out-skirts, who upon all Occasions, and especially in cold Weather, will plunder of whatever is combustible'.

The residential squares of London, Edinburgh, Bath and other cities that developed in the seventeenth and eighteenth century became a uniquely British urban landscape form (Lawrence 1993; Johnston 2015, 112–114). Their origin can be traced back to the public plazas of the seventeenth century and most were initially created as private gardens for the residents of the surrounding houses. The first London residential square was Covent Garden, begun in 1630, and was an exclusively hard landscape without trees or any horticultural features of note. Some other London squares, such as Bloomsbury, St James's and Leicester, were established just before the Great Fire of 1666. The open spaces of the squares changed from open plazas in the seventeenth century to enclosed private parks at the end of the eighteenth century, a trend based on the social values of the aristocracy, later adopted by the middle class. Public rights of access were steadily extinguished through parliamentary enclosure acts similar to those used at the same time on rural estates. The first was passed in 1726 to enable the residents of St James's Square to restrict access to the square and to levy a rate on themselves for its improvement and maintenance.

In 1722, Thomas Fairchild (1667–1729) published his influential book *The City Gardener*, the first British text on urban horticulture, which included information about trees, shrubs and other plants that thrived best in London (Fairchild 1722). Fairchild also had advice for laying out squares, which is all about extensive tree planting and the benefits this brings. He says:

> If a Square was planted in the Manner of a Wilderness, it would be a Harbour for Birds. Secondly, the Variety of Trees would be delightful to the Eye. Thirdly, Groves and Wildernesses would be new and pleasant in a London Prospect. Fourthly, The Walks, tho' regular as the Walks in the common Squares, would be more shady and private, and the Hedges and the Groves of Trees in every Quarter would hide the Prospect of the Houses from us; every House would command the Prospect of the Whole, as well as if it was lay'd out in plain Grass Platts and Walks.
>
> Fairchild 1722, 12–14

Fairchild (1722, 43) recommends various trees and shrubs that he believes would flourish under such conditions, pointing out that 'the plants that will grow in a Square or large Place, will not always grow in a Street or a Court'. Trees and shrubs became increasingly common in squares and initially they were kept far enough apart and sufficiently low to allow views across and through the open space. Small trees such as laburnum were frequently planted, and when larger ones such as limes and hornbeams were used they were usually tightly pruned and pleached.

Another location for many urban trees was in churchyards. After the Great Fire many London parishes took the decision to replant trees in most burial yards, even in the smallest urban sites. Ironically, this large-scale tree planting, particularly noticeable from the 1670s onwards, would also have rapidly reduced light levels in many churchyards, in spite of the evidence for regular pruning and trimming. In David Marsh's (2004, 232–234) research on London churchyards at this time, at least forty-nine parishes provide evidence of the existence of trees in their churchyards; thirty-four recording payments for planting and thirty-three recording payments for tree care. Some parishes also chose to plant trees in boxes. At St Mary Magdalene, Old Fish Street, five boxes were bought in 1710 and then painted for use in the churchyard, while at St Thomas the Apostle records show the carpenter, John Kent, was paid to repair such boxes in 1709, with new ones being bought eventually in 1718. This parish also used tree boxes on neighbouring streets, Kent being paid 14*s* for 'mending the Boxes for the Trees on the Hill' (*ibid.*, 236).

Pioneers of the new landscape in the early Georgian era

The year 1714 saw the end of the War of the Spanish Succession and the start of the reign of George I. The long-awaited peace and improving economic situation signalled a marked revival in garden-making. However, royal leadership in the art of garden design began to decline after the accession of George I. He did not share the Stuart passion for the arts in general or gardens in particular and artistic leadership passed to the nobility. As garden-making received a boost, it also witnessed the early development of a new style of garden design in Britain that would hold sway for several decades to come.

The criticism of some influential commentators of what they viewed as the excesses of the Baroque formal garden prompted an exploration of more naturalistic styles. A series of essays was placed by Addison in *The Spectator*, one of the leading magazines of the day, which promoted much discussion on this topic. In one particularly scathing attack on the formal style, Addison declared:

> Our British gardeners, on the contrary, instead of humouring nature, love to deviate from it as much as possible. Our trees rise in Cones, Globes and Pyramids. We see the marks of the scissors upon every plant and bush. I do not know whether I am singular in my opinion, but, for my own part, I would rather look upon a tree in all its Luxuriancy and Diffusion of Bough and Branches, than when it is thus cut and trimmed into a Mechanical figure.
>
> Addison 1712, 101–102

FIGURE 33. Stephen Switzer's plan for an ideal forest garden in his *Ichnographia Rustica* (1718). While this still shows a strong central axis, many of the walks are serpentine and meander through the garden.

While those objecting to formal gardens focused mostly on practical considerations of cost and aesthetic concerns about their unnatural appearance, there was also a political dimension to this in England that was associated with the Whig party. The formal garden was increasingly viewed as an expression of some despotic and decadent tendency and the French style in particular was seen as symbolising the absolute power and presumed cultural supremacy of Louis XIV. Some English politicians and garden designers were now looking for a more natural style of landscape that symbolised English national values, its constitutional monarchy and 'love of liberty'. One of the pioneers of this new naturalistic approach was the garden designer Stephen Switzer (1682–1745) and he named it the English Forest Style. In 1715 Switzer wrote *The Nobleman, Gentleman, and Gardener's Recreation*, the first British gardening book to advocate the new landscape style (Switzer 1715). He followed this a few years later with his major work on the subject, the three-volume *Ichnographia Rustica*, which focused on 'the general Designing and Distributing of Country-Seats into Gardens, Woods, Parks, Paddocks, etc., which I therefore call Forest, or, in a more easy style, Rural Gardening' (Switzer 1718, 1, xv). Switzer believed that his style was more economical and beautiful than the previously fashionable formal garden style. He thought more money should be spent on forest planting and that it should be obtained by reducing the size of parterres or by laying them to grass. Since few avenues could be made by cutting through existing woods in England, massive tree planting was essential to create any semblance of the French style. These new outward plantations would help to merge the woodland area with the inward garden once its wall was removed. However, Switzer was not entirely abandoning the French formal style. The plan for an ideal forest garden in *Ichnographia Rustica* shows a strong central axis and extensive tree planting, although many of the walks through this are serpentine and meander through the garden (Figure 33). Switzer wrote of outward plantations that 'these kinds of Woods, as they are more Natural and Rural than the set Wildernesses and Groves, so much us'd of late amongst us, yet they are a great deal less expensive [to establish], an Acre of this being made full five times cheaper than the other' (Switzer 1718, 1, 272–273). There were also significantly reduced maintenance costs due largely to a much less intensive pruning regime along the sides of the walks.

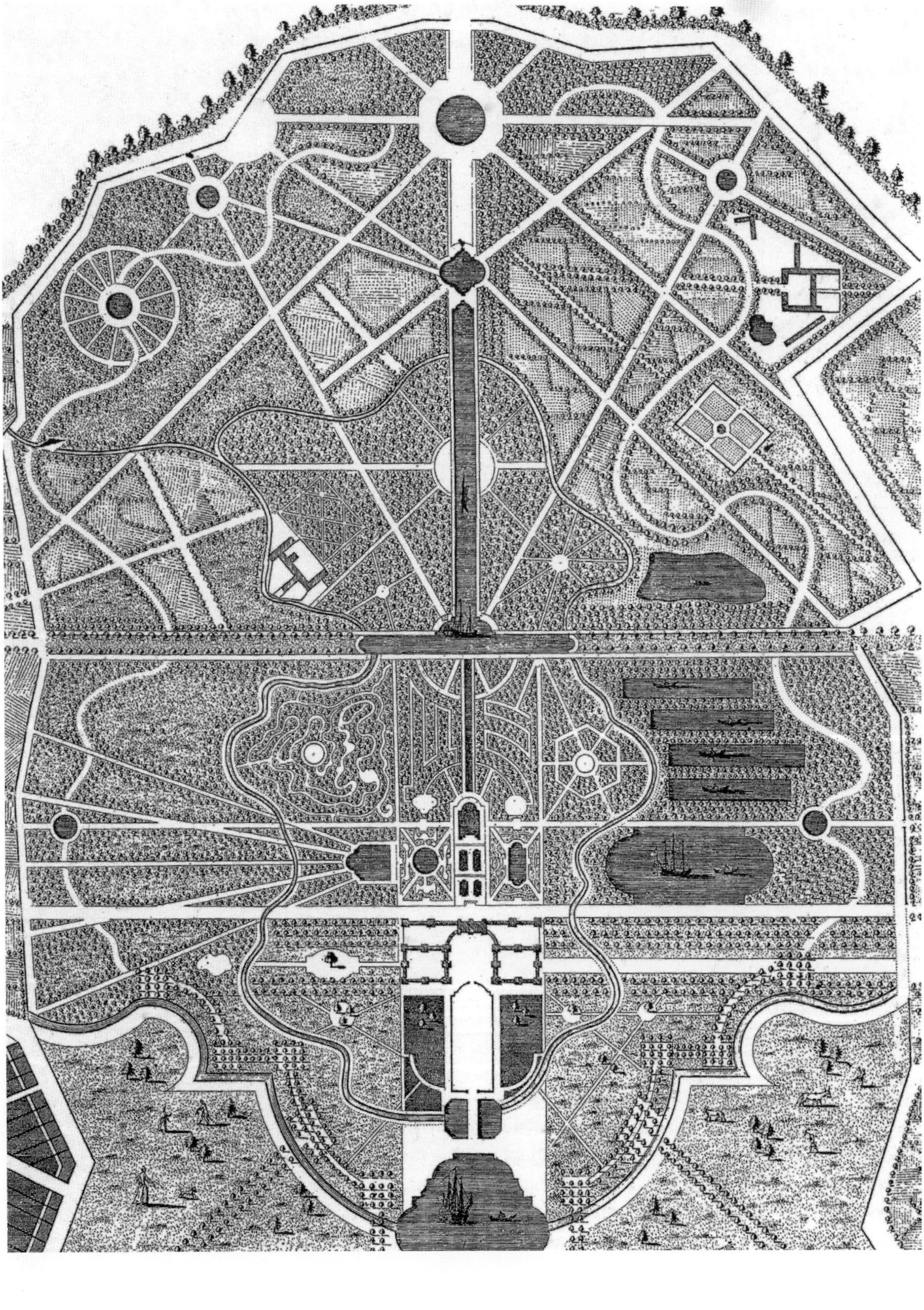

One of the few surviving examples of the English Forest Style is at Cirencester Park, Gloucestershire. This extensively wooded park with radiating avenues was planted by the first Earl of Bathurst from 1714 with the help of his friend Alexander Pope and much influenced by Bathurst's association with Switzer. The scale of the tree planting was vast and prompted Pope to ask, 'Who plants like Bathurst?' and then invest £4,000 in his friend's forest venture (Hussey 1967, 80). However, in Switzer's view it was Lord Carlisle's treatment of Wray Wood at Castle Howard, Yorkshire, which represented 'the highest pitch that Natural and Polite Gardening can possibly ever arrive to' (Switzer 1718, 1, 87). Switzer related how in 1699 'Mr London desgn'd a Star' for Wray Wood, but the Earl of Carlisle, with his 'superlative Genius', vetoed the plans 'which would have Spoil'd the Wood' (Switzer 1718, 2, 198). The initial proposal for Wray Wood caused Switzer to criticise similar designs that he considered were too rigidly geometric and not appropriate to the location. He complained of those 'who call themselves Gardeners, who having wrought a little while at some or other great Works of this Kingdom, immediately put on an Apron, get a rule and pair of Compasses, with other things belonging to this work; thus equipt, what Wonders are we not to expect from so profound a set of Mathematicians and Designers!' (Switzer 1718, 1, xxiii). This approach had led to many ancient trees being felled and Switzer complained, 'Thus do we see many a noble Oak, or sometimes whole lines of these and other umbrageous Trees, fell'd, to humour the regular and delusive Schemes of some Paper Engineer' (*ibid.*, xxvii). What was eventually fashioned at Wray Wood was a network of meandering walks creating a sort of labyrinth woodland. While this was a radical departure from earlier formal gardens, it echoed the work done at Cassiobury some thirty-five years earlier by Cook and the Earl of Essex, who was actually Carlisle's father-in-law.

In the late 1720s, Batty Langley (1698–1751), a garden designer from Twickenham, began writing on gardening and architecture. His book *New Principles of Gardening* contains designs for gardens and wider landscapes that have much in common with the work of Switzer (Langley 1728a). While Langley devoted a major section of his book to 'Forest trees, their Culture, etc.', this did not exhaust his knowledge on the subject. In the same year, he wrote *A Sure Method of Improving Estates by Plantations of Oak, Elm, Ash and Other Timber-Trees* to further promote the planting of trees for pleasure and profit (Langley 1728b). Another pioneer of the naturalistic landscape style was Charles Bridgeman (1690–1738) who also favoured meandering walks through his plantations. Despite the influence of more naturalistic styles, wildernesses did not suddenly become 'rural'. Plantations in geometrical blocks reminiscent of formal *bosquets* with tall palisades and gravel alleys were still being created in the 1720s. Throughout that decade, however, symmetrical wildernesses were less common as more naturalistic arrangements became increasingly fashionable. Nevertheless, as with their 'greens', the nobility and gentry were often reluctant to destroy what they had recently striven to acquire, and fine examples of high palisades continued to excite admiration into the 1730s.

Arboricultural practices

The tremendous increase in the number of amenity trees across Britain during the second half of the seventeenth century and early eighteenth century required an unprecedented amount of arboricultural work to achieve this. At the same time, information on the planting, maintenance and management of amenity trees was being published in a growing number of books. While these books did not use the term arboriculture to describe these activities, the terms arborist or arborator were used by some authors to describe the people involved in this work. Indeed, Worlidge (1683, 73) describes Evelyn as 'our best Arborist'.

In introducing these topics some authors were keen to highlight what they regarded as 'errors' that had been handed down from previous generations. These were a mix of old superstitions, misguided practice and incorrect science. Ralph Austen (1653, 78–97) was one of the first to do this when he listed a number of 'Errors discovered' in his *Treatise of Fruit-Trees*. He wrote, 'Let not men think that Ancient and Learned Authors have discovered all Truths, or that all they have said is truth: they are but men and have their Errors' (*ibid.*, 78). In terms of the source of these errors, Austen identifies the authors from the ancient world. However, with those authors from recent times he does not give names, although he makes the exception of naming Francis Bacon as the source of a few errors. He is especially keen to challenge Bacon's belief in the 'descention of sap in Autumne' (*ibid.*, 91). In order to avoid these errors Austen argues for using the 'Scale of Reason' and the 'Touchstone of Experience'.

Moses Cook also lists a number of 'errors' in common arboricultural practice in his book *Forest-Trees* (Cook 1676, To the Reader). These include boring a hole in a tree and pouring in honey to make its fruit sweet, and watering seeds with coloured water so that they would produce flowers of that colour. In refuting these errors Cook explains how he had learned about trees and gardening with his use of the experimental method:

> I alwayes took Notes of what I did set or sow, the Times, and on what Ground, etc., and when it proved well, I noted it so; but when ill, I did endeavour as much as I could to know the Reason; which when once I found, I noted it well: I also alwayes was very wary of taking things upon trust; for many Learned men have abused their Works by so doing; and if any man told me any thing, unless he had sufficient experience of it, or could give very good Reasons why it was so, I always was incredulous of it, unless my Judgement told me it were possible, or he by Discourse made it plain to me.
>
> Cook 1676, To the Reader

In a fascinating insight into Cook's standing among aristocrats who professed a knowledge of trees and gardening, he says, 'I have often been blamed by Noble-Men, for not consenting to the Opinion of some of their Favourites; for when their Notions were not grounded on Reason, or had not been proved by experience, (though never so new) I could not entertain them' (Cook 1676, To the Reader). Evelyn, by contrast, highlights few 'errors' by past and present authorities. Despite the efforts of Austen, Cook and others, some popular misconceptions continued for many years and were repeated in some learned works, especially if they had originated with classical authors who continued to be revered.

One popular belief shared by most contemporary authors, and one that still has currency today, is the belief that the moon and other celestial bodies have a major influence on plant growth. Evelyn (2001, 55–59) in the draft of his *Elysium Britannicum* has a chapter entitled 'Of the Celestial Influences, particularly the Sun, Moon, and of the Climates'. Although aware of the criticism in some quarters, he still thought 'there is doubtless some regard to be had' for these beliefs (Evelyn 1664, 91). Many authors specifically mentioned the phases of the moon in relation to the timing of arboricultural practices such as planting, pruning and felling. By the early part of the eighteenth century the influence of Bacon's writings and growing recognition of the scientific method contributed eventually to the decline of astrology in arboricultural and horticultural practice.

Planting and post-planting maintenance

In deciding what trees to plant and where these should be located, future maintenance considerations begin to feature in this era of mass amenity planting. An example is the recognition of the impact of the leaves falling in autumn. Evelyn was an early critic of the sycamore (*Acer pseudoplatanus*) and states:

> The Sycomor is much more in reputation for its shade then it deserves; for the Leaves which fall early (like those of the Ash) turn to a Mucilage, and putrifie with the first moisture of the season; so as they contaminate and mar our Walks, and are therefore (by my consent) to be banish'd from all curious Gardens and Avenues.
>
> Evelyn 1664, 29

Aware of these considerations, Cook recommends planting pine trees as 'they be fine to set round a Garden, or Bowling Green, for the Leaves will not do any harm' (Cook 1676, 86). Tamsin Chambers (2008, 64–76) has undertaken a fascinating analysis of the different tree species recommended by several post-Restoration writers, giving the reasons why they favoured these species for specific purposes.

Austen (1653, 95) stresses the need to select planting sites that have suitable soil and recommends always planting trees in better soil than the soil they came from. He notes that many trees are purchased from nurseries around London where the soil is very fertile and they are then planted in locations where the soil is poor, resulting in weak growth and even death. Ground preparation in advance of planting is also highlighted by some authors as a crucial first stage in the planting process. Cook (1676, 22) remembers the old saying, 'A thing once well done, is twice done: And those that are resolved to plant, that they may make their Ground fit for those Trees before they set them, and not bury them in a hole like a dead dog, as too many do'. When planting trees in the ground, Austen highlights what he perceives as a number of common errors. These include planting trees too close together and planting them at the wrong time. He recommends planting 'when trees have done growing, and that is divers weeks before the Leaves fall', which ideally should be at the end of September (Austen 1653, 60). Even considering the adjustment of the Gregorian calendar, this is generally much earlier than most professionals, then and now, would recommend. Unlike many authors of the time, Austen does not recommend pruning back the branches of young trees before they are planted. Most authors agreed that planting in the dormant season was best for deciduous trees, as long as the weather and ground conditions were favourable. Most also agreed with Evelyn (1932, 67), 'But plant not Ever-greenes 'til the beginning of April, when you perceive them begin to shoote'.

The debate continued about whether to establish woodland by seeding or by transplants. In the early seventeenth century Church and Markham had both recommended the use of transplants. However, Evelyn was strongly in favour of establishing woodland by seeding and was forthright in giving his reasons. These were:

> First, because they take soonest. Secondly, because they make the straightest, and most uniform shoot. Thirdly, because they will neither require Staking, nor watering (which are two very considerable Articles) and lastly, for that all transplanting (though it much improve Fruit-trees) unless they are taken up the first Year, or two, is a considerable impediment to the growth of Forest-trees.
>
> Evelyn 1664, 3

Cook was of the view that using transplants was the better method and of all the authors around this time he goes into the most detail on his planting technique. He relates how the trees at Cassiobury were raised from seed in nurseries and then transplanted to the site where they were planted in well-prepared pits, their bruised roots pruned and their branches pruned back (Cook 1676, 21–22). Cook also took care to plant the tallest of the young trees in the worst ground, rather than reducing them all to the same height. In addition to nursery training to select a strong leader, it was common practice to cut off the crown of young trees when they were planted out in avenues or walks. This 'heading' of the young trees was undertaken for a variety of reasons: because the trees were transplanted bare-rooted and this compensated for root loss, to encourage branching and crown growth, or to ensure all the young trees started off at the same height. By the early eighteenth century, it was widely accepted that transplants were the best method for establishing ornamental groves rather than direct seeding which took too long. However, with the growing trend for more naturalistic landscapes, the pendulum started to swing back again towards direct seeding (Switzer 1718, 1, 272). Switzer led the way in believing that the growth of these trees grown from seed could soon overtake those established as transplants and the cost of establishing woodland by seed was also much less.

Opinions differed on planting arrangements for ornamental woodlands. Evelyn favoured the quincunx pattern, while others favoured a square or rectangular arrangement. Cook preferred his trees planted in a triangular pattern: 'Your Best way is to plant them Triangular and not square, as some doe; for you can plant them in no form or order whatsoever to be more pleasing to the most Noble Sense, than to have every three Trees to make an Equilateral Triangle' (Cook 1676, 104). In orchard planting it was widely accepted that the quincunx arrangement was best, for ease of management and appearance. In the 1720s, with interest growing in more naturalist garden designs, some authorities argued that planting arrangements should reflect this. Langley recommended, 'In the Planting of Groves, you must observe a regular Irregularity; not planting them according to the common Method like an Orchard, with their Trees in straight Lines ranging every Way, but in a rural Manner, as if they had received their Situation from nature itself' (Langley 1728a, 202). Being more of a designer than practical gardener, he was probably unaware that trees can be planted in a regular pattern for ease of weeding and initial maintenance, and then thinned out later to achieve an irregular effect.

The need to stake newly planted trees to ensure their root systems were secure in windy conditions was stressed by some authors. While most gave little detail on how to do this, Le Gendre (1660, 103) noted that 'this stake must be shorter than the stem of the tree, lest it wry [distort] the branches that shoot forth'. Timothy Nourse (1700, 134) had a novel way of countering the effects of high winds when there was no stake: 'Plant the tree leaning towards the prevailing wind so that the force of the wind will bring it upright'. Nevertheless, when stakes were used, he stressed that care should be taken to ensure they did not bruise the bark or branches. Some form of protection was used for trees planted in locations where they might be damaged by cattle and other animals. Evelyn recommended a form of staking and protection that is still used in parkland planting today:

> But where they are expos'd in single rows, as in walks and avenues, the most effective course is to empale them with three good quartet-stakes of competent length, set in a triangle, and made fast to one another by short pieces above and beneath; in which a few brambles being stuck, secure it abundantly without that choaking or fretting, to which trees are obnoxious that are only single staked and bushed, as the vulgar manner is: Nor is the charge of this so considerable as the great advantage, accounting for the frequent reparations which the other will require.
>
> Evelyn 1706, 1, 23

As well as the focus on good planting techniques during this era, there was also more awareness of the importance of post-planting maintenance, such as watering, mulching and weeding. Austen (1653, 66) talks about mulching around the base of trees and believes that applying various types of dung will increase growth and fruiting. Cook (1676, 22) states that once his trees were planted 'I then laid round every Tree … wet Litter taken off from the Dung hill, a good Barrow-full to every Tree, and covered that with a little Mould'. Cook (*ibid.*, 111) also emphasises the need to ensure post-planting maintenance when he says, 'In the Spring, walk over the Ground you planted in Winter, and set your Trees to right'. He then outlines the maintenance activities that should be considered.

The importance of watering newly planted trees, particularly during a dry spring and summer, is stressed by many writers. However, the difficulties of doing this in large plantations at some distance from the house are rarely considered. This would have required some form of mobile water tank or bowser, possible horse-drawn, that would have got its supply from a nearby lake or pool. In more formal ornamental gardens, watering of trees and other plants was often written into the maintenance contract. In 1699, in Henry Wise's estimate for works at Hampton Court there is an item listed for 'The Charges of watering all ye Trees that have been planted this year and ye Last' (Chittenden 1939, 466). This indicates a two-year watering regime for newly planted trees. In 1702, in a comprehensive contract for most of the royal parks, Wise undertook to provide all implements including wheelbarrows and 'a sort of Engine to sprinkle the several trees and plants' (Green 1956, 55). He further undertook 'At his own proper cost and charges to make good the several Plantacons of Fruit trees, forest trees, flowering shrubs and hardy Evergreens that shall at any time happen to dye, by planting others in their places'.

Transplanting large trees

While the moving of large trees was undertaken occasionally in earlier times in Britain, in this era it became relatively common in some prestigious gardens. John Evelyn wrote in some detail on this topic. In *Sylva* he writes:

> My Lord Hopton planted Oaks as big as twelve Oxen could draw, to supply some defect in an Avenue to one of his houses in Devonshire; as the Right Honourable Sir Charles Barclay, Treasurer to His Majesty's Household, assur'd me; who had himself likewise practiced the Removing of great Oaks by a particular address extremely ingenious, and worthy the communication.
>
> Evelyn 1664, 13

In the 1679 edition of *Sylva* Sir Charles Barclay is referred to as Lord Fitzharding and his technique of rootball preparation one or two years in advance of moving is subsequently mentioned by Steuart (1828, 48) and others. Evelyn then refers to Pliny and Roman transplanting practices before describing his preferred technique:

> Chose a Tree as big as your thigh, remove the earth from about him; cut through all the collateral Roots, till with a competent strength you can enforce him down upon one side, so as to come with your Axe at the Tap-root; cut that off, redress your Tree, and so let it stand cover'd about with the mould you loosen'd from it, till the next year or longer if you think good; then tak it up at a fit season. … A little before the hardest Frosts surprize you, make a square Trench about your Tree, at such

> distance from the Stem as you judge sufficient for the Root; dig this of competent depth, so as almost quite to undermine it; by placing blocks, and quarters of wood, to sustain the Earth; this done, cast in as much Water as may fill the Trench, or at least sufficiently wet it, unless the ground were very moist before. Thus let it stand, till some very hard Frost do bind it firmly to the Roots, and then convey it to the pit prepar'd for its new station; but in case the mould about it be so ponderous as not to be remov'd by an ordinary force; you may then raise it with a Crane or Pully hanging between a Triangle, which is made of three strong and tall Limbs united at the top, where a Pully is fastned, as the Cables are to be under the quarters which bear the earth about the Roots: For by this means you may weigh up, and place the whole weighty Clod upon a Trundle to be convey'd, and Replanted where you please, being let down perpendicularly into the place by the help of the foresaid Engine. And by this Address you may Transplant Trees of a wonderful stature, without the least disorder; and many time without topping, or diminution of the head, which is of great importance where this is to be practis'd to supply a Defect, or remove a Curiosity.
>
> Evelyn 1664, 13–14

Evelyn recommends preparing the root system for at least a year before moving the tree, as in the technique developed by Fitzharding. Once the move has been completed, he describes some aftercare such as watering, staking and protecting against damage from animals. He also gives information specifically about the transplanting of large elms:

> Of all the Trees which grow in our Woods, there is none which does better suffer the Transplantation than the Elm; for you may remove a Tree of twenty years growth with undoubted success: It is an Experiment I have made in a Tree almost as big more as my waste; but then you must totally disbranch him, leaving only the Summit intire; and being careful to take him up with as much Earth as you can, refresh him with abundance of water. This is an excellent and expeditious way for great Persons to plant the Accesses of their Houses with; for being dispos'd at sixteen, or eighteen foot Interval, they will in a few years bear goodly heads, and thrive to admiration.
>
> Evelyn 1664, 18

Moses Cook mentions transplanting large trees but refers his readers to the method described by Evelyn as being the most satisfactory. In contrast to those experts who were content to move large trees in one season, John Reid's method required two years of preparation and may have been common practice in Scotland:

> The Rule for removing old large Trees out of woods or other places, who was never before Transplanted, is to make a trench at two sides of the Trees, distance considerable, till you can Inforce the Tree upon one side: then cut the top Root through, saving as many collateral Roots as you can; lessen its Head, or lop it if it can suffer, and to set up the Tree again, and tread in the earth about it, as it was; let it stand 2 years to emit fibres or feeding Roots to nurse it when Planted out. But to my nursed trees again. When you remove, as is directed, carry them as quickly to their new quarters as you can.
>
> Reid 1683, 80

Some of the most impressive examples of transplanting large numbers of trees occurred in the royal gardens. Much of this work was undertaken by the Brompton Park Nursery established in 1681. This company will be discussed later but here we are concerned with their tree transplanting work. When George London, one of the partners in the business, was made Deputy Superintendent of the Royal Gardens, he secured the services of the company to undertake the considerable alterations at Hampton Court, supervised by his fellow partner Henry Wise. A major part of this work was the transplanting of one of the rows of lime trees which formed the avenue by the semicircular canal (Amherst 1896, 224). In 1703 the trees on the northern banks were taken up and replanted on the south of what had been the most southern row. A total of 403 large lime trees, some thirty years old, were then replanted into prepared holes 10 ft (3 m) in diameter. The work must have involved many men and taken a considerable amount of time. In the first few months of Anne's reign, the company undertook another major tree transplanting project in St James's Park. This time 350 limes needed to be planted 'in ye roome of those that are dead', and for that purpose trees 'of ye Largest size that can be procured out of Nurseryes, being about 7 inches [18 cm] in Circumference' were sought (Green 1956, 90). Wise needed to look no farther than Brompton Park, and the trees he chose were more than 13 in. (33 cm) round. He also undertook the replanting of '48 large old Lyme trees of a large size … which if they had stood, most of them would have dyed'. At Hampton Court and St James's Park Wise had demonstrated his mastery of transplanting limes. Later at Blenheim for the Duke of Marlborough he successfully repeated the exercise, with elms.

Tree pruning

With the increasing number of books dealing with tree-related topics, much was being written on the subject of pruning. However, most of this related to either forestry or fruit trees and there was little that was directly about ornamental trees. In a section of Evelyn's *Sylva* devoted to pruning, which is taken mostly from other authors, the emphasis is on pruning for the 'Improvement of our Timber and Woods' and there is almost no reference to pruning for aesthetics. There is, however, more recognition of this in Evelyn's *Directions for the Gardiner* when he declares, 'The principal Art of a Gardner consists in pruning' (Evelyn 1932, 73–76). Evelyn was often critical of tree work standards and had this to say on poor pruning:

> For 'tis a misery to see how our fairest Trees are defac'd, and mangl'd by unskilful Wood-men, … hacking and chopping off all that comes in their way; by which our Trees are made full of knots, boils, cankers, and deform'd bunches, to their utter destruction; Good Husbands should be asham'd of it. As much to be reprehended are those who either begin this work at unseasonable times, or so maim the poor branches, that either out of laziness or lack of skill, they leave most of them stubs,

> and instead of cutting the arms and branches close to the bole, hack them off a foot or two from the body of the tree, by which they become hollow and rotten.
>
> Evelyn 1664, 74

Cook (1676, 116) reproduces this quotation in his own book, although he says of Evelyn that 'he quotes from an ancient author'. Evelyn then follows with a passage that is identical to one in Cook's book:

> By this Animadversion alone it were easy for an ingenious man to understand how Trees are to be govern'd, which is (in a word or two) by cutting clean, smooth, and close, making a stroke upwards, and with a sharp Bill, so as the weight of an untractable bough do not splice, and carry bark with it, which is both dangerous and unsightly.
>
> Evelyn 1664, 74; Cook 1676, 117

What is curious about this passage in Cook's book is that it follows the previous 'quotation' from Evelyn, given in the usual italics, but the text for this is presented as his own words. Considering that Cook gave advice to Evelyn on trees before he wrote *Sylva*, we can only speculate about the origin of those words on pruning. In *Sylva* Evelyn says that he could have included a lot more on pruning but this had already been 'set down by our Country-man honest Lawson' in his *New Orchard and Garden* (Evelyn 1664, 74). He then reproduces an extensive excerpt from Lawson set in Old English typeface. It is interesting that Evelyn does not quote Lawson on wound treatment after pruning and has nothing to say on this himself. With regard to the timing of pruning, most authors agreed this should be in the dormant season. Like many of his contemporaries, Evelyn also considers the astrological dimension: 'The fittest time of the Moon for pruning is (as of Graffing) when the sap is ready to stir (not proudly stirring) and so to cover the wound. Old Trees would be prun'd before young Plants' (*ibid.*, 77).

Moses Cook has plenty to say on the subject of pruning, much of it original. He discusses the benefits of pruning trees early in their life to ensure good timber (Cook 1676, 42–43). He also stresses the importance, for all trees, of formative pruning when they are young (*ibid.*, 114–115). When removing major branches he stresses the need to undertake this correctly:

> Great Boughs ill taken off, as I have many time said before, spoyls many a Tree; therefore take them off close and smooth, and not parallel to the Horizon; Cover the wound with Loom, or Clay, well mixed with Horse-dung, and keep it so covered; for if the wet fall perpendicular on such a place, it will in little time make the whole Tree hollow, but the wet getting in there, which comes by great Lops and bad Lopping; as you may see in too many Elms, ashes, Hornbeams, etc. spoyl'd by such carelessness.
>
> Cook 1676, 118

Different tree experts recommended different recipes for the preparations they used to treat pruning wounds. Many included some type of soil or animal dung. More 'exotic' preparations were often taken from the translations of continental books on pomology. For example, Nicholas Venette (1685, 30) advises, 'Care must be taken to cover the great Wounds of Trees with a Plaister made of a pound of *Drogue a flambeau*, four Ounces of Rosine, and two Ounces of Sheeps Sewet'.

While Cook recognised that pruning trees to produce good timber needed to follows some simple rules, he was also aware that pruning ornamental trees and fruit trees was a more sophisticated matter (Cook 1676, 112). On this subject he echoes the words of Le Gendre:

> It is a Thing very rare among Gardners to Prune Trees well; for the doing of it well depends more upon their Ingenuity than upon their Hand. It is also very hard to give Instructions for it, because it consists not in certain and general Maxims, but varies according to the particular Circumstances of each Tree, so that it depends absolutely upon the Gardners Prudence, who ought of himself to judge what Branches must be left, and which are to fit to be cut away.
>
> Le Gendre 1660, 113

Neither Evelyn nor Cook was in favour of pollarding trees as a general rule. Cook was horrified when this was done to oaks:

> But our Yeomen and Farmers are too much subject to spoyl such Trees as would make our best Oaks, by heading them, and making them Pollards. I wish there were as strict a Law as could be made, to punish those that do presume to head an Oak, the King of the Woods, though it be on their own land.
>
> Cook 1676, 42

However, Cook recognised that careful and timely pruning could enhance the ornamental impact of many species of tree and praised those that responded well to pruning.

Many writers drew parallels between the care of trees and the care of people, something that continued for centuries to come. Talk of treatments to limbs, wounds and the use of plaisters were just some of the anthropomorphic references. While repeating some of these, Thomas Fairchild gives an eloquent appeal to avoid further weakening already sick trees with harmful pruning:

> Some learned Men say that whatever can be made agreeable to a sick Man will help his Cure, or contribute to his Health, but whatever is the contrary increases his Distemper and might even cause his Death: And sure nothing could be more tending to his Detriment than wounding him when his Body was already weak and low? But the Cutting off a Limb would surely give so great a Change of Circulation of Juices that a sick man could not bear.
>
> Fairchild 1722, 39–40

The pruning of fruit trees was recognised as a specialised task for the gardener or arborist that required different skills to pruning ornamental or forest trees. It was also something that increasingly took place in designated orchards rather than in ornamental gardens or pleasure grounds. In larger gardens it was often left to the head gardener, assisted by one of his staff, or a fruit specialist was engaged temporarily for the work. The aim was to increase the quality and quantity of fruit, and while the basic techniques of branch removal remained broadly the same, the 'when', 'what' and 'why' of pruning was often quite different to ornamental pruning.

As the clipping, shaping and pruning of all sorts of trees became a routine and time-consuming task in formal gardens, so it was mentioned increasingly in gardening books and in the house and garden account books of larger gardens. Regular pruning was viewed as necessary to keep vigorous trees in check and bring good order to the garden. John Laurence (1727, 294) declared, 'The luxuriancy and vigour of most healthful trees is like the extravagant sallies of youth, who are apt to live too fast, if not kept within due bounds and restrained by seasonable corrections'. Regular pruning kept 'all in order, which would otherwise be perfect anarchy and confusion' (*ibid.*, 205). Sometimes trees would have their side branches severely pruned back, leaving just the branches at the top, giving a 'tufted' appearance. The epitome of close pruning was seen in the palisades that lined the walks through many groves, wildernesses and other plantations. Maintaining all those high palisades or green 'walls', often stretching out into the distant landscape, necessitated a programme of intensive and regular pruning by a small army of gardeners. The work to create this effect was sometimes referred to as plashing or pleaching, although those terms were also used in other contexts. Specially designed mobile wooden platforms could make the work easier but it was still very labour-intensive. When Evelyn visited the Luxembourg Palace in Paris he commented on the high level of maintenance in the gardens, due largely to the intensive pruning regime. He remarked, 'What is most admirable, you see no gardeners, or men at work, and yet all is kept in such exquisite order, as if they did nothing else but work; it is so early in the morning, that all is dispatched and done without the least confusion' (Evelyn 1901, 1, 63).

In the smaller formal garden this clipping and close pruning work was still very demanding as there was often only one full-time gardener and some occasional assistance. For Thomas Ackres, the gardener at Knole in 1710, among his many other duties it was stipulated that he should look after existing plantations (O'Halloran 2013, 180). He was also required 'to prune and clip the furr Walk' in the Dwarf Ground, which referred to the central avenue of silver fir trees. In 1724, the gardener Gervase Whitehead who was pruning these trees was paid £1 3*s* 'for shriping up ye firr trees in ye Dwarfe Garden'; the term 'shriping' meaning to lop, prune or trim. With the trend towards more naturalistic landscapes in the 1720s, there were significantly less maintenance costs. In such informal woodlands the cost and toil of clipping palisades and espaliered trees would be unnecessary as only rough trimming of trees along the walk was required. As Switzer noted: 'Instead of that prodigious deal of Clipping in Espaliers, a Scythe fix'd into a Pole or Handle will do this Work, and retrench the Extravagances of Nature in a more expeditious and less expensive Method' (Switzer 1718, 1, 272).

The presence of more trees in the urban environment also required gardeners and arborists skilled in the techniques of pruning. While there is little archival evidence, the research of David Marsh (2004, 237) again provides an insight into this with churchyards in London. Payments for tree pruning, lopping or trimming are a regular occurrence in those church accounts around this time, the bulk of entries beginning after the 1690s. Much of the work is likely to have been crown lifting and crown thinning, for access and for light, as well as the removal of some deadwood and other branches for safety. The prunings appear to have been treated as 'rubbish' and were not burnt or composted but 'carried away' by the gardener or sexton.

Tree health and tree care

This era saw much interest in how to promote healthy and vigorous trees and various terms were used to refer to the threats to this. In Evelyn's *Sylva* there is a chapter on the 'Infirmities' of trees, which includes weeds, small mammals and ivy (Evelyn 1664, 68–70). Snails and insect 'pests' such as caterpillars and wasps are also mentioned. While the word 'disease' appears, this is used to embrace all the 'infirmities' of trees. Cankers and 'hollowness', which we now regard as aspects of disease, are mentioned but without any reference to biological processes. The only cure proposed for this was to cut out the 'bad bits'. Cook (1676, 117–119) has a chapter entitled 'Of the Diseases of Trees' that covers similar ground to Evelyn. Richard Bradley, writing some fifty years later, offers little more. In his *New Improvements of Planting and Gardening* he has a chapter on what he calls the 'Blights' of trees, which includes all the known harmful agents (Bradley 1718, 3, 53). During this period advances in tree biology and pathology were often achieved more by chance than scientific means. Nevertheless, the foundation of the Royal Society in 1660 gave major impetus to scientific enquiry in horticulture, arboriculture and related fields. In its early days, many of its 'leading lights' were not only passionately interested in plants but also dedicated to its aim of working 'for the improvement of natural knowledge by experiment'. Part of this included encouraging research into the many agencies that were a threat to the health of trees. While positive outcomes were slow to emerge, at least a start had been made.

In the absence of good science, even the most revered tree experts resorted to folk remedies to revive ailing trees. For example, Evelyn recommends:

> For the improvement of the speedy growth of Trees, there is not a more excellent thing then the frequent rubbing of the Boal or Stem, with some piece of hair-cloth, or ruder stuff, at the beginning of Spring; … the more rugged bark with a piece of Coat of Maile, which is made of small wyres; this done, when the body of the Tree is wet.
>
> Evelyn 1664, 77

Other practices, which may have seemed like folk remedies at the time, were many years later to prove beneficial to the trees. Ralph Austen, in the second edition of *A Treatise of Fruit-Trees*, talks about directly feeding the roots by puncturing the soil with an iron bar and placing fertiliser in the holes (Austen

1657, 71). This must be the first British reference to tree feeding via the roots. Austen also cautions against over-fertilisation using this method.

As the population of many British towns and cities increased substantially, health problems arose for both people and plants as a result of the obnoxious smoke, mainly from the domestic and industrial burning of coal. In 1661, Evelyn published *Fumifugium: Or the Inconvenience of the Aer and Smoake of London Dissipated* (Evelyn 1661). This was a proposal to solve London's smoke problem that included establishing a series of plantations around the city, 'supplied with such Shrubs as yield the most fragrant and odoriferous Flowers, and are aptest to tinge the Aer upon every gentle emission at a great distance' (*ibid.*, 14). While this was a ground-breaking text, the plantations were designed solely to act as a scented buffer zone to counter the bad smell. Some sixty years after Evelyn's *Fumifugium*, Thomas Fairchild's book *The City Gardener* would give some useful practical information about managing plants in a polluted urban environment. He was the first British writer to observe that not only did evergreens and conifers not thrive in the polluted atmosphere of cities, but any amount of pruning usually made their condition worse (Fairchild 1722, 39).

Although authors such as Evelyn and Cook were highlighting the value of Britain's ancient trees and stressing the need to preserve these magnificent specimens, there was little mention of preventative or remedial tree surgery in the contemporary literature. If mentioned, this was usually taken from the classical Roman authors or what appear to be folk treatments. However, not all tree care practice looked back to the past. In another early reference to an accepted modern tree care technique, Cook (1676, 48) describes bridge grafting to overcome ring-barking or girdling of the trunk, something that can truly be called tree surgery.

Concern about the loss of countless trees and woodlands across Britain was a constant theme of this age, regardless of the true extent of this. As with modern times, nothing focused attention on the fragility of the nation's tree cover more than the destruction caused by dramatic weather events. As noted earlier, the severe winter of 1683–1684 killed almost every Mediterranean cypress in England, which highlighted the risk in trying to establish species that were only marginally hardy in Britain. The Great Storm of 26–27 November 1703 was another such event. At the time, Evelyn wrote in his dairy:

> The effects of the hurricane and tempest, of wind, rain and lightening, through all the nation, especially London, were very dismal. Many houses demolished, and many people killed. As to my own losses, the subversion of woods and timber, both ornamental and valuable, through my whole estate … is almost tragical, not to be paralleled with anything happening in our age.
>
> Evelyn 1901, 2, 362

When amenity trees die or are damaged beyond repair the normal course of action is to fell them. In forestry operations the felling of trees is a routine activity in the production of timber and other wood products. However, in the literature of this era, whether it is concerned with trees for pleasure or profit, there is little information on the practical techniques used to fell trees. This is surprising given the level of detail on other aspects of practical arboriculture, such as branch removal. There is, however, an interesting passage from Cook that concerns testing for decay in the trunk of a tree to determine if it should be felled. He states:

> When trees are at their full growth, there are several signs of their decay, which give you warning to fell it before it be quite decayed: As in an Oak, when the top boughs begin to die, then it begins to decay: In an Elm or Ash, if their head dies, or you see they take wet at any great Knot, which you may know by the side of the tree being discolour'd below that place before it grows hollow, you may know by knocking it with the head of an axe, of which you may be the surer satisfied by boring into the middle of it with a small auger, or you see the Nighills make holes in it: these be certain Signs the Tree begins to decay, but before it decays much, down with it, and hinder not yourself.
>
> Cook 1676, 126

An aspect of tree surveys and inspections that Cook spends much time discussing is the measurement of trees (Cook 1676, 126–134). His grasp of mathematics really comes to the fore over several pages of text where he gives various methods for measuring the height of trees and calculating their timber volume.

Tools and equipment

Many of the tools and items of equipment used in practical arboriculture during this era were already in use in Tudor and early Stuart times. This is confirmed by the descriptions and illustrations of these that appear in gardening books from this era.

In Leonard Meager's popular book *The English Gardner* there is an illustration of a range of tools used for pruning and grafting (Meager 1670, pl. 1). This includes the pruning knife, the primary tool for both pruning and grafting, along with different sizes of saws, chisels and files. These 'instruments' are little changed from those illustrated by Mascall in 1569. In the draft of Evelyn's *Elysium Britannicum* there is a chapter entitled 'Of the Instruments belonging to a Gardiner, and their various uses', which includes two full-page illustrations that depict a wide range of tools and equipment for gardening (Evelyn 2001, 83–92). Many of these would have been used in arboricultural work, especially pruning. They include 'a paire of Reachers … which are sheares fixed upon a staff of 10, 15, or 20 foote long [operated with string and pulley] … and wonderfully usefull for the clipping of hedges, arbours, trees and tall palisades, without the trouble of a ladder' (*ibid.*, 87). Also for working at height but closer to the ground there is 'a pruning hooke (set on a handle 4 foote in length) made somewhat like a Battell-axe with edges to cut both ways, upwards thrusting, and downeward pulling, especially such branches as grow from the main stem, and that without the use of the Ladder which many tymes prejudices the boughs' (*ibid*,. 88). This appears to be similar to the 'Scythe fix'd into a Pole' that Switzer later described for 'rough cutting' the sides of walks. Although these tools could be used to prune trees at height while remaining on the ground, ladders were still much used for high pruning work. Evelyn describes 'Ladders of severall sizes for length and strength, one made with a stage to plash, bind and doe high-worke upon, also to serve for the gathering of fruite'. With regard to the 'one made with a stage', the illustration of this given by Evelyn (*ibid.*, 91) shows it to be the same as that illustrated by Lawson in 1618 and Parkinson in 1629. Evelyn believed this equipment was also appropriate 'to the pruning of the taller trees', although this does not seem very practical.

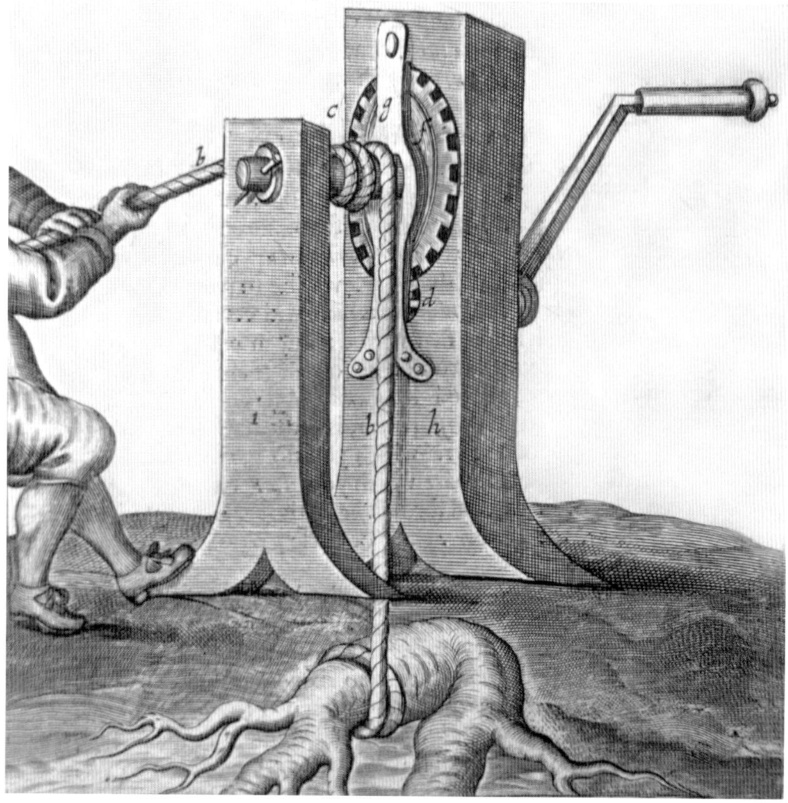

FIGURE 34. The German-devil, illustrated in the third edition of John Evelyn's *Sylva* (1679). Evelyn believed this 'might effect wonders, not only for the extirpation of Roots, but the prostrating of huge Trees'.

With regard to working at height in trees, Evelyn's *Elysium Britannicum* does not mention any tree climbing or the use of ropes. It also has no mention of large saws and axes for the heavier tree work such as felling. In his *Directions for the Gardiner*, which also has a list of tools and equipment, the situation is the same. However, this does have two interesting additions to the *Elysium* list with regard to watering on a large scale, 'A Water-barrow and Tub' and 'A Tree-pump' (Evelyn 1932, 105). The latter machine might have been similar to the equipment used by Henry Wise in the royal parks in 1702 (Green 1956, 55). In the third edition of *Sylva* Evelyn describes and illustrates an apparatus called the German-devil (Evelyn 1679, 23–24). According to Evelyn this 'might effect wonders, not only for the extirpation of Roots, but the prostrating of huge Trees' (Figure 34). With regard to other machinery used in heavy tree work, apart from tree moving machines, there is a dearth of information.

Most authors stressed the need to keep all tools and equipment in good condition so that they operated effectively and lasted a long time. Pruning tools are most often mentioned with plenty of good advice about sharpening and cleaning. Evelyn instructs his gardener accordingly: 'The Tooles are to be carried into the Toole-house, and all other instruments set in their places, every night when you leave work: And in wett weather you are to cleanse, sharpen, and repair them' (Evelyn 1932, 99). In some cases, the tools and equipment would belong to the gardener and not the owner. The gardener would supply them as part of his terms of employment and be paid an additional sum to cover their use and replacement. Tools and equipment would often be purchased, as in earlier times, from local sources such as the blacksmith or carpenter. However, with the steady increase in commercial nurseries that were supplying gardens with seeds and plants, many now stocked a wide range of tools and equipment for the gardener and arborist.

Nurseries and tree production

As orchards and fruit production became increasingly separated from the ornamental gardens, so this work came to be regarded more as the province of commercial horticulture, as it is today. As commercial horticulture is beyond the scope of this book, our coverage here focuses on nurseries and tree production in terms of ornamental and forest trees rather than fruit trees. However, since the principles of propagation are common to most trees, and the nurseries of this era often produced trees of all types, there is inevitably some overlap.

The dramatic increase in tree planting that came with the popularity of the Baroque garden prompted a similar increase in tree nurseries and tree production to satisfy that demand. This in turn stimulated some important advances in propagation techniques, especially in the area of grafting and cuttings. Many tree nurseries were located on landed estates and mainly met the demand of their own gardens and grounds. There were also commercial nurseries aiming to supply many different customers and frequently offering a wide range of species to meet the growing demand for more exotic trees. Evelyn gave momentum to this after seeing the range of tree planting at Charles II's gardens at Theobalds House in Hertfordshire. Prompted by that visit he wrote in *Sylva*: 'Methinks it should rather encourage our Countrymen to add yet to their Plantations other Forreign and useful Trees, and not in the least deter them' (Evelyn 1664, 58).

Tree and shrub propagation

Most late seventeenth-century books on plant propagation describe grafting and budding (inoculation) techniques in sufficient detail that they could be faithfully executed today with reasonable expectation of success. However, misinformation continued to abound on the issue of the genetic limits of compatibility, something that went back to Roman times and before. As with other aspects of horticultural and arboricultural practice, there was widespread belief that astrology played an important role. Ralph Austen was one of the first to attack the common belief that the phases of the moon were significant in deciding when to graft: 'Men (we see by Experience) graft in all seasons of the Moon, and find no such difference in the bearing of the trees' (Austen 1653, 90). Most contemporary books dealing with the vegetative propagation of trees and shrubs concentrated on grafting. Not only was this a common method of propagation, it was also something frequently undertaken by garden owners either on their own or with their gardeners. For example, Sir Courtenay Pole at Shute in Devon recorded working with his gardener on two days in April 1658 when together they grafted about 170 trees of mainly apple varieties (Gray 1994, 35).

Some of the most important advances in tree propagation during this era were made by Georg Andreas Agricola (1672–1738), a German physician and botanist, who experimented with plant cuttings and grafting. His book on propagation enjoyed an enthusiastic reception in the horticultural world and was promptly translated into Dutch, French and English. It was recognised as the definitive work on the propagation of woody plants for many decades after. The English translation of the book, entitled *A Philosophical Treatise of Husbandry and Gardening*, was edited by Richard Bradley and published in 1721 (Agricola 1721). Of particular interest was Agricola's special method of preparing grafting wax, or 'mummy' as it was often called (*ibid.*, 159). Commercial nurseries in Britain and on the Continent were much influenced by Agricola and followed his advice on propagation for many years to come (Figure 35). A huge advance in horticultural science came in 1716 when Thomas Fairchild conducted the first successful experiment in plant breeding, although this was not widely applied to woody plants until much later. By the simple act of taking pollen from a carnation and inserting it into a Sweet William, Fairchild produced an entirely new variety that became known as 'Fairchild's Mule' (Leapman 2000, 163).

Many tree species continued to be propagated from seed rather than by vegetative reproduction. Cook (1676, 6) was one of the first authors to stress the need to ensure seeds were collected from healthy trees with good form. Evelyn advised to 'sow all stony, and hard kernels and seeds' in September and recommended that acorns were 'laid in sand', which was common practice, so that 'the rigour of the Winter, may least prejudice them' (Evelyn 1664, 5).

Tree and shrub nurseries

Many treatises on gardening published in the second half of the seventeenth century advised garden owners to set up their own nurseries to supply their trees. For many landowners, buying from commercial nurseries either in Britain or overseas was generally not an option until the late seventeenth century, especially for those located in more remote districts. The English translation of Le Gendre's classic book specifically advises British landowners engaged in extensive planting 'to have Nurseries of their own, in which they may bring up trees wherewith to plant in their Gardens: it being certain, that Trees taken from such domestick Nurseries, thrive much better then others procured from abroad' (Le Gendre 1660, 1–2). Trees from overseas were raised in 'foreign' soils; they often suffered in the long journey and many subsequently died.

FIGURE 35. This illustration from Agricola's *Philosophical Treatise of Husbandry and Gardening* (1721) shows gardeners taking root cuttings in a tree nursery. Many nurseries in Britain and on the Continent were much influenced by Agricola's pioneering work.

Thomas Langford also argued for British landowners to raise their own trees. He declared, 'The trouble and expense of buying young plants and getting them home (many times from places very remote) and the prejudice they often receive in the carriage, will be wholly prevented' (Langford 1681, 2). The landowner would also be sure of getting the right species and varieties. Gentlemen with smaller gardens and estates who would be planting on a smaller scale would aim to source young ornamental trees and shrubs from their neighbours, fellow gardeners and any commercial growers they could find. Where saplings of native forest trees were required, many would obtain a supply from those growing naturally in local woodlands.

When landowners raised their trees from seed, the area of ground where the seed was sown and germinated was called the 'seminary' (Evelyn 1664, 6). The young trees were then lifted and grown on in another nearby plot of ground called the 'nursery'. In modern terminology these areas are all part of the nursery. Sometimes the terms plumps or clumps were adopted for tree nurseries that were located in open parkland (Jacques 2017, 218).

Commercial nurseries had long been used as a source of trees and shrubs by the more prestigious gardens, especially when this involved obtaining trees in large sizes and in great numbers. However, when Charles II laid out the new avenues at Hampton Court as a wedding gift for Catherine of Braganza it was simply not possible to acquire a sufficient number of similar sized large trees from British sources and they had to be imported from Dutch nurseries (Marsh 2004, 394). Nevertheless, with the ever-increasing demand for trees, entrepreneurs with an interest in plants and gardens began to realise the opportunities in establishing their own commercial nurseries. By 1690 there were at least five commercial nurseries in London and just a few located in other regions of the country. From 1700 onwards the growth of the nursery trade across Britain indicates that many landowners did not raise sufficient stock for their needs and it was now preferable to buy in stock where possible. Garden writers such as Switzer (1718, 1, 201) were arguing that there was little sense for someone within 40 miles (64 km) of London or a good regional nursery to raise their own trees.

The popularity of the Dutch style of garden had given a tremendous boost to topiary and the use of 'evergreens' in Britain. Previously, suitable evergreens, either as developed topiary or plants ready for training, would usually have been imported from Dutch nurseries, renowned for their skill in the art. Aware that this business would otherwise go overseas, a number of British nurseries, many of them in London, were encouraged to produce saleable specimens. Some of these London nurseries were located in Hoxton, in the East End of the capital, and had been established for many years. The largest London nursery in most of Charles II's reign was owned by Captain Leonard Gurle and records show he supplied trees to much of the nobility (Harvey 1974, 45). In some respects Gurle was a pioneer in the way he marketed his business, being one of the first to provide not only a catalogue but also a guarantee that he would replace any plant that died. Also in Hoxton was William Darby, regarded as an outstanding nurseryman at the end of the seventeenth century. While he only had a small nursery, he was renowned for stocking 'curious greens' and propagating variegated hollies by budding them. He is also remembered as one of the first to specialise in exotic trees, including among his stock large numbers of tulip trees (*Liriodendron tulipifera*). At this time, Captain Foster of Lambeth was said to have the finest striped holly in England, which he raised from grafts. Besides Darby and Foster, a nurseryman named Wrench in Fulham was famed for variegated hollies and he gave rewards to people who spotted remarkable kinds (Hadfield 1979, 147). Yet another Hoxton nurseryman was Thomas Fairchild, famed for his book *The City Gardener*.

In Leonard Meager's (1670) *The English Gardner* he gives a catalogue of ornamental and fruit trees then available from nurseries in the London area. The main sections in Meager's catalogue are those that are still largely in use today. There are also some categories for trees that reflect the specific uses of the time, for example, 'Trees fit to plant by Out-Walk sides'. Around 1700 Robert

Furber founded his nursery at Kensington, London, having acquired some of the exotic plants introduced by Bishop Compton to Fulham Palace. In 1727, Furber published *A Catalogue of English and Foreign Trees: Which will Thrive in the Natural Ground in England* (Furber 1727). At the end of this catalogue is a section entitled 'A Collection of English and Foreign Trees mentioned in my Catalogue, which are placed in the Plan as their natural Growth and Size require'. There was also another section entitled 'Sorts of Exotics that require a warm situation, with some Care, in the Winter, to preserve them whilst Young'. All this was intended to give good guidance on tree selection for various situations, and for a nursery catalogue this was quite advanced for its time. It seems likely that for professional gardeners and arborists these early catalogues, and the advice of the nurserymen themselves, were their main source of information for tree selection. Books were scarce and expensive and more for the landed aristocracy.

While there are good records for London nurseries at this time that sold ornamental trees and shrubs, much less is known about these nurseries elsewhere in Britain. With regard to the north of England, there were two great nurseries at York and Pontefract that gave a major boost to estate and garden planting in Yorkshire (Harvey 1972, 32). The former belonged to the Telford family of York, where John Telford was said to be 'one of the first that brought our northern gentry into the method of planting and raising all kinds of forest trees, for use and ornament' (Drake 1736, 275). The latter was owned by the Perfect family and this achieved prominence due partly to its location close to the Old Great North Road and through the supply of trees to John Aislabie at Studley Royal (Harvey 1974, 66). However, many parts of the north were not so well served and it was not until a generation later that customers in Tyneside, for example, were able to buy trees locally instead of importing them by ship from London. While nurserymen were operating in Edinburgh shortly after the Restoration, little information exists on the nature and extent of their business and substantial quantities of ornamental trees and shrubs were often sourced from London (Cox 1935, 67).

As with the early development of many industries, the growth of commercial nurseries did have some 'teething problems'. Common complaints from customers related to the quality of the nursery stock, especially that which was purchased at markets and less reputable nurseries, and also that those selling the trees and shrubs had little or no knowledge of gardening. Thomas Fairchild, a nurseryman himself, put this very graphically and also proposed a solution. He declared:

> Another Disappointment these Lovers of Gardens meet with, is from the Markets; for most of those People who sell the Trees and Plants in Stocks and other Markets, are Fruiterers, who understand no more of Gardening than a Gardener does of making up the Compound Medicines of an Apothecary. They often tell us the plants will prosper, when there is no Reason or Hopes of their growing at all; for I and others have seen plants that were to be sold in the Markets, that were as uncertain

> of Growth as a Piece of Noah's Ark would be; but when such Plants are bought at the Gardens where they were raised, there can be no Deceit, without the Gardener who sold them loses his Character.
>
> Fairchild 1722, 68

From the perspective of the nurserymen, they had a range of problems to contend with and one of the most significant concerned logistics. Getting trees and shrubs to their customers quickly and in good condition was not easy, especially to estates and gardens in more remote districts that were often a considerable distance from the nursery. Because many roads were in poor condition, use was made of water transport where possible as this was generally cheaper as well as quicker (Willson 1982, 5). Nevertheless, packing must have been of a high standard and the cost considerable. Plants travelled packed in straw and mats, sometimes with moss round the roots or in wicker baskets. The other major problem for nurserymen was financial. Many of their customers, especially the monarchy but also many of the nobility, were notoriously late in making payment and in some cases failed to make any payment at all.

In many respects, those reputable nurserymen who specialised in the supply of trees were the tree experts of the latter part of this era. Many of them not only sold a wide range of trees, with much of this stock already formatively trained and shaped, but they also gave advice on planting, maintenance and management and in some cases undertook these activities themselves on a contract basis. The most outstanding example of this combination of nurserymen and contractor was the Brompton Park Nursery.

The tree experts of Brompton Park Nursery

The Brompton Park Nursery was founded as a commercial business in 1681 by four head gardeners on 100 acres (40 ha) of land at Brompton, a hamlet in Kensington, London. The partners and their former positions were George London (1640–1714), gardener to the Henry Compton, Bishop of London, at Fulham Palace; Moses Cook (*c.* 1640–1714), gardener to the Earl of Essex at Cassiobury; Roger Looker (?–1695), gardener to Queen Catherine of Braganza at Somerset House; and John Field (?–1687), gardener to the Earl of Bedford at Woburn Abbey.

Within just a few years the business was thriving, not only selling nursery stock but also offering to design, create and maintain gardens for their clients. As early as 1682, the Brompton Park partnership was commissioned to plan and create a prestigiously expensive English Versailles for Viscount Weymouth at Longleat in Wiltshire (Mowl 2000, 50). This project would earn them £30,000 and set up the business very nicely. However, the partnership itself was set to undergo some major changes. Looker, who may have been instrumental in forming the partnership, died in 1685 and then Field passed away in 1687. While both these men had played important roles in establishing the business, the next change was especially significant when Cook retired in 1689. With Cook's renowned skills in propagation it is likely that the nursery depended on him in its early years to establish a wide range of high-quality stock. As well as his experience of nursery practices, Cook was supremely skilled at laying out major gardens and landscapes involving extensive tree planting. As the nursery steadily expanded into a garden design and contracting business, Cook's expertise was invaluable. For example, the Earl of Exeter sought the opinion of 'Mr Cooke and Mr London about ye Gardens' at Burghley in 1683 (Till 1991, 136). In 1684, Cook and London were undertakers for the Office of Works in forming and planting the King's Avenue (today's Long Walk) linking Windsor Castle to the Great Park, when 2½ miles (4 km) of the southern section was planted with elms (Jacques 2017, 125). When Cook retired he sold his share in the partnership to Henry Wise (1653–1738), then London's apprentice.

When James II was deposed in 1688, the fortunes of the business were improved dramatically by an accident of history. At the outbreak of the Glorious Revolution it was Bishop Compton who secretly spirited Princess Anne out of London and away from her Catholic father and stepmother; and the bishop's chief agent in the melodramatic escape was his gardener George London (Harvey 1972, 26). Consequently, the Brompton Park Nursery was doubly recommended to the newly crowned William III who was now keen to promote the Dutch style of gardening in Britain, something that London and his nursery were very able to satisfy. This also avoided the necessity of sourcing plants and expertise from the Netherlands. From the outset, the partners had always intended to take away from the Dutch as much of the trade in evergreens as they could. Not only did their royal connections help to achieve this, their efforts were also supported by the high standards of their nursery stock. A major consideration of gardens purchasing plants to follow the Dutch style was the condition of the 'greens', and those supplied by Brompton Park were always reliably 'bright'. A significant figure at this stage in the business was Willem Bentinck, a Dutch intimate of William III, who was appointed Superintendent of the Royal Gardens. London, fortunate in his loyalty to Compton and the Glorious Revolution, was appointed under him as Master-Gardener and Deputy Superintendent, and in a position to ensure Brompton Park secured some important work for the royal gardens. Although London lost his post as royal gardener when Anne came to the throne in 1702, Wise was appointed to replace him, so the firm's prestige and patronage did not suffer. While Anne did, however, decide to cut back on the overall cost of the royal gardens, the partnership was still engaged on a number of major projects.

From 1689, the Brompton Park business was carried on by London and Wise and the remarkable partnership of these two men came to dominate the garden scene in Britain until London's death in 1714. London was the main creative force, riding around the country from one prestigious commission to the next. Switzer says, 'It will perhaps be hard to believe, in Time to come, that this one Person actually saw and gave Directions, once or twice a year, in most of the

Noblemens and Gentlemens Gardens in England, … it was common for him to ride 50 or 60 Miles in a Day' (Switzer 1718, 1, 81). Wise was concerned with financial matters, planning the projects and running the nursery with its staff of around twenty men and a few women.

As London was being asked for advice, he found himself being requested to devise designs for many garden projects. This now developed into a significant part of their business, as confirmed by Evelyn in his advertisement to London and Wise's edition of De la Quintinie's *The Compleat Gard'ner*. Evelyn states, 'They have a numerous Collection of the best Designs, and I perceive are able of themselves to Draw, and contrive other, applicable to the places' (De la Quintinie 1699, Advertisement). London and Wise were very happy to take commissions throughout the country from owners wanting gardens reflecting French and Dutch designs. This was the 'age of the tree' in garden design and the Brompton nursery had a massive amount of stock to satisfy the demand. It always had ready for sale huge quantities of young limes, elms, hornbeam and other favourites, at various stages of growth, and every possible variety of fruit tree. Their designs also called for vast quantities of clipped trees and shrubs and the nursery really excelled at this with many thousands of young plants, carefully prepared and trimmed, lined up in row after row of different sizes and species, ready for dispatch. The second edition of Langford's book *Fruit-Trees* includes a catalogue not only 'of Choice Fruits' but 'also of Greens and Blossoming Shrubs: To be had at Brompton Park, near Kensington' (Langford 1696, 210–220). Some trees in the Brompton Park Nursery were grown on to a considerable size as these 'instant trees' were often required in some of the most prodigious projects. The prowess of London and Wise in the specialist skill of moving large trees has already been noted.

As a young man London had been able to study at first hand some of the most prestigious French gardens when, as the apprentice of John Rose, he had been sent to France to further his horticultural studies. Some decades later, in 1698, London was chosen to accompany Willem Bentinck, now the first Earl of Portland, on another tour of gardens in France, which meant that he was out of the country for six months. During this time he met De la Quintinie at Versailles and discussed design with Le Nôtre. These experiences undoubtedly fired London's passion for the Baroque style of grand gardening which was so eloquently expressed by the Brompton partnership in Britain. A good account of the many garden projects undertaken by London and Wise is given in David Green's excellent biography of Henry Wise (Green 1956). Some of these have already been mentioned and only a few additional highlights can be given here.

On London's return from France one of the first clients to allow him to express his renewed passion was the politician Thomas Coke of Melbourne Hall, Derbyshire. In October 1699, Coke drafted a design and sent it to Wise, who then forwarded it to London (Green 1956, 43). The design was approved and the following month a huge consignment of trees, shrubs and other plants were supplied by the Brompton Park nursery and transported, mainly by water, to

Melbourne. Included were 1,000 small Dutch elms, 600 large elms and 2,000 hornbeams in addition to vast numbers of shrubs and bulbs. The hornbeams were to hedge straight wilderness walks; the elms and limes filled in the resulting geometrical shapes between the walks. After a pause in the development of the gardens, work continued in 1704 when Brompton Park sent Coke two alternative designs and he chose the one 'to suit with Versailles' (Mowl 2000, 51). This now survives as one of the best examples of the classic Baroque style of garden in Britain. One of their greatest projects was at Chatsworth, the seat of the Duke of Devonshire, where their involvement began in 1688 and continued until at least 1704. Huge quantities of trees and other plants were supplied and planted there, and the illustration by Knyff and Kip (1707) in *Britannia Illustrata* gives an indication of the staggering extent of this (Figure 36). In 1705, London and Wise were the obvious choice to design the gardens of Blenheim Palace, Oxfordshire, for the Duke of Marlborough. No other gardeners in Britain at that time were equipped to manage such a huge project from design to implementation (O'Halloran 2013, 169). Remarkably, it was Wise who took on the works without any involvement from London, presumably because he was committed to other projects.

FIGURE 36. This view of Chatsworth is from Knyff and Kip's *Britannia Illustrata* (1707). It shows the vast quantities of trees that were supplied and planted there by London and Wise to create this extraordinary garden in the classic Baroque style.

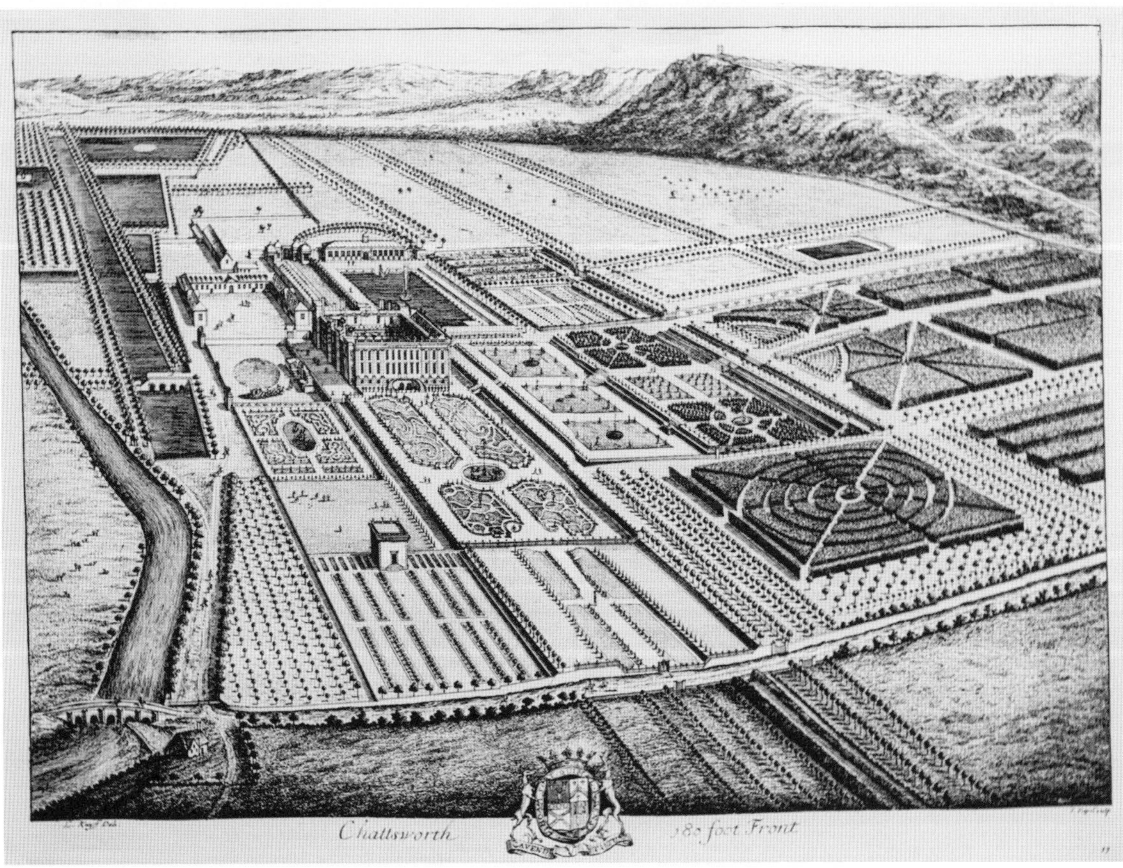

While almost none of the gardens created by London and Wise still exist, we are fortunate in having a remarkable record of many of their achievements in the stunning illustrations of Knyff and Kip. Not only was the Brompton Park partnership directly involved in designing and creating many of those landscapes, it was also responsible for supplying the nursery stock to many more. It is interesting to note that these views show a remarkable diversity of designs that do not always conform to our modern conception of what constitutes the formal French or Dutch style. Among the nearly seventy views in *Britannia Illustrata* there are many that do not have parterres or an axial layout. However, there is barely a single garden that does not have vast swaths of tree planting or extensive new avenues stretching out towards the horizon. These landscapes not only symbolised the landowner's power and possessions, they were also a significant investment for the future that would benefit his heirs when the trees had matured and were ready to harvest.

The Brompton Park Nursery's unrivalled position as a supplier of nursery stock was reflected in the prices they were able to charge. George Clarke (1985, 75), in his study of the planting at Stowe, shows that London and Wise could command up to thirty times the price that local sources could ask, even for common trees. In 1715, Switzer estimated the value of plants at Brompton Park Nursery at between £30,000 and £40,000, 'perhaps as much as all the nurseries of France put together' (Switzer 1715, 78). Roderick Floud (2019, 13) has calculated this value as between £56 million and £75 million in 2019. Brompton Park's commanding position in the nursery trade only started to be challenged in the 1720s with the rise of provincial nurseries that could begin to provide some serious competition.

By the 1720s, the dominant position of London and Wise in the area of landscape design was also being challenged by new attitudes and tastes. Influential commentators, such as Stephen Switzer and Batty Langley, were calling for a move away from what they viewed as the excesses of the formal gardens and were pioneering the development of a more naturalistic approach. In those attacks on the old style, the work of London and Wise was an obvious target. Horace Walpole's censure of the Brompton Park approach to garden design is well known: 'Improvements [in formal gardening] had gone on till London and Wise had stocked our gardens with giants, animals, monsters, coats-of-arms and mottoes in yew, box and holly. Absurdities could go no further, and the tide turned' (Walpole 1995, 41). Langley was even more pointed in his criticism of London and Wise, suggesting that the 'thick Planting' of their garden designs may have been the result of the desire for 'greater Sales of their Plants' (Langley 1728a, xi). One important feature of the many gardens laid out by Brompton Park partners was their cost. Not only were the designs very costly to realise, especially in terms of the numbers and types of trees and shrubs required, they were also very expensive to maintain, being extremely labour-intensive. This was a major factor why almost all were eventually destroyed, most of them within fifty years of their creation.

Although there is little of the legacy of the Brompton Park partnership in terms of surviving gardens, it had a lasting impact in many respects on professional arboriculture and horticulture in Britain. The Brompton business was a training ground for several individuals who would go on to make a major contribution of their own. Leonard Meager was the first foreman at Brompton and he was succeeded by Stephen Switzer, who went on to establish his own great reputation, having been taken on as an apprentice at Brompton around 1698 (Green 1956, 31). Later, probably before 1709, London and Wise were joined at Brompton Park by Charles Bridgeman, who was to become an eminent garden designer in the 1720s and 1730s. Among the lesser recruits of the Brompton enterprise were Joseph Carpenter and William Smith, who would eventually try to run the nursery themselves. By the late seventeenth century the reputation of Brompton Park meant that, regardless of the expense of living in London, an apprenticeship there was highly sought after. Those young men received the highest standard of training and when they had completed their apprenticeship they were much in demand. By the eighteenth century London and Wise were supplying gardeners to employers throughout the country (O'Halloran 2013, 35). Other employees were engaged on their many different contracts around the country, with several of them acting as the foreman responsible for running those contracts. London and Wise also passed on their knowledge as popular writers. As well as the translation of French works already mentioned, they added notes to these from their own vast experience.

George London died in 1714 and within a year Henry Wise was disposing of Brompton Park Nurseries and the great business soon went downhill. Wise retired in 1727 to Warwick Priory and died a very wealthy man in 1738. He insisted on a simple funeral and the money saved was given to the poor. On the site of the Brompton Park Nursery now stands the South Kensington Museums complex.

Status of gardeners and tree-related professionals

We have noted the labour-intensive nature of the formal Baroque gardens, and for the most prestigious this necessitated a huge complement of gardeners. For example, in the 1690s, the gardens at Hampton Court employed eighty men and women and at Kensington there were seventy-two, not least because of the intricate wilderness and parterre work (Jacques 2017, 173). Much of the work involved the creation and maintenance of various tree-related features, something confirmed by the images of Knyff and Kip which vividly highlight the prominence of trees in these landscapes. While much of this would now be called arboricultural work, that term was not used in this era and it came under the umbrella of horticultural and landscape work. Nevertheless, many of the gardeners employed in this era would have required a range of arboricultural skills. How and where would they acquire those skills?

Crucial to the professionalisation of gardeners during this era was the importance of apprenticeships, as these were the only source of formal education available to them. There is, however, little record of these apprentices outside of London where they were not controlled by the Worshipful Company of Gardeners, or legally obliged to the Statute of Labourers and Apprentices (O'Halloran 2013, 16–21). Apprentices were trained by a master, who had to be a householder, at least twenty-four years of age, and exercise 'an art mystery or manual occupation'. This meant that a father who was a professional gardener could train his son; while this occasionally happened on country estates, most would move to London and be trained by a non-relative. The apprentice had to be over ten years and under eighteen years of age, and he gradually acquired skills over a period of several years until he was at least twenty-four. Once selected, the chosen master, in exchange for a sum of money, agreed to instruct the apprentice for a set term of years and this was recorded in a legal document known as the apprenticeship indenture. Seven years was the average length of service for apprentices in London, but without legal standing this varied elsewhere.

Knowledge was transmitted through observation, imitation and practice, and books were very rarely used. The early years of apprenticeships were devoted to acquiring practical gardening skills. However, the extent of what was taught depended on the level of knowledge of the master and his ability to convey this clearly. Many gardeners, once trained, were often resistant to new ideas and practices, perhaps because they were not encouraged in this direction by their employers. As Agricola put it, 'the old beaten Road they have been bred up in stands them in lieu of reason' (Agricola 1721, Preface of French Translator). The learning of skills such as the supervision of others, the organisation of planting schemes and overall management of the garden usually began at the 'journeyman' stage when the apprentice took up a paid position, which came with added responsibilities. Due to the high demand for good gardeners after the Restoration a well-trained journeyman would have little difficulty in finding employment. Those with good tree-related skills would certainly be in demand on estates with extensively treed landscapes. Because owners found themselves in competition to obtain and keep them, they could be tempted to poach talent, and a select number of highly skilled journeymen would have found themselves moving frequently.

The most highly skilled and experienced gardeners obtained positions on the wealthiest estates or royal gardens where the owners took a keen interest in gardening. In order to stay at the 'cutting edge' of horticultural practice, some gardeners were given the opportunity to travel overseas for brief periods and not having this overseas experience limited a gardener's opportunities and perspectives (Switzer 1718, 1, x). Those apprentices who were able to train at locations where their superiors or employers had overseas experience were at a distinct advantage. Nevertheless, despite the excellent training afforded to those apprenticed at the most prestigious locations, some gardening writers called for alternative educational structures that would benefit more trainees. Agricola (1721, 60) was one of the first to call for the establishment of 'Academies for the Benefit of Gardeners', where they would be 'instructed in the Science of Gardening, and get a thorough Knowledge of the Fundamentals of their Profession'. Richard Bradley (1724, 129) agreed and recommended 'proper Academies erected for the better Instruction and Information of the Persons who intend to profess the Art'. However, these views did not bring about change and apprenticeships were often seen as a source of cheap labour and remained in place throughout this era (O'Halloran 2013, 42). Self-education through studying books was not a viable option for prospective gardeners. The cost of even the cheapest books on gardening was beyond the means of almost every apprentice and also many professional gardeners.

As with any trade throughout history, particularly when there is a shortage of skilled workers, many who offered their services as gardeners had neither the skill nor the experience required. Many were itinerant workers who had left the north of England and Scotland to seek employment in the wealthier gardens and estates further south. Regardless of their actual ability, they were viewed with much suspicion by respected gardeners and garden writers. Switzer declared:

> There are likewise several *Northern* lads, which whether they have served any time in this Art, or not, very few of us know anything of; yet by the help of a little Learning, and a great deal of Impudence, they invade these *Southern* Provinces and … pretend to know more in one twelve-month than a laborious, honest *South* Countryman does in seven years.
>
> Switzer 1718, I, xxiv

Langley was similarly scathing:

> There is a certain Kind of People in England, who having been employ'd in the labouring Part of Gardening, perhaps but for one Month only, will then put on Aprons and call themselves Gardeners, tho' at the same time, they in fact are no more than Wheel-barrow Men, and very often hardly worthy of that Office.
>
> Langley 1729, ix

In an age of gardening that often required a high degree of skill, avoiding the cost of this could turn out to be an expensive mistake. Langley (*ibid.*, xi) cites a prominent Middlesex garden where unskilled labourers who did not recognise the difference between flower buds and leaf buds were employed to prune fruit trees, with the result that the owner had scarcely any fruit.

One area of gardening and tree work that was now expanding was in the towns and cities. Not only were there a growing number of small gardens for residential properties but also more trees were being planted in churchyards and residential squares, as well as in the newly created public walks. In time these would all require regular maintenance. Rather than seeking out gardeners with experience of working in the urban realm, residential property owners especially seem to have been bedevilled by 'jobbing' gardeners calling at their houses who had little practical experience of any kind (Fairchild 1722, 66–67).

On the big estates the head gardener was probably the only member of the gardening team on the permanent payroll. The rest of the gardening staff were likely to have been casual labour, employed for specific tasks at certain times of the year, especially in the spring when the workload increased considerably. For example, with grafting, the work was usually done by the permanent gardener but with casual workers often sealing the graft with clay. Casual labour was also likely to have been the only opportunity for women to gain employment in gardening work. It is hard to be precise about this because the names of gardeners were not always recorded in household and estate accounts. However, no mention could be found of a woman in the position of head or permanent gardener. As in the past, the casual work for women would have involved menial tasks such as weeding.

The wages and working conditions for gardeners varied greatly depending on the circumstances of their employment (O'Halloran 2013, 190–230). The main gardener received an annual salary similar to the other servants and domestics, typically paid in instalments throughout the year. In return for their salary, they signed contracts for one year for 'keeping the gardens', the written contract detailing the maintenance and other work required. The gardener was housed in the grounds of the estate and accommodation became part of the contract negotiations by the late seventeenth century. Many gardeners in Scotland at this time were very badly paid, which would have contributed to their desire to migrate south (Cox 1935, 72).

By the latter half of the seventeenth century the interest in gardens and gardening among the landed class was reflected in greater recognition of the gardening profession and a rise in the status of individual gardeners within the social hierarchy of the estate where they worked. As owners looked more to gardeners for advice, they sometimes developed a personal relationship, which was not common or necessary with other servants. Leonard Meager (1670, Dedication) describes his employer as more like an 'indulgent father than a Master' because he was so supportive of him. The most remarkable example of an owner/gardener relationship was that between the Earl of Essex and Moses Cook. Not only were the two men genuine friends, despite the gulf in their social status, but Essex was a keen and willing assistant to Cook. This is confirmed by Cook in his book's dedication to Essex when he reminds his employer 'that you have not been onely a Spectator but an Actor in most of what is treated in the ensuing lines' (Cook 1676, Epistle). The rising status of the gardener is evident from contemporary books, with many writers giving examples of outstanding professional gardeners of the time. Some more prominent gardeners were also given the prefix 'Mr' as a mark of their professional status. In Switzer's *The Practical Husbandman and Planter* there is a list of 290 subscribers, which includes thirty-five professional gardeners all listed as 'Mr' (Switzer 1733, Subscribers). As the status of gardeners improved, some began to show a more independent spirit towards their employers. Writing in the reign of Charles II, Lady Reresby of Thrybergh in Yorkshire complained to her husband that while he was in London she 'cannot get the Gardener to worck as ye ordered' (Cliffe 1999, 56). She was then obliged to seek the assistance of a kinsman who sent his own gardener to prune the trees and vines.

This era saw the rise of the gardener acting as a contractor. As head gardeners have never made their fortunes with this work, one path to financial improvement was to move into trade. The size of companies selected by garden owners and their stewards would depend on the scale of the project and their particular requirements. Large garden construction projects might involve an owner in much tedious administration, the use of itinerant labourers and numerous suppliers, and were often beyond the capability of his own men and steward. The alternative was to turn to 'undertakers' (contractors), to deliver a specified piece of work for an agreed cost and relieve the owner of much of the trouble. After the Restoration the use of undertakers became increasingly common and some companies, especially the Brompton Park Nursery, acquired numerous contracts all over the country. As well as construction projects, these contractors also delivered maintenance contracts. While most rural seats would have their own head gardener, the smaller gardens could not always justify a full-time gardener. The solution was a contract for upkeep entered into with a gardener or nurseryman who looked after several properties.

As well as implementing the designs of others as contractors, some notable professional gardeners got involved in producing their own designs. From the 1680s, the Brompton Park partnership moved into the field of landscape design in a major way. Encouraged by their success, other professional gardeners of repute also offered this service. Previously, the landed aristocracy would have consulted someone of their own social class with gardening expertise. Some garden owners opted to follow design templates set out in books, although this would prove a disappointment if they or their gardening staff did not have the necessary expertise. However, there were now professional gardeners with a proven track record in design prepared to offer advice for a fee, and many who would also be able to implement those schemes. Some had particular specialisms such as water gardens or parterres or, in the case of Brompton Park, they would design any landscape or garden feature, especially if it involved trees and shrubs. However, most garden designers were based in London, with very few working outside of the capital. Switzer described the normal design process rather sarcastically: 'The Method commonly taken in this Affair, is, Gentlemen have their Ground survey'd, and perhaps the Levels taken, and then 'tis brought to London, where there are a great many Drafts-men, and Paper Engineers, so a regular fine Scheme is made' (Switzer 1715, xii).

With the focus on trees in the formal gardens of this age, those gardeners with arboricultural skills were at a premium, although the term 'arboricultural' was not yet used. Nevertheless, the term 'arborist' was used by some writers to identify somebody who had specialist tree-related skills. William Lawson (1618) had referred to arborists in the early part of the seventeenth century and the term was used in this era by authors such as Evelyn. In his various editions of *Sylva* Evelyn refers to arborists frequently and this is always in the context of tree-related work. Evelyn also occasionally uses the Latin term *arborator*, which he defines as 'Pruner, or one that has care of the trees' (Evelyn 1679, Advertisement). It seems likely that Evelyn just copied the term *arborator* from the Roman writings of Pliny, without translating it as arborist or forester, but then felt obliged to define this. Interestingly, Cook does not use the term arborist or any similar name, despite being one of the foremost tree experts of his age. He is also reluctant to refer to himself as a gardener, largely because he viewed his work as concerned mainly with forest trees and outside the enclosed garden (Cook 1676, 98).

While many garden owners recognised the importance of tree-related skills in the gardens of the day, many did not and therefore would engage gardeners who had little knowledge or experience of working with trees and shrubs. Not recognising the value of retaining a specialist tree worker and replacing them with a succession of casual and inexperienced staff usually resulted in very poor quality work. Samuel Collins (1717, Introduction) found this a particular problem with pruning: 'Gentlemen are often … apt to imagine their Gardeners fit to be removed, when indeed they are not'. The result was that 'This year's Gardener prunes contrary to the last, and the next to both of them'. Another common failing among garden owners was not to appreciate the long-term nature of arboricultural work. It can often take several years for practical tree work to have the desired result and years of careful tree work can often be ruined by one thoughtless action. Writing from his experience with trees, Cook puts this very eloquently:

> A good cook can dress you several dishes of meat very well in half a day, and if one miscarry, they can in a little time make another; but the gardiner must have several months or years, to bring some things to perfection; and if he miscarry, he cannot begin again when he will, but he must wait his time with patience; therefore he ought to be more careful.
>
> Cook 1676, 62

Conclusions

The age of the formal garden was also an age of trees. The popularity of those grand Baroque gardens with their French and Dutch influences resulted in tree planting on a scale not previously seen in Britain. Not only did the basic framework of these gardens and their wider landscape rely heavily on trees but they also included many tree-related garden features. This all required a vast amount of arboricultural work in planting, establishing and maintaining these landscapes and much of that work, particularly in the more formal elements, involved quite intricate and skilled pruning. This in turn led to a substantial increase in the numbers of professional gardeners and those with tree-related skills were always much in demand. Consequently, in comparison to the previous era, professional arboriculture in Britain made great strides in a relatively short period of time.

The dramatic increase in the volume of arboricultural work also prompted some significant developments in arboricultural practice. The growing number of books on gardening and husbandry during this era often include a substantial amount of material on the planting and care of trees. While much of this focused on fruit trees and growing trees for timber, some of those basic practices were also applicable to amenity trees. Nevertheless, arboricultural practice was still heavily influenced by ancient Roman and Greek authors and by age-old superstition. There were also few developments in the range of tools and equipment used in tree work.

The huge increase in planting gave a major boost to the establishment of new commercial and estate nurseries specialising in tree and shrub production. However, many commercial nurseries were located in the London area, leading to supply problems to other parts of Britain, and it would be some time before there was better coverage across the country. Another major development was the establishment of some commercial companies to undertake landscaping work, usually on a contract basis, involving not only the design and creation of those gardens and landscapes but also their ongoing maintenance. In some instances, the supply of the trees and plants, along with the design, construction and maintenance of the landscape, would be undertaken by the same company. The most famous example of this was Brompton Park Nursery, which held an unrivalled position for some forty years.

This era also saw the emergence of some notable tree experts. These include Ralph Austen, John Evelyn, Moses Cook, George London, Henry Wise and Stephen Switzer. However, in some cases their reputations today do not reflect the real extent of their contribution to the development of arboriculture in Britain.

References

Addison, J. (1712) Article on Gardening. *The Spectator*, 414, 25 June, 98–102.

Agricola, G.A. (1721) *A Philosophical Treatise of Husbandry and Gardening. The whole Revised and Compared with the Original by Richard Bradley.* P. Vaillant: London.

Amherst, A. (1896) *A History of Gardening in England.* Second edition. Bernard Quaritch: London.

Aubrey, J. (1982) *Aubrey's Brief Lives.* Edited by Richard Barber. The Boydell Press: Woodbridge.

Austen, R. (1653) *A Treatise of Fruit-Trees.* Thomas Robinson: Oxford.

Austen, R. (1657) *A Treatise of Fruit-Trees.* Second edition. Thomas Robinson: Oxford.

Beale, J. (1657) *Herefordshire Orchards: A Pattern for all England.* Roger Daniel: London.

Blith, W. (1649) *The English Improver.* John Wright: London.

Blith, W. (1653) *The English Improver Improved.* Third impression, much augmented. John Wright: London.

Bonnefons, N. de (1658) *The French Gardiner.* Translated by John Evelyn. Thomas Basset: London.

Boyceau, J. (1638) *Traité du Jardinage, selon les raisons de la nature et de l'art.* Michel Vanlochom: Paris.
Bradley, R. (1718) *New Improvements in Planting and Gardening.* A. Bettesworth: London.
Bradley, R. (1724) *A General Treatise of Husbandry and Gardening.* Volume 1. J. Peele: London.
Bradley, R. (1731) *New Improvements in Planting and Gardening.* Sixth edition. A. Bettesworth: London.
Campbell-Culver, M. (2006) *A Passion for Trees: The Legacy of John Evelyn.* Eden Project Books: London.
Chambers, T. (2008) *The Arboriculture of West Country Parks and Gardens, 1660–1730.* PhD thesis, University of Bristol.
Chittenden, F.J. (1939) Henry Wise and the Royal Gardens in Queen Anne's time. *Journal of the Royal Horticultural Society* 64, 464–479.
Clarke, G. (1985) Where Did All the Trees Come From? An Analysis of Bridgeman's Planting at Stowe. *The Journal of Garden History* 5(1), 72–83.
Cliffe, J.T. (1999) *The World of the Country House in Seventeenth Century England.* Yale University Press: New Haven and London.
Collins, S. (1717) *Paradise Retriev'd: Plainly and Fully Demonstrating the Most Beautiful, Durable, and Beneficial Method of Managing and Improving Fruit-Trees.* John Collins: London.
Cook, M. (1676) *The Manner of Raising, Ordering, and Improving Forest-Trees.* Peter Parker at the Leg and Star: London.
Cotton, C. (1675) *The Planters Manual: Being Instructions for the Raising, Planting, and Cultivating all sorts of Fruit-Trees.* Henry Brome: London.
Couch, S.M. (1992) The Practice of Avenue Planting in the Seventeenth and Eighteenth Centuries. *Garden History* 20(2), 173–200.
Cox, E.H.M. (1935) *A History of Gardening in Scotland.* Chatto and Windus for Flora and Sylva: London.
Defoe, D. (1742) *A Tour through the island of Great Britain.* Third edition. Volumes 1–4. J. Osborn: London.
Defoe, D. (1962) *A Tour through the Island of Great Britain.* Volume 1. Everyman's Library, Dent: London.
De la Bédoyère, G. (ed.) (1995) *The Writings of John Evelyn.* The Boydell Press: Woodbridge.
De la Quintinie, J.-B. (1693) *The Compleat Gard'ner, or, Directions for cultivating and right ordering of fruit-gardens and kitchen-gardens.* Made English by John Evelyn. M. Gillyflower: London.
De la Quintinie, J.-B. (1699) *The Compleat Gard'ner by Monsieur de la Quintinye. Now compendiously abridg'd, and made of more use, with very considerable improvements. By George London and Henry Wise.* M. Gillyflower: London.
Desmond, R. (1977) *Dictionary of British and Irish Botanists and Horticulturists.* Taylor and Francis: London.
Dezallier d'Argenville, A.-J. (1709) *La Théorie et la Pratique du Jardinage.* Jean Mariette: Paris.
Drake, F. (1736) *Eboracum: Or the History and Antiquities of the City of York.* In Two Books, Book 1. William Bowyer: London.
Drope, F. (1672) *A Short and Sure Guid[e] in the Practice of Raising and Ordering Fruit Trees.* Printed for Ric. Davis: Oxford.

Evelyn, J. (1661) *Fumifugium: Or the Inconvenience of the Aer and Smoake of London Dissipated*. Gabriel Bedel and Thomas Collins: London.

Evelyn, J. (1664) *Sylva; or, A Discourse of Forest-Trees, and the Propagation of Timber in His Majesties Dominions.* John Martyn and James Allestry: London.

Evelyn, J. (1679) *Sylva; or, A Discourse of Forest-Trees, and the Propagation of Timber in His Majesties Dominions.* Third edition. John Martyn: London.

Evelyn, J. (1706) *Silva; or, A Discourse of Forest-Trees, and the Propagation of Timber in His Majesties Dominions.* Fourth edition in two volumes. Robert Scott: London.

Evelyn, J. (1901) *The Diary of John Evelyn*. Edited in 1901 from the original manuscript by William Bray in two volumes. Volume 1: 1620–1664, Volume 2: 1665–1706. M. Walter Dunne: New York and London.

Evelyn, J. (1932) *Directions for the Gardiner at Says-Court*. Edited by Geoffrey Keynes. Nonesuch Press: London.

Evelyn, J. (2001) *Elysium Britannicum, or The Royal Gardens*. Edited by John E. Ingram. University of Pennsylvania Press: Philadelphia, PA.

Fairbrother, N. (1956) *Men and Gardens*. The Hogarth Press: London.

Fairchild, T. (1722) *The City Gardener. Containing the most experienced method of cultivating and ordering such evergreens, fruit-trees, flowering shrubs, exotick plants, etc, as will be Ornamental, and thrive best in London Gardens.* Printed for T. Woodward: London.

Fiennes, C. (1888) *Through England on a side-saddle in the time of William and Mary.* Field and Tuer, The Leadenhall Press: London.

Floud, R. (2019) *An Economic History of the English Garden.* Allen Lane: London.

Furber, R. (1727) *A Catalogue of English and Foreign Trees: Which will Thrive in the Natural Ground in England.* Robert Furber: London.

Gray, T. (1994) 'Their Idols of Worship': Fruit Trees in the Walled Garden of Early Modern Devon. In: Pugsley, S. (ed.) *Devon Gardens: An Historical Survey*. Sutton Publishing: Dover.

Green, D. (1956) *Gardener to Queen Anne: Henry Wise (1653–1738) and the Formal Garden*. Oxford University Press: London.

Hadfield, M. (1979) *A History of British Gardening*. Third edition. John Murray: London.

Hadfield, M., Harling, R. and Highton, L. (1980) *British Gardeners: A Biographical Dictionary*. A. Zwemmer: London.

Hanmer, T. (1933) *The Garden Book of Sir Thomas Hanmer.* Edited by E.S. Rohde. Gerald Howe: London.

Harding, V. (1990) Gardens and Open Space in Tudor and Early Stuart London. In: Galinou, M. (ed.) *London's Pride: The Glorious History of the Capital's Gardens*. Anaya Publishers: London.

Hartlib, S. (1652) *A Designe for Plentie: By an Universall Planting of Fruit-trees*. Richard Wodenothe: London.

Harvey, J. (1972) *Early Gardening Catalogues*. Phillimore and Co.: London and Chichester.

Harvey, J. (1974) *Early Nurserymen*. Phillimore and Co.: London and Chichester.

Henrey, B. (1975) *British Botanical and Horticultural Literature before 1800. Vol. 1: The Sixteenth and Seventeenth Centuries.* Oxford University Press: Oxford.

Horwood, C. (2010) *Gardening Women: Their Stories from 1600 to the Present.* Virago Press: London.

Hussey, C. (1967) *English Gardens and Landscapes 1700–1750*. Country Life: London.

Ingram, J. (1998) John Evelyn's 'Elysium Britannicum': Provenance, Condition, Description. In: O'Malley, T. and Wolschke-Bulmahn, J. (eds) *John Evelyn's 'Elysium Britannicum' and European Gardening*. Dumbarton Oaks Research Library: Washington, DC.

Jacques, D. (1989) 'The Chief Ornament' of Gray's Inn: The Walks from Bacon to Brown. *Garden History* 17(1), 41–67.

Jacques. D. (2002) Who Knows What a Dutch Garden Is? *Garden History* 30(2), 114–130.

Jacques, D. (2017) *Gardens of Court and Country: English Design 1630–1730*. Yale University Press: New Haven, CT.

James, J. (1712) *The Theory and Practice of Gardening*. Done from the French Original. George James and Maurice Atkins: London.

Janson, H.F. (1996) *Pomona's Harvest: An Illustrated Chronicle of Antiquarian Fruit Literature.* Timber Press: Portland, OR.

Johnston, M. (2015) *Trees in Towns and Cities: A History of British Urban Arboriculture.* Windgather Press: Oxford.

Johnston, M. (2017) *Street Trees in Britain: A History.* Windgather Press: Oxford.

Knyff, L. and Kip, J. (1707) *Britannia Illustrata or Views of Several of the Queen's Palaces, etc.* David Mortier: London.

Langford, T. (1681) *Plain and Full Instructions to raise all sorts of Fruit-Trees.* Richard Chiswell: London.

Langford, T. (1696) *Plain and Full Instructions to raise all sorts of Fruit-Trees.* Second edition. Richard Chiswell: London.

Langley, B. (1728a) *New Principles of Gardening: Or, The Laying out and Planting parterres, groves, wildernesses, labyrinths, avenues, parks, etc.* A. Bettesworth and J. Batley: London.

Langley, B. (1728b) *A Sure Method of Improving Estates by Plantations of Oak, Elm, Ash and Other Timber-Trees.* Francis Clay and Daniel Browne: London.

Langley, B. (1729) *Pomona: Or, The Fruit-Garden Illustrated.* Printed for G. Strahan: London.

Laurence, J. (1716) *The Gentleman's Recreation: or the second part of the art of gardening improved.* Bernard Lintott: London.

Laurence, J. (1727) *A New System of Agriculture: Being a Complete Body of Husbandry and Gardening.* J. Hyde and E. Dobson: Dublin.

Lawrence, H.W. (1993) The Greening of the Squares of London: Transformation of Urban Landscapes and Ideals. *Annals of the Association of American Geographers* 83(1), 90–118.

Lawson, W. (1618) *New Orchard and Garden.* Alsop for Roger Jackson: London.

Leapman, M. (2000) *The Ingenious Mr. Fairchild: The Forgotten Father of the Flower Garden.* Headline Book Publishing: London.

Le Gendre, A. (1660) *The Manner of Ordering Fruit-Trees.* Humphrey Moseley: London.

Le Sueur, A.D.C. (1954) The Industrious Mr Cook. *Country Life* 115, 2985, 1 April, 964.

Linnard, W. (2000) *Welsh Woods and Forests: A History.* Gomer Press: Llandysul.

London, G. and Wise, H. (1706) *The Retir'd Gardner.* Translated from the French in two volumes. Jacob Tonson: London.

Macky, J. (1732) *A Journey Through Scotland.* Second edition. J. Pemberton: London.

Marsh, D. (2004) *The Gardens and Gardeners of Later Stuart London.* PhD thesis, Birkbeck College, University of London.

Mascall, L. (1569) *A Booke of the Arte and Maner Howe to Plant and Graffe all sortes of Trees.* Henry Denham and John Charlewood: London.

Meager, L. (1670) *The English Gardener, or, A Sure Guide to Young Planters and Gardeners.* P. Parker: London.

Milledge, J. (2009) *The Restoration Garden at Cassiobury Park Created by the Earl of Essex and his Gardener Moses Cook.* Unpublished notes for a presentation to Birkbeck Garden History Study Day, February 2009.

Mollet, A. (1651) *Le Jardin de Plaisir.* Henry Kayser: Stockholm.

Mollet, A. (1670) *The Garden of Pleasure Containing Several Draughts of Gardens, etc.* John Martyn and Henry Herringham: London.

Mowl, T. (2000) *Gentleman and Players: Gardeners of the English Landscape.* Sutton Publishing: Stroud.

Nourse, T. (1700) *Campania Fœlix, Or, a discourse of the benefits and improvements of husbandry.* Thomas Bennet: London.

O'Halloran, S. (2013) *The Serviceable Ghost: The Forgotten Role of the Gardener in England from 1630–1730.* PhD thesis, Department of Landscape, University of Sheffield.

Parkinson, J. (1629) *Paradisus in Sole Paradisus Terrestris, or a Garden of All Sorts of Pleasant Flowers.* Humfrey Lownes and Robert Young: London.

Pope, A. (1956) *The Correspondence of Alexander Pope.* Edited by George Sherburn. Volume 2. Clarendon Press: Oxford.

Reid, J. (1683) *The Scots Gard'ner.* David Lindsay and his Partners: Edinburgh.

Sharp, L. (1975) Timber, Science and Economic Reform in the Seventeenth Century. *Forestry* 48(1), 51–86.

Smith, J. (1670) *England's Improvement Reviv'd: In an Treatise of all Manner of Husbandry and Trade By Land and Sea.* Benjamin Southwood and Israel Harrison: London.

Steuart, H. (1828) *The Planter's Guide.* Second edition. John Murray: London.

Stout, F. (2015) Before Evelyn: Trees, Tree Planting and Tree Management in Sixteenth- and Early Seventeenth-century England. *Arboricultural Journal* 37(3), 150–165.

Switzer, S. (1715) *The Nobleman, Gentleman and Gardener's Recreation.* B. Barker and C. King: London.

Switzer, S. (1718) *Ichnographia Rustica: or, The Nobleman, Gentleman, and Gardener's Recreation.* Volumes 1–3. D. Browne, etc: London.

Switzer, S. (ed.) (1733) *The Practical Husbandman and Planter. Volume 1 containing April, May and June.* S. Switzer: London.

Thomas, K. (1983) *Man and the Natural World: Changing Attitudes in England 1500–1800.* Allen Lane: London.

Till, E.C. (1991) The Development of the Park and Gardens at Burghley. *Garden History* 19(2), 128–145.

Turner, J. (1978) Ralph Austen, an Oxford Horticulturist of the Seventeenth Century. *Garden History* 6(2), 39–45.

Venette, N. (1685) *The Art of Pruning Fruit-Trees, with an Explanation of some Words which Gardiners make use of in speaking of Trees.* Thomas Basset: London.

Walpole, H. (1995) *The History of the Modern Taste in Gardening.* Introduction by John Dixon Hunt. Ursus Press: New York.

Williamson, T. (2018) Seeing the Wood for the Trees: The Long-term Aesthetics of

Woodland in England. In: Woudstra, J. and Roth, C. (eds) *A History of Groves*. Routledge: Abingdon.

Willson, E.J. (1982) *West London Nursery Gardens*. The Fulham and Hammersmith Historical Society: London.

Worlidge, J. (1683) *Systema Horti-culturae: Or, the Art of Gardening*. The Second Edition with Large Additions. Thomas Dring: London.

Woudstra, J. (2018) The History and Development of Groves in English Formal Gardens (1600–1750). In: Woudstra, J. and Roth, C. (eds) *A History of Groves*. Routledge: Abingdon.

CHAPTER SIX

Arboriculture in the English Landscape Garden

The period covered by this chapter includes much of the Georgian era. This lasted from 1714 to 1830 and is named after the Hanoverian kings George I, George II, George III and George IV. Some historians extend the Georgian era to include the relatively short reign of William IV, which ended with his death in 1837. The Georgian era includes the Regency period when the Prince of Wales (later George IV) assumed the role of Prince Regent when his father, George III, was declared unfit to rule due to illness. This was a time of continued military conflict, with France as the main adversary. There were also mixed fortunes for the Empire with the loss of the American colonies but also major expansion and consolidation of its territories in Asia. The Industrial Revolution, which began around the 1780s, saw Britain emerge as the world's first industrialised nation. This prompted a rapid expansion of the towns and cities as workers and their families migrated from the depressed countryside to find work in the new industrial centres, which were mainly in the north. There was a huge increase in the population of new industrial towns like Manchester, Leeds, Birmingham, Bradford, Sheffield and Nottingham. While this industrialisation brought unimaginable wealth to a very small number of people, the bulk of the population at the bottom of the social scale suffered severe reductions in their living standards. These glaring inequalities led to episodes of significant social and political unrest with calls to overturn the established order, encouraged by the revolutionary movements in America and France. However, any agitation by working people in Britain to obtain reforms, such as extending the right to vote, were often met with brutal suppression.

The social, economic and political developments during the Georgian era were also reflected in the landscape of Britain. This included its amenity landscapes in the form of gardens and landscape parks, which were also dominated by trees but in much less formal arrangements than the previous era. The period covered by this chapter begins around 1730 with developments in the new landscape style pioneered by garden designers such as Charles Bridgeman and William Kent. The chapter ends around 1810, which saw the conclusion of the career of Humphry Repton, generally regarded as the last great landscape designer of the eighteenth century.

Sources of information

Those sources of information relevant to the previous chapter are also significant to this period, and often to an even greater extent due to a substantial increase in the volume of material. With regard to surviving gardens and landscapes, there are many authentic examples, such as Stowe, Chatsworth, Painshill and Stourhead, which are now internationally famous. This period also witnessed a huge increase in relevant written material. During the eighteenth century there were around 600 books on horticultural and botanical subjects published in England and Wales (Henrey 1975, 3) and many of these have sections dealing with amenity trees. This figure does not include those published in Scotland or the different editions of various works. Neither does it include works on forestry or general husbandry, many of which have some relevance to amenity trees. A number of dictionaries of gardening were also published that contained definitions and descriptions of horticultural practices, some of which are tree-related. However, following a detailed examination of numerous books from this period it was noticeable that many do not say much about practical arboriculture, apart from the usual topics of planting and pruning. Furthermore, many books about amenity landscapes tended to focus on design and tell us little about the practicalities of how those landscapes were created. In terms of information about the work and status of professional gardeners and arborists, contemporary household and garden account books have again been useful, as have the diaries and garden books of some garden owners. The latter half of the eighteenth century also saw the publication of the first trade and commercial directories, covering specific urban centres, districts or counties. Some of these give information about the range of trades and occupations involved in gardening and landscape work. However, these directories were less helpful in identifying anyone who might have specialised in tree-related work since the particular branches of horticultural or landscape work were rarely specified. Indeed, in many cases these individuals or businesses would probably have been willing to take on any suitable work. Lastly, the eighteenth century saw the establishment of a number of newspapers, with their average readership increasing dramatically from 1750 onwards (Barker 1999, 256). These early newspapers have been a source of information about relevant events and activities and also of advertisements. While the first magazine or journal of botanical and horticultural interest, *The Botanical Magazine*, was published by William Curtis in 1787, this contained little on practical arboriculture.

Visual sources have played a useful role in the research for this chapter. This has related mainly to the depiction of landscapes and gardens and specific tree-related features within them. In some cases these are still in existence and have been examined first hand by the author with a garden visit. Contemporary images from books and other sources have been less informative in providing information about arboricultural practices and tools and equipment.

Charles Bridgeman and William Kent

In the early eighteenth century the British aristocracy and gentry were reacting against the formal Baroque garden style that was largely French-influenced. The popularity of these landscapes with their parterres, long radiating avenues of trees and straight-edged canals was fading. Now, a more simple and natural approach was gaining favour that involved the extensive planting of trees as part of the 'improvement' of estates. From this emerged the English landscape movement with its emphasis on creating green parklands with belts, clumps and groves of trees and serpentine lakes. This generated a revolution in the appearance of British gardens and estate landscapes throughout the eighteenth century that also impacted on urban landscapes.

In the previous chapter two pioneers of the naturalistic landscape style were mentioned, Stephen Switzer and Batty Langley. They were among the first to set out the basic principles that underpin the England Landscape Garden, even if they did not always follow these in their own designs. Others, including Charles Bridgeman and William Kent, took up the mantle and added their own early contributions to the emerging style with the creation of some outstanding examples on the ground.

In Horace Walpole's influential book *The History of the Modern Taste in Gardening*, published in 1780, he refers to Charles Bridgeman (1690–1738) as 'the next fashionable designer of gardens' after London and Wise. He summarises Bridgeman's main achievements and states, 'He banished verdant sculptures … and distained to make every division tally to its opposite, and though he still adhered much to strait walks with high clipped hedges, they were only his great lines; the rest he diversified by wilderness, and with loose groves of oak, though still within surrounding hedges' (Walpole 1995, 42). Although little is known of Bridgeman's early life, he trained at Brompton Park under London and Wise, probably before 1709, and became involved with their work at Blenheim. Unlike many of his predecessors who were primarily nurserymen-contractors, Bridgeman went into business as a garden designer rather than as a garden-maker. Also, unlike some of the leading designers of this era, Bridgeman was not an author; therefore, any understanding of his design ideas must rely on an analysis of his surviving plans and other meticulous research (Willis 2002). In 1714 he began his long association with Viscount Cobham at Stowe House, Buckinghamshire. In many respects Stowe was Bridgeman's masterpiece, although his work there was subsequently modified by some of the great names in landscape design. Nevertheless, he established the framework of the new landscape park at Stowe and in 1735 planted between 60,000 and 70,000 trees (Clarke 1985, 74). Bridgeman is credited with popularising the garden device known as the ha-ha, a ditch with a vertical retaining wall that is invisible at a distance, which divides the garden from the pasture beyond and keeps out grazing animals. This was employed by Bridgeman at Stowe and elsewhere, although it had been first described by Dezallier d'Argenville in 1709.

William Kent (1685–1748) is regarded by many garden historians as the first to successfully undertake the English Landscape Garden style. Kent was forty-six in 1730 when he turned his talents to garden design, having previously spent eight years in Rome studying to be a painter. In some respects he was the successor to Bridgeman and at first made gardens with a similar layout. However, he soon left the formal garden behind and substituted it for the new naturalistic landscape style. In the words of Walpole, 'At that moment appeared Kent, painter enough to taste the charms of landscape, bold and opinionative enough to dare and to dictate, and born with a genius to strike out a great system from the twilight of imperfect essays. He leaped the fence, and saw that all nature was a garden' (Walpole 1995, 43). One of Kent's ruling principles was 'nature abhors a straight line', so avenues, straight walks and neatly clipped hedges were distasteful to him, a feeling that was increasingly shared by other landscape gardeners. This passion for the imitation of nature was part of a general reaction in British society at that time, not only in gardening, but in the world of letters and fashion, where the extremely artificial French taste was definitely out. However, on an artistic level, Kent and the new generation of landscape designers drew inspiration from the paintings of Arcadian landscapes by French artists such as Claude Lorrain and Nicolas Poussin.

It was at Rousham in Oxfordshire between 1738 and 1741 that the essential style of Kent's gardening revolution was to be seen and which, remarkably, remains much as he left it. His client there was General James Dormer, a veteran of Blenheim, who engaged him to enhance and develop a garden that had been created by Bridgeman. What had been a rectangular enclosure was transformed into a seemingly natural landscape, incorporating the distant River Cherwell, with the formal terraces made into a slope. Among the rolling parkland landscape were clumps of trees and shady glades interspersed with statuary and architectural set pieces, all slowly revealed along a circuit walk. His other projects included Stowe, where he worked with Bridgeman and succeeded him as garden designer. His masterpiece at Stowe is the Elysian Fields with its Temple of Ancient Virtue that looks across to his Temple of British Worthies.

One curious feature of Kent's landscape work was his practice of planting large dead trees. The idea behind this was to help create an immediate 'natural' impact and romantic mood. However, it attracted some ridicule, especially when he did this at Kensington Gardens. Walpole commented, '[Kent] had followed nature, and to imitate her so happily, that he began to think all her works were equally proper for imitation. In Kensington-garden he planted dead trees, to give a greater air of truth to the scene – but he was soon laughed out of this excess' (Walpole 1995, 49).

Uvedale Price, in his *An Essay on the Picturesque*, has some interesting thoughts on the difficulty of moving large dead trees and on the need to consider their placement. He states:

> I may here observe that it is almost impossible to remove a large old tree, with all its branches, spurs, and appendages; and without such qualities as greatness of size, joined to an air of grandeur, and of high antiquity, a dead tree should seldom be *left* in a conspicuous place; to entitle it to such a situation, it should be 'majestic even in ruin': A dead tree which could be moved, would from that very circumstance, be unfit for moving. These dead trees of Kent's were probably placed where they would attract the eye; for it is rare that any improver wishes to conceal his efforts.
>
> Price 1796, 252–253

Increased tree planting in the early eighteenth century

The impetus given to extensive tree planting by John Evelyn, Moses Cook and others in the latter part of the seventeenth century continued into the eighteenth century as a wave of 'improvement' swept over the country. Many of the landed nobility and gentry vied with one another in planting large expanses of lowland and mountain slopes with thousands of trees, both deciduous and coniferous. They were encouraged by writers such as Batty Langley (1728) and James Wheeler (1747), who not only raised concerns about the strategic and economic implications of dwindling timber resources but urged extensive tree planting as a patriotic and financially sound investment to remedy the problem. An Act of Parliament in 1714 'to Encourage the Planting of Timber-trees, Fruit trees and other Trees, for Ornament, Shelter or Profit', was followed by a similar one in 1719, and, together with other factors such as the enclosure of large areas of land, this encouraged landowners to enter into the 'spirit of planting'. Not only were deciduous trees established but also an increasing number of coniferous plantations, a reflection of the steadily expanding range of trees becoming available for landscape planting. In just the 1730s, twenty-four new trees and forty-five shrubs were introduced into Britain (Loudon 1838, 1, 81). Among the many landowners engaged in extensive tree planting in the early part of the eighteenth century, two stand out for their imaginative use of newly introduced exotic species.

In 1722, Archibald Campbell (1682–1761), then Lord Islay but later the third Duke of Argyll, acquired the Whitton Park estate in Twickenham, Middlesex. Argyll was an enthusiastic gardener and tree connoisseur, interests he had developed in his native Scotland, and he quickly established a tree nursery at his new property (Symes *et al.* 1986, 140). Through his friendship with the amateur botanist, Peter Collinson, he was part of a small circle of estate owners who were able to acquire a wide range of rare and exotic trees and shrubs, many of them collected by John Bartram in the eastern American colonies. From the catalogue of Argyll's trees, compiled later in 1765 by his gardener Daniel Crofts, we have some idea of the enormous range of species that were raised in the nursery and then planted out in the large pleasure ground that was laid out near the house. Contemporary images (Figure 37) show the house surrounded by a wide mixture of deciduous and coniferous trees. Argyll was nicknamed the 'Tree-monger' by Horace Walpole who claimed that 'the introduction of foreign trees and plants,

FIGURE 37. The third Duke of Argyll was nicknamed the 'Treemonger' for his role in introducing many exotic trees and for the extensive planting on his Whitton Park estate. This print from 1757 shows his house surrounded by a marvellous mix of deciduous and coniferous trees.

which we owe principally to Archibald Duke of Argyle, contributed essentially to the richness of colouring so peculiar to our modern landscape' (Walpole 1995, 46). Due to the efforts of Argyll and his friends during the 1720s and 1730s the ornamental plantings on their estates gave new life, shape and colour to the previously mainly native appearance of trees in landscape gardens (Symes *et al.* 1986, 148). On the death of Argyll in 1761, many of the trees in his garden, including mature specimens, were moved to the Princess of Wales's new garden at Kew, which later became the Royal Botanic Gardens.

Robert Petre (1713–1742), eighth Baron Petre of Writtle, came to gardening later than the Duke of Argyll but made up for it in the scale of his enterprise. In 1729, having been interested in horticulture from a young age, he took over the management of his grandmother's estate at Thorndon Hall in Essex. This was a property of over 1,000 acres (405 ha) where he was not only able to establish a series of nurseries but the landscape park was also sufficiently extensive to accommodate his hugely ambitious planting schemes. Petre was another member of Collinson's circle of landowners who took advantage of John Bartram's American plant collecting exploits. By 1740, only five years after he began subscribing to Collinson's scheme, some 10,000 American trees were planted. After the very harsh winter in 1739–1740, Collinson was delighted to record that all the plants at Thorndon had survived in good condition (Wulf 2008, 89). This was a major achievement since many of the recently introduced trees from North America were initially feared to be quite tender.

As a young aristocrat Petre had become familiar with French Baroque gardens, but it appears he was far more captivated by the exciting new designs of Switzer's (1718) *Ichnographia Rustica*. At Thorndon Petre was blessed with an essentially blank canvas on which to paint his vision of a radically different arboreal landscape and thanks to Bartram he had a wonderfully broad palette to play with. Bartram's trees came in a wide array of shapes and sizes, many of them retaining a neat and compact appearance without the use of pruning shears. These trees also exhibited a beautiful kaleidoscope of colours and textures at different times of the year. In a letter written to Philip Southcote in 1752, Collinson praised Petre's planting schemes as 'painting with Living pencils … the Mixture of Trees and their various Shades of Green' (Chambers 1993, 107). The landscape at Thorndon with its clumps of assorted trees growing in smooth grass was in marked contrast to the complex geometric patterns of Baroque gardens. Those gardens were best viewed from the elevated position of a window high in the house, while Thorndon could only be appreciated by walking along its meandering gravel paths.

In 1742, Petre died suddenly of smallpox aged just thirty-five. The contents of his splendid nursery were then sold to fellow tree enthusiasts the Duke of Richmond, Lord Lincoln and Sir Hugh Smithson, later first Duke of Northumberland (Elliott *et al.* 2011, 20). The death of Petre at such a young age was not only a great loss to his family and friends but also a severe blow to British horticulture and arboriculture. Given what he had achieved in little over a decade at Thorndon, there is no telling what else he might have accomplished in a few more decades.

The English landscape movement

As noted in the previous chapter, landscapes can be both symbols and instruments of cultural power and to understand the nature of those landscapes we must understand the social, economic and political forces at work behind them. A purely aesthetic appreciation of the English landscape movement would be a quite limited perspective without first placing this in the context of those wider influences. As this landscape movement gathered pace during the eighteenth century some quite profound changes were sweeping though British society at every level.

The social, economic and political context

The eighteenth century was a golden age for landowners in Britain. As more and more land was enclosed and let to tenant farmers for money rents, the landowners' incomes rose and they were able to build themselves substantial mansions in every fertile part of the country. They were now able to plan ahead with considerable confidence that wealth would always be there to support their gracious lifestyle. Later in the eighteenth century, as the Industrial Revolution gathered pace, there was a dramatic increase in wealth among the upper echelons of society, not only among the old landed aristocracy but also the *nouveaux riches* of industrialists, financiers and other entrepreneurs. The mines, foundries, textile mills and other industries, employing a rapidly growing working class, generated enormous profits for their owners. This was also a period of economic development in the countryside where Britain's growing population fuelled greater demand for agricultural produce and stimulated agricultural innovation. For those wealthy individuals who benefited from all this, one very potent way of displaying their social status was through the symbol of the estate and the garden. The essential requirement was an extensive country estate together with a town house for the Season. There was much competition among the landed class to be seen at the forefront of the latest innovations and taste. Trees provided much of the expression of this landscape movement, linking the man-made garden with its surrounding countryside.

While a beautifully landscaped park was an impressive status symbol, this had to be seen for people to talk about it. In the mid-eighteenth century a few great houses opened their grounds to select members of the public who could come and admire the views and, hopefully, tell their friends. To fully appreciate the garden, visitors needed information and a helpful plan, and so the very first garden guidebook was produced for Stowe in 1744 (Seeley 1744). In those days the implications of garden tourism in terms of risk management would hardly have occurred to the owners who could never have imagined that over 200 years later vast numbers of people would visit these same properties every year.

During the latter part of the eighteenth century the landscape fabric of much of Britain underwent a series of striking transformations. Of particular relevance here was the parliamentary enclosure of common fields and the amalgamation of small farms into more productive units, which helped to facilitate estate 'improvements'. These developments also dispossessed large sections of the peasantry and led to much hardship among the rural poor. Other radical transformations to the landscape of Britain included major improvements in transport infrastructure, rapid expansion of resort towns and commercial cities, the growth of vast new areas of industrial, manufacturing and commercial activity, and the development of London into a world metropolis. Taken together, these major advances signalled Britain's arrival into the modern world.

While a significant proportion of the money that financed the English landscape movement came from the landowners' profitable exploitation of their land and from the profits of the Industrial Revolution, a substantial proportion came from another source that has only recently received much recognition. It is an uncomfortable truth that those beautiful naturalistic landscapes so admired as part of English culture and artistic achievement were to a large extent financed by the slave trade. Initially, this was from the very lucrative trade in human beings forcibly taken from their homes on the west coast of Africa and transported in appalling conditions across the Atlantic Ocean to the Americas. Of the European countries engaged in this trade, Britain was the most dominant. Once there, these enslaved men, women and children were forced to work on plantations growing cotton, sugar, tobacco and other crops. Many of these plantations were British owned and the enormous profits generated went back to Britain to help fund the extensive house and estate improvements, as well as the opulent lifestyles, of the owners and their families (Dresser and Hann 2013).

The creation of the landscape park could have unpleasant consequences for the peasantry who were unfortunate enough to stand in the way of the landowner's plans for 'improvement'. Through a series of parliamentary acts, the majority being passed between 1750 and 1860, open fields and 'wastes' were closed to use by the peasantry. Open fields were large agricultural areas to which a village population had certain rights of access in order to pasture animals, collect firewood or otherwise benefit. With enclosure the land became the private property of the landowner and completely free of any control or use by the local community. Countless rural labourers and their families had come to depend on this land to fend off starvation. As compensation for their loss, the displaced people were often offered alternative land of smaller size and quality, sometimes with no access to water or wood. Many of those who were driven off the land ended up working in the new factories and mills. Removing housing could be more difficult but the total, or partial, removal of villages for emparkment is a common theme in eighteenth-century landscape history, highlighted by Oliver Goldsmith in 1770 with his poem *The Deserted Village*, and research has identified various examples (Neave and Turnbull 1992, 9). What received less contemporary attention was the impact of enclosure and emparkment on the local road network. Roads, tracts and footpaths were frequently terminated or diverted when parks were laid out. It is extremely difficult to estimate the number of roads and footpaths affected but the total must have amounted to several thousand miles (Williamson 1995, 105).

The extent to which the major landowners continued to dominate the political life of the country in the middle and later decades of the eighteenth century is a matter for debate among historians (Williamson 1995, 17). What is unquestionably true is that the nature of their relations with the local gentry and the mercantile and professional classes underwent a profound change. Increasingly, they sought to play down differences of status and hierarchy between them, emphasising instead a collection of shared cultural values often referred to as 'politeness'. This was characterised by easy and affable behaviour, knowledge of 'taste' or current fashions, and the acquisition of a particular set of social skills, not least an ability to hold cordial conversation. 'Polite society' embraced not only the landed class but also the upper echelons of the middle class. They were thought of as a group distinct from the 'common people' or working class, whose fortunes were generally in decline throughout the second half of the eighteenth century. The development of garden design during this time was inextricably bound up with the emergence of 'polite society', and especially in its relations with the increasingly impoverished and alienated world around it. It was out of this complex and potentially explosive mix of social, economic and political circumstances that the landscape park emerged.

Continuing influence of the formal garden

In the second half of the eighteenth century the English landscape style was the height of fashion among many British landowners who could afford the necessary modification to their existing parks and gardens. However, to describe the greater part of the eighteenth century as a period in which the 'reign of nature' held sway in English gardens, let alone those in the rest of Britain, would be misleading. As previously noted, in the first few decades of the century the formal garden remained the dominant style among landowners of all social levels. The 1730s and 1740s witnessed a slow but growing number of landowners incorporating more naturalistic elements into their gardens, but it was only after 1750 that the formal geometric garden really began to go out of fashion, and then mainly in England. Nevertheless, as with many fashions, what was written about garden style in the popular magazines and discussed in polite society did not necessarily match the reality on the ground. Those square and level lawns, straight walks flanked by topiary and avenues leading directly to the house continued to be popular with some landowners, even into the second half of the eighteenth century.

While the English landscape movement gathered pace in England, its progress in Scotland was much slower, with many formal gardens remaining largely unaltered throughout the eighteenth century (Cox 1935, 84). Various factors were involved in the survival of these formal gardens in Scotland. There was a general dislike of wasting money by destroying what had previously been considered money well spent, which must have also concerned some English landowners. There was a degree of conservatism in matters of fashion among many Scottish landowners and reluctance by some to adopt an overtly English style of garden. Furthermore, in Scotland there does not seem to have been any identification of the formal garden with the despotic and decadent regime in France or, conversely, the naturalistic garden with freedom and liberal philosophy that was a major factor in the gardening tastes of the English Whigs (Tait 1980, 4).

Even in the 1750s, many gardens throughout Britain incorporated both formal and naturalistic elements and this continued for some years. Those who made their living by designing and creating gardens were generally not wedded to any particular style and were happy to work with features in either style, as and when their clients requested these. Typical of this all-embracing approach was Henry Ellison, a gardener and land surveyor based in Norwich. In an advertisement in the *Ipswich Journal* in 1753 he announced, 'he undertakes and performs all Sort of Gardening Works, Wood Works, Water Works, Forest Planting, and Planting in all Branches whatsoever' (*Ipswich Journal* 1753). While

Ellison may have had his own personal preferences in garden styles, he saw his role as doing his best to deliver what his clients wanted. However, three or four decades on and everything had changed. Almost all members of the English and Welsh landed class now had their home set within a landscape park, an open landscape of grass and irregularly scattered trees. The formality of geometric planting had largely disappeared and the enclosed gardens around the house had gone, so that grass flowed up to its very walls.

Characteristics of the landscape park

Some of the most notable gardens in the new English landscape style were initially laid out in the late 1730s and 1740s, although some took many years to complete and often involved various garden designers. Some of these more famous gardens include those at Claremont, Holkham Hall, Stowe, Painshill, Wilton, Hagley Park and Wroxton Abbey. Some of these will be mentioned later but first we will outline the characteristics of the English landscape garden.

The new landscape parks were generally larger than the formal gardens they replaced, not least because they often included a substantial amount of new land that had recently been enclosed. Many had also incorporated ancient deer parks; indeed, the basic style of the landscape park owed much to the decorative wood-pasture landscape of the post-medieval deer park. The landscape parks had a far more open appearance than the formal gardens, with their extensive tracts of grassland and irregularly scattered groups of trees. While these parks varied considerably in appearance, some features were considered important, if not essential, to achieve the overall landscape effect. Trees formed the basic framework, a point emphasised by Uvedale Price who asserted, 'It is in the arrangement and management of trees, that the great art of improvement consists' (Price 1796, 285). However, linear planting, such as avenues or geometric vistas, was now shunned. All blocks of woodland, a crucial feature of these designs, had to have curvilinear, serpentine outlines. Two forms of tree planting were particularly distinctive and popular. The first was the placing of strips of woodland, generally referred to as 'belts', on the periphery of the park. These sometimes ran all the way around the boundary, so that the park was effectively sealed off from the surrounding landscape. The second common form of planting involved the placing of small groups of trees, known as 'clumps', dotted in a seemingly casual arrangement across the interior of the park. This type of landscape was now more suited to the owner's sporting pursuits as the change from stag to fox hunting encouraged the replacement of straight rides in woodlands by expanses of turf interspersed with coverts. Almost all parks had some type of ha-ha that separated the pleasure ground near the house from the park beyond and kept out grazing animals. There would also be a serpentine path that meandered around the pleasure ground and possibly even further afield. Many parks, especially the larger ones, also had lakes. Ideally, this was positioned in the middle distance when viewed from the windows of the house.

Although industrialisation had played a major role in funding these landscape parks, owners did not want to be reminded of this as they looked out across their private pastoral idyll. Any unpleasant image of industrial activity, such as spoil heaps, smoke belching from chimneys or the humble dwellings of an emerging industrial working class, had to be excluded from the scene. This, of course, was the primary function of the perimeter belt of trees. At Horsford Hall in Yorkshire, the home of the Stanhope family, the grounds to the north of the mansion rose sharply to the colliery on Hunger Hills. From the 1780s extensive tree planting was carried out there by the landscaper Richard Woods, so that the presence of industry would not impinge on the scene (Sheeran 1990, 128). Even the sight of agricultural buildings was considered inappropriate. The great tree expert William Hanbury (1770, 1, 71) recommended, 'Whenever any disagreeable objects present themselves to view, such as dove-cotes, old barns, etc., such places are always to be hid with trees'. Having obscured anything on their own property that was unsightly, owners generally did not want to gaze out to somebody else's land beyond, even if this was an attractive and 'tasteful' view. The landscape gardener William Marshall declared that parks should only be open to the surrounding world if all the land in view belonged to the owner, in which case the prospect would gratify his 'love of possession [which] is deeply planted in every man's breast' (Marshall 1785, 614). If, however, someone else's property entered the view from the house, then it had to be obscured by thick belts of trees.

As should now be apparent, one of the most important functions of the park was to act as a landscape of polite exclusion (Williamson 1995, 107). The park's ability to supply its owner and his family with their precious privacy may in part explain not only its popularity but also certain aspects of its style. An insulating and segregating landscape had by definition to be an extensive one and the belts of trees around the perimeter not only concealed the unwanted world beyond but the slowly maturing timber trees did this in a very cost-effective way.

Tree-related features of the English landscape garden

Having outlined the main characteristics of the landscape garden or park, we take a closer look at some of its features that made significant use of trees. Some tree-related features that were popular in the earlier formal gardens were no longer fashionable and had either disappeared from the garden or been modified or renamed. Arbours are an example of this. According to Philip Miller's *Gardeners Dictionary*, 'These were formerly in great Esteem with us than at present; few gardens were without cover'd Arbours and shady seats; but of late they have been much rejected, and that not without good Reason' (Miller 1737, 'Arbours'). Arbours were now considered expensive to erect and maintain, while the damp and shady conditions rotted the woodwork and left the seats continually damp and 'unhealthy'. Nevertheless, arbours still featured in the commercially run pleasure gardens of the eighteenth century. Dwarf-trees, those neatly clipped fruit trees that were a particular feature of gentry gardens, were also no longer in fashion. Again, according to Miller, 'These were formerly in much greater Request than they are at present. Since the introducing of Espaliers into the English Gardens, Dwarf-Trees have been destroyed in most good gardens' (Miller 1737, 'Dwarf-trees'). Parterres were one of the most characteristic and certainly one of the most expensive features of the formal garden. According to Mawe and Abercrombie (1778) in their *Universal Gardener and Botanist*, 'Parterres in general are becoming in disrepute in England, their regularity being thought too stiff and formal, their figures often of a trifling appearance, and their making and keeping in proper order attended with considerable trouble and expense'. As for topiary or clipped evergreens in any form, once so popular in gardens, these are hardly mentioned in the literature after the 1730s. While these earlier tree-related features went out of fashion and disappeared, others came to the fore in the new landscape style. One particular difficulty in giving an account of these features is the different terms that were used at that time, and even today, to describe what are basically the same or very similar features. According to Thomas Whately (1770, 36) in his hugely influential *Observations on Modern Gardening*, 'Every plantation must be either a *wood*, a *grove*, a *clump*, or a *single tree*'. We will take these categories and add some additional ones.

Belts of trees

As already noted, belts of woodland around the boundary were an essential feature of the landscape park, contributing much to its character and its role as a landscape of polite exclusion. However, belts of trees of some description that marked out the boundaries of property were an established feature of the countryside in many regions of Britain, especially in England. Pehr Kalm remarked in 1748, 'The country was everywhere divided into small inclosures, with hawthorn and other hedges around them, so that one could only suppose he was travelling all the way through a garden. Here and there appeared small woods of all sorts of leaf-trees' (Kalm 1892, 54). What was also noticeable were the wide hedgerows growing in strips some 30 to 40 ft (9.1 to 12.2 m), giving the fields the appearance they were reclaimed from more extensive woodland. These belts of woodland not only marked out boundaries and acted as shelterbelts giving animals and crops some protection from inclement weather, they could also be harvested for timber and fuelwood by managing the trees as coppice or pollards. This economic value of belts of woodland continued when they became a regular feature of the landscape park.

John Reid, the author of *The Scots Gard'ner* in 1683, had already noted the ornamental potential of wide hedgerows and belts of trees as places for a pleasant walk (Reid 1683, 85). The man attributed by some historians for developing the ornamental potential of the woodland belt is Philip Southcote in the 1730s

(Stroud 1975, 33). His original *ferme ornée* (ornamental farm) at Woburn Farm, near Chertsey, greatly influenced the English landscape movement and was regarded by his contemporaries as the first to demonstrate the value of belts for providing a woodland walk around and overlooking the park.

While most tree planting in belts and other locations in the landscape park comprised mainly native or naturalised trees, more exotic species could occasional be found. For example, 'Capability' Brown used many exotics such as London plane and evergreen oak, as well as planting a wide range of conifers, principally Scots pine, spruce and larch, as Kent had done before him. Indeed, the cedar of Lebanon is known to be Brown's favourite tree and he often planted one at a focal point in the landscape park. He also used conifers occasionally to vary the margins of his perimeter belts. Many landscape designers used conifers as 'nurse' crops in plantations of hardwoods, before removing them as thinnings when the hardwoods were well established. However, not all advocates of the English landscape style were favourably disposed towards conifers. George Mason remarked:

> The greatest fault of modern planners is their injudicious application of Fir-trees. A quick growth and perpetual verdure have been the temptations for introducing them; but these advantages are very insufficient to justify the prevailing mode, which gives them universal estimation. Trees of conic figure are by nature unsociable – not to be allowed a place amidst the luxuriant heads of oak, or other noblest progeny of the forest.
>
> Mason 1795, 80

Clumps of trees

The use of clumps of trees in landscape settings for both ornament and profit had already been mooted long before the idea was considered an innovation by William Kent. In the Commonwealth era Walter Blith had advised that 'several Groves or Plumps of Trees may be Erected about any Manour, House or Place', for both profit and pleasure (Blith 1653, 153). In 1728, Batty Langley advised 'Ashes planted with Oaks, Chestnuts, etc. in Plumps, on the tops of little Hills in Parks, etc. have a very good Effect' (Langley 1728, 85). Stephen Switzer was of a similar mind when he states:

> To conclude this Part, the raising or planting of Woods in Hills rather than Vales, adds greatly to the Beauty of any place, when beheld from afar, which is visible in several Groves, Clumps or compartments of Wood, many of them but small, which one sees at a distance … it being certain that one Tree on a Hill adds more Beauty to a Seat, and looks better than six, I have almost said ten, in a Bottom.
>
> Switzer 1733, iii, 22

In Lord Petre's plans for Thorndon and Worksop, made around 1740, he laid out the peripheral distribution of trees in the manner called for by Switzer (Chambers 1993, 103). Of the great garden designers of the English landscape movement, Kent was probably the first to introduce clumps of trees, as evidenced in a sketch by him for a group of clumps that is preserved at Holkham (Clark 1948, pl. 4). While landscape designers of this era often had different views on the design and use of clumps, we refer to Whately's definition. He stated, 'A clump … may be either close or open; when close, it is sometimes called a thicket; when open, a group of trees, but both are equally clumps, whatever may be the shape or situation' (Whately 1770, 36). He also regarded clumps as 'beautiful objects of themselves' (*ibid.*, 56).

In the landscape park clumps served a number of purposes, apart from their aesthetic role. As with most park plantations, the early thinnings would provide wood for fencing, fuel and other uses. Later in their management clumps could produce timber. Clumps also provided shelter for grazing animals as well as being used to hide pits and disturbed ground. Clumps could be used as tree nurseries, providing a supply of young trees that would then be planted out elsewhere in the park, although some authors were concerned about the loss of roots in the process (Kennedy 1777, 1, 40). Nevertheless, Repton (1803, 46) observed that 'Capability' Brown intended trees to be moved out of his clumps when big enough to resist the sheep, and the resultant drift of planting would give a suitably indistinct appearance to the woodland edge. While this may have applied to Brown, many other landscapers surrounded their clumps with stone walls and hedges, rather than wooden fences, so their dimensions would be determined in lasting form by the designers (Phibbs 2017, 100). Also, in many cases the boundaries of the clumps were fixed by the views they were designed to contain. While landscapers anticipated that many trees in clumps would be transplanted or thinned out, this often failed to happen. The architect John Buonarotti Papworth pointed out that the mature appearance of Brown's clumps was often accidental. He stated, 'These were to be thinned, and the best selected for standing in the manner common to parks; but the richness obtained by the concealment of the stems, and by the increased foliage and shadow of the branches, caused them to be preserved as clumps, and Brown did not object to be complimented on the contrivance' (Papworth 1818, 67).

The wilderness, grove and shrubbery

Despite its name, the wilderness was a highly ornamental feature of the Baroque formal garden and was basically a designed grove or *bosquet* with paths cut through it. In the 1730s and 1740s there was a trend towards the simplification of wildernesses that involved not only a reduction in the number and complexity of paths, but also in many cases the removal of the underwood between them and the clipped hedges beside them. Writing about wildernesses in 1733, Thomas Hamilton noted:

> They are too formal, and Stiff, besides they make Every thing so Busy, that they croud the ground too much. Were I to plant a Wilderness I should have nothing in it, but Ever Greens Trained up as much to Trees as I could, Flowering Trees, and

> them that have fine Blossoms, with a kind of Willow, that has a Bark of a Bright Yellow.
>
> Hamilton 1953, 58

As hedges were removed from the perimeter of the wilderness quarters, attention focused on planting flowering shrubs within them. This trend seems to have signalled the evolution of the wilderness into the shrubbery. Writing in 1770, William Hanbury still referred to the wilderness rather than the shrubbery, but from his description it is clear how much had changed from the previous era (Hanbury 1770, 1, 68–70). With regard to walks within the wilderness, these 'ought to be serpentine, winding in an easy manner throughout the whole works, and should on a sudden … lead into a large grand walk, which commands a view of the circumjacent country. This would not fail to strike the imagination, and fill the mind both with pleasure and surprise'.

With regard to planting in the wilderness, this should be dense, graduated and natural:

> With respect to the quarters that are to be planted, they should be of a size large enough to hold a sufficient number of trees to intercept the view of any two walks, i.e. there must be so many trees as to prevent a person walking in one from seeing the company walking in another. The largest-growing trees should always be planted in the middle of the quarters, the lowest next to the walks; and the different growths should be so placed, one next the other, as that the heads of the lowest, and the next, may hide the stems of the tallest, forming at the same time a beautiful amphitheatre. And here let me observe, that all the trees must retain their natural growth, and are by no means to be either clipped or sheered, but should naturally form their waving heads with the pleasing luxuriancy and beauty in growth.
>
> Hanbury 1770, 1, 69

By the 1750s, the straight paths and neatly trimmed palisades of the wilderness had largely disappeared, to be replaced by graduated planting along the winding paths through the shrubbery. In many cases the old close grove or wilderness was not simply uprooted but slowly evolved into an open grove that resembled a shrubbery once the forest trees had matured and suppressed the understorey (Laird 1999, 8).

What Thomas Whately (1770, 36) considered a grove was essentially an open grove or 'trees without underwood', consisting of single trees 'dotted' over grassland without an understorey, giving the appearance of a kind of savannah. The word grove itself is ancient, and medieval groves of a kind did survive into the eighteenth century (Phibbs 2009, 178). Nonetheless, despite the claims of the promoters of the English landscape garden to be pursuing a naturalistic effect, it seems unlikely that the grove of parkland and pleasure ground was directly copied from anything in English nature. As Thomas Whately (1770, 229) observed, 'A fine open grove is unusual, except in a park or a garden'.

One consistent theme of the composition of the later eighteenth century shrubberies was that the plants should be arranged in a graduated array down to the paths, to create Hanbury's 'beautiful amphitheatre'. Initially, this was done in straight rows but then designers favoured planting in irregular drifts to give a more natural and picturesque composition. The graduated arrangement was also an ideal way to display the increasing number of exotic flowering trees and shrubs coming into Britain, especially from North America. The resulting planting scheme would be 'covered with verdure down to the very front, in an easy theatrical manner, and in summer scarce a stem visible' (Meader 1779, 1). As time went by, the angle at which the planting was graduated got increasingly steeper to the point where it became quite extreme. The steeper the slope, the more conspicuous were the plants, and the greater the skill required by the gardener. It reached the stage where this was mocked in magazine cartoons that drew parallels between the 'vanity of the shrubbery' and the extremes of rakish graduation then fashionable in women's headgear (Phibbs 2017, 161). It is interesting to note the similarities of this horticultural style with the supposed excesses of the Baroque garden and how far it was removed from naturalistic planting.

One consistent theme of the planting in shrubberies was the grouping of species. The debate centred around whether there should be 'mingling' or mixing of plants of different species, or 'massing' or planting a group of the same species. This usually focused on whether evergreen species should be segregated from other plants. According to Philip Miller (1737, 'Wilderness'), there were two good reasons for segregating evergreens. The first was cultural, as he believed the evergreens would be 'starved' if planted too near some kinds of deciduous trees. The second reason was aesthetic, because in winter the bare stems of deciduous trees and shrubs looked unattractive when combined with evergreen verdure. In the early years of this debate most gardening authorities, such as Hanbury (1770, 1, 68) and James Meader (1779, 1), upheld the principle of segregation. However, it later became increasingly common to undertake 'mingling' or intermixed planting.

One aspect of planting design and species composition that is surprisingly absent in most eighteenth-century literature is the use of autumn colour. This seems strange in a period when the North American maples, oaks, sumacs and liquidambar were becoming common in British gardens. For example, that great horticultural authority, Philip Miller, mentions eight species of *Acer* in his *Gardeners Dictionary* of 1737, and although some flowers are identified there is no mention of autumn colour. One of the few writers to comment positively on this is William Hanbury (1770, 1, 155) when he praises the Carolina poplar (*Populus x canadensis*), not least for its light green leaves which are 'heightened in the autumn ... turning to a red colour'.

By the 1750s the term 'shrubbery' was commonly used, not just by many gardeners but also by the general public. By the 1770s, horticultural manuals adopted a new section, 'Shrubbery', to replace Philip Miller's long-standing entry of 'Wilderness', even though Miller himself was never convinced. By 1800, most gardens of any size would have been furnished with a shrubbery, as this horticultural embellishment was now the favoured spot for private walks and intimate conversation.

The pleasure ground

The pleasure ground was a term used to describe the most extensive area of ornamental planting within the wider landscape garden. It was essentially distinct from the less intensively planned and maintained landscape park and was often separated from this by a ha-ha or other physical barrier. It was also separated from utilitarian enclosures, such as the kitchen garden. Very often it contained one spot primarily devoted to shrubs, the shrubbery, as well as another spot principally dedicated to flowers, the flower garden.

Pleasure grounds, which had featured during the previous era of the formal garden, continued to be created and maintained at most country houses during the era of the English landscape garden, but they were now usually hidden away from the main facade of the house and placed to the sides or rear. Geometric layouts were still occasionally employed, where the pleasure ground was arranged in regular circles or ellipses. More usually they featured curvilinear, relatively narrow gravel paths, which ran through areas of grass and shrubbery planted with varying densities of trees. These provided a circuit walk, mirroring the circuit drive within the park. An examination of contemporary images of pleasure grounds suggests that it might have been common practice to prune some of the large trees in a manner that would now be termed crown lifting (Figure 38). This involves the removal of the lower branches of the tree to a specified height, leaving a clear stem. The purpose of this pruning in the pleasure ground was probably to give greater visibility to the shrub layer or understorey and to allow more light to this.

Some contemporary gardening authorities define the pleasure ground to include features beyond the ha-ha, such as plantation walks or groves, pieces of water, even an entire small park (Mawe and Abercrombie 1778, 'Pleasure-garden'). In the work of landscape gardener Richard Woods, the pleasure ground and the landscape park were often merged into what might be described as a 'pleasure park' (Cowell 2009, 40). However, of the forty-seven estates that Woods is known to have been associated with, only six were of a size to have even the potential for planting on a truly Brownian scale. With most landscape designers of this era the pleasure ground was commonly taken to be a particular area of mainly ornamental garden.

Coppice, pollards and shreds

Areas of woodland throughout the landscape garden and park were often managed as coppice. Some of this coppice pre-dated the 'improvements' and had been incorporated from the previous landscape, while others areas had been planted recently specifically with the intention of managing them as coppice. This coppice woodland had a number of benefits, not least as a source of wood and wood products that could have many different uses around the estate. It also gave a steadier and quicker commercial return than timber trees, taking

FIGURE 38. The practice of removing the lower branches of mature trees in the pleasure ground was probably to give greater visibility to the shrub layer and to allow more light to this. This print from 1779 shows an example of this at Ranston House in Dorset.

perhaps ten to twenty years before harvesting, as opposed to 100 years or more for oak timber. Some landscape designers also valued coppice for the interest it gave to walks through the garden and park, and for the cover it provided for game. Areas of coppice were sometimes retained or planted around the pleasure ground, giving the appearance of natural woodland from the outside and acting as shelter from the wind on the inside. While there are examples where Brown incorporated and maintained extensive coppice regimes in his landscapes, he more often grubbed these out and replaced them with plantations (Phibbs 2017, 26). Indeed, many areas of coppice in and around landscape parks were replaced by timber trees in the second half of the eighteenth century. One of the shortcomings of coppice from an aesthetic perspective was that it needed continual fencing from deer and other grazing animals, which gave it an unnatural appearance.

With regard to pollards, most of the seventeenth-century writers on forestry and gardening had criticised these for being unsightly and a ruination of timber trees. This continued into the eighteenth century with influential writers, such as Switzer (1718, I, 259), echoing these views. William Marshall (1785, 487) spoke for many authorities on trees and landscape at that time when he proclaimed,

'We declare ourselves enemies to Pollards; they are unsightly; they encumber and destroy the Hedge they stand in (especially those whose stems are short), and occupy spaces which might, in general, be better filled by timber trees; and, at present, it seems to be the prevailing fashion to clear them away'. Pollards were not just viewed as unsightly and mutilated trees, they were also decrepit, hollow and decayed, and for those creating the English landscape garden anything that indicated deformity and decay should be removed. As a result, a great many pollards were felled during the eighteenth century. These negative attitudes towards pollards seem incongruous from a modern perspective as these trees are now much admired for their antiquity and historical associations, for the richness of their natural history and for their remarkably gnarled and picturesque appearance.

While pollards were not welcome in the landscape park, large mature trees were often retained from the previous landscape. When hedges were removed to make way for the new landscape, any majestic hedgerow trees were left standing, although usually not in such numbers as they clearly stood in lines. Retaining these old trees gave an air of antiquity to the newly created scene. As William Gilpin said of Cadland in Hampshire, the 'abundance of old timber gives the house, tho lately built, so much the air and dignity of an ancient mansion that Mr Brown, the ingenious improver of it, used to say, "It was the oldest new place he knew in England"' (Gilpin 1794, 2, 202). Almost all eighteenth-century parks contained at least a few large mature trees that had been retained from the previous landscape.

Similar to pollarding is the ancient practice of shredding trees. This is like 'vertical coppicing', where the side branches on the main stem, usually a single stemmed tree, are regularly cut back to the trunk, leaving just a small tuft of crown at the very top of the tree. According to Arthur Young, the late eighteenth-century writer on agriculture, across much of the south of England 'the trees are all … stripped up like Maypoles, which they nearly resemble when fresh stripped, with only a little tuft of leaves on the top. This is a most barbarous custom; destroys the beauty of the country, and is absolutely ruinous for timber. Norfolk, Suffolk and a part of Essex, are, however, exempt from this detestable practice' (Young 1769, 308).

It is difficult to assess the extent of this practice in other parts of Britain around this time as the terms used to describe these ancient forms of tree management can vary in different regions. However, it seems unlikely that shreds (shredded trees) would have been incorporated into the landscape park, and certainly not in areas where they would have been noticed, since they would have been regarded, like pollards, as unsightly.

Avenues

While the avenue was a major feature of estates and gardens in the seventeenth century, from the mid-eighteenth century the straight line was definitely out of fashion and the avenue was rarely included in garden design. Nevertheless, many writers continued to give advice on planting avenues well into the latter half of the century. The edition of Miller's *Gardeners Dictionary* published in 1737 has an entry for 'Avenues', although he recognises they are no longer much planted. He states, 'The old Method of planting Avenues with regular Rows of Trees, is what has generally been practic'd till of late Years; since the old Manner of laying out Gardens being with good Reason disused, they have lately introduced a more magnificent Method of planting Avenues; which is to plant the Trees in Clumps, or Platoons' (Miller 1737, 'Avenues').

However, even clumps and platoons for avenues were no longer being planted by the mid-eighteenth century. In purely aesthetic and practical terms, contemporary writers raised various objections to the avenue as a landscape feature, focusing mainly on its 'sameness' and that it cut the scenery in two and reduced the view to a narrow vista (Whately 1770, 139). Where avenues were part of an existing landscape that was now undergoing 'improvement', they would often be felled. Of all the leading British landscape gardeners of the time it was 'Capability' Brown who was most associated with the felling of many fine avenues on country estates. This assault on the avenue was so widespread that William Chambers (1773, xi) was moved to comment, 'Our virtuosi have scarcely left an acre of shade, nor three trees growing in a line, from the Land's-end to the Tweed'.

With a straight avenue leading directly to the house considered inappropriate, the idea now was to have an approach specifically designed to show off the park as a prelude to showing off the house. This would wind its way around the landscape park among trees and clumps of woodland carefully positioned along the route to give tantalising glimpses of the building as it came into view and then was hidden again.

Costs of the landscape style

There has been much debate among historians regarding the costs of creating and maintaining the English landscape garden in comparison to the formal style that it replaced. While we will not dwell on that debate, the issue does have a bearing on our subject, since much arboricultural work was involved and it was undertaken by landscape professionals. Linked to that debate is the question of just how productive was the English landscape garden or park?

Many authorities on gardening in the early part of the eighteenth century state that the decline of the formal garden was not only due to objections on matters of style and ideology, but was also about the high cost of establishing and maintaining these gardens. Furthermore, many of the advocates of the new naturalistic style of gardening were keen to point out the reduced cost of this in comparison to the formal garden style. These issues were highlighted in the previous chapter and the position was summarised by Stephen Switzer:

> The Chief of our Design in this Book, being the Decoration and Embellishment of the whole Estate, or at least, that Part of it that lies most contiguous to the Mansion House, instead of such vast and expensive Gardens, that the Folly of this and the past Ages hath run into, whereby that most innocent and harmless Employ, is become a Burden too great for the biggest Estate, and the Gardens themselves not at all answerable to the needless Expense that is laid out upon them.
>
> Switzer 1718, 3, 45

Looking back on this today, it does seem reasonable to assume the costs involved with the simple naturalistic style were normally less than the meticulously manicured and painstakingly geometric formal style. However, there are many factors involving both expenditure and income to take into account before rushing to any judgement. First, many of the new landscape parks were much larger than the earlier formal gardens because they had incorporated major areas of new land, often as a result of enclosure or the acquisition of small farms. This could also involve considerable additional costs such as the replacement of diverted roads and demolished buildings, and the laying out of new carriage drives. Furthermore, immense sums could be expended in removing all traces of the old landscape, such as avenues, pollards, woodland and any remaining formal gardens, which was typically heavy and labour-intensive work. Major earthworks were often involved in the creation of the landscape park, especially in making a new lake or the extensive contouring of the ground. These earthworks could often be the most expensive part of the project. Otherwise, the main expense of the new landscape design could come from the planting component. While many trees might be retained from the existing landscape, many new ones would need to be planted, especially for the perimeter belts. There is also the question of the extent of planting in the pleasure ground, in comparison to that in the landscape park. Because of its essentially ornamental nature and mix of exotic trees and shrubs, the pleasure-ground planting was invariably more expensive than planting a similar area of the landscape park, which usually involved the use of cheaper, mainly forest or native trees. Also, the plantings in the pleasure ground, especially where there this was steeply graduated, would have been infinitely more expensive to maintain than those in the landscape park.

While contemporary figures such as Switzer highlight the high maintenance costs of the formal garden as a significant factor in the change of style, finding firm evidence to support this can be difficult (Sheeran 1990, 40). Where estate accounts do exist, they are not always sufficiently complete to give a reliable picture. Where expenditure on the maintenance of landscape parks was considerable, this was often focused on the pleasure ground and ornamental gardens around the house, including any glasshouses stocked with costly exotics. Where there was a substantial increase in gardeners' or labourers' wages, this was often due to a substantial increase in the area of more intensively maintained garden. However, taking all these potential variables into account, it does seem reasonable to conclude that, on balance, the formal garden style was more expensive to maintain than the naturalistic style, simply because it generally required less intensive and less demanding upkeep.

Some recent historians have characterised the landscape park of 'Capability' Brown and his contemporaries as a largely unproductive expanse of grass and scattered trees where the emphasis was on aesthetics and little else. While there may be some truth in this, landscape parks could also have a significant economic role. Many wealthy landowners were also interested in agricultural improvements and some economic benefit from the landscape park would have appealed to them. This economic aspect also encouraged the popularity of the landscape park among the less wealthy ranks of the landed class, who could not afford to dedicate large areas of their land to purely aesthetic purposes.

It has been mentioned that the park was used for shooting deer and game, and while this was essentially a recreational activity, it also had an economic benefit in providing meat for the table. However, it is doubtful whether the form of the landscape park was critically shaped by such considerations, although it almost certainly played some role in the emergence of the landscape style (Brown and Williamson 2016, 123). The grassland of the park provided grazing for the deer as well as for sheep and cattle. Park deer had some advantages as they could be bred and fattened much quicker than in the forest. Nonetheless they were regarded by Brown and most other improvers as pests. Some sections of the park were also kept and mown as hay meadows.

While most of the parkland trees in belts, clumps and other plantations were not managed primarily as an economic resource, they could still generate a significant income. However, woodland in more remote areas of the larger parks would certainly have been managed mainly as commercial forestry and income could be generated in various ways. Areas of coppice, either incorporated into the park or newly planted, would provide a regular source of fuelwood and other wood products. There would also be the repeated thinnings from recent plantations, generating a regular income to the landowner or also providing fencing material and other wood products for the estate. It is likely that some of the forest trees in the less visible part of the perimeter belts would also be felled for timber. However, this would have been done on a very selective basis and it would only have started as those trees were approaching 100 years old.

'Capability' Brown and the emerging landscape industry

Of the many landscape designers involved in the English landscape movement one name stands out, that of Lancelot 'Capability' Brown. While he was not the originator of the English landscape style, he was its most successful proponent, undertaking more commissions than any of his rivals and achieving national fame in his lifetime. This section of the chapter examines the contribution of Brown and his contemporaries to creating a landscape industry in Britain. The contribution of Humphry Repton, widely regarded as the successor to Brown, is considered later.

Lancelot Brown (1716–1783) was born at Kirkharle in Northumberland, the son of a prosperous farmer. In 1739 he moved south where he gained a reputation as an engineer with experience in water management. In 1742 he took up a post at Viscount Cobham's gardens at Stowe in Buckinghamshire and his career in landscape gardening really began. While still at Stowe he undertook several commissions for other clients and after the death of Cobham in 1749 Brown moved to Hammersmith to establish himself as an independent designer (Stroud 1975, 53–55). His career took off remarkably quickly and by 1760 he had undertaken perhaps thirty major commissions. However, it was the 1760s and 1770s that really brought him success and public recognition as the most important landscape designer of his age (Brown and Williamson 2016, 9). He gained the nickname 'Capability' because he would tell his clients that their property had 'capability' for improvement.

Brown's landscape style evolved over time and was not always the same as some have suggested. His early commissions were focused on transforming the immediate setting of the house, where walls were removed and fashionable pleasure grounds were laid out; and on creating lakes (Brown and Williamson 2016, 75). It was only with designs like Bowood in Wiltshire in 1763 that his fully developed style emerged, with its circuit rides, perimeter belts and extensive planting. By the late 1760s his designs were invariably envisaged on a large scale, embracing entire landscapes and detailing the planting a mile or more from the house. While Brown destroyed many trees to create his new landscapes, he was also responsible for a colossal amount of planting, especially elm, oak, beech and ash. On just one project for Lord Donegall at Fisherwick, Staffordshire, he planted 100,000 trees. The eminent garden historian Christopher Hussey (1975, 35) calls Brown the 'Director-General of tree planters'. In a much-quoted passage Brown acknowledges his reliance on trees and explains how there was 'so much Beauty depending on the size of the trees and the colour of their leaves to produce the effect of light and shade so very essential to the perfecting of a good plan' (Stroud 1975, 157). Brown had no particular supplier of his trees. Some of his clients would have had their own tree nursery to meet the demand of his plan for their own estate. Unlike London and Wise in Anne's reign, who used trees from their own nursery at Brompton Park, Brown relied on different commercial tree growers, many of which were in the London area.

Brown's landscape style was not only something that had evolved over time but was also now widely shared by other landscape designers. Some were former contractors and associates of Brown, most notably Nathaniel Richmond, Robert Robinson and Thomas White. Nathaniel Richmond (1723/24–1784) appears to have begun working for Brown in 1754, executing his designs and supervising the work. He then started taking commissions of his own from 1759 and was renowned for his design of pleasure grounds, achieving much success in the south of England. Robert Robinson (1734-1794) worked with Brown in the early stages of his career before moving to Edinburgh in 1757 where he entered into partnership with William Boutcher, a Scottish nurseryman and garden designer. They were successful in the lowlands of Scotland and Robinson openly promoted himself in the local newspapers as the 'late Draughtsman and Executor of the Designs of Lancelot Brown' (*Caledonian Mercury* 1760). Thomas White (1739–1811) is notable for establishing the most successful landscaping business of this era in Scotland. He was born in Shropshire and gained experience of landscape gardening and design working as a foreman for Brown before branching out on his own at the age of twenty-six (Turnbull 1990, 402). White prepared improvement plans for some seventy estates throughout Scotland, mainly in the Lowlands, and was known for his distinctive planting style that made frequent use of larch (House and Dingwall 2003, 137). He later shared the landscape practice with his son, Thomas junior.

There were also very successful landscape designers who had little or no connection with Brown. One was William Emes (1729/30–1803), who was head gardener at Kedleston Hall, near Derby, before he moved to live in Bowbridge House, Mackworth, where he developed his private practice, working mainly in the Midlands and north Wales (Brown and Williamson 2016, 118). Another independent designer of note was Richard Woods (1715–1793), who began his career as a nurseryman at Chertsey in Surrey in the 1740s, subsequently working with Philip Southcote on the design and planting of the *ferme ornée* at nearby Woburn Farm (Cowell 2009, 12–19). After setting up as an independent designer, he worked on commissions in southern England from the late 1750s, in Yorkshire in the 1760s, before moving to Essex and working on estates there. Although living in Essex, he was still prepared to travel considerable distances for work. The career of Richard Woods is of interest for a few reasons. His progression from nurseryman to landscape gardener and surveyor, involving the abandonment of his nursery, was rather unusual, but probably due to the risky finances of the plant trade (Harvey 1974, 97–98). Wood was uncommon among his contemporaries in being brought up as a Catholic, which was not conducive to advancement in social and business circles. Nevertheless, his religion may also have influenced some of the Catholic patrons, such as with his early commission for Sir Robert Throckmorton at Buckland House in Oxfordshire from 1758 (Cowell 2009, 10). Woods was also a landscape designer with a style quite distinct from his contemporary Brown. He was essentially a pleasure ground designer and a highly ornamental mix of trees was typical of his planting style (*ibid.*, 57) (Figure 39). He was also noted for his more pronounced use of flowers than most of his contemporaries.

Until the late 1760s, Brown, Emes, Richmond and Woods were the only significant names in the landscape improvement business (Cowell 2009, 57). Thereafter, there was an explosion of improvers as a new generation, and in many respects a new type, came on the scene. Many of these combined landscaping with another livelihood, such as surveying or the nursery trade. None of these improvers would usually create entire landscapes from scratch. While all could occasionally be asked to do this, they were more often commissioned to expand,

FIGURE 39. Richard Woods's elevated Elysian Walk on the opposite side of the canal at Hare Hall, Essex, involved a highly ornamental mix of trees that was typical of his planting style. This print dated 1791 also shows some of the associated planting near the house.

update or modify an existing park or garden, the scope of their work limited to a specific area or element of the landscape by the brief drawn up by the client. Sometimes a well-known designer would draw up the broad plan and then a less expensive designer would provide the detail.

While historians often differ over claims that Brown was an artistic genius, there is no doubt about his remarkable abilities as a businessman. There are various reasons for his phenomenal success (Brown and Williamson 2016, 152). One obvious feature was his flexibility, where different contracts and different projects were organised in different ways. Sometimes, for example, estate labour was used to implement some or all of a design; sometimes local labour was recruited and directed by one of Brown's associates. Sometimes the risk of inflated costs were shared between client and contractor and written into the contract, for example, if wages had to be increased for some reason. Another major factor was his ability to choose the right people to develop and execute his design outlines. Such men were recruited from varied backgrounds. Some came from the expanding nursery industry, since nurseries not only provided plants but often offered design advice and labour to local customers. Then there was Brown's gregarious and likeable personality. He was extremely well connected, in terms of both trade contacts and client networks. He was also blessed with the social skills necessary to converse easily with his most elevated clients, many of whom would comment on his charm and wit as well as his abilities as a designer.

Due to some recent research, we now have some idea of the scale of Brown's success and can place that in context. By the time of his death in 1783 he owned a substantial estate at Fenstanton, Huntingdonshire, and had probably worked on at least 250 sites (Brown and Williamson 2016, 9). The level of Brown's earnings as recorded in his archives is stupendous. At the height of his career between 1762 and 1779, his annual income was always above £20 million in modern values and in two years surpassed £50 million (Floud 2019, 96). However, while Brown was responsible for more individual projects than anyone else, he nevertheless produced a minority of the total number of designed landscapes of the eighteenth century, perhaps as few as 5 per cent (Brown 2000, 1). From these statistics it is clear that by the time Brown died there was now a landscape *industry* in Britain. This had grown out of all recognition from the days of London and Wise at the start of the eighteenth century when this one company had dominated the scene. There was now a myriad of businesses ranging from the substantial enterprises of Brown and his competitors engaged on commissions across the country through to countless individuals working on a small scale locally, often alongside some other business interest. Furthermore, a major focus of this industry was on tree planting and management.

Linked to the emergence of a landscape industry in the second half of the eighteenth century was the substantial increase in forestry planting on many estates. While this tree planting was motivated primarily by the economic interests of timber production, this was often viewed under the general umbrella of estate 'improvement'. A source of encouragement in this planting effort was the Society of Arts (now the Royal Society for Arts) which from 1758 offered premiums for the successful establishment of tree plantations. Between 1758 and 1835 the Society awarded numerous gold and silver medals to stimulate large-scale plantations, resulting in some 60 million trees being planted. This included some enormous forestry projects that now extended outside the bounds of landscape parks into the countryside beyond. They were initiated by some far-sighted individuals, such as Thomas Johnes at Hafod, Cardiganshire, who planted almost 5 million trees in the period 1782–1816 (Linnard 1971, 153). In Scotland, the landscape of Perthshire was transformed by three successive Dukes of Atholl, who between 1740 and 1830 planted 14 million larches. Regardless of the incentive of medals, planting on such a scale required a deep purse and for many major landowners this came with the dramatic growth in economic activity that occurred nationally from the 1760s and especially from the 1770s. In the countryside this was characterised by profound changes in agricultural practice leading to substantially increased incomes. This was also a time of relative political stability, which, together with the security of a well-established system of inheritance, gave landowners the confidence to undertake such a long-term investment as forestry with deciduous trees. Where landowners wanted a quicker return on their investment and growing conditions were favourable, conifers were now becoming acceptable and their planting increased appreciably in late Georgian era on the heaths and commons then being rapidly enclosed. On the sandy soils of Welbeck the Duke of Portland nurtured his larches and pines 'with as much pride as his oaks' and Thomas Coke, Earl of Leicester, was nicknamed King Pine for his extensive planting of conifers on his Holkham estate. Of course, there was always a potential conflict between forestry planting and landscape aesthetics, particularly on those estates that had areas dedicated to both within the landscape park. However, in these situations the commercial areas of woodland were invariably located in the more obscure or peripheral areas of the park, where clear felling and replanting could occur without ruining the view.

Many landscape companies that were planting trees in parks and pleasure grounds were also involved in large-scale commercial forestry planting, which usually included drainage and fencing work. One of the most notable was Thomas White, the man who had formed his successful landscape gardening business after working for Brown. White and his son were now equally successful in delivering large-scale forestry planting projects in Scotland and northern England (White 1813, 94; Turnbull 1990, 333). In recognition of their extensive and high-quality forestry schemes, the Whites received the largest number of medals from the Society of Arts, eleven in all, with eight of these for planting at their own Butsfield estate in County Durham.

The Picturesque movement and appreciation of ancient trees

As with various garden styles, after many years of unchallenged dominance there was a major reaction against the naturalistic approach of Brown and his contemporaries. When the reaction against Brown set in the main charges laid against him were his excessive formalism and lack of 'naturalness'. Taste was moving on and some authoritative voices in garden design were now appreciating more wild and windswept scenery, such as in the Lake District. The once-praised 'natural' style of Brown now appeared 'artificial' and even 'formal'. The first serious attack on Brown came from Sir William Chambers in 1772 in his *Dissertation on Oriental Gardening*. Chambers (1773, v) criticised Brown's approach to landscaping where 'our gardens differ little from common fields, so closely is vulgar nature copied in most of them'. The historian John Phibbs (2017, 269–271) gives a good summary of the main objections to Brown's work from his contemporary critics such as Uvedale Price and Richard Payne Knight.

The driving force behind this reaction against Brown and his followers was the rise of a new movement in landscape and art that became known as the Picturesque. If anyone can be described as the Father of the Picturesque in landscape design it has to be the clergyman William Gilpin (1724–1804). While Gilpin never owned or designed a park, he was a hugely influential author based in the New Forest in Hampshire. In 1782 he published *Observations on the River*

Wye and several parts of South Wales, etc. relative chiefly to Picturesque Beauty, although he wrote this in 1770 (Gilpin 1782). Other books followed, including *Remarks on Forest Scenery*, which contains some inspired descriptions of trees and woodlands in their natural state (Gilpin 1794).

The term 'picturesque', meaning literally 'in the manner of a picture', needs to be understood in relation to two other aesthetic ideals: the 'sublime' and the 'beautiful'. The sublime refers to that which is wild, rough, dangerous and abandoned, such as mountainous regions and virgin forests. The beautiful refers to something physical that generates feelings of pleasure, which is both sensual and safe. The picturesque is a hybrid aesthetic category, representing a way of viewing or describing nature, which can be inspirational but also controllable. Its key features are asymmetrical and sinuous shapes, the rugged, the irregular, light and shade, contrast and variety. In Gilpin's view the picturesque wild landscape contrasted dramatically with the fashionable garden landscapes of the day. He remarked, 'How flat, and insipid is often the garden scene, how puerile, how absurd! The banks of the river how smooth, and parallel! The lawn, and its boundaries, how unlike nature!' (Gilpin 1792, 57).

Two of the most prominent advocates of the picturesque approach to landscape design were Uvedale Price and Richard Payne Knight, the men who had been at the forefront of the attack on Brown. Both published important contributions to the literature of the Picturesque movement. Knight published *The Landscape: A Didactic Poem* in April 1794 (Knight 1794). Price's *Essay on the Picturesque* was published a few months after Knight's contribution (Price 1794). Both were taking Gilpin's ideas about natural 'picturesque' scenery and applying the principles to landscape gardening. Both not only attacked Brown and his 'imitators' but also Humphry Repton, the leading landscape designer of that time, who we will discuss shortly. There was much heated debate surrounding the ideas of these men as this 'picturesque controversy' became a matter of considerable interest among polite society. Rather than discuss that in detail here, we are fortunate that Knight summarised the differences between the 'Brownian' and the 'Picturesque' landscape in two illustrations for his book *The Landscape* (Figure 40). The images were derived from drawings by Thomas Hearne, for which he was given written instructions, parodying Repton's technique of showing developments by contrasting scenes. The bottom plate shows the scene 'dressed in the modern style'. The whole scene is typical of the landscape park with its smooth, rather lifeless look. The trees have been thinned to a few clumps dotting shaven lawns. The stream has been channelled to make a placid, serpentine stretch of water over which there is a Chinoiserie bridge. The top plate shows an unimproved scene, pleasingly picturesque and full of life. The view looks out from the edge of an ancient forest across the boulder-strewn and fern-covered foreground. There are some dead branches and a small tree that has fallen over. The stream has a rippling, abandoned look, over which is a rustic wooden bridge. The message of these two images could not be clearer: the cold and lifeless scene of the landscape park needs urgent transformation to restore the landscape and give it back its natural energy and grandeur. Both Knight at his own home of Downton Castle, Herefordshire, and his near neighbour Price on the family estate of Foxley, tried to put their ideas about landscape into practice. They also inspired a great many other landowners to do the same.

FIGURE 40. These two images of the same imaginary property appear in Richard Payne Knight's book *The Landscape* (1794). The bottom plate shows the scene 'dressed in the modern style', while the top plate shows an unimproved scene, pleasingly picturesque and full of life.

Regardless of the differences in the overall appearance of the picturesque landscape compared to the Brownian park and pleasure ground, both relied equally on the presence of trees and woodland to create their effect. Uvedale Price was very clear about the crucial role that trees played in the 'improvement' of landscapes as he imagined this should always be done. He declared:

> It is in the arrangement and management of trees, that the great art of improvement consists: earth is too cumbrous and lumpish for man to contend much with, and its effects when worked upon, are flat and dead like its nature. But trees, detaching themselves at once from the surface, and rising boldly into the air, have a more lively and immediate effect on the eye. They alone form a canopy over us, and a varied frame for all other objects; which they admit, exclude, and group with, almost at the will of the improver. In beauty, they not only far excel everything of inanimate nature, but their beauty is complete and perfect in itself, while that of almost every other object requires their assistance.
>
> Price 1796, 285–286

As the Picturesque movement was emphasising the sublime and beautiful in the natural landscape, there was also evidence of a growing appreciation of the value of ancient trees. Furthermore, this was not just about their value as private property, although that was also very apparent. The legislation relating to the protection of trees and forests remained extremely severe throughout the eighteenth century. The Black Act of 1723 decreed the death penalty for a number of offences in royal forests or chases, including cutting down or otherwise destroying 'any trees planted in any avenue, or growing in any garden, orchard or plantation, for ornament, shelter or profit' (Thompson 1975, 271). This law was enforced on many occasions throughout the eighteenth century and even as late as 1814 when an Essex labourer was hanged for cutting down an orchard (*ibid.*, 255). In July 1766, a new Act of Parliament came into force 'for the better Preservation of Timber Trees, Woods, Underwoods, etc.' that were outside royal forests (*Manchester Mercury* 1766). Punishment for destroying or damaging timber trees ranged from a fine of £20 plus costs for a first offence to seven years' transportation for a third offence. The Act's definition of timber trees included most trees commonly grown in parks and gardens, so trees did not have to be in a forestry plantation to be covered by the Act. For stealing or damaging shrubs, plants and bulbs, the punishment was less, only £2 for a first offence, but for a third offence the punishment was the same as for trees, seven years' transportation. In the highly materialistic and unequal Georgian world, all private property was sacrosanct and trees were no exception. However, there is evidence towards the end of the century that the ruling aristocracy and gentry were also developing a genuine appreciation of the beauty and majesty of our

ancient trees. In the excellent book by Keith Thomas (1983, 212–223) entitled *Man and the Natural World*, the author goes into detail about the connections then being made in literature and the arts that highlighted the association between ancient trees and family histories, seasonal celebrations, religion, eternity and much more. Of interest here is the publication of some early books that celebrate ancient trees. One of the first of these focused on trees that stood in the park at Welbeck, Nottinghamshire, seat of the Dukes of Portland.

In 1740, the third Duke of Portland engaged Francis Richardson to restyle the 900 acre (364 ha) park at Welbeck with lawns, clumps of trees and a lake as its centrepiece. This large-scale and long-term programme of landscape improvements was directed by William Speechly, his resident gardener from 1767 to 1804. The pride of Welbeck was its trees, which included many oaks that were remnants of ancient woodland in Sherwood Forest that had been incorporated into the improved landscape. These ancient trees were to become famous thanks to the work of Hayman Rooke, a Londoner who settled in Nottinghamshire after retiring from military service. Major Rooke took an interest in local natural history and was fascinated with the nearby ancient oaks. This interest in trees led Rooke to become friendly with the Duke of Portland, who agreed to fund a survey of the old oak trees on his estate. As a result of this survey Rooke published a book entitled *Descriptions and Sketches of Some Remarkable Oaks in the Park at Welbeck* (Rooke 1790). Rooke was a talented artist and the book contains some wonderful images of the trees. Towards the end of the book, Rooke illustrates one of the trees that was known locally as the Cockpen Tree, although his caption just refers to this as 'An Ancient Oak in Birchland Wood' (Figure 41). This tree is still standing today and is probably the most famous tree in England. Because the tree was unnamed in Rooke's book, visitors began referring to it as the Major's Oak, although some also called it the Queen Oak. However, in the twentieth century it became known worldwide simply as the Major Oak.

What is especially noteworthy about Rooke's book is that it highlights the ancient trees themselves and not the overall landscape. Of course, Robert Plot had been the pioneer of recording ancient trees back in the mid-seventeenth century, and John Evelyn and Moses Cook had utilised his work. However, Hayman Rooke is probably the first to publish an entire book on this subject and in the process began our popular fascination with the subject of ancient trees. It is interesting that, from the literature and landscape art of this time, there is no indication of any tree surgery on these ancient trees, such as the propping or bracing of heavy branches. It seems likely that this would have been viewed as unnatural and unsightly and not in keeping with the values of the English landscape garden or picturesque movement.

FIGURE 41. This illustration by Major Hayman Rooke appeared in his book *Descriptions and Sketches of Some Remarkable Oaks in the Park at Welbeck* (1790). The caption just refers to this as 'An Ancient Oak in Birchland Wood' but it would eventually become known as the Major Oak and is now probably the most famous tree in England.

Humphry Repton, the consummate professional

Humphry Repton (1752–1818) is widely regarded as the last great landscape designer of the eighteenth century and the worthy successor to 'Capability' Brown.

Born in Bury St Edmunds, Suffolk, he was the son of a collector of excise and as a child spent some time in the Netherlands. He embarked on his landscape career in 1788, aged thirty-six, after a series of unsuccessful ventures. He quickly established his name as the leading landscape designer in Britain, which had lacked a national figure in this field since the death of Brown in 1783. In a career spanning nearly thirty years he carried out over 400 commissions, many of which involved more than one consultation, some lasting for years. He worked at various locations throughout England and a few in Wales but never in Scotland.

With regard to Repton's landscape style, a few observations should be made. He was not dismissive of his predecessor's achievements and wrote in 1806 of 'Brown's style corrupted'. He continued, 'Mr. Brown, after his death, was immediately succeeded by a numerous herd of his foremen and working gardeners' (Repton 1840, 328). He viewed many of these as having 'introduced all that bad taste which has been attributed to [their] great master, Brown'. At the same time as defending Brown, Repton objected to the aesthetic theories of Price and Knight, the 'wild men' of the Picturesque, as he believed they did not take into account the differences between painting and horticulture, differences that were essentially in the materials employed. Repton's early views on tree planting tended to be formalistic. He recommended conifers as complements to classical architecture for the same qualities of shape, size and lighting that he recommended deciduous trees as complements to Gothic architecture (*ibid.*, 56–60). By the 1800s, Repton had become more flexible and imaginative with the use of deciduous trees in his designs, although increasingly intolerant of conifers and of those who planted them (Figure 42). Repton's contribution to landscape design marked the end of Brown's etiolated Arcadia and provided the essential link with Loudon's Gardenesque style in the early nineteenth century. Regardless of the debate around Repton's place in the history of landscape design, what is significant here is his enormous contribution to the development of professional landscape gardening.

Repton refashioned landscape gardening as a profession from a range of paid skills and amateur accomplishments concerned with the portrayal and improvement of landed property (Daniels 1999, 1). Instead of following Brown's 'design and build' approach, Repton detached the contracting side of the business and supplemented the consulting side. To understand his approach to landscape design and the methods he used, we can not only examine his surviving landscapes but more informatively read about this in the books that he published. In seeking to raise the status of landscape gardening further as a polite art Repton felt obliged to publish various treatises on its theory and practice. His first book, *Sketches and Hints on Landscape Gardening*, was published in 1795, and his second, *Observations on the Theory and Practice of Landscape Gardening*, was published in 1803. He published his last treatise in 1816 entitled *Fragments on the Theory and Practice of Landscape Gardening*. All these books were later republished by Loudon as Repton's collected works (Repton 1840).

FIGURE 42. Humphry Repton depicted at the height of his career sketching trees. He was always fascinated with the different shapes, colours and texture of trees, which he expressed in his love of sketching them.

While Brown and his competitors had drawn large-scale plans for their clients, Repton provided watercolour drawings and detailed written instructions, often bound in morocco in the form of so-called Red Books. The texts of the Red Books were theoretical as well as practical, making observations on landscape aesthetics as well as actual landscape on site. Landscape gardening for Repton was never merely a business or a vocation concerned with the improvement of grounds or the design of gardens, it was a profession that gave him access to the best circles of polite society and recognition of his undoubted abilities from his most distinguished clients. Repton certainly had a good grasp of how to promote his image and his services to his potential clients. In the front of each Red Book he pasted his trade card (Figure 43). Engraved to Repton's design, it depicts him as a surveyor, at his theodolite and taking a bearing. The card also gives him the title 'landscape gardener', a term he coined himself. In the introduction to his first book, Repton explains that he adopted the term landscape gardening 'because the art can only be advanced and perfected by the united powers of the *landscape painter* and the *practical gardener*' (Repton 1840, 29). He added, 'The former must conceive a plan, which the latter may be able to execute'.

In managing to elevate landscape gardening to the level of a profession and working essentially as a landscape consultant, Repton was assisted by some of the dramatic technological advances of his era. In travelling around the country quite rapidly and visiting many sites in quick succession, he was assisted by new and improved systems of road travel. Between 1750 and 1811 journey times for the fastest coaches between London and major cities were cut by two-thirds

FIGURE 43. Humphry Repton's trade card, engraved to his own design, depicts him as a surveyor, at his theodolite and taking a bearing. The card also gives him the title 'landscape gardener', a term he coined himself.

(Chartres and Turnball 1983, 64–99). People, goods and information were on the move, at faster speeds, at increasing distances. Those travelling professionally, such as Repton, were joined on the road by those travelling for pleasure, for example, visiting friends, country houses, resort towns and race meetings. Repton's standard charge of 5 guineas a day and expenses was revised according to the distance he travelled (Daniels 1999, 37). His correspondence from 1807 records rates of 10 guineas for a first visit to within one stage of London, 50 guineas up to 100 miles (161 km), 70 guineas up to 140 miles (225 km) and so on in proportion. This was broadly consistent with the travel expenses charged by the architects he worked with. He charged return visits at half these rates. Repton's task of supplying his clients with work-related documents without delay was made possible by a reformed and more extensive postal system. This operated with greater speed and security for small packages and accelerated the circulation of information via newspapers, magazines and correspondence of all kinds (Gregory 1987).

While Repton dominated the profession he set out to redefine, his career was troubled by controversy and disappointment. In one of the final passages of his last book he declares, 'I have lived to see many of my plans beautifully realised, but many more cruelly marred; sometimes by false economy; sometimes by injudicious extravagance' (Repton 1816, 237). He goes on to state, 'I have also lived to reach that period when the improvement of Houses and Gardens is more delightful to me, than that of Parks or Forests, Landscapes or distant prospects'. This last remark reflects the changing nature of the commissions that Repton undertook and the type of client that engaged him towards the end of his career. Noting in 1802 that he had earned just half the amount he had in 1799, he blamed the impoverishment of prospective clients on heavy war taxation (Repton 1816, viii). As his workload on large estates was diminishing, his clients now were more likely to be located in and around the major cities and the properties themselves were urban and suburban villa gardens. Very few of these commissions later in Repton's career were commemorated in a Red Book, and he gives the impression he would rather forget these urban assignments. He clearly laments the situation when he comments:

> It seldom falls to the lot of the improver to be called upon for his opinion on places of great extent … while in the neighbourhood of every city or manufacturing town, new places as villas are daily springing up, and these, with a few acres, require all the conveniences, comforts, and appendages, of larger and more sumptuous, if not more expensive places. And … these have of late had the greatest claim to my attention.
>
> Repton 1816, 69

In the twilight of his career Repton found himself designing mostly small-scale urban landscapes for clients who had neither the resources nor the land to make the most of his talents to work on a grand scale. However, these urban commissions did not necessarily mean his clients were the *nouveaux riches.* Occasionally, Repton was asked to work on commercial sites that were subject to various urban pressures, such as speculative housing and public recreation.

In doing this, he got a glimpse of the type of work that would engage many landscape gardeners of the nineteenth century and beyond.

Pleasure gardens and the urban realm

The study of eighteenth-century urban gardens is often overlooked in garden history, where the country landscapes of the elite have been the focus of attention. However, there are notable exceptions, such as the work of Todd Longstaffe-Gowan (1990, 2001, 2012) and Joy Uings (2013). During the eighteenth century the ownership or use of a small private town garden became a common expectation and popular amenity for much of Britain's urban middle class. In London the expansion of these town gardens was particularly marked after 1750 and many were located to the rear of the rows of newly built terraced houses in fashionable areas such as Bloomsbury and Belgravia. The end of the eighteenth century saw the emergence of tiny front gardens in some of these more expensive urban developments. However, the vast majority of the urban poor in towns and cities throughout Britain continued to have no access to a garden of any kind.

Where urban and especially suburban gardens contained some mature trees, this tended to add to the desirability of the property. This is evidenced by the fact that ornamental trees and specimen fruit trees feature prominently in the sale particulars for suburban properties from the early eighteenth century (Longstaffe-Gowan 2001, 86). Nevertheless, trees only appear to have been a valuable commodity if the garden was relatively large, the fruit trees in a 'bearing state' and the ornamental trees were 'of sufficient size to produce considerable effect'. The value of established trees in residential properties was reflected in some of the new suburban developments where it became the practice to plant trees in advance of the main building work (Johnston 2015, 111).

One form of urban greenspace that was prominent in the eighteenth century was the pleasure garden, although not to be confused with the pleasure garden or pleasure ground on a country estate (Wroth 1896; Coke and Borg 2011; Conlin 2013). After the restoration of the monarchy in 1660, pleasure gardens emerged as a uniquely British form of public open space in many towns and cities. They can be defined as privately owned enclosed ornamental grounds, open to the public as resorts or amusement areas and operated as a business. They are widely regarded as the forerunners of today's public parks, although there were some significant differences. While pleasure gardens were generally open to all members of the public, they had to pay an admission charge and entry could be refused to those regarded as 'undesirable' for whatever reason. While the admission charge was usually low, significant additional costs could be incurred, such as travel expenses to get there and to use the dining facilities and bars, which would preclude many working-class patrons. The gardens were mainly a summer phenomenon, usually open for business around May to September, which was understandable given their reliance on outdoor entertainment. Furthermore, the most famous and popular venues, such as Vauxhall and Ranelagh in London, only really came to life in the evenings. Pleasure gardens were usually located on the fringe of urban centres, where land values were lower but where the gardens were still accessible to large numbers of people.

Trees, shrubs and other forms of greenery were a universal feature of the pleasure gardens and had both aesthetic and functional roles. The larger pleasure gardens usually had extensive tree cover in quite formal arrangements. There were avenue plantings along the main walks which were set out in straight lines or gently curving to ensure plenty of visibility for those promenading. At the main entrance to Vauxhall, the Grand Walk lined with elm trees led away for more than 300 yards (274 m) to the eastern perimeter of the gardens (Downing 2009, 23). These open avenues contrasted with the denser tree planting along paths that led off the main thoroughfares where couples could find some seclusion and privacy. In Vauxhall these secluded paths were known as the Dark Walks. Trees along the main paths were used for the attachment of lamps that not only illuminated the way at night but also lit up the trees and highlighted their natural beauty, shape and texture in an atmospheric glow. Large forest-type trees such as plane, lime and elm were used to mark entrances to the gardens, such as at Vauxhall where the main entrance was said in 1767 to be 'planted on each side with very lofty trees' (Anon 1767, 177). Trees also provided the framework and backdrop to the clearings where open-air displays and entertainments took place, such as for fireworks, music, dancing, theatre and the occasional ascents in a hot-air balloon. The larger gardens were keen to promote their collections of trees, shrubs and other plants as a main attraction in their own right. Because these were mostly laid out in formal arrangements their designs were largely resistant to the English landscape style of naturalism that was popular on country estates. Many pleasure gardens included landscape features from the era of Baroque gardens that were now out of fashion elsewhere. For example, the maze was a popular feature of some gardens and provided much amusement, particularly when darkness descended and alcohol took effect. Many gardens also had arbours located along their shady walks, which served as discreet resting places for couples and locations where small parties could gather, such as those noted by the tourist Resta Patching at Birmingham's Vauxhall Gardens in 1754 (Mitchell 2018, 84). Topiary shapes and figures were often dotted at regular intervals along the sides of the walks. As might be expected, this unfashionable formality in the gardens did not please everyone and to provide a contrast some larger gardens created wilderness areas to give an impression of the countryside. As well as providing contact with plants and nature, formal or otherwise, the pleasure gardens also packaged the most exciting aspects of city living, such as encounters with the elite and the latest music, fashions and novelties.

The creation of pleasure gardens in many towns and resorts throughout Britain created major new opportunities for those involved in the design, creation and maintenance of gardens. Since much of this involved the use of trees and shrubs, arboricultural work formed a major element of this. Information on tree work in pleasure gardens comes largely from contemporary images of a few well-known gardens. It also comes from some enlightening contemporary accounts of visitors to the gardens. The juxtaposition of large crowds of people with extensive tree and shrub cover presented the garden managers with significant challenges. This was particularly the case as the busiest times were in the evening and at night when the visitors were in high spirits, there were ongoing programmes of sometimes risky entertainments and large parts of the garden were intentionally poorly lit. The combination of these factors could be a health and safety nightmare. Ensuring that the trees did not make the situation worse would have required some careful thought to tree management and a regular programme of tree inspections.

Much routine tree work is best done in winter and conveniently this was when the pleasure gardens were closed. The formal features, such as the topiary, arbours and mazes, would have required fairly intensive pruning regimes to keep them looking neat and tidy, and this would have been done in the mornings before the gardens were open. Ensuring that the trees did not present a hazard or obstruction to the visitors would have required routine pruning. Trees and shrubs were frequently damaged, which might require broken branches being pruned back or complete removal. Damage could occur when drunken visitors swung on branches or tried to climb trees, or just the result of some routine activity or accident. For example, all the illuminations hanging from the trees at Vauxhall had an adverse effect on the trees themselves, when leaves were scorched or oil spilt (Wroth 1896, 314). At Vauxhall the trees in the Grove and other busy areas were crown-lifted to provide clearance for plenty of activity at ground level (Figure 44). This pruning was also done to the trees at the sides of the Grand Walk and other main avenues to avoid the company tearing their fine clothes on low branches. In the Dark Walks at Vauxhall the lower branches of trees and shrubs were retained, although cut back from the paths, to ensure dense cover provided sufficient privacy for those using these areas. On some walks the higher branches on either side of the path were trained to form a closed canopy, creating particularly shady and secluded conditions and plunging the walk into shadow even before dusk. Creating these shadowy conditions was a deliberate policy of the owners as they knew that one of the inherent attractions of the gardens was the opportunity for sexual freedom offered by its more secluded areas (Coke and Borg 2011, 75–76; Corfield 2012, 16). However, this did present some challenges for management. Sexual harassment and abuse of unaccompanied women who strayed into these areas was commonplace. Pickpockets and footpads (muggers) were a constant problem in and around the gardens but the secluded walks provided them with ideal cover. To police inside Vauxhall Gardens, constables were employed who were given distinctive blue uniforms, but they were not able to watch everybody all the time. Local magistrates were sufficiently concerned about these threats

posed to public safety and morals that in 1764 the owner of Vauxhall Gardens was forced to install artificial lighting in these areas, but this was done above the closed canopy, thus ensuring the gloomy conditions largely remained. One very unromantic aspect of these secluded areas was that some visitors relieved themselves under the trees, even when toilets were provided within the site (Coke and Borg 2011, 202). This must have made working in these areas most unpleasant for the gardeners and arborists.

FIGURE 44. Many commercial pleasure gardens had extensive tree cover, which required careful management to avoid this becoming a hazard or obstruction to the visitors. This print of Vauxhall Gardens from 1737 shows the trees in the Grove with their lower branches removed to provide clearance for all the activity at ground level.

Many pleasure gardens became less tree-covered over time. This happened at Birmingham's Vauxhall Gardens where the gradual removal of tree cover is evident from a comparison of various maps (Mitchell 2018, 84). London's Vauxhall also saw a gradual felling of trees over the years, again evidenced by maps and plans (Coke and Borg 2011, 417–422). A substantial amount of the felling of trees at Vauxhall occurred in 1735, to accommodate major improvements to the gardens that included the widening of walks and the construction of new built facilities (*ibid.*, 59). The tree cover in many gardens, which was initially very dense, may also have been reduced to cut back on costs and potential liability.

The composition of staff at the larger pleasure gardens would have included many service workers and artisans (Coke and Borg 2011, 182). There would also have been gardening staff, some of whom would have been employed throughout the year. These permanent gardeners would be supplemented to a greater or lesser extent by temporary staff, called upon when there was a heavy workload or a specialist task. It is likely that major tree work, such as tree felling or tree moving, would have been done by specialist contractors.

At the beginning of the nineteenth century, pressures were mounting on pleasure gardens and many had already closed. Their popularity and viability was waning and they would be eventually replaced by the public parks, owned and managed by local authorities (Johnston 2015, 148–150).

Some tree experts and their books

Before discussing the development of arboricultural practices in this era, a brief account will be given of some of the individuals who were recognised as tree experts, particularly in relation to amenity trees. Mention will also be made of their more important published works. The eighteenth century was the age of the scientific gardening book, where rigorous experiment and personal experience came to be increasingly valued as the only reliable basis for sound practical knowledge. The old truths passed down from the classical authors and those long-held folk beliefs were being challenged on many fronts in this Age of Reason. A number of gardening books were now published in calendar form, following the different activities, month by month, that should occupy the gardener throughout the year. In 1752, when Britain changed over from the Julian to the Gregorian calendar, omitting eleven days in September, new editions of these books had to make some slight adjustments.

This account begins with an outstanding horticulturist who had no special expertise with trees but considerable influence on arboricultural practice. Philip Miller (1691–1771) is thought to have been born in the London area and was of Scottish descent. He was Curator of the Chelsea Physic Garden for almost fifty years and has been described as the 'the greatest British gardener of the period' (Hadfield 1960, 158). His most famous work is *The Gardeners Dictionary*, which was the standard authority on gardening for much of the eighteenth century. It was first published in 1724 as *The Gardeners and Florists Dictionary* and then went into several subsequent editions, including an abridged version in 1735. What is significant here is that anyone seeking information about trees and tree care would probably consult Miller's *Gardeners Dictionary*, often as the first book they explored. However, as time went by, it became increasingly dated as it was slow to incorporate changing terminology, tastes and ideas. Nevertheless, it remained a much-revered text during his lifetime.

One of the great tree experts of the eighteenth century was William Hanbury (1725–1778), although he is little known today among arboriculturists and horticulturists. Hanbury was rector of Church Langton in Leicestershire from 1753 and used his genius for planting and gardening to develop extensive plantations and gardens in his parish. In 1758, he published *An Essay on Planting* that set out his moral and religious views on working with trees (Hanbury 1758). In 1770, he published his seminal work entitled *A Complete Body of Planting and Gardening*, originally appearing in sixpenny weekly numbers, and then in two folio volumes of nearly 900 pages each (Hanbury 1770). This remains one of the outstanding works on horticulture, arboriculture and forestry in the English language. Unfortunately, the initial cost of this, even in weekly parts, and its limited circulation and sheer size, ensured it was never widely read. Also, for today's researchers, it is not available on the internet, only a small number of specialist libraries have original copies and these are rare and very expensive to purchase.

What is especially commendable about Hanbury is that he put his practical knowledge and religious beliefs about trees and arboriculture into a nursery enterprise that would benefit not only his parishioners but many other people. As a young man Hanbury had acquired land and in 1751 began sowing seeds from distant countries, particularly North America. In 1758, now rector of Church Langton, Hanbury put forward proposals for an annual sale of trees, shrubs and plants from his gardens and plantations, and to put the money thus made into a trust for the benefit of his church and parish (Hanbury 1758, 36). At this time his stock was valued at £10,000. When the trust was formed and the trustees appointed, Hanbury advertised a sale of 'Forest Trees of all Sorts, and of the best Breed; both deciduous and Evergreen; Great Variety of American Plants; Flowering Shrubs; of these I have a curious Collection' and much more (*ibid.*, 33). Around 1760 a large sheet catalogue was printed, inviting all gentlemen 'who will favour the Society with their orders' to direct them to any of the trustees, to Hanbury, or to his gardener (Harvey 1974, 96). His annual sales were so successful that by 1765 he was able to build an organ, install three new bells and erect a gallery in his church from the profits. It is also probable that a significant portion of the profits was used by Hanbury to fund the publication of *A Complete Body of Planting and Gardening*, which would have been a very expensive project. The trust's income at Hanbury's death in 1778 was £190 a year, and had risen to £900 in 1864 when the gardens and nurseries were closed, so it must have carried on a substantial trade by the standards of the time (*ibid.*, 105).

One of the leading figures in Scottish forestry in the early eighteenth century was Thomas Hamilton, sixth Earl of Haddington (1680–1735). While he has received much recognition for the extensive tree planting and landscaping on the family estate at Tyninghame, East Lothian, research has now highlighted his wife's leading role in this. In 1696 at the age of eighteen Hamilton married his cousin, Helen Hope. The Countess, who died in 1768 at the age of ninety-two, has been credited with being the real driving force behind the tree planting (Anderson 1953, xviii). In addition to the planting of the 800 acre (324 ha) estate with over fifty species of tree, the Countess oversaw the development of a 'pleasaunce' or wilderness area, and a bowling green that had fourteen walks emanating from its centre (Horwood 2010, 78). Her arboreal enthusiasm was much appreciated by her husband and he became a tree expert himself, and very supportive of her vast tree-planting schemes. Haddington will always be remembered for the publication in 1761 of *A Treatise on the Manner of*

Raising Forest Trees (Hamilton 1953). This was written in the form of a letter by Haddington to his grandson and, although the manuscript is not dated, it must have written before his death in 1735. Soon after its publication as a book, this became one of the most influential texts of the eighteenth century on tree planting and propagation. The extent to which the Countess contributed to it will remain a matter of speculation.

Towards the end of the eighteenth century, three further books on tree planting were published in Scotland that also had relevance to arboriculture. The first in 1775 by William Boutcher (*c.* 1725–1780), a nurseryman and landscaper from Edinburgh, was entitled *A Treatise on Forest-Trees* (Boutcher, 1775). Boutcher has already been mentioned as the partner of Robert Robinson in their Edinburgh-based landscape business. Boutcher also ran a successful nursery at Comely Bank in Edinburgh that he inherited following his father's death in 1738. He was particularly interested in tree propagation and tree care, prompting him to undertake numerous experiments to establish what was good practice. He was also very interested in extensive tree-planting schemes and their role in major landscape projects. Boutcher's experience of working with forest trees was set out in his *Treatise on Forest-Trees*. The list of some 400 subscribers to this publication is an impressive collection of the Scottish nobility and gentry. It is clear from his book that Boutcher was not an enthusiast for the Brownian style of landscaping, especially where many trees were destroyed in the name of 'improvements' (*ibid.*, 255).

The second major book on tree planting to be published in Scotland in the late eighteenth century was by James Anderson and entitled *Miscellaneous Observations on Planting and Training Timber-Trees* (Anderson 1777). Anderson (1739–1808) was a leading figure in the Scottish Enlightenment and wrote mainly on agricultural topics. This book on trees, written under the pen name of Agricola, takes the form of a series of letters that deal with different aspects of the planting and management of forest trees. Nevertheless, there are some important passages of arboricultural interest. The third Scottish text on tree planting published in the late eighteenth century was by Walter Nicol (1769–1811) entitled *The Practical Planter* (Nicol 1799). Nicol was born near Edinburgh and his father managed the gardens of the Raith Estate, near Kirkcaldy. Walter also became a gardener, firstly at Raynham Hall in Norfolk in 1787, although two years later he returned to Scotland where he was employed on a number of estates. He later became interested in large-scale tree planting and landscaping and his knowledge and experience of this is set out in his book, which many regarded as one of the best texts of planting from this era. Nicol calls himself a 'surveyor and planner' and at the start of the book is an advertisement that states:

> The Author surveys, and gives Designs for Gardens, Hot-Houses, Shrubberies, Park, Approaches, etc. in the most approved taste. Also, gives Advice and Examples for Thinning and Pruning of Neglected Plantations, in every Stage of their Growth, according to Locality of Situation; with proper Memorandums, relative to their

> future Culture. Terms, One Guinea per Day; with Travelling Charges on Horseback, or by Stage Coach.
>
> Nicol 1799, advertisement

In the mid-eighteenth century the first publication on forest trees by a Welsh author appeared (Linnard 2000, 137). It was *A Treatise on Forest-Trees*, published in 1753 by William Watkins, curate of Hay-on-Wye (Watkins 1753). This useful little booklet records the results of Watkins's own experience in the planting of trees and also gives quotations from other writers on the subject.

In 1776, John Evelyn's classic work *Sylva*, now spelt *Silva*, was published in a new edition edited by Alexander Hunter (1729–1809), a Scottish physician (Evelyn 1776). Hunter seems to have been in awe of Evelyn's venerable reputation and made no serious attempt to trace, correct or elaborate Evelyn's innumerable references (De la Bédoyère 1995, 177). Nevertheless, he did add to the text a great many notes of his own, giving a substantial amount of original information. One further book on forest trees should be mentioned. In 1794, Samuel Hayes, the Irish MP and amateur architect, wrote the first book on tree care and management in Ireland. Entitled *A Practical Treatise on Planting and the Management of Woods and Coppices*, it was based on his experience at Avondale, his estate in County Wicklow (Hayes 1794). While the book deals mainly with forestry, there are also arboricultural topics such as tree pruning, wound dressing and moving large trees. The book was not just directed at Irish landowners but also to the larger audience of their counterparts in Britain.

Two important books dealing with tree planting in the context of landscape and garden design should be mentioned. The first is by James Meader, published in 1779 and entitled *The Planter's Guide: or, Pleasure Gardener's Companion* (Meader 1779). Almost nothing is known of Meader's early life and little about his career. The title page of this describes him as 'late Gardener to his Grace the Duke of Northumberland' and it is known that he worked at the Duke's gardens at Syon Park when 'Capability' Brown extensively redesigned this. When the book was published Meader was then gardener to Empress Catherine of Russia, in charge of her palace gardens at Peterhoff, near St Petersburg (Henrey 1975, 359). Meader's book lists the trees and shrubs suitable for planting in parks and pleasure grounds. There are several columns of information on each page with great detail on various plant attributes, such as size, shape, soil requirement, flowers, fruit and method of propagation. The book also includes two folding plates showing suggested planting arrangements, one for deciduous trees and shrubs and the other for evergreens. The plates and accompanying text had a significant influence over how shrubberies and other plantations were designed.

William Marshall (1745–1818) was born in Yorkshire and was a planter in the West Indies before returning to England around 1774 to manage a farm in Surrey. He later settled at Stafford and became an eminent writer on agriculture and the rural economy. After a time in London he returned to his native Yorkshire where he purchased a large estate in the Vale of Cleveland. His book *Planting and Ornamental Gardening* was published 1785, although later editions were called *Planting and Rural Ornament* (Marshall 1785). In researching this Marshall used Hanbury as his chief authority on planting in preference to Philip Miller (*ibid.*, viii). Indeed, substantial sections are quoted directly from Hanbury and this probably contributed to the success of the book. In the Advertisement at the start Marshall not only promotes his own arboricultural expertise but also gives his views of other writers on this subject. His comments on Evelyn are typical of his forthright style: 'His book was written in the wane of his life, at Greenwich, during a long and painful fit of the gout. His *Sylva* contains many practical rules, valuable, no doubt, in his day, but now superseded by modern practice' (*ibid.*, vi). By contrast, Marshall is full of praise for Moses Cook, who he describes as 'our first writer on planting'.

Many important eighteenth-century books on trees and arboriculture were also published on the Continent. Perhaps the most significant was by Henri-Louis Duhamel du Monceau (1700–1782), the eminent French physician, engineer and botanist. Duhamel was interested in the cultivation and use of trees and his *Traité des Semis et Plantations des Arbres et de leur Culture* is his main work on the subject (Duhamel du Monceau 1760). The book discusses the soil, the climate and the choice of trees, together with their propagation, nursery gardens, planting, tree maintenance and the management of forests, with all instructions based firmly on the author's long experience of forestry and arboriculture. Duhamel's book was published in Paris in 1760 with a second edition in 1780 and although it was very advanced, it was never translated into English. The extent to which it influenced contemporary British tree experts is impossible to determine but the very limited reference made to him in their works suggests this was not great.

Arboricultural practices

The eighteenth century was a time of contradiction with regard to the practical techniques used in planting and caring for amenity trees and shrubs. This Age of Reason was the era of scientific gardening where rigorous experiment and personal experience came to be increasingly valued as the only reliable basis for sound practical knowledge. However, this often seems to have been forgotten as the old truths passed down from the classical authors and those long-held folk beliefs continued to hold much sway in horticulture and arboriculture.

Planting and post-planting maintenance

In the third edition of Miller's *Gardeners Dictionary*, published in 1737, the entry entitled 'Planting' has some sound information on the subject. This includes taking care when lifting trees in the nursery to avoid damage, good ground preparation before planting, planting at the same depth as in the nursery, staking trees against the wind and ensuring post-planting watering but not over-watering. However, there is also some surprisingly poor practice. Miller states:

> And first, as to the Roots; all the small Fibres are to be cut off, as near to the place from whence they are produc'd, as may be (excepting such Trees as are to be replanted, immediately after they are taken up) otherwise the Air will turn all the small Roots and Fibres black; which if permitted to remain on, when the Trees is planted, will grow mouldy and decay, and thereby spoil all the new Fibres which are produced; so that many times the Trees miscarry for want of duly observing this. … The next thing is the Pruning of their Heads [crown], which must be differently perfor'd in different Trees, and the Design of Trees, must be consider'd; for if they are intended for Walls or Espaliers, it is better to plant them with the greatest Part of their Heads.
>
> Miller 1737, 'Planting'

To investigate the effect of pruning the heads or crowns of trees before planting, Miller describes an experiment he undertook with two standard almonds, of 'equal Strength and Age', where one was severely pruned immediately before planting and the other was not pruned, except for removing small broken and damaged branches. He then describes the planting and the results:

> These two Trees I planted in the same Soil, and to the same situation, gave them both equal Attendance, and managed them both as nearly alike as possible; yet, in the Spring, when these Trees began to shoot, that whose Branches were intirely preserved, came out early, continued to shoot longer, and is at present much larger and in better Health than the other. And since this, I have made several other Experiments of the like Nature, which have constantly succeeded in the same manner; from whence it is reasonable to conclude, that the shortening of the Branches is a great injury to all newly-planted trees.
>
> Miller 1737, 'Planting'

It is interesting that Miller uses the experimental method of repeatable results to reach his conclusions. However, he does not appear to have undertaken similar experiments with root pruning before planting, as his recommendation to remove 'all the small Fibres' of roots before planting would result in many trees subsequently dying. Some fifty years later, William Marshall was taking the opposite approach to Miller. While not advocating root pruning before planting he is in favour of severely pruning their tops (Marshall 1785, 27–28). Some arboriculturists today still follow Marshall's approach and recommend top pruning of bare-rooted trees before planting. This is in the belief that it will help to maintain a suitable root/shoot ration by compensating for fibrous roots lost in the transplanting process.

As already noted, much of the tree planting in the eighteenth century involved the mass planting of forest trees to create extensive plantations, often for mainly economic objectives. Consequently, much of the contemporary literature relating to tree planting needs to be read in this context. In Haddington's *Forest Trees* he gives five different techniques for forestry planting: pitting, notching, screefing, mound and dibbling (Hamilton 1953, xxxv). These are methods still in general use, although the common method of turf planting is not mentioned.

The debate that began in the early seventeenth century about whether to establish woodland by seeding or by transplants continued into the latter part of the eighteenth century. In Hunter's edition of *Silva* his notes include an account of how oak plantations are formed on the Duke of Portland's Welbeck estate by his gardener, Mr Speechly, who many regarded as an authority on planting. This involves fencing off the designated area of land called a 'nursery wood' and sowing acorns, having previously spread lime and grown a crop of cabbages or turnips (Evelyn 1776, 90–100). Speechly also described a division of labour for tree planting: 'It will be proper to inform you that in the planting season we divide our hands into four classes, which we term Takers-up, Pruners, Carriers, and Planters' (*ibid.*, 95). As soon as the seedlings are pruned by the *pruners*, they are taken to the *planters* by the *carriers*, 'who are generally a set of boys or some of the worst of the labourers'. The *planters* go in pairs; one makes the holes and the other sets and treads around the seedlings. Other authorities on planting had a range of views on the seeding versus transplants debate. For example, Hanbury (1770, 1, 14) favoured raising a wood by planting seedlings rather than directly from seed, while Miller preferred the latter. Samuel Hayes was also not in favour of sowing seeds directly and preferred seedlings raised in his own nursery. He declared, 'I am confident, from my own experience, that the decided opinions of Miller, and Sir James Justice, in favour of the above practice [direct seeding], has retarded the growth of many plantations in this kingdom, and deadened the ardour of many a Planter' (Hayes 1794, 6).

Thomas White and his son specialised in undertaking extensive tree planting schemes and could be expected to know the best methods of establishing trees since their business depended on this. They recommended transplants and on exposed sites these should be no more than 18 in. (46 cm) high, otherwise they could be damaged by the wind before having a chance to establish (White 1813, 94). Using such small trees was not common practice at that time. White junior noted the reaction of some visitors to their Woodlands estate who came to inspect the planting but who, 'after riding through the plantations, began to enquire where they [the trees] were, never supposing that the little pigmies under their horses' feet were trees' (*ibid.*, 94). On sheltered sites the Whites were happy to use transplants up to 5 ft (1.5 m) high, as long as they had previously been transplanted a few times in the nursery to encourage good root growth. In all their planting schemes the trees were planted in holes, and according to White junior 'the holes were made with a hollow spade, invented with many other tools, by my late father' (*ibid.*, 93). It seems that this spade allowed for much speedier planting.

Some form of staking was usually regarded as essential when planting larger trees in landscape park situations. Speechly commented: 'The fastening a tree well is a material article in planting: for if it once becomes loose, the continual motion which the wind occasions, is sure to destroy the fibres as fast as they are procured, which must end in the destruction of the plant if not prevented' (Evelyn 1776, 96). Some form of protection was also essential to protect against browsing animals. Normally the whole plantation would be fenced, but where single trees were planted in groves they would be protected individually. 'Capability' Brown preferred wooden tree guards for these grove trees, often with three posts as recommended by Evelyn (1706, 1, 23), which did not give an overly intrusive appearance. When single trees were planted at the edge of clumps or belts in full view of the house, thorny shrubs such as hawthorn were often used to protect the young trees.

Some modern authorities have stated that during the seventeenth and eighteenth centuries a method of tree planting, now called 'bundle planting', was sometimes used for landscape effect (Green 1994). This is said to have been recommended by Evelyn and was a technique known to landscapers (Read 2000, 21). It apparently involved planting a bundle of trees in the same hole and as the trees grow up they blended together and eventually formed an impressively wide crowned tree. However, no historical evidence could be found in the literature to support this being a recognised method of tree planting undertaken for landscape effect.

Transplanting large trees

The practice of transplanting large trees was relatively common by the early eighteenth century. Evelyn, Reid and others had described various techniques for this in the late seventeenth century and in the 1700s London and Wise at the Brompton Park Nursery had undertaken some impressive tree-moving projects.

The practice of moving large trees became increasingly common as the eighteenth century progressed. According to Peter Collinson, 'Lord Petre … removed, in the spring of 1734, twenty-four fully-grown elms about 60 ft [18.2 m] high, and 2 ft [0.6 m] in diameter: all grew finely, and now [1764] are not known from old trees they were planted to match' (Loudon 1838, 1, 55). Petre also undertook several other transplanting projects on his estate at Thorndon Hall and it is thought he may have been inspired by André Le Nôtre's tree-moving exploits at Versailles. Around the same time as Petre's projects, Frederick, Prince of Wales, was organising the landscaping of his new Carlton House. He did not want to wait for the trees and shrubs to mature and insisted on the creation of an 'instant' landscape. To this end some 15,000 trees were transplanted wholesale from Robert Furber's nursery at a cost of £1,239 (£2.2 million today) (Floud 2019, 35). These included an 18 ft (5.5 m) Virginia black walnut and a 25 ft (7.6 m) tulip tree (*Liriodendron tulipifera*) that alone cost £21 (£38,120 today) (*ibid.*, 57).

Although 'Capability' Brown is sometimes incorrectly credited with the innovation of moving large trees, he nevertheless made some advances in this. When he was working at Stowe there had been no major developments in tree-moving techniques for over fifty years. He was charged with the task of moving some of Bridgeman's trees to create a more irregular vista to the lake. Instead of following the accepted convention that large trees should be moved upright,

Brown had the novel idea that they should be laid flat. He then designed a machine to do just that, which Henry Steuart later described in his *Planter's Guide*:

> It consisted of a strong Pole of considerable length, with two high Wheels, and acting on an iron Axle, which was placed at right angles to it. At the extremity of the Pole there was a smaller Wheel, that turned on a pivot, and was used for trees of more than ordinary magnitude. These two Wheels were of great strength, made nearly upright, or, in the workman's phrase, very little 'dished'. A strong Cross-bar was bolted on the axle, with a space in the middle, gradually hollowed out for receiving the stem of the Tree, of whatever size it might be. According to Brown's method, the top was pretty severely lopped and lightened, and sometimes quite pollarded. The roots were next cut round, to the depth of the fibres, and only two or three feet out from the body, and the Machine was brought up upon its wheels, as close as possible to the Tree. The pole was set upright, and applied to the stem, to which it was then lashed in the firmest manner. By a rope fixed to the top of the Pole, it was, last of all, forcibly drawn down by several men's strength; and thus the stem and the pole of the Machine, forming a lever of great power, forced or tore up the roots from their under-bed, with as many fibres adhering to them as escaped laceration; leaving the Tree suspended horizontally on the cross-bar, and ready to be drawn away root foremost, as the artist may require it. … This, we must own, was not very gentle treatment of the roots, any more than of the branches. Yet Brown's improvement possessed the double advantage of contributing, by the height of the wheels, to the safety of the Tree during the transmission; and, by materially increasing the dispatch, it proportionally diminished the cost of the process.
>
> Steuart 1828, 43–44

The source of Steuart's information about Brown's method was two of Brown's assistants, Thomas White and James Robertson (Steuart 1828, 46). The great advantage of his rather crude method was that it was quick and cheap. It dispensed with the crane and pulleys used by Evelyn and also the careful root preparation over at least a year recommended by both Evelyn and Reid. While this enabled Brown to move many trees at a significantly reduced cost, his method probably reduced the tree's chances of survival, or at least significantly reduced its vigour. However, this cannot be confirmed as there is no record of the overall success of Brown's tree-moving exploits.

Thomas Hale, the respected writer on husbandry, was not in favour of moving large trees immediately and recommended careful root preparation over one year, similar to that recommended by Reid over two years. He also suggested moving it just after a good frost and furthermore, 'Let the hole in which it is to be set be got ready; and the earth about the root well wetted before the frost, that it may bind together in a firm lump about the root: in this condition let it be removed' (Hale 1758, 296).

It was clear to Hale that moving the tree with a sizeable rootball was crucial to its chances of surviving and thriving. He also restated the accepted practice of orientating the tree in its new location in the same way it was before moving. In contrast to Hale and some other authorities, the Irish tree expert Samuel Hayes did not believe there was any benefit in preparing the root system. Based in his experience of moving large trees on his estate in Ireland, he declares:

> I venture to caution the Planter against relying too implicitly on the receipt for removing large trees given by Evelyn and others, however specious it appears in theory, viz: cutting the tree all round and filling the trench up, and so leaving it for one year before removal: I can say from my own experience on many beech trees so managed that those of the same age and upon the same ground, which had been removed at once without such preparation in a very few years surpassed them.
>
> Hayes 1794, 22

It is interesting that Hayes goes on to say that 'if the Planter wishes to try Evelyn's method, he should by all means let the tree stand two years at least before its removal', which echoes Reid's method (Hayes 1794, 22–23). However, he then states that after the two years, just before the tree is moved, 'he may prune off the smallest fibres of the last year; and carry only the more hardened and woody parts of the roots to the new site'. This removal of the vital fibrous roots would probably kill the tree and it raises questions about how successful Hayes was in moving any trees this way. Hayes describes a machine for transplanting trees that he says was introduced by 'Mr Robinson, a Scottish Engineer' (*ibid.*, 42–45). He also gives an illustration of this machine in action (Figure 45). From both the description and the illustration, the machine closely resembles that used by Brown, and later described by Steuart.

Tree pruning and felling

In the early eighteenth century there was vigorous reaction against intensive 'pruning and clipping' and greater appreciation of the natural form of trees that came with the development of the English landscape movement. Stephen Switzer, a pioneer of that movement, seems to have regarded pruning later in his career as something inherently bad, declaring:

> In the Second Place (as to Pruning) there is still much less Occasion for it, than there is for Weeding. For since Nature (or rather the great Director of Nature) never design'd any Thing in vain, the collateral Branches, or Side-Shoots, which many People are so fond of hacking, cutting, or pruning off, should always remain, and never be cut off till they grow very large, or give visible Marks of their Decay.
>
> Switzer 1734, v, 96

With the emergence of the Picturesque movement, this aversion to pruning continued. To William Gilpin even clipped hedges in the countryside were not picturesque and he complained that 'no sooner is a hedge in perfection, than it is destroyed' by 'clipping and making' (Gilpin 1794, 2, 79–80). While the popular view of the fashionable eighteenth-century writers on gardening was to avoid pruning at all costs, professional gardeners and arborists largely continued to restate the principles and practice of pruning from the previous century. Miller, the respected authority on most things horticultural, regarded

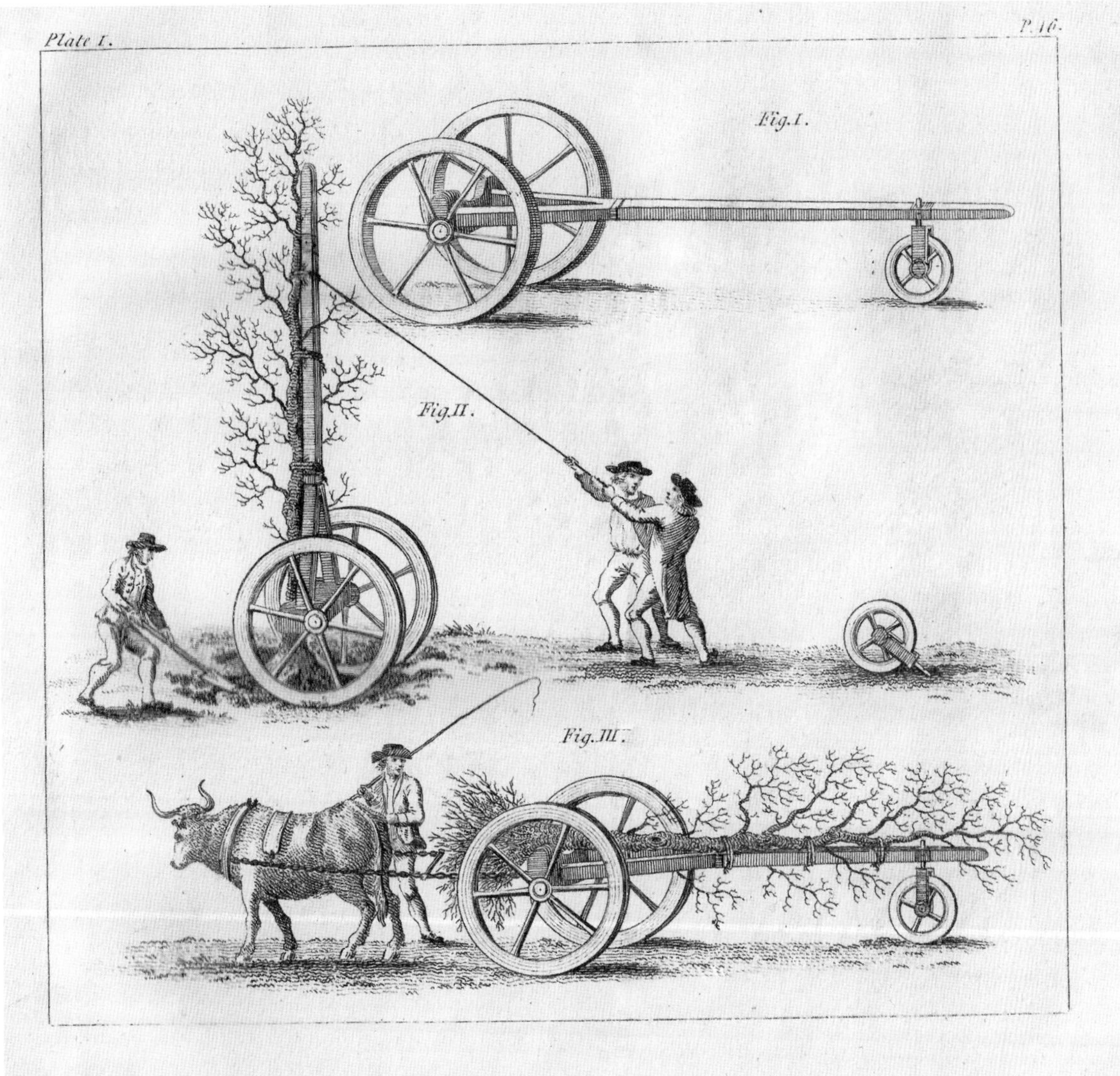

FIGURE 45. This illustration by Samuel Hayes (1794) shows a tree transplanting machine in action on his Avondale estate in County Wicklow, Ireland. Hayes says this machine, which closely resembles that used by 'Capability' Brown, was introduced by 'Mr Robinson, a Scottish Engineer'.

tree pruning as essential work, as long as it was done correctly. In his *Gardeners Dictionary* he states:

> It is very observable, that most old Trees are hollow within; which does not proceed from the Nature of the Trees, but is the Fault of those who have the Management of them, who suffer the Tops to grow large before they lop them [and] don't consider that the cutting off great Tops do endanger the Life of a Tree, or at best, wound it so that many Trees decay more yearly in their Bodies, than the yearly Tops come to.
>
> Miller 1737, 'Lopping'

Miller is referring here to pollarding, a practice that was out of favour in the landscape gardening world. By contrast, Hale (1758, 284) recognises the role of pollards in the rural economy, calling their branches 'shrowds' and the lopping of them as 'shrowding'. With regard to pruning young trees, Miller recognises the benefits of early tree training. With the pruning of mature trees, he notes that an inexperienced arborist can cause much damage by removing large branches unnecessarily. For Miller, pruning must always be undertaken in the correct manner and at the right time:

> Great Boughs ill taken off, often spoil many a Tree; for which Reason they should always be taken off close and smooth, and not parallel to the Horizon, and cover the Wound with Loam and Horse-dung mix'd, to prevent the Wet from entering the Body of the Tree. … All sort of Resinous Trees, or such as abound with a Milky Juice, should be lopped very sparingly, for they are subject to decay when often cut. The best season for Lopping these Trees is soon after Bartholomew-tide [24 August], at which Time they seldom bleed much, and the Wound is commonly heal'd over before the cold Weather comes on.
>
> Miller 1737, 'Lopping'

Much of the eighteenth-century literature on tree pruning is concerned with forest trees grown for timber or with fruit trees. Those books on forestry that include sections on tree pruning are almost exclusively concerned with the early pruning of trees to increase the quality of the timber crop and produce knot-free timber. As this is commercial forestry it is understandable that there is invariably no mention of the treatment of pruning wounds. Of the many books published on pomology, these often include extensive accounts of the best methods of pruning to increase the quantity and quality of fruit. As this is outside the scope of this book only two popular titles are mentioned. Thomas Hitt (1690–1770), a gardener who later became a nurseryman, published *A Treatise of Fruit-Trees* in 1755 which long remained the best practical work on the art of pruning and training fruit trees (Henrey 1975, 357). It is likely this book's account of tree pruning was consulted by many gardeners working with amenity trees. While it gives much information on this, there is no mention of applying any composition to pruning wounds. Thomas Skip Dyot Bucknall (1734–1806) was MP for St Albans, Hertfordshire, and an amateur horticulturist with an interest in fruit trees. For his pioneering work on pruning orchards he received a silver medal from the Society for Arts in 1793 and its gold medal in 1796. Bucknall's book *The Orchardist* was published in 1797 and is a collection of essays and testimonials describing his methods. His 'medication' for pruning wounds was a mixture of corrosive sublimate, gin and vegetable or common tar (Bucknall 1797, 14).

In contrast to tree pruning, there is surprisingly little reference to tree-felling techniques in the relevant eighteenth-century literature. This might be expected with books on gardening as heavy tree work could be regarded as beyond their scope, but the same could not be said of forestry or landscape texts. This omission is puzzling since one of the consequences of the English landscape style was the felling of many mature trees, especially in avenues and other formal arrangements, to make way for the new landscape parks. It is also

FIGURE 46. In every landscape park mature trees were sometimes removed as these died, were blown over or were deemed to be in the wrong place when parks were redesigned. This 1793 print shows workers removing a large fallen tree with an axe and saw at Caversham Park, Oxfordshire.

apparent that in every park some mature trees were felled from time to time as these died, were blown over or were deemed to be in the wrong place when parks were redesigned (Figure 46). In a rare reference to felling techniques, Hayes describes a method for felling trees in plantations, normally done as part of a late thinning. He states:

> The axe must do its office, taking care at the same time, that the falling of one tree does not injure another, which is best prevented by a young labourer attending on the feller, with a long pole, whose top is furnished with an iron fork and a hook on one side of it, by which he either draws the falling tree towards him, or pushes it off, so as best to keep clear of the branches or stem of those trees which are designed to remain.
>
> Hayes 1794, 71

Tree health and tree pathology

This era witnessed a significant growth of interest in tree pathology and how to improve tree health and vitality. A good standard of cultivation in the early years of trees and shrubs was always recognised as an important factor in promoting health. There was also a greater awareness of the impact of climatic factors, if not an understanding of the underlying biological processes behind this. For example, Switzer (1734, v, 82) writes about how the foliage of trees in avenues growing within several miles of the sea can still be damaged by salt, although he does not recognise how the damage is caused. When it came to treating various pests and diseases, a lack of knowledge of tree pathology and biological processes often resulted in some rather drastic actions. Young trees and shrubs that were infected in some way were often pulled up and burned as soon as any sign of infection was noticed. This sanitary measure has much merit and still continues today. With large trees, however, it was the custom to cut out any infected parts. Washes of various kinds were used to get rid of such things as mildew and 'blight'; some were fairly effective while others could do more harm than good. Occasionally, someone would announce that they had found a cure for one or other of the common ailments of trees and this would always have aroused interest whether it was of any value or not. One such instance caused a huge amount of debate and controversy in the late eighteenth century.

William Forsyth (1737–1804) was a Scotsman who had worked under Philip Miller in the Apothecaries' Garden in Chelsea. In 1784, at the age of forty-seven, Forsyth was appointed gardener to George III at Kensington and St James's and thus became one of Britain's foremost horticulturists. However, towards the end of his career he became involved in a horticultural 'scandal' that did not help his reputation. In 1789 he created a 'plaister' (protective coating) that he claimed was able to remedy defects in trees and, most especially, cure diseased and decayed trees 'as though they had never been injured' (Forsyth 1802, 310). Britain was then at war with France and there was an urgent need for sound timber to build warships, as much of the Royal Forests were in poor condition. Understandably, Forsyth's plaister interested the British government and they approached him to find out more. He informed them that its recipe was known only to him but this could be given to the government for a suitable fee. A government committee was then formed with instructions to pay particular attention to the effect of Forsyth's special preparation on forest trees, especially oak. When their investigations were complete the committee reported: 'We have every reason to believe … that Mr Forsyth's Composition is a discovery which may be highly beneficial both to individuals and the public' (Webber 1968, 103). On receiving this report the Treasury recommended that Forsyth should be given £1,500 at once and another £1,500 after an extended trial of the special preparation. The recipe for the plaister or 'composition' was:

> Take one bushel of fresh cow dung, half a bushel of lime rubbish of old buildings (that from the ceilings of rooms is preferable), half a bushel of wood ashes, and a sixteenth part of a bushel of pit or river sand: the three last articles are to be sifted fine before they are mixed; then work them well together with a spade, and afterwards with a wooden beater, until the stuff is very smooth, like fine plaster used for the ceilings of rooms.
>
> Forsyth 1802, 321

The instructions for using his plaister were quite precise. It involved removing all the decay back to sound wood, leaving the area very smooth and then applying a layer of the plaister one-eighth of an inch thick (3 mm) (Forsyth 1802, 321). A quantity of powdered wood ashes, mixed with powdered ashes of burnt bones, was then applied over this. This was repeated until the whole area became a smooth dry surface. Forsyth later discovered from experience that the best way of applying the composition was in a liquid state. He therefore reduced it 'to the consistence of pretty thick paint, by mixing it up with a sufficient quantity of urine and soapsuds' and applying it with a painter's brush (*ibid.*, 323). This refinement and the original recipe were included in Forsyth's book published in 1802 entitled *A Treatise on the Culture and Management of Fruit-Trees*. Although the details of his 'cure' were now made public, few had the knowledge or the time to challenge it. However, Thomas Andrew Knight (1759–1838), a Herefordshire squire, was in a position to do this as a knowledgeable horticultural scientist who had written his own book on fruit growing. In the second edition of this he added a criticism of Forsyth's 'plaister', saying he had little confidence in it. Accusations and counter-accusations were then traded between the two men and their supporters and opponents were drawn into an increasingly acrimonious debate (Webber 1968, 103–110). Despite the controversy Forsyth's *Treatise* sold well and the first edition of 1,500 copies was cleared within a year. In the third edition he took the opportunity to publish a letter signed by eight eminent medical men supporting his claims for his plaister. Far from calming the heated debate, this outraged Knight, who promptly offered 200 guineas if Forsyth or his friends could produce a single foot of timber restored from a previous state of decay. He also offer 100 guineas if a traverse section of a tree could be produced that substantiated Mr Forsyth's claim to restore hollow trees. While all this was going on, Forsyth died, but the dispute continued for another year, mainly in the pages of the *Gentleman's Magazine* and among some leading gardeners and scientists of the day.

As modern science has established, decayed wood cannot be restored. However, it seems that Forsyth was not deliberately deceiving everyone with his plaister and that he genuinely believed the claims he made for it. Regardless of the damage this episode may have done to his reputation, he was undoubtedly a very knowledgeable horticulturist and his *Treatise* remained a bestseller for many years. Nevertheless, this well-publicised controversy, and especially the implacably opposed positions of the protagonists, must have done little for the public's confidence in the tree experts of the day.

Arboricultural tools and equipment

With regard to the tools and equipment used in arboricultural operations, little progress seems to have been made during most of the eighteenth century. Those items that were in common use in the first few decades of the century, and were described in the previous chapter, were very much the standard items in use at the end of it. For working at height in trees, there is reference once again to the use of reachers, pruning hooks, ladders and specialist platforms. However, there is still no mention of any tree-climbing equipment.

Tree work accidents and risk management

With the advent of newspapers in the eighteenth century, accounts were occasionally published of tree work accidents that resulted in serious injury or death. Such accidents in agriculture or forestry would not have been uncommon as this was long before modern safety measures such as personal protective equipment and risk assessment. Most tree work accidents would have required some unusual perspective to be reported in the newspapers, for example, if the accident occurred in an urban area or if the public were involved. A search of contemporary newspapers has identified several such cases involving fatalities and three are referenced here. The first is an accident among arborists, while the second and third were accidents involving members of the public.

The accident involving arborists is particularly interesting because it relates to somebody climbing up a tree rather than just working on the ground. The incident occurred in 1767 and an account of it was reported in the *Derby Mercury*:

> On Tuesday last the following melancholy Accident happened in a Piece of Ground Called Goose-Green Orchard, near St. Mary's Bridge in this town: Three men were employed in felling some Trees in the said Orchard, and as they were at work upon one of them whose Root had been wash'd by the River Derwent, it was thought proper that a Rope should be tied as high as possible, in order to direct its course from falling in the Water; accordingly one of the men, named Joseph Nicholas, went up and ty'd the Rope, but before he could get down again the Tree fell, and broke a large Branch from a Tree adjoining, which falling upon the said Joseph Nicholas, killed him on the Spot. The Coroner's Inquest having sat on the Body Yesterday, brought in their Verdict Accidental Death. He has left a disconsolate Wife and one Child.
>
> *Derby Mercury* 1767a

The second tree work accident involved members of the public, and, although it occurred in New Hampshire in the United States, it was deemed sufficiently unusual and shocking to be reported in the British press. It is particularly interesting as one of the earliest accounts in British newspapers of a tree falling across a highway and causing death to people travelling in a vehicle. It is also unusual because the tree fell prematurely during the course of tree work. This account was published in the *Ipswich Journal*:

> The following melancholy and extraordinary accident happened in Chesterfield, in the county of Hampshire; as a man was felling a tree near a road, and when it was near falling, a traveller passing by told him he had better leave chopping for there was a sley [carriage] coming with two women and a child in it; ... upon which the man left off chopping, and walked on the road to meet them, and as they drew near the tree, the man went back to see if there was any danger, but not perceiving

> any till the sley had got just where he supposed the tree would fall, when he saw it agoing, and called to the women to drive on their horses; but they moving but slowly, the tree fell on the women, by which the child was killed instantly, the young married woman expired in half an hour, and the old lady in about two hours after.
>
> *Ipswich Journal* 1775

The third fatal accident also involved the public and occurred in Nailsworth, Gloucestershire, in 1772. It was reported in a number of British newspapers and the account in the *Kentish Gazette* read as follows:

> On Friday morning a poor woman from Nailsworth, picking up chips near a tree, which some men were hewing down, had laid her child at the bottom of the tree; the woman seeing the tree falling ran to catch up her child, but before she could get clear of it the tree fell and killed her. She had the presence of mind to throw the child from her, by which means it escaped unhurt.
>
> *Kentish Gazette* 1772

There is no doubt that the reports of tree work accidents involving the public would have raised concerns among many people. Landowners would have been made aware of their potential liability when their staff were working on trees in close proximity to people or the highway. Professional gardeners and arborists would also have been made aware of the potentially fatal consequences of not ensuring the safety of the public. Lastly, the public themselves would be alerted to the fact that they might be killed or seriously injured if they happened to be nearby when tree work operations went badly wrong. These considerations, which are obvious today, would not have concerned most people in the late eighteenth century without such incidents being reported in the newspapers.

It is interesting that the whole issue of tree risk management and public safety only begins to come to the fore in the late eighteenth century. It was noted earlier that this was an issue in the pleasure gardens and had an influence on tree management there. However, although this era witnessed the beginnings of garden tourism with some of the larger parks and pleasure grounds opening their gates to the public, the numbers were not yet sufficient for tree safety to be an important consideration for the owners or their staff.

Nurseries and tree production

Loudon calculated that in the course of the eighteenth century no fewer than 143 species of tree and 311 species of shrub were introduced into Britain (Loudon 1838, 1, 80–84). This wave of new introductions dramatically increased the range of trees and shrubs that were propagated and offered for sale by the nursery industry, especially by those nurseries specialising in unusual plants.

The demand for trees and shrubs for planting in parks and gardens continued much as it had in the seventeenth century, even though many of the planting designs were now quite different. The emerging landscape industry largely drove the tremendous demand for large quantities of high quality nursery stock.

Furthermore, the huge surge in forestry planting, particularly in the second half of the century, promoted the creation of many nurseries specialising in the production of vast numbers of forest trees. These nurseries offered trees for sale by the thousand and they were usually one-year or two-year seedlings. Larger forest trees, which had been transplanted at least once to develop their root systems, were often sold by the hundred and were available in a variety of heights. Large estates with extensive forestry plantations that wanted forest trees for the forthcoming planting season would place orders for many thousands of trees. By contrast, specimen ornamental trees in larger sizes were often priced and sold individually and orders for these were generally much smaller than for forest trees.

As in the seventeenth century, the debate continued about whether estates and gardens should have their own nursery or whether they should rely on buying in stock, or perhaps a combination of both. Miller was quite decided on this: 'There is no such thing as having a fine Garden or Plantation, without a Nursery both for Trees and Flowers' (Miller 1737, 'Nursery'). In England and Scotland, as the nursery industry expanded rapidly to meet the growing demand, many estates chose to obtain most of their stock from commercial nurseries. However, they may also have retained a small nursery of their own, where it was still cheap and convenient to satisfy some of their own requirements. On some estates where there was an extensive forestry enterprise, it was common to periodically establish a 'nursery wood' or 'nursery plantation', as referred to earlier at Welbeck. This was a patch of land, within the area dedicated to forestry, which was marked out, fenced and then directly sown with tree seed. As the young trees grew up they were gradually thinned out to the desired final density by removing surplus plants to a nursery where they could be held as further planting stock until required.

Operating a commercial nursery was a skilled but also risky business where good financial management was needed as much as horticultural knowledge and experience. The historical archives are littered with examples of nursery businesses in this era that folded, some within just a few years and others after many decades of success. Just as today, there were many challenges to overcome, including inclement weather, plant disease, theft, clients failing to pay and competition from other nurseries. Since most could not afford to buy enough land, enforced moves occurred when leases expired, resulting in the quick sale of stock. Many were family businesses that had been running for several generations, which can have both its advantages and disadvantages. Many of those establishing new nurseries were formerly gardeners in private service or had worked with one of the numerous firms of landscape gardeners. As we have noted, some people continued operating in both spheres. Indeed, it can often be difficult to make any clear distinction between nurserymen and gardeners at this time as many individuals moved freely from one sphere to the other. Furthermore, the two spheres depended greatly on each other. While the nurserymen depended largely on recommendations from clients and their gardeners, the clients in turn looked to the nurserymen to recommend to them reliable gardeners.

The eighteenth century was a period of steady growth for the nursery trade. When the Swede Pehr Kalm commented on the situation in 1748 most nursery businesses were still on a small scale. He noted, 'In England there is the advantage that in every town and large village there is one or more nurseryman, whose principal occupation is to sow and plant the seeds of a number of different kinds of trees … so that they can sell a number of all kinds … for a reasonable price to one who requires them' (Kalm 1892, 316). Around 1760, the Society of Gardeners estimated that England contained 100 nurserymen, already quite a substantial number (Cowell 2009, 66). Then, from the middle of the eighteenth century onwards, there was a tremendous increase in the number of nurseries in Britain (Harvey 1974, 94). Although not part of the Industrial Revolution and on a much smaller scale, it was an accompanying and parallel phenomenon. By the end of the eighteenth century, as with the landscape sector, there was clearly a nursery stock *industry* in Britain.

As in the previous century, London continued to be the centre of Britain's nursery trade. This was largely due to its good transport links and because the landed aristocracy, in London for the Season, could select plants for their parks and gardens at the same time. While the Brompton Park Nursery no longer existed, a few other established firms continued. Robert Furber had probably founded his nursery at Kensington Gore around 1700 and by the 1730s he had gained a formidable reputation for being able to supply a huge amount of high-quality plants (Harvey 1974, 78). The other long-established nursery flourishing in the 1730s was that of Christopher Gray at Fulham. Gray has long been famous for having issued his catalogue in 1740, under the mistaken impression that this was the first of its kind (*ibid.*, 79). As a highly skilled plantsman Gray claimed in 1755 that his nursery had a greater variety of trees, shrubs and other plants not only of any nursery in England 'but also in any part of Europe'. Two of the greatest London nurseries were founded around 1760 (*ibid.*, 88). One of these was the largest in extent and probably the biggest economically, that of John Russell at Lewisham. The other, in Hackney, was founded by the German, John Busch, and later became world famous under Conrad Loddiges.

In the rest of England and especially in the north, some excellent nurseries were flourishing by the end of the century. In South Yorkshire an important nursery had been founded before 1779 by John Littlewood at Handsworth, near Sheffield (Harvey 1974, 117–118). Littlewood also undertook tree planting on a grand scale and one of his clients was Doncaster Corporation. Another Yorkshire nursery, owned by the Telford family, was mentioned in the previous chapter. Under the leadership of John Telford senior this became the most distinguished nursery in the north (*ibid.*, 64). Telford was an improver as well as a nurseryman and was involved in a major land drainage and planting project for the Corporation of York at the Knavesmire racecourse. The brothers

John and George Telford succeeded their father John in 1762 and distinguished themselves by producing what seems to have been the first catalogue of nursery stock to be priced throughout, at any rate in England. It is dated 1775, but may possibly have been anticipated by a few months by the catalogue of Robert Anderson of Edinburgh.

According to Loudon (1835, 361), the first two tree nurseries in Scotland were established at Edinburgh, around the beginning of the eighteenth century, by William Malcolm at the Watergate and by James Gordon at Fountainbridge. Also in Scotland was one of the most significant nurseries in the whole of Britain (Harvey 1974, 117–118). This was founded by Robert Dickson in 1728 at Hassendeanburn near Hawick. The business was continued by Archibald Dickson who was selling large numbers of forest trees in the 1770s at prices that represented a big undercutting of the current price of the Yorkshire nurseries. In time they secured a substantial amount of business south of the border. In 1795, Nicholas Ashton of Woolton Hall, near Liverpool, received a Gold Medal of the Society of Arts for having planted 133 acres of enclosed land with forest trees. He had employed Dicksons of Hassendeanburn, who in November and December 1795 planted 487,040 trees on his land near Northwich in Cheshire.

In Wales, it was common throughout much of the eighteenth century for large estates to operate their own nurseries as there were few commercial nurseries in the country (Linnard 2000, 139). However, as demand for nursery stock increased, coupled with the active promotion of nursery activities by agricultural societies, this led eventually to the formation of many small and a few larger commercial nurseries in Wales towards the end of the eighteenth century.

Establishing a reputation for quality plants and satisfied customers was a major factor in the success of a nursery. Once the confidence of customers had been lost, it was very difficult to recover this. Within a few years of the death of George London and the retirement of Henry Wise, the Brompton Park Nursery went into decline because of the incompetence of their successors, Joseph Carpenter and William Smith (Harvey 1974, 10). While it is impossible to know for sure, it seems that as the nursery industry expanded, most companies enjoyed a satisfactory reputation. There were, however, still plenty of nurseries who consistently provided nursery stock that failed to live up to expectations, and their most strident critics were often other nurserymen. William Boutcher (1775, x) delivers a stinging attack on some of his fellow nurserymen for their poor quality plants, sold cheaply 'under no meaner pretence than serving their country'. He continues, 'To appear more formidable too in this quackery, they have published pompous catalogues of they know not what, collected I know not where, and strangely jumbled together nobody knows how'. He appeals to honest nurserymen to take action, since 'to stop the fast increasing evil of nurserymen (or those who call themselves so) selling bad plants, to the general discouragement of improvements, would show a truly public spirit' (*ibid.*, xiii). However, he believes that words from people like him will have little impact; it is only the public who can change the situation.

One of the major reasons for poor quality nursery stock arriving at planting sites was, and still is, delays in transportation and poor packaging of plants when in transit. Walter Davies, rector of Manafon in Montgomeryshire, described his personal experience of the problems in the 1790s that arose with buying stock from distant nurseries:

> Proprietors planting upon a large scale, and not raising trees from seed in their own nurseries, were formerly used to procure seedlings of larch, firs, and pines, etc., from Scotland; but owing to their heating in close bundles, and otherwise damaging upon the road, not above one half, and frequently not above one fourth, of the number, could be expected to grow.
>
> Davies 1810, 243

Fortunately, over much of the rest of Britain there had been major improvements in transport infrastructure. To speed up the delivery of plants the nursery industry in the eighteenth century was able to take advantage of an expanding road network, better road surfaces and an excellent network of canals. Superior methods of packaging of plants were also being employed, especially by the larger or more successful nurseries, to help ensure that plants arrived at their destination in good condition. Thomas Hitt devoted a chapter in his *Treatise of Fruit-Trees* on 'How to pack up trees so that they may be carried safe to places remote from whence they are raised' (Hitt 1757, 71–72). He advocated the provision of large hampers, especially for trained espalier trees, but had the honesty to admit that simpler packing, often straw and mats, was not necessarily harmful.

All the individuals mentioned above were nurserymen – quite literally, they were male. Now, thanks to some recent research by Dianne Barre, we can include the contribution of a woman (Barre 2018, 107–109). Ann Sandys was the wife of Charles Sandys (1725–1761) who had a nursery near Ashbourne, Derbyshire. After the death of Charles in 1761, Ann was left with at least three young children. Determined to support herself, she continued the business and after some initial cash-flow problems this went from strength to strength. In 1767 she was advertising: 'To be sold: A Large Quantity of Forest Trees of various Sizes. Likewise all sorts of Fruit Trees, Ever-Greens, Flowering shrubs' and other plants (*Derby Mercury* 1767b). By 1773, Sandys had the confidence to describe herself as 'Nursery and Seeds-woman', offering a wide range of trees, shrubs and bulbs 'with a great Variety of American Trees and Shrubs' (*Derby Mercury* 1773a). All of these were included in her printed catalogue (*Derby Mercury* 1773b). Sandys last advertised in February 1786, when her prices were reduced, so she may have retired soon afterwards (*Derby Mercury* 1786). She had run a successful nursery for twenty-five years, testament to her skills as a businesswoman and horticulturist.

Status of gardeners and tree-related professionals

Some coverage has already been given in this chapter on the status of gardeners and tree-related professionals. An account was given of the expansion of both the landscape and nursery sectors during the eighteenth century where eventually these could be described as industries. This included some of the leading figures in the landscape sector who helped drive its expansion to the point where Humphry Repton was able to elevate landscape gardening to the status of a profession rather than just a trade. In this final section of the chapter we give some further information about the status of professional gardeners and also highlight some of those who specialised in tree-related work.

As in previous the era, the term gardener in the eighteenth century embraced professionals who often had a wide range of horticultural and arboricultural skills. There was, however, a distinction emerging between those who focused on the ornamental aspects of plants and those who were concerned with their commercial aspects, such as nursery stock production, pomology and market gardening. A calculation from the London Gardeners' Company around 1760 gives the approximate numbers of gardeners of all kinds in England and Wales (Harvey 1974, 6). This estimated there were 10 garden designers, 150 noblemen's gardeners, 400 gentlemen's gardeners, 100 nurserymen, 150 florists, 20 botanists and 200 market gardeners. This gives a total of just over 1,000 principals in professional gardening in England and Wales, not counting their employees or temporary staff.

In 1747 a book was published entitled *A General Description of All Trades*, which lists these in alphabetical order along with a brief description of each (Anon 1747). It has a section covering 'Gardeners' with the expected subdivisions of ornamental and commercial horticulture. However, for 'Nursery-men' it states, 'these are also Designers, Drafts-men, and Surveyors, who draw plans, lay out Ground, direct the Plantation, and furnish everything necessary for either Use or Beauty, according to the Humour or Taste of the Proprietor' (*ibid.*, 101). The book has no listing for forester, arborist or landscaper. Also appearing in 1747 was a similar publication entitled *The London Tradesman*, which only covered the capital (Campbell 1747). This has a section headed 'Of the Gardener, etc.', which is enlightening. It states, 'The Gardener is a Country Business, but mentioned here it is a City Company. It is a healthful, laborious, ingenious, and profitable Trade. A Gardener ought to have a good Notion of laying out Grounds to Advantage, but this I rather chuse to treat under the Head of Land-Surveyor, which many of them are' (*ibid.*, 274). Again, there is no listing for forester, arborist or landscaper. However, its category called 'Land-Surveyor' has a description of this that is similar to the category of 'Nursery-men' given in the previous publication. It states:

> The Land-Surveyor is employed in measuring Land, and laying out in Gardens, and other kinds of Policy about Gentlemen's Seats. To have a good Taste this Way he ought to travel to France and Italy, and to have a Liberal Education; but especially

> a thorough Knowledge of Geometry and Designing. They may earn a Guinea a Day when employed in laying out, and are always esteemed above a Mechanic.
>
> Campbell 1747, 275

From the 1760s, various trade directories were published, usually focused on an individual town or city, which listed those involved in different trades in that area. During the 1790s, volumes of the *Universal British Directory* covering the whole country were produced, recording the names of all gardening firms of any standing. However, research has yet to be undertaken to analyse this, which may be a difficult task given the lack of differentiation and confusing terminology between the types of gardeners listed. Furthermore, as already noted, many gardeners who offered their services on the open market routinely operated under various 'hats', such as nurseryman, landscape gardener or surveyor. While offering a range of services, some stressed their skill in arboricultural work. For example, as well as running a tree nursery and garden construction service in the Newcastle area, John Pringle's newspaper advertisements specifically mention 'trees pruned and trained in the neatest and best form' (*Newcastle Courant* 1784). Also, with the rapidly expanding landscape industry, some nurseries and gardening businesses decided to diversify into the retail trade, selling tools, equipment and sundries. One example was John Adams, a nurseryman and landscaper from Aberdeen. As well as his tree and shrub nursery and garden construction business, he also sold tools and equipment, including arboricultural items such as 'pruning irons for trees and pruning knives' (*Aberdeen Journal* 1798).

The emerging landscape industry in the second half of the eighteenth century enabled a few very ambitious and talented men to make a lot of money. As we have seen, men like Brown and his immediate competitors were able to make huge profits. Also, those individuals who focused mainly or exclusively on offering a design service were also able to generate a substantial income. In Jane Austen's novel *Mansfield Park* the character Mr Rushworth mentions Humphry Repton and states, 'His terms are five guineas a day'. This is equivalent to £3,800 today (Floud 2019, 107–108). In fact his rates were usually more expensive and similar to the architects he worked with. For his Red Books Repton charged his clients between 6 guineas (£6,000 today) and £42 (£40,000 today). Around the same time Richard Woods was charging 2 guineas (£2,200) a day for his design and consultancy services. However, this scale of income was just a distant dream for the vast majority of people working on 'improvement' projects.

The landscape industry was a major employer of people with horticultural and arboricultural skills. It also employed large numbers of unattached labourers, many of whom moved from job to job and lived on the margins of society, a class of worker that now comprised a substantial proportion of the population. From the 1760s the development of Britain's canal system required a vast workforce of labourers and this initially affected the balance of the floating labour market in some regions where these were recruited locally. This often had the effect of driving up the cost of labour in the area and creating a shortage for local landscape improvers (Burton 1972, 157). However, from around 1780 these labourers (known as 'navigators' and then 'navvies') had become a body of specialised men who followed the canal-building around the country and therefore had less of an impact on local labour supply.

It was common practice for the larger landscape companies to subcontract much of their work to other smaller companies. There were definite advantages in this. If one of the subcontractors was unsatisfactory, for whatever reason, they could easily be replaced. Also, if there was some highly specialised work involved in the contract and a company was available with the right skills, it could make sense to make them a subcontractor. This might be the case with some heavy tree work, such as the sectional felling of a mature tree in a confined space or transplanting a large tree. Most of the major landscape companies had foremen who were responsible for the successful execution of its designs. A large part of that work would often involve the supervision of various subcontractors. The foreman was a key figure who had to establish a good working relationship with the head gardener or steward on the estate if the project was to proceed smoothly and be delivered on time.

Just as the creation of new landscape parks and pleasure grounds involved a large workforce, usually supplied by the landscape contractor, so the estate owner needed people to maintain these in the future. This would be led by the head gardener, although not necessarily with that precise title. As in the previous era, this was a position of considerable responsibility that took years of apprenticeship to achieve, with only a very limited number of gardeners rising to those lofty heights. Many head gardeners were competent in a wide range of horticultural and arboricultural skills and would also have acquired supervisory experience. Some were very conscious of their elevated status among gardening professionals and would make this apparent. This was the case with Thomas Mawe, whom one visitor to the Duke of Leeds in 1764 found 'so bepowdered and so bedaubed with gold lace' that he thought he was in the presence of the Duke himself (Thomas 1983, 225). While some head gardeners were relatively well paid, most received less than many other skilled household staff. William Boutcher complained: 'A great man bestows from fifty to a hundred pounds a year on a French cook; for a British gardener, seldom more than twenty to forty' (Boutcher 1775, vii). Gardeners were male, usually from a rural working-class background, and to rise up the ranks required considerable self-discipline and motivation to engage in many years of self-improvement and study. If they had not learned how to read and write at school, it was important to acquire these abilities if they were to progress to head gardener. While the eighteenth century saw a boom in horticultural publishing, books were often very expensive and beyond the means of ordinary gardeners. These books were invariably purchased by the landowner, who then passed on the knowledge to their head gardener or let him have sight of the book.

As in the previous era, Scotland continued to provide many of the gardeners working in England and Wales. There was undoubtedly more work south of the border, offering the chance to gain a wider range of experience and enhanced opportunities for advancement. As before, this influx of Scottish gardeners caused widespread resentment among their fellow gardeners in England. However, estate owners were happy to continue to employ Scottish gardeners and some went on to make a name for themselves in British horticulture (Cox 1935, 201).

The great majority of gardeners working on estates would remain in poorly paid and lowly positions. As there was little communication among ordinary gardeners in different localities to compare their working conditions, most accepted their lot. Consequently, the variation in wages for gardeners was considerable. Furthermore, there were no labour unions of any kind to agitate for better pay and conditions. While the Worshipful Company of Gardeners continued to represent professional gardeners in London, there were not yet any comparable organisations in the rest of Britain. Living conditions for gardeners could be atrocious and on big estates they usually had to live next to the gardens, meaning they were effectively 'on the job' at all times. If the gardener had a family, his wife and children might bring in supplementary income by menial tasks such as weeding.

Working-class women were often engaged as labourers in eighteenth-century gardens, which included physically exhausting jobs (Hoyles 1995, 80). As male labourers mowed the grass with scythes, women would follow them and sweep up the cuttings. Women also helped in nursery work and were employed to pick insects from plants. These very limited opportunities for horticultural work were not just restricted to working-class women. Until the twentieth century it was virtually impossible for a woman of any class to have what might be described as a career in horticulture, although this was not a situation unique to gardening.

One new type of gardener that appeared during this era was the jobbing gardener. This followed the emergence of town gardening as a commercialised leisure activity and was a product of the eighteenth-century consumer revolution (Longstaff-Gowan 1990, 125). Although many middle-class urban residents were happy to pursue gardening as a leisure activity, they were seldom prepared to do all the work involved. They would frequently employ jobbing gardeners for cheap and quick gardening services. By the middle of the eighteenth century these jobbing gardeners were a feature in London and other major cities. They were self-employed, itinerant operatives or tradesmen gardeners freed from professional indenture. They secured work either by knocking door to door or sometimes through advertisements in local papers. Jobbers catered to an urban clientele that needed occasional 'professional' gardening services and advice, but had neither the quarters nor the inclination to keep a full-time gardener.

Thomas Fairchild was one of the first to urge people to be wary of these jobbing gardeners. He declared:

> There are many ignorant Pretenders, who call at Houses where they know there is any ground, let it be in Season or out of Season, and tell the owners it is a good

> Time to dress and make up their Gardens; and often impose on them that employ them, by telling them that everything will do, when perhaps it is the wrong Season.
> Fairchild 1722, 67

Fairchild believed these men brought the gardening trade into disrepute and discouraged garden owners from employing any gardeners. Some jobbing gardeners would specifically target tree work as a source of income, in the same way that this happens today. They would spot trees growing in gardens and then tell the owners these should be pruned or removed, whether this was necessary or not. As tree work can often be difficult, demanding and potentially dangerous, it can be tempting for owners to have the work done while these 'tree experts' are in the neighbourhood, especially as other gardeners may be reluctant to undertake the work. By the late eighteenth century, jobbers who had previously found plenty of work, especially in London, were now feeling the squeeze of the commercialised services provided by large, organised contracting firms and their professional gardeners (Longstaffe-Gowan 1990, 125).

In terms of tree-related practical skills required by the professional gardener or landscaper, there was a shift during this era away from the more horticultural skills towards those more associated with forestry. With the decline in popularity of the formal garden, the demand for people skilled in intricate tree pruning and training seems to have diminished. Instead, with the rise of the English landscape garden and landscape park, the emphasis was more on basic forestry skills, such as the mass planting of forest trees, tree felling and stump removal. While much of the work of the professional 'improvers' was forestry work, they mostly identified themselves as landscapers rather than foresters. It was only towards the end of the eighteenth century with the introduction of coniferous trees for planting on wasteland that what we would recognise as the modern forester came into existence (Forbes 1906, 315–316). This individual probably originated in Scotland where, due to the far greater extent of wasteland, planting was conducted on a far larger scale and was therefore more rapidly specialised. Before the emergence of the modern forester, forestry on most country estates was the responsibility of the land steward, although on a practical level this came under the woodman or gardener. In the 1740s on the Midlothian estate of the Scottish politician John Cockburn, his gardener Charles Bell directed the forestry as well as the garden work (Colville 1904, 3). However, from his letters to Bell it is obvious he had little confidence in the forestry or arboricultural skills of the rest of his estate staff (*ibid.*, 14).

While much of the work of professional gardeners and landscapers involved what we now call arboricultural work, it was not called this at that time. Furthermore, those who specialised in tree-related work were only occasionally identified in this role by a tree-specific title. While the term 'arborist', meaning a tree expert, continued to be used throughout the eighteenth century, it was not common. Switzer (1734, v, 89) uses the term, as does Hayes (1794, vii), and it appears in the work of some other authors. However, an extensive search of the literature has indicated that the term 'arborist' may have been no more widely used in this era than the previous one. It is interesting that some modern historians have described some of the famous improvers as an 'arboriculturist'. For example, Herbert Edlin (1970, 196) calls 'Capability' Brown an arboriculturist and Deborah Turnball (1990) uses the same term for Thomas White. While both these men undertook huge amounts of arboricultural work and were eminent tree experts, they would not have used any tree-specific term to describe themselves.

Conclusions

The decline of the formal garden and the rise of the English landscape garden probably made little difference to the amount of arboricultural work in the amenity landscapes of the landed class. However, there was a major change in the type of work undertaken. While there was much less demand for the more horticultural skills, such as intricate tree pruning and training, there was a corresponding increase in demand for basic forestry work such as tree planting and felling. This era did, however, witness a significant increase in arboricultural work in urban areas, with a steady rise in residential gardens owned by the middle class and in the well-treed commercial pleasure gardens. With this noticeable increase in urban trees came the first signs of a growing awareness of the safety implications for trees and tree work in close proximity to people and property.

While this era was preoccupied with great debates about landscape styles, and trees and shrubs were central to this, there was only limited progress in the development of practical arboriculture. Of course, this was the Age of Reason that witnessed the rise of scientific gardening when rigorous enquiry and experimental methods came to the fore. However, this seems to have had little impact on arboricultural practices. While some advances were made, common practices relating the planting and care of trees still relied heavily on the teachings of ancient authors, folk remedies and pure superstition. Similarly, there was little development in tools and equipment for tree-related work. Much of the literature dealing with practical tree work was about timber or fruit trees and had limited relevance to amenity trees.

The eighteenth century witnessed some important advances in the development of professional gardening and landscaping. By the end of the century there were two distinct but closely related industries in Britain associated with horticulture and landscape. The first was concerned with the design, construction and maintenance of amenity landscape and much of the work involved trees and shrubs. The second was the nursery industry that supplied the trees, shrubs and other plants for all these gardens and landscape parks.

While the previous era had several notable individuals who were renowned as tree experts, this was not the case in this era. While there were outstanding figures in the English landscape movement who had considerable expertise in what would now be called arboriculture, this was always viewed within the context of their landscape work and not as some separate specialism. This continued into the nineteenth century and it would be some time before the celebrated tree expert re-emerged. Furthermore, this would only come with the growing recognition of arboriculture as a separate discipline, distinct from horticulture, landscape and forestry.

References

Aberdeen Journal (1798) Advertisement for John Adams, Nurseryman and Seedsman in Aberdeen. *The Aberdeen Journal*, 23 January, 1.

Anderson, J. (1777) *Miscellaneous Observations on Planting and Training Timber-Trees; particularly calculated for the climate of Scotland.* Charles Elliot: Edinburgh.

Anderson, M.L. (1953) Introduction. In: Hamilton, T., *Forest Trees: Some Directions about raising Forest Trees by Thomas Hamilton 6th Earl of Haddington.* Edited by M.L. Anderson. Thomas Nelson and Sons: London.

Anon (1747) *A General Description of All Trades, Digested in Alphabetical Order, etc.* T. Waller: London.

Anon (1767) *A Companion to Every Place of Curiosity and Entertainment in and about London and Westminster.* Printed for J. Lawrence: London.

Barker, H. (1999). *Newspapers, Politics and English Society 1695–1895.* Routledge: Abingdon.

Barre, D. (2018) Enterprising Women: Shaping the Business of Gardening in the Midlands, 1780–1830. In: Dick, M. and Mitchell, E. (eds) *Gardens and Green Spaces in the West Midlands since 1700.* West Midlands Publications: Hatfield.

Blith, W. (1653) *The English Improver Improved.* Third impression, much augmented. John Wright: London.

Boutcher, W. (1775) *A Treatise on Forest-trees.* Printed by R. Fleming and sold by the author: Edinburgh.

Brown, D. (2000) *Nathaniel Richmond (1724–1784) 'Gentleman Improver'.* PhD thesis, University of East Anglia.

Brown, D. and Williamson, T. (2016) *Lancelot Brown and the Capability Men.* Reaktion Books: London.

Bucknall, T.S.D. (1797) *The Orchardist: Or, A System of Close Pruning and Medication for Establishing the Science of Orcharding.* G. Nicol: London.

Burton, A. (1972) *The Canal Builders.* Methuen Publishing: London.

Caledonian Mercury (1760) Designing, Drawing, Planting, etc. *Caledonian Mercury*, 8 March, 3.

Campbell, R. (1747) *The London Tradesman, being a Compendious View of All the Trades, Professions, Arts, both Liberal and Mechanic.* T. Gardner: London.

Chambers, D. (1993) *The Planters of the English Landscape Garden: Botany, Trees, and the Georgics.* Yale University Press: New Haven and London.

Chambers, W. (1773) *A Dissertation on Oriental Gardening.* Second edition. W. Griffin: London.

Chartres, J.A. and Turnball, G. (1983) Road Transport. In: Aldcroft, D. and Freeman, M. (eds) *Transport in the Industrial Revolution.* Manchester University Press: Manchester.

Clark, H.F. (1948) *The English Landscape Garden.* Pleiades Books: London.

Clarke, G. (1985) Where Did All the Trees Come From? An Analysis of Bridgeman's Planting at Stowe. *The Journal of Garden History* 5(1), 72–83.
Coke, D. and Borg, A. (2011) *Vauxhall Gardens: A History*. Yale University Press: New Haven and London.
Colville, J. (ed.) (1904) *Letters of John Cockburn of Ormistoun to His Gardener, 1727–1744*. Scottish History Society: Edinburgh.
Conlin, J. (ed.) (2013) *The Pleasure Garden, from Vauxhall to Coney Island*. University of Pennsylvania Press: Philadelphia, PA.
Cook, M. (1676) *The Manner of Raising, Ordering, and Improving Forest-Trees*. Peter Parker at the Leg and Star: London.
Corfield, P. (2012) *Vauxhall: Sex and Entertainment, London's Pioneering Urban Pleasure Garden*. History and Social Action Publications: London.
Cowell, F. (2009) *Richard Woods (1715–1793): Master of the Pleasure Garden*. The Boydell Press: Woodbridge.
Cox, E.H.M. (1935) *A History of Gardening in Scotland*. Chatto and Windus for Flora and Sylva: London.
Daniels, S. (1999) *Humphry Repton: Landscape Gardening and the Geography of Georgian England*. Yale University Press: New Haven and London.
Davies, W. (1810) *General View of the Agriculture and Domestic Economy of North Wales*. Richard Phillips: London.
De la Bédoyère, G. (ed.) (1995) *The Writings of John Evelyn*. The Boydell Press: Woodbridge.
Derby Mercury (1767a) Thursday's Post: Derby 9th April. *The Derby Mercury*, 10 April, 4.
Derby Mercury (1767b) Advertisement for Ann Sandy, Sale of Plants. *Derby Mercury*, 30 October, 4.
Derby Mercury (1773a) Advertisement for Ann Sandy, Nursery and Seeds-woman. *Derby Mercury*, 12 February, 3.
Derby Mercury (1773b) Advertisement for Ann Sandy, Nursery and Seeds-woman. *Derby Mercury*, 22 October, 4.
Derby Mercury (1786) Advertisement for Ann Sandy, Nursery and Seeds-woman. *Derby Mercury*, 16 February, 4.
Downing, S.J. (2009) *The English Pleasure Garden 1660–1860*. Shire Publications: Oxford.
Dresser, M. and Hann, A. (2013) *Slavery and the British Country House*. English Heritage: Swindon.
Duhamel du Monceau, H.-L. (1760) *Traité des Semis et Plantations des Arbres et de leur Culture*. Guerin and Delatour: Paris.
Edlin, H.L. (1970) *Collins Guide to Tree Planting and Cultivation*. Collins: London.
Elliott, P., Watkins, C. and Daniels, S. (2011) *The British Arboretum: Trees, Science and Culture in the Nineteenth Century*. Pickering and Chatto: London.
Evelyn, J. (1664) *Sylva; or, A Discourse of Forest-Trees, and the Propagation of Timber in His Majesties Dominions*. John Martyn and James Allestry: London.
Evelyn, J. (1706) *Silva; or, A Discourse of Forest-Trees, and the Propagation of Timber in His Majesties Dominions*. Fourth edition in two volumes. Robert Scott: London.
Evelyn, J. (1776) *Silva or a Discourse of Forest-Trees, and the Propagation of Timber in His Majesties Dominions*. With notes by Alexander Hunter. J. Dodsley: London.
Fairchild, T. (1722) *The City Gardener*. Printed for T. Woodward: London.
Floud, R. (2019) *An Economic History of the English Garden*. Allen Lane: London.
Forbes, A.C. (1906) *English Estate Forestry*. Second edition. Edward Arnold: London.

Forsyth, W. (1802) *A Treatise on the Culture and Management of Fruit Trees*. T.N. Longman and O. Rees: London.
Gilpin, W. (1782) *Observations on the River Wye and several parts of South Wales, etc. relative chiefly to Picturesque Beauty*. R. Blamire: London.
Gilpin, W. (1792) *Three Essays: On Picturesque Beauty; On Picturesque Travel; and On Sketching Landscape*. R. Blamire: London.
Gilpin, W. (1794) *Remarks on Forest Scenery, and other Woodland Views*. In three books [2 volumes]. Second edition. R. Blamire: London.
Green, T. (1994) Bundle Planting. *Quarterly Journal of Forestry* 88, 243–244.
Gregory, D. (1987) The Friction of Distance? Information Circulation and the Mails in Early Nineteenth-century England. *Journal of Historical Geography* 13(2), 130–154.
Hadfield, M. (1960) *A History of British Gardening*. John Murray: London.
Hale, T. (1757) *Eden, Or, a Complete Body of Gardening*. T. Osborne: London.
Hale, T. (1758) *A Complete Body of Husbandry*. Volume 1. Second edition. Thomas Osborne: London.
Hamilton, T. (1953) *Forest Trees: Some Directions about raising Forest Trees by Thomas Hamilton 6th Earl of Haddington*. Edited by M.L. Anderson. Thomas Nelson and Sons: London.
Hamilton, T. (1761) *A Treatise on the manner of raising Forest Trees*. G. Hamilton and J. Balfour: Edinburgh.
Hanbury, W. (1758) *An Essay on Planting, and a Scheme for Making It Conducive to the Glory of God and the Advantage of Society*. S. Parker: Oxford.
Hanbury, W. (1770) *A Complete Body of Planting and Gardening*. [In Two Volumes.] Published for the author by E. and C. Dilly: London.
Harvey, J. (1974) *Early Nurserymen*. Phillimore and Co.: London and Chichester.
Hayes, S. (1794) *Practical Treatise on Planting and the Management of Woods and Coppices*. Published privately and printed by William Slater: Dundalk, Ireland.
Henrey, B. (1975) *British Botanical and Horticultural Literature before 1800. Vol. 2: The Eighteenth Century History*. Oxford University Press: Oxford.
Hitt, T. (1755) *A Treatise of Fruit-Trees*. The Author: London.
Horwood, C. (2010) *Gardening Women: Their Stories from 1600 to the Present*. Virago Press: London.
House, S. and Dingwall, C. (2003) 'A Nation of Planters': Introducing the New Trees, 1650–1900. In: Smout, T.C. (ed.) *People and Woods in Scotland: A History*. University of Edinburgh Press: Edinburgh.
Hoyles, M. (1995) *Bread and Roses: Gardening Books from 1650 to 1950, Volume Two*. Pluto Press: London.
Hussey, C. (1975) Introduction. In: Stroud, D., *Capability Brown*. Faber and Faber: London.
Ipswich Journal (1753) Advertisement for Henry Ellison, Norwich, Gardener and Land Surveyor. *The Ipswich Journal*, 23 September, 4.
Ipswich Journal (1775) Sunday's Post: 14. Boston, Jan. 16. *The Ipswich Journal*, 18 March, 1.
Johnston, M. (2015) *Trees in Towns and Cities: A History of British Urban Arboriculture*. Windgather Press: Oxford.
Kalm, P. (1892) *Kalm's Account of his Visit to England: On his way to America in 1748*. Translated by J. Lucas. MacMillan: New York.
Kennedy, J. (1777) *A Treatise Upon Planting, Gardening, and the Management of the Hot-House*. Second edition, in two volumes. Hooper: London.

Kentish Gazette (1772) Extract of a letter from Exeter dated May 19. *The Kentish Gazette*, 26 May, 419, 3.

Knight, R.P. (1794) *The Landscape: a didactic poem in three books, addressed to Uvedale Price, Esq*. W. Bulmer and G. Nicol: London.

Laird, M. (1999) *The Flowering of the Landscape Garden: English Pleasure Grounds, 1720–1800*. University of Pennsylvania Press: Philadelphia, PA.

Langley, B. (1728) *A Sure Method of Improving Estates by Plantations of Oak, Elm, Ash and Other Timber-Trees*. Francis Clay and Daniel Browne: London.

Linnard, W. (1971) Thomas Johnes (1748–1816): Pioneer of Upland Afforestation in Wales. *Forestry* 44(2), 135–143.

Linnard, W. (2000) *Welsh Woods and Forests: A History*. Gomer Press: Llandysul.

Longstaffe-Gowan, T. (1990) Gardening and the Middle Classes. In: Galinou, M. (ed.) *London's Pride: The Glorious History of the Capital's Gardens*. Anaya Publishers: London.

Longstaffe-Gowan, T. (2001) *The London Town Garden 1700–1840*. Yale University Press: New Haven and London.

Longstaffe-Gowan, T. (2012) *The London Square*. Yale University Press: New Haven and London.

Loudon, J.C. (1835) *Encyclopaedia of Gardening*. A new edition. Longman, Rees, Orme, Brown, Green and Longmans: London.

Loudon, J.C. (1838) *Arboretum et Fruticetum Britannicum or the Trees and Shrubs of Britain*. In eight volumes. Longman, Brown, Green and Longmans: London.

Manchester Mercury (1766) Abstract of the New Act for the better Preservation of Timber Trees, Woods, Underwoods, etc. *The Manchester Mercury*, 8 July, 4.

Marshall, W. (1785) *Planting and Ornamental Gardening; A Practical Treatise*. J. Dodsley: London.

Mason, G. (1795) *An Essay on Design in Gardening, First Published in 1768*. Benjamin and John White: London.

Mawe, T. and Abercrombie, J. (1778) *The Universal Gardener and Botanist, or, A General Dictionary of Gardening and Botany*. G. Robinson and T. Calder: London.

Meader, J. (1779) *The Planter's Guide: or, Pleasure Gardener's Companion*. G. Robinson: London.

Miller, P. (1724) *The Gardeners and Florists Dictionary: or a Complete System of Horticulture*. Charles Rivington: London.

Miller, P. (1737) *The Gardeners Dictionary: containing the methods of cultivating and improving the kitchen, fruit and flower garden*. Third edition. The Author: London.

Mitchell, E. (2018) Duddeston's 'shady walks and arbours': The Provincial Pleasure Garden in the Eighteenth Century. In: Dick, M. and Mitchell, E. (eds) *Gardens and Green Spaces in the West Midlands since 1700*. West Midlands Publications: Hatfield.

Neave, D. and Turnbull, D. (1992) *Landscaped Parks and Gardens of East Yorkshire*. Georgian Society for East Yorkshire: Hull.

Newcastle Courant (1784) Advertisement for John Pringle, Nursery and Seedsman, Chester-le-Street. *Newcastle Courant*, 24 January, 1.

Nicol, W. (1799) *The Practical Planter, Or, A Treatise on Forest Planting*. Printed for the Author: Edinburgh.

Papworth, J.B. (1818) *Rural Residences*. R. Ackermann: London.

Phibbs, J. (2009) The Persistence of Older Traditions in Eighteenth-century Gardening. *Garden History* 37(2), 174–188.

Phibbs, J. (2017) *Place-Making: The Art of Capability Brown.* Historic England: Swindon.
Price, U. (1796) *An Essay on the Picturesque: As Compared with the Sublime and the Beautiful.* A new edition. J. Robson: London.
Read, H. (2000) *Veteran Trees: A Guide to Good Management.* English Nature: Peterborough.
Reid, J. (1683) *The Scots Gard'ner.* David Lindsay and his Partners: Edinburgh.
Repton, H. (1803) *Observations on the Theory and Practice of Landscape Gardening.* J. Taylor: London.
Repton, H. (1816) *Fragments on the Theory and Practice of Landscape Gardening.* J. Taylor: London.
Repton, H. (1840) *Landscape Gardening and Landscape Architecture of the Late Humphry Repton.* A new edition by J.C. Loudon. Longman and Co.: London.
Rooke, H. (1790) *Descriptions and Sketches of some Remarkable Oaks, in the Park at Welbeck.* J. Nichols: London.
Seeley, B. (1744) *Stowe: A Description of the Magnificent House and Gardens.* Printed by W. Dicey: Northampton.
Sheeran, G. (1990) *Landscape Gardens in West Yorkshire 1680–1880.* Wakefield Historical Publications: Wakefield.
Steuart, H. (1828) *The Planter's Guide.* Second edition. John Murray: London
Stroud, D. (1975) *Capability Brown.* New edition. Faber and Faber: London
Switzer, S. (1718) *Ichnographia Rustica: or, The Nobleman, Gentleman, and Gardener's Recreation.* Volumes 1–3. D. Browne: London.
Switzer, S. (ed.) (1733) *The Practical Husbandman and Planter. Volume 1 containing April, May and June.* S. Switzer: London.
Switzer, S. (ed.) (1734) *The Practical Husbandman and Planter. Volume 2 containing July, August and September.* S. Switzer: London.
Symes, M., Hodges, A. and Harvey, J. (1986) The Plantings at Whitton. *Garden History* 14(2), 138–172.
Tait, A. A. (1980) *The Landscape Garden in Scotland 1735–1835.* Edinburgh University Press: Edinburgh.
Thomas, K. (1983) *Man and the Natural World: Changing Attitudes in England 1500–1800.* Allen Lane: London.
Thompson, E.P. (1975) *Whigs and Hunters: The Origins of the Black Act.* Allen Lane: London.
Turnbull, D. (1990) *Thomas White (1739–1811): Eighteenth Century Landscape Designer and Arboriculturist.* PhD thesis, University of Hull.
Uings, J. (2013) *Gardens and Gardening in a Fast-changing Urban Environment: Manchester 1750–1850.* PhD thesis, Manchester Metropolitan University.
Walpole, H. (1995) *The History of the Modern Taste in Gardening.* Introduction by John Dixon Hunt. Ursus Press: New York.
Watkins, W. (1753) *A Treatise on Forest-trees.* John and James Rivington: London.
Webber, R. (1968) *The Early Horticulturists.* David and Charles: Newton Abbot.
Whately, T. (1770) *Observations on Modern Gardening, Illustrated by Descriptions.* T. Payne: London.
Wheeler, J. (1747) *The Modern Druid, Containing Instructions Founded on Physical Reasons.* James Wheeler: London.
White, T. (1813) Various letters to the Editor from Thomas White, Junior. *Transactions of the Society, Instituted in London, for the Encouragement of the Arts, Manufactures and Commerce* 31, 81–105. R. Wilks: London.

Williamson, T (1995) *Polite Landscapes: Gardens and Society in Eighteenth-century England.* Sutton Publishing: Stroud.

Willis, P. (2002) *Charles Bridgeman and the English Landscape Garden*. Elysium Press Publishers: Newcastle upon Tyne.

Wroth, W. (1896) *The London Pleasure Gardens of the Eighteenth Century*. Macmillan and Company: London and New York.

Wulf, A. (2008) *The Brother Gardeners: Botany, Empire and the Birth of an Obsession.* William Heinemann: London.

Young, A. (1769) *A Six Weeks Tour through the Southern Counties.* W. Strahan: London.

CHAPTER SEVEN

Heroic Arboriculture in the Nineteenth Century

As Georgian Britain entered the nineteenth century the nation had just become embroiled in the Napoleonic Wars (1799–1815). This was a series of major conflicts between several European countries, led and financed by Britain, and the French Empire and its allies led by Napoleon Bonaparte. The epic and protracted power struggle cast a long and bloody shadow over much of Europe and beyond until it eventually concluded with the Battle of Waterloo in 1815. This was followed by the Congress of Vienna that resulted in the redrawing of the borders of Europe and a period of relative peace. However, it also heralded a time of severe economic hardship in many rural areas of Britain. Grain prices tumbled after the inflated values of scarcity and improved agricultural methods increased production. Enclosure drove workers and small-scale tenant farmers off the land. There was severe unemployment in many districts as the British economy struggled to absorb the thousands of men who were discharged from the armed forces at the end of the wars. Average rural labourers' wages fell dramatically and many families found themselves on the verge of starvation. Parliament, dominated by the interests of the landowning elite, did little to address this desperate situation. This was also a time of serious social and political unrest across much of Britain as vast numbers of working people agitated for improved conditions and democratic representation. By the end of the Georgian era, Britain had recovered from the war and once again entered a period of economic growth, driven by the Industrial Revolution at home and by the expansion of its overseas empire.

Queen Victoria ascended the throne in 1837 and reigned for sixty-three years until her death in 1901 in a period known as the Victorian era. This was a time of unprecedented economic growth with increasing industrialisation at home, especially in the second half of the nineteenth century. Britain's military might secured the position of the British Empire as the largest empire in world history and a source of immense wealth. The growth of Britain's transport infrastructure that began in the Georgian period continued at an even greater pace, most significantly with the development of the nation's railway network. During its boom years of 1836 to 1847 Parliament authorised 8,000 miles (12,875 km) of railway lines.

The Victorian era is notable for some dramatic changes in Britain's social and political system. The political agenda was increasingly liberal, with gradual political and social reform and the widening of the franchise. From the 1830s, significant advances were made in the development of local government in Britain with increasing powers for local authorities. The balance of political power was also shifting away from the nobility and gentry towards an emerging middle class, especially in the expanding industrial centres where they had considerable control of the newly formed borough councils and boards of health. There were also unprecedented demographic changes across Britain with the population of England and Wales almost doubling, from 16.8 million in 1851 to 30.5 million in 1901, and Scotland's population also rising rapidly, from 2.8 million in 1851 to 4.4 million in 1901. Alongside this population increase was the dramatic expansion of Britain's towns and cities with vast areas of new suburban housing engulfing great swathes of the surrounding countryside.

From the 1870s all was not well in the countryside as farming began to slide into a long depression, prompted by the importation of cheap American grain. Having experienced a few decades of continuing prosperity, the series of agricultural depressions between 1873 and 1896 came as a great shock to the rural idyll of the Victorian landed estate. While some British landowners were always interested in industry and mining, the shift of wealth from land to business and other forms of investment became more pronounced during the second half of the nineteenth century. Although landed families continued to present an appearance of wealth and gained considerable sums from agriculture and rents, much more income came from business ventures. The Great Depression of British Agriculture lasted until the late 1890s and led to high levels of unemployment among rural workers, which forced many families to abandon the countryside to seek work in Britain's towns and cities.

The Victorian era was an age of invention and innovation, with Britain leading the world in engineering and technology. Great advances were made in many different fields of science as well as in medicine and public health. This tremendous progress was also reflected in some rapid advances in the development of horticulture, landscape gardening, forestry and related activities. It was also a time when arboriculture emerged as a clearly defined field much as we understand it today.

Sources of information

There are many surviving examples of parks and gardens with the different landscape styles that were popular during this period. With the dramatic increase in public greenspace from the 1840s onwards, many renowned examples of these landscapes are now maintained by local authorities and other public bodies. There are also detailed written records on how some of those landscapes were created and maintained. However, many landscapes from this era have now been redeveloped to meet the needs and tastes of their modern users. This has been especially evident with many Victorian public parks and avenues of street trees.

In the nineteenth century there was a huge increase in books on tree-related subjects compared to those published in the previous century. It was also a time when many relevant magazines and journals were founded, which have proved a particularly rich source of research material. The Victorian gardening and professional periodicals frequently published correspondence from their readers, which give an invaluable insight into the experiences, ideas and concerns of those working in horticulture and related fields. With regard to the development of professional arboriculture in Britain, these periodicals provide a unique insight into the range of practical work that was undertaken, the reasons for this, the terms used, and much more. Also of value, although fairly scarce, are the surviving notebooks of head gardeners and garden owners. Lastly, nineteenth-century newspapers and popular magazines have provided excellent research material.

Visual images have played a vital role in the research for this chapter, with many of the same sources as in the eighteenth century. There are, however, some significant differences. With the huge increase in relevant publications there was a corresponding increase in the number of visual images such as illustrations, diagrams and plans. From the 1880s, photographs became a feature in horticultural books and magazines, replacing engravings and other forms of illustration.

The most prolific writer on horticulture, landscape and arboriculture in the nineteenth century was John Claudius Loudon (1783–1843) and his work is a major source of information on these topics for the first half of the nineteenth century. Loudon was a Scotsman born in Lanarkshire in 1783, and as a young man he studied biology, botany and agriculture at the University of Edinburgh. After moving to England and then spending time travelling abroad, he founded a horticultural publishing empire in London, producing a host of gardening magazines and books. Through his publications he spread his ideas on public open spaces and city planning, as well as many aspects of horticulture, arboriculture and related subjects, to both amateur enthusiasts and professionals. Unlike many of his prominent contemporaries, Loudon was never a head gardener but his writings became essential reading for relevant professionals. He was especially interested in trees and shrubs and during his lifetime made a huge contribution to British arboriculture. Loudon was also one of the first authors to appreciate the unique role that trees could play in human welfare and development. He declared, 'Trees are not only, in appearance, the most striking and grand objects of the vegetable creation; but, in reality, they are those which contribute the most to human comfort and improvement' (Loudon 1838a, 1, 1). Loudon is undoubtedly the greatest British tree expert of the nineteenth century.

Nineteenth-century gardens

As the Georgian era entered a new century, interest in horticulture continued to increase. This was not just a fascination of the upper class on their country estates but also a growing preoccupation of the expanding middle class of Britain's towns and cities. While garden design in all locations was to undergo some fundamental changes in the nineteenth century, this did not happen immediately and certainly not in a consistent way across Britain. As in previous eras, an assessment is made here of the extent to which these landscape designs used trees and shrubs, the role that these played and the amount of practical arboriculture that might have been involved in their creation and maintenance.

From the mid-eighteenth century a few great houses had begun opening their grounds to select members of the public and this trend continued into the nineteenth century. The coming of the railways made garden visiting an option for the wider public and the number of visitors to some famous gardens increased dramatically. By the late 1860s, Chatsworth was welcoming holiday crowds of men and women from the cotton factories of Manchester. However, some garden owners were horrified at the prospect of such 'hordes', and by Edwardian times the gardens of the wealthy were overwhelmingly out of bounds. Happily, with the development of public parks from the 1840s onwards, working-class families were at last able to freely enjoy greenspace, trees and horticultural wonders in their own towns and cities.

Exotics, grouping and colour in private gardens

In the early nineteenth century the Picturesque continued to exercise a strong influence over garden design in the parks and gardens of many country estates. The social and economic stability of the landowning class was reflected by many in attitudes towards their houses and gardens where the late eighteenth-century landscape designs were often maintained. There was a sound economic reason for this. Many of these landscape gardens had been laid out between 1760 and 1780, and were only coming to maturity in the mid-nineteenth century. They represented a major investment in the future and it was the heirs to the estate who would enjoy such gardens at their best.

While many landscape parks and gardens remained largely as they had been envisaged, especially in the first half of the nineteenth century, tastes in landscape design among the gardening elite were starting to change. These changing tastes were expressed first in the books published on the subject and then in the horticultural press as this emerged from the late 1820s onwards. What became known as the Victorian garden arose out of a rebellion against the eighteenth-century landscape park and a conscious move away from naturalistic landscapes. In 1808, the great landscape gardener Humphry Repton had declared that 'gardens are works of art rather than of nature' (Repton 1840, 374).

The component parts of the landscape began to be freed from subordination to the overall scheme.

During the 1820s large garden landscapes were still mainly characterised by a 'free and easy' style poised somewhere between 'nature' and 'art' (Elliott 1986, 59). The style was dominated by the flowering shrubbery, a group of shrubs designed for display. This had long been a feature of many gardens and was now promoted by the first book on the subject, *Silva Florifera* by Henry Phillips (1823). By the 1830s, it was increasingly appropriate for the garden to display the unmistakable signs of artifice. Furthermore, it was widely accepted that there was no constant in terms of taste as everyone could have their own view and even this could change over time.

The revival of more formal styles was the most obvious way of making the garden as a whole into a work of art (Elliott 1986, 33). In order to achieve this, Loudon in particular was keen on the use of exotics in landscape planting (Loudon 1838a). Indeed, exotic planting was to become one of the major determinants of Victorian gardening as it so clearly demonstrated the supremacy of art over nature. Trees of unusual shape or habit became immediate signals of art when planted so as to attract attention. Many of the well-known weeping tree cultivars, such as those of elm and beech, initially appeared in the first quarter of the nineteenth century (Elliott 1986, 33). The fastigiate Lombardy poplar was also much planted at this time.

By the 1830s, the isolation of trees as individual specimens was a major feature of gardens (Elliott 1986, 34). This involved not merely separation, but allowing each tree to 'arrive at perfection' by displaying to the maximum its most aesthetically pleasing characteristics. However, Loudon believed there should still be some aesthetic cohesion to this specimen planting. He deserves recognition as the man who introduced the concept of artistic grouping of trees into nineteenth-century British landscapes. He frequently protested in his own *Gardener's Magazine* against the 'meaningless dotting' of single trees in places such as Regent's Park.

Colour in the garden became hugely popular in the nineteenth century and the contribution of trees and shrubs took a variety of forms. Colour groupings depended largely on flowering shrubs, especially rhododendrons and azaleas. Rhododendrons started to become popular in the 1840s, partly as a result of an article by Philip Frost of Dropmore in the *Gardeners' Chronicle*. Oblivious of its long-term consequences, he wrote about the ability of *Rhododendron ponticum* to grow in ordinary English soil, unlike the other rhododendrons then available, which, as American swamp plants, needed peaty conditions (Frost 1841). Perhaps the garden most noted in the 1840s and 1850s for its rhododendrons was Trentham in Staffordshire, due to the initiative of the head gardener George Fleming. In the 1860s, the nurseryman William Paul attempted to take arboricultural colour a step further. In a series of articles in the *Gardeners' Chronicle* in 1864 he urged gardeners to cast their eyes beyond just flower colour to the use of trees with purple, golden or silver leaves or with coloured bark for winter effect. Along with trees of pyramidal or weeping effect, he labelled these together as 'pictorial trees' (Paul 1892, 243–278). While conifers had often dominated earlier plantings, the excitement generated by Paul's ideas led to a revival of interest in deciduous trees, especially with the new introductions of Japanese maples and the purple-leaved plum (*Prunus x pissardii*) (Elliott 1986, 183). Variegated foliage became particularly popular with many late Victorian gardeners, although this trend towards 'colourful trees' did meet with some significant opposition. Probably the most influential critic of arboricultural colour was Alexander Dean, who at the turn of the century launched an attack on positive colour in trees, in favour of green (*ibid.*, 184). Dean had powerful opponents but he also had much support. As a result of this controversy the planting of purple beech nearly came to a halt in the 1890s.

The first garden to attract attention for its massing of trees and shrubs by colour was Waddesdon Manor in Buckinghamshire, whose grounds were laid out in the late 1870s and early 1880s (*Gardeners' Chronicle* 1885, 820). The planting of trees and shrubs in masses of single colours was a short-lived movement, characterising the last quarter or third of the nineteenth century; but it left behind the varied legacies of the woodland garden and flowering shrub garden (Elliott 2018, 100).

Revival of the formal garden

While the formal garden had never completely disappeared, it had been out of fashion for many decades during the popularity of the English landscape garden and the Picturesque. Now it was time for the formal style to make a return. Britain's economy was booming and the landed aristocracy and great industrialists were enormously wealthy on the profits of successful enterprises, urban expansion and overseas investments. As with previous eras, this wealth was poured into creating or renovating their great mansions and laying out impressive new gardens. As the naturalistic landscape park was yesterday's fashion, what was needed was a grand and opulent garden style that contrasted dramatically with this. The grandiose Italianate schemes of architect Sir Charles Barry (1795–1860) were the perfect solution. Barry had spent time in Italy studying antiquities and on his return based his designs on what remained of the great Renaissance gardens. His vast architectural schemes at Trentham Park in Staffordshire and Shrubland Park in Suffolk remain, although the latter has now lost much of its original horticultural intricacy. Another notable exponent of what became known as the 'Italianate' or 'architectural style' was William Andrews Nesfield (1793–1881). His trademarks were the terrace and the intricate parterre, the latter marked out in box and often with an infill of coloured minerals rather than flowers. His designs can still be seen in London at Regent's Park and Kew. At this time the bedding system of planting with ornamental flowers became very popular. Carpet bedding was a variation of this that appeared like groundcover with a very uniform surface, achieved through careful clipping to give a 'carpet-like' appearance. The majority of country houses from the 1870s boasted some carpet beds and the fashion spread into suburban villas and town gardens.

As the formal or architectural garden became ever more popular, opposition to this also grew. During the 1890s a battle raged that was to have a profound effect on the future of garden design. It was not only about whether a garden should be 'natural' or 'formal', but also who should be primarily responsible for its design, the gardener or the architect. The chief protagonists were the gardener William Robinson, who disliked the formal and favoured a naturalistic style, and the architect Reginald Blomfield, who enthusiastically promoted formalism and challenged Robinson in his influential book *The Formal Garden in England* (Blomfield 1892). His ideas struck a chord with many wealthy garden owners since their affluence could accommodate the considerable expense of the formal garden with its long herbaceous borders, many yards of clipped hedges, extensive kitchen gardens and the large teams of gardeners this required. The popularity of the formal garden also reflected a general revival of Classicism that had been prevailing in all aspects of the arts since the late Victorian period. Somewhere between the directly opposing views of Robinson and Blomfield was the contribution of John Dando Sedding with his influential book *Garden-Craft Old and New* (Sedding 1891). Like Blomfield, Sedding had trained as an architect, but later developed an interest in garden design. While Sedding agreed with many of Robinson's views, he was still firmly on the formalist side of the fence.

Topiary had long been out of fashion but it underwent an astonishing revival along with the fortunes of the formal garden. This began with the remarkable garden created by the fourth Earl of Harrington at Elvaston in Derbyshire. In 1830 he invited the gardener William Barron to create a remarkable garden that would be 'the biggest wedding present in the world' for his bride-to-be, the actress Maria Foote. Over the next twenty years Barron created a mature landscape of formal gardens that included several miles of evergreen hedges. Some of these hedges were used to divide the garden into a series of garden rooms, which established a trend in garden design that has continued to the present time. Barron steadily turned the Earl's grounds into a grafter's playground, with avenues of grafted cedars and specimens of weeping beech. Training shrubs into standards was another of his practices, although some regarded this particular distortion as unsightly. Perhaps the most outstanding feature of the garden at Elvaston was its extraordinary collection of architectural and sculptural topiary, the fame of which helped promote a revival of interest in topiary (Figure 47). The fact that Barron was able to create this mature garden so quickly was due to his unique expertise in transplanting large trees, which will be discussed later. Although no visitors were allowed into the garden before the Earl's death in 1852, a series of articles about Elvaston were published in the *Gardeners' Chronicle* in 1849 and 1850. These spread the word about these remarkable gardens and helped to encourage huge public interest when they eventually opened. Other prominent gardens picked up on this interest and

FIGURE 47. The outstanding feature of the garden created by William Barron at Elvaston was its extraordinary collection of architectural and sculptural topiary. Its fame helped promote a revival of interest in topiary and this 1881 print shows some of the finest examples of that work.

topiary once again became a regular feature of the larger British gardens. The acceptability of topiary in the late Victorian garden was succinctly expressed by Sedding (1891, 180) who declared: 'I have no more scruples in using scissors upon tree or shrub, where trimness is desirable, than I have in mowing the turf of the lawn that once represented the virgin world'.

Leading the opposition to the formal garden was the Irishman William Robinson (1838–1935). Although brought up in Ireland, Robinson spent most of his life in England where he laid much of the foundations of the modern British approach to horticulture. He was also to have a huge influence, both directly and indirectly, on the development of British arboriculture. Robinson was dismissive of many elements of the formal garden and topiary was one of his main targets. He clearly despised topiary and had little time for those who promoted it, commenting sarcastically, 'It is probable that Mr Blomfield has never noticed what a beautiful tree a yew in its natural form is' (Robinson 1892, 54). He was equally dismissive of mazes, stating, 'The maze is one of the notions about gardening which arose when people had very little idea of the dignity and infinite beauty of the garden flora as we now know it' (Robinson 1900, 344). He also refers to mazes as 'ugly frivolities'. They should be left, he says, 'for the most part to places of the public tea-garden kind'.

While Robinson failed to put gardeners off topiary, bedding out and other elements of the formal garden, he exposed some Victorian pretensions and introduced a beautiful free style of planting. Two of his books, *The Wild Garden* (Robinson 1870) and *The English Flower Garden* (Robinson 1883), were especially influential and spoke to a new anti-industrial generation, many of whom were the followers of the Arts and Crafts movement. In contrast to the formal garden, Robinson and his allies, who included Gertrude Jekyll, favoured the image of the English cottage garden, with its roses, honeysuckle, foxgloves and primroses. However, despite Robinson's reputation for dogmatism and his vociferous opposition to the formal garden, he was also a man of contradictions. For example, he expresses the view in *The Garden Beautiful* that 'any style may be right where the site fits it' (Robinson 1906, 40).

In 1858, a large folio volume by Edward Adveno Brooke was published entitled *The Gardens of England* (Brooke 1898). This gives an account of what Brooke considered to be nineteen of the finest gardens in England. While there is some interesting text, the illustrations are outstanding and show the role of trees and arboriculture in these gardens. The images are based on Brooke's own paintings and are widely regarded as some of the best evocations of the spirit of Victorian gardens.

Arboretums, pinetums and woodland gardens

The steady introduction of new species of trees, many from North America, that began in the early eighteenth century now continued in the Victorian era. In the past, these trees were often displayed in the pleasure ground or similar location near the house where they could be readily admired by the owner and their visitors. Those areas where collections of trees could be displayed to maximum effect now became known as an arboretum. While Linnaeus has used the term arboretum to denote published lists of national trees, it was then adopted in British and Irish botanical gardens for collections of trees and later came to be applied to country estates (Elliott *et al.* 2011, 6). The garden of the Botanical Society at Glasnevin in Dublin, established in 1789, included an area designated as an arboretum, possibly the first time this term was used in that context. The term was then picked up by Loudon and promoted through his *Gardener's Magazine* and other writings.

One of the first of the nineteenth-century arboretums was established in the 1830s at Bicton in Devon where the head gardener, Robert Glendinning, collaborated with Loudon on the design and planting (Loudon 1840a, 626). Another early nineteenth-century arboretum was at Chatsworth, Derbyshire, where planting commenced in 1835 under the direction of Joseph Paxton (1803–1864), who would later become the leading landscape gardener of the Victorian era (Paxton 1835). One of Loudon's pet projects was to persuade the landowners in different parts of Britain to 'undertake to collect and cultivate each a single genus; and let the species and varieties be examined by the botanical world, on certain days throughout the year' (Loudon 1837). This was an early proposal to establish national collections of tree species, something which exists now but was not formally taken up in Loudon's day.

A pinetum is similar to an arboretum and features a collection of coniferous trees. The first pinetum in Scotland was formed at Methven Castle, near Perth, in 1830 and this encouraged other Scottish landowners to follow suit (Cox 1935, 144). In the mid-nineteenth century, the discovery of many spectacular conifers along the west coast of North America and in parts of South America really promoted the idea of the pinetum. Landowners were especially excited by the monkey puzzle or Chile pine (*Araucaria araucana*) introduced in 1844 and the giant sequoia (*Sequoiadendron giganteum*) in 1853, named the Wellingtonia by the Victorians, although this imperialist tribute was never recognised by the rest of the world. Thomas Appleby, touring the north in 1852, reported that many prominent industrialists' estates were being planted with conifers, often in arboretums or in newly designed pinetums (Appleby 1852). Coniferous trees have a close association with the Victorian garden, not just in the grounds of large country estates but also in smaller suburban gardens.

Another garden feature of arboricultural interest that was gathering momentum from the 1840s was the floral embellishment of woodlands and wasteland by naturalising exotic plants (Elliott 1995, 77). The culmination of these various trends – the exotic tree collection, colour massing, woodland embellishment – was the woodland garden. In some respects this now replaced the shrubbery of previous garden styles. In 1870, William Robinson's *The Wild Garden* recommended a range of exotic plants that would naturalise themselves in the woods and waste areas of an estate. One of the first examples to gain the attention of the horticultural press was Lord Armstrong's garden at Cragside, Northumberland, where a forest filled with rhododendrons was planted during the 1870s and 1880s. The discovery of the flora of the Himalayas, and later China, with its almost limitless supply of new rhododendron species would keep Britain's woodland gardens refreshed for many years to come.

Elsewhere on country estates, usually in the more remote or upland areas, commercial forestry was flourishing. This continued to be supported by the Society of Arts which offered gold and silver medals for those who had planted the greatest number of trees or the largest area of ground during a given year (Hudson and Luckhurst 1954, 86–89). Landed estate forestry took two main forms in the nineteenth century: traditional woodland management, mainly coppice woodland or coppice with standards; and plantation forestry, which had developed its own strong traditions from the early eighteenth century onwards. Many of the developments in British forestry from the 1850s onwards were to come from Germany which was then leading the world in what was termed the 'new forestry' based on scientific principles (Simpson 1900).

Trees in the urban realm

The nineteenth century was a time of unprecedented population growth in Britain, particularly in England and Wales. This was due not only to an increase in the birth rate but also a fall in the death rate through medical advances. The 1851 census showed that the population of England and Wales was for the first time mostly urban (some 54 per cent), concentrated into a relatively small number of great cities and conurbations. This had resulted from vast numbers of people leaving the depressed countryside to find work in the new industrial centres, which were mainly in the north. There was a huge increase in the population of new industrial towns like Manchester, Leeds, Birmingham, Bradford, Sheffield and Nottingham. The growth of most towns was not planned; they grew by expansion outwards and by infilling within existing boundaries. The population explosion, coupled with the desire of many middle-class families to abandon the depressing and unsafe city centres, led to a suburban house-building boom that began in earnest in the 1840s. This continued steadily before culminating in the greatest wave of Victorian suburbanisation in the last few decades of the nineteenth century, which not only catered for the fast expanding middle class but also for the 'superior' working class (Ravetz 2001, 9–10).

The most significant developments in the planting and care of amenity trees in the nineteenth century came in the rapidly expanding towns and cities. Important changes in urban administration encouraged a huge increase in urban tree planting and management by improvement commissions, urban enclosure committees, corporations, boards of health and their successor local authorities. This was prompted by public health concerns, civic pride and altruism, and was financed mainly from local rates with additional funding coming through the generosity of local businessmen (Johnston 2015, 46–54). The growing cultural significance of trees and shrubs was reflected in the public's increasing interest in urban arboriculture, at least among the upper and middle classes, and the topic began to feature regularly in the national and local press. With the emergence of the horticultural press, from the 1830s onwards, articles about urban arboriculture also began to feature regularly in gardening magazines, generating much debate among their readership. These debates were often initiated and led by a small number of horticulturists and foresters who were now gaining a reputation as tree experts.

The nineteenth century, especially from the 1840s onwards, witnessed some major new opportunities for tree planting in our towns and cities. The development of public parks, the growth of suburban residential districts and the new vogue for street tree planting collectively resulted in a huge increase in the extent of urban tree cover across Britain. These developments have been extensively covered elsewhere (Johnston 2015 and 2017) and only a brief summary is given here.

Small private gardens and residential squares

Many middle-class residential properties in the nineteenth century had rear gardens, although the extent of these varied considerably, not just with the value of the house but also with its style and location. The larger gardens were able to accommodate some trees, including forest-type trees in the most extensive, but in more modest gardens this could be limited to one or two fruit trees or small ornamentals. More houses also had front gardens in the nineteenth century, although these were almost exclusively in middle-class areas and only in the more expensive properties did these give much scope for tree planting. The monkey puzzle was often a popular choice for Victorian front gardens.

Early in the nineteenth century Loudon (1812, vi) had pointed out the absurdity of trying to make small villa gardens in the landscape style, with miniature lawns and 'clumps and strips of trees'. Instead, he recommended designs of a more formal style, laid out in accordance with his 'gardenesque' principles, something which he promoted through his writing. In the December 1832 issue of the *Gardener's Magazine* he introduced the gardenesque as a new style of planting design in which each individual plant was allowed to develop its natural form and habit, without the constraints of the 'picturesque' conditions so admired by painters (Loudon 1832, 701). In his classic text *The Suburban Gardener and Villa Companion*, first published in 1838, Loudon makes clear the distinction between the picturesque and the gardenesque. The picturesque was 'the imitation of nature in a wild state, such as the painter delights to copy', while the gardenesque was 'the imitation of nature subjected to a certain degree of cultivation and improvement, suitable to the wants and wishes of man' (Loudon 1838b, 164). Loudon was very conscious of the growing desire of many middle-class families to relocate to the suburbs and enjoy a more relaxed and 'greener' home life. His *Suburban Gardener* provided them with all the information they needed for their new suburban homes. The book sets out a hierarchy of four types of suburban garden that Loudon considered might be classed in the same manner as their houses. These ranged from 'first-rate' gardens, with not less than 50–100 acres (20.2–40.5 ha), to 'fourth-rate' gardens of no more than one acre (0.4 ha). Loudon planned the garden to illustrate the social position of the landowner. Even with a small villa the trees and shrubs were as important as the house architecture in achieving general effects and in demonstrating the 'means and taste' of the owner. Loudon's emphasis was always on the ornamental value of trees and shrubs and trying to accommodate as many different species as possible in the garden. He classified his own garden at Bayswater as 'fourth rate' and he devotes more space in his book to this class of garden than any other. Considering the limited size of these 'fourth rate' gardens, it is remarkable how many different features and how much variety of planting is included in the proposed designs, which Loudon managed to achieve over the years with different designs in his own garden.

As the Victorian era progressed, a range of garden styles became acceptable in urban and suburban gardens. These were often miniature versions of the gardens that were fashionable with the nobility and gentry. The Italianate villa, through the work of Loudon and Nash, became the dominant style for the suburban villa in the mid-nineteenth century. These gardens had an emphasis on parterres, terracing, statues and the use of gravel instead of grass (Elliot 1986, 74). Considerable importance was accorded to evergreen trees. The popularity of the Italian style also prompted a demand for fastigiate trees to produce the effect of the Mediterranean cypresses. Japanese gardens also became fashionable and Japanese trees and shrubs were a requirement of the planting design of these gardens (*ibid.*, 200). Later, William Robinson's naturalistic approach to planting and garden design became popular with many middle-class garden owners who were no longer enamoured with the more formal and traditional styles of the Victorian age.

To meet this interest in urban and suburban gardens, many gardeners followed Loudon in publishing books and magazine articles giving guidance on how these could be created. One of the most popular in the 1850s was Edward Kemp's *How to Lay Out a Small Garden* (Kemp 1850). The second and subsequent editions omitted the word small from the title. In a section of the book entitled 'What to Avoid' there is a short passage about cutting down large trees. Kemp advises:

> Where a garden is to be made on land that has been planted at some previous period, and *trees of considerable magnitude* exist upon it, especial care should be used in reference to the removal of any of these, so as not to render the place too open and bare. … There is no subject on which greater deliberation is demanded than the cutting down and removal of large trees, as nothing changes the character of a place more.
>
> Kemp 1850, 32

The residential square and its private garden continued to form the central elements in the landscape of elite districts of London and many provincial towns until at least the mid-nineteenth century. The landscape of these squares had developed from open to closed spaces as trees and shrubs became an increasingly dominant feature. This trend had been promoted by Loudon who published his first article in the *Literary Journal* in 1803 on his plans to plant the residential squares of London with trees and shrubs (Loudon 1803). Loudon criticised the gardens for their overuse of evergreen trees and shrubs and lack of 'picturesque beauty'. There were still many squares with just grass and very few trees, and Loudon demanded this be remedied with more planting and a much wider selection of species. However, the claim that he actively promoted the planting of plane trees in these squares is incorrect (Fricker 1974). By the 1830s the need for public open space was becoming severe in London and other major cities and these private residential squares were now being eyed longingly by nearby residents who were denied access to them.

Pleasure gardens

The commercial pleasure gardens that had blossomed in the eighteenth century continued into Victorian times, although their popularity was now in decline. Those that remained open underwent some fundamental changes as they attempted to keep up with shifting attitudes and demands. The notorious excesses of the Georgian pleasure gardens, especially the drunken behaviour and sexual freedom, were viewed as alarming by the respectable Victorian middle class. Those densely wooded paths and corners where couples could find some seclusion and privacy were no longer acceptable. The wooded areas of many famous pleasure gardens were now in retreat. This was not just about promoting moral behaviour; it was primarily about generating more income. Not only did wooded areas encourage unwanted activities that put off many people from visiting, the removal of trees created more space for mass entertainments such as popular exhibitions and displays. The economic need to pack in the crowds had taken precedence and lovers had to find other more discreet places to go. In the case of Vauxhall, many of its mature trees were felled in 1810 and 1811, notably decreasing its once renowned woodland effect (Corfield 2012, 3). Other pleasure gardens experienced a similar pattern of decreasing tree cover as the Victorian era progressed. Some trees at Vauxhall were removed for safety reasons and this is likely to have occurred in other gardens. A watercolour by George Scharf entitled *View of the Exterior of the Gardens* shows trees at the edge of Vauxhall Gardens bordering Kennington Lane (Coke and Borg 2011, 290). Painted around 1859 just before the gardens closed, it shows the trees as heavily topped, undoubtedly as a safety measure due to the increase in pedestrians and vehicles now using the road.

By the 1840s, pressures from many directions were mounting on the pleasure gardens and many had already closed. Their role as an 'oasis' of greenspace and horticultural splendour in the grey and depressing urban fabric was being superseded by the new municipal parks, which were open all year and did not charge for admission.

Municipal public parks

The massive expansion of towns and cities in the wake of the Industrial Revolution was one of the main drivers for the creation of new public parks as substitutes for the increasingly unobtainable countryside. The conditions in which many families now lived in Britain's larger towns and cities were a matter of grave concern and growing disquiet. The cholera epidemic of 1832 was devastating in the crowded slums of the capital and tens of thousands of people died. One of the cures recommended was the provision of many more public open spaces that would open up the urban fabric to let in light and air. With pressure mounting on the government it established the Parliamentary Select Committee on Public Walks (SCPW) in 1833 with its members 'appointed to consider the best means of securing Open Spaces in the Vicinity of populous Towns, as Public Walks and Places of Exercise calculated to promote the Health and Comfort of the Inhabitants' (Hansard 1833, 1049–1059). When it presented its report to Parliament this contained the first official recognition of the need for public walks or parks (Conway 1991, 2). The term 'public walk' was synonymous with 'public park' until as late as 1875, indicating that these facilities were primarily intended for walking and not for any other recreation or games. The motivation behind the promotion of new parks was not driven entirely by the altruism of the public health agenda. The need for parks in the 1830s was also identified against a background of severe social unrest and increasing class polarisation (*ibid.*, 35–36). The park promoters saw the creation of parks as part of the political process; otherwise 'great mischief must arise'.

In 1833, the year of the SCPW, Moor Park in Preston became the first municipal park freely open to the public in an industrial town (Conway 1991, 18). Although the Royal Victoria Park in Bath had opened as a public park in 1830, this was not a municipal park as the Corporation only leased the land and did not own it (*ibid.*, 15). In 1839, Loudon was commissioned to design the Derby Arboretum, a pioneering public arboretum and park, on 11 acres (4.5 ha) of land donated to the town by Joseph Strutt, a former Mayor and wealthy local industrialist.

After 1837, largely due to Joseph Hume in the House of Commons, although initially proposed by Loudon, all parliamentary enclosure acts were required to contain provision for some public open space, representing a significant step forward in park promotion. Around 1845, the Victorian park movement became recognisable as such due to the growing number of initiatives that had been undertaken or were in the planning stage (Conway 1991, 39). One of these was in direct response to the SCPW which placed special emphasis on the crowded East End of London where there was almost no public open space. It prompted the creation of Victoria Park, which opened in 1845 as the East End's first major public park. It rapidly became a very popular venue, particularly at weekends, as tens of thousands of working-class people flocked to enjoy its attractions that included an arboretum (*ibid.*, 165).

The pioneering period of park development in the 1830s and 1840s was marked with various initiatives from central and local government, from benefactors and from the community. Park and garden design was no longer an affair for the wealthy alone; its impact was now extending to the working class and was becoming a matter of public policy (Elliott 1986, 54). However, even the most prosperous municipal corporations were not yet empowered by a general act of Parliament to raise their own funds to create and maintain public walks and parks. In 1847, Birkenhead Park opened with the distinction of being the first municipal park in Britain established as a result of an application to Parliament for powers to use public funds for that purpose (Conway 1991, 49). It became known as 'the people's park' and there is evidence that local residents had a clear sense of ownership of it (Elliott 1986, 53–54). Joseph Paxton was chosen to design the park and Edward Kemp as the works supervisor because both had previously worked on the successful redesigning of the gardens at Chatsworth House.

By the 1880s most towns had at least one public park or recreation ground, often on the town's outskirts or a short distance from it. However, William Robinson voiced his concern that many of the new public parks were not where they were most needed and that efforts should be focused more on the poorer and densely populated parts of cities (Robinson 1869, xviii). Some modern historians have cautioned against overstating the contribution of many of these new 'people's parks' to improving the lives of the working class precisely because they were often on the periphery of urban areas (O'Reilly 2013). For working-class families in the heart of the city, visiting their public park could involve travelling some distance and at some expense by tram.

While Victorian public parks varied in size from tiny patches of greenspace to extensive landscapes of over 100 ha, they were usually quite similar in design (Johnston 2015, 159–162). These rather formal landscapes comprised a network of paths through closely mown lawns with some built features interspersed among groups of ornamental specimen trees and islands of flowerbeds. Built structures included bandstands, palm houses, shelters, statues, drinking fountains and commemorative monuments. Most Victorian parks were extensively planted with trees, with plenty of individual specimens, avenues, belts and woodland. In some cases trees were labelled with their name and some basic information, much as in the manner of an arboretum, to promote the park's educational role. Planting around the periphery of the park could be particularly dense. Although most eighteenth-century writers on park and garden design focused on private gardens, some were notable for expressing their views on the design of public parks. Apart from Loudon, other writers on this subject included Joshua Major (1852), Charles Smith (1852), Edward Kemp (1850) and Henry Milner (1890), who set out the design approach of his late father, Edward Milner.

Many of these early public parks were titled 'arboretum'. This reflected their emphasis on trees, not just for their beauty and stature, but also as living botanical specimens to be studied as part of a programme of educational and moral improvement (Elliott *et al.* 2007a). They were likened to 'living museums', with the trees and shrubs clearly labelled, numbered plans, displays and guides. While many landscape gardeners such as Loudon advocated public urban arboretums as alternatives to exclusively aristocratic gardens, others such as Joseph Paxton argued that public arboretums were not usually suitable as public parks.

Urban botanical gardens

The dramatic increase in plant introductions that flooded into Britain in the nineteenth century led to the founding of many new botanic gardens (Johnston 2015, 216–221). In the first fifty years of the nineteenth century a total of seventeen botanic gardens were established, many of them in towns and cities. The first of these was Liverpool Botanic Garden which opened in 1802, followed by other large towns and cities such as Hull (1812), Manchester (1831), Birmingham (1832), Bath (1834), Sheffield (1836) and Leeds (1840). Many of these gardens were located in rapidly expanding industrial centres and were often created thanks to generous private donations and subscriptions. In the early nineteenth century, before the establishment of public parks, these subscription botanic gardens provided a vital recreational and educational resource that was especially appreciated by the expanding middle class (Elliott 2016, 62). The exotic tree and shrub collections were usually the most important features of the botanical society gardens, providing an unusual opportunity for members and visitors to encounter new species and varieties that they would only have seen before in reference books or gardening periodicals.

Despite their relatively small size, the Liverpool and Hull botanic gardens both had impressive tree collections. The Manchester Botanic Garden opened in 1831 and, after a chequered history that included national fame and financial disaster, effectively ceased to exist in 1908 (Brooks 2011). Fortunately, the Birmingham Botanical Gardens that opened in 1832 fared much better (Ballard 2003). This was laid out by Loudon, who prepared a plan for the 16 acre (6 ha) site at Edgbaston, and although there have been some modern improvements, the overall appearance of the gardens has changed little. As might be expected of Loudon, the planting plan included a wonderful selection of trees and shrubs, all detailed in the catalogue of plants produced in 1836 (*ibid.*, 131–147). Over the years, the gardens have trained many talented individuals, most notably the eminent arboriculturist and plant collector, Ernest 'Chinese' Wilson, who trained there from 1893 to 1897.

In the 1820s and 1830s, the botanical gardens at Kew were in recession due to declining financial support from government (Desmond 2007, 137). Meanwhile, its rival, the Horticultural Society of London, formed in 1804, was in the ascendant. Then, a threat to Kew's destiny as the country's National Botanic Garden came from the Royal Botanic Society. In 1838, it purchased the Inner Circle of Regent's Park for landscaped gardens, a conservatory, hothouses, a museum and a library (*ibid.*, 139). Eventually, in 1840 the responsibility for Kew was transferred from the Crown to the government. Sir William Hooker was appointed as director and the gardens were finally opened to the public. In 1843, with the gardens on a more secure financial foundation, Hooker finally acquired much-needed additional space for the garden's arboretum.

Cemeteries

In the early nineteenth century, urban churchyards and burial grounds were becoming increasingly overcrowded with the growth of urban populations. The situation was generating widespread public concern and shocking reports appeared in newspapers that used lurid phrases to describe their overcrowded and dilapidated condition, such as 'dangerous masses of corruption' and 'injurious to the living'. The foul 'miasmas' that escaped from the 'seething corruption' of graves was thought to cause cholera and other diseases. It was believed that trees and shrubs would help to absorb these harmful gases and generally improve public health. In the 1830s and 1840s, with mortality rates in cities at horrific levels, there was also a growing clamour to defeat overcrowding in burial grounds by establishing new cemeteries outside city boundaries. These ideas came together in the concept of the 'garden cemetery' and prompted the development of a cemetery movement in Britain that proposed to locate new attractively landscaped burial grounds on the outskirts of towns and cities (Johnston 2015, 204–216). In contrast to churchyards and many burial grounds, cemeteries are principally secular spaces where ownership is almost always by local authorities or private sector concerns.

Liverpool was among the first British cities to establish a garden cemetery. In 1829, St James's Cemetery was opened as a commercial venture on the site of an exhausted quarry occupying 12 acres (4.8 ha), funded through public subscription (Rutherford 2013, 13). The romantic landscape design was a major factor in attracting customers to the cemetery. By the 1830s, the idea of new landscaped cemeteries in the urban fringe was being considered by civic leaders in many towns and cities (*ibid.*, 14–15). In response, further commercial cemeteries were established on a business model that was later to be used for municipal cemeteries. Among the cities to gain cemeteries at this time were Birmingham, Newcastle, Sheffield, Nottingham and Manchester. The Glasgow Necropolis opened in 1832 as the first major cemetery in Scotland. One Victorian cemetery that deserves a special mention for its trees is the Abney Park Cemetery in Stoke Newington. This was one of London's 'Magnificent Seven' and opened in 1840 as the first wholly non-denominational garden cemetery in Europe (Curl 2000, 100–108). The site was fortunate in already having some lush planting and the cemetery company wisely retained as much of this as possible. According to George Collison, one of the promoters of the cemetery, the presence of many sizeable and beautiful trees was a major attraction of the site, providing a mature look to the landscape lacking in many other new cemeteries. In addition to the existing treecover, it was the new plantings of trees and shrubs that were to radically transform the landscape and ensure its reputation as a pioneering example of the garden cemetery. Abney Park was laid out as an extensive arboretum with the help of the celebrated Loddiges nursery in Hackney, which also supplied the wonderful selection of trees and shrubs.

One of the most passionate campaigners for attractively landscaped cemeteries was John Claudius Loudon. In 1843, he published his seminal work on the topic entitled *On the Laying Out, Planting, and Managing of Cemeteries* (Loudon 1843). It was one of the most exhaustive books on cemeteries ever written and was to have a lasting impact on how these were laid out.

Trees in streets

The Victorian street tree movement has close parallels with the Victorian park movement (Johnston 2017, 79–116). This was also driven by a public health agenda and by the desire to bring trees and greenspace close to urban populations. Trees had always featured to a very limited extent in the streets and roads of Britain's towns and cities, as remnants of past landscapes before urbanisation. There were also trees that had lined public walks and promenades, some dating back to the seventeenth century, which effectively became street trees when the sides of the walks were lined with shops, houses and other built development. However, the deliberate and systematic planting of trees in Britain's streets only started in the 1820s. One of the earliest examples was at Pittville, an exclusive residential estate in Cheltenham, which was probably completed by 1827. As Britain's seaside resorts and spa towns flourished in the Victorian era, street tree planting was a popular addition to the amenities. In Eastbourne's Terminus Road around 1840, trees were planted along this prominent thoroughfare leading to Grand Parade and the seafront. This could be the earliest example of street tree planting in a seaside resort in Britain and other resorts and spa towns quickly followed.

When systematic street tree planting finally arrived in London it was in Chelsea, then a new suburb on the western fringes of the capital. The project was conceived by local scholar and antiquarian, Dr John Samuel Phené, who was an early advocate of street tree planting. In 1850 he developed a nearby tract of vacant land as a new suburban residential development that is now Margaretta Terrace (named after his first wife), Oakley Street and Phene Street. In 1851, in accordance with his belief in the value of urban greenspace, he planted plane trees along both sides of the streets. The first major street tree scheme to be undertaken in central London was along the Victoria and Chelsea sections of the Thames Embankment. The first trees to appear alongside the roadway of the new Embankment were planted in 1869 in the Victoria Embankment section. Reaction in the Victorian horticultural world to the tree planting on the Thames Embankment was a mixture of relief and pride. When the announcement was made that the Embankment project was going ahead the gardening magazines issued repeated calls for trees to be planted along the side of the new roadway. When the trees were eventually planted, the gardening press was generally delighted with the results, as were much of the national press. The success of this very high-profile project did much to expel any public doubts about the wisdom of putting trees in streets and was a major factor in encouraging the extensive street tree planting that followed in towns and cities throughout Britain.

Three types of tree came to dominate the scene in the Victorian street tree movement and these were the plane, lime and horse chestnut, all large forest-type trees. The first two were by far the most widely planted and many of our urban streets dating from the late Victorian and Edwardian eras still have a legacy of these trees. The over-reliance on these three trees was often the subject of adverse comment and ridicule. The type of plane tree most commonly planted in British streets is the hybrid London plane (*Platanus x acerifolia*). It has long been regarded as an ideal street tree because it can withstand air pollution, drought and compacted soil. However, in the Victorian and Edwardian eras it was often planted in narrow streets where it was allowed to outgrow its situation and cause a variety of problems.

Opportunities for professional arboriculture in the urban realm

The dramatic increase in the number of trees in towns and cities in the Victorian era would have a huge and lasting impact on the development of arboriculture in Britain. Mature trees, in contrast to other vegetation in amenity landscapes, often require specialist skills and equipment to maintain and manage them in any setting, but especially in the urban realm. As Victorian Britain became increasingly urbanised, trees were integrated into the urban fabric in ever-closer proximity to built development and urban infrastructure. This presented major new challenges for those responsible for these trees, not just in ensuring their successful establishment and healthy growth, but also in avoiding many potential conflicts with the urban environment, especially those that might result in harm to people or damage to property (Johnston 2017, 244–271). Many of these new urban trees were owned by local authorities and as planting gathered pace throughout the Victorian era this resulted in ever-greater numbers of trees requiring maintenance and management. This meant that local authorities now needed specialist workforces with appropriate equipment to deal with arboricultural matters, or to obtain those services from contractors, or a combination of both. This was really the start of municipal arboriculture in Britain, giving a tremendous boost to the world of professional arboriculture and dramatically increasing the number of people employed in this work. In the late nineteenth century it would lead to the development of arboriculture as a distinct trade or occupation, separate from horticulture and forestry.

Ancient trees

The rise in public interest in ancient trees that became apparent in the late eighteenth century grew significantly in the nineteenth century. One of the driving forces for this was the horticultural and landscape press. There was a noticeable increase in the number of books focusing wholly or partly on ancient trees and the topic was featured regularly in periodicals devoted to horticultural and tree-related subjects.

An early contribution came from Edward Kennion (1744–1809) with his book *An Essay on Trees in the Landscape*, published after his death by his son Charles (Kennion 1815). Kennion had worked as manager on a relative's estate in Jamaica before returning home to England and establishing himself as an artist. His art focused on rural landscape, particularly trees, and he was much influenced by the Picturesque movement. Although the fifty-four plates in his book show trees of various species and ages, he believed that it was 'old or fully grown trees, which have the strongest character' (*ibid.*, 15). Natural appearance was also crucial, so pollards and other trees 'where the axe has been exercised' cannot be picturesque. An influential book exclusively on ancient and remarkable trees was *Sylva Britannica; or, Portraits of Forest Trees*, with text and fifty illustrations by Jacob Strutt (1784–1867), the landscape artist (Strutt 1830). First published in 1826 in a folio edition, the subscribers included George IV, the Duke of York and many figures from the aristocracy. When it was published again in 1830 in a smaller format, it sold very well and got excellent reviews. Strutt's record of some of Britain's most notable ancient trees clearly struck a chord with the public. While many of the portraits show trees in a very dilapidated state there is no evidence of any tree work such as propping or bracing, which in accordance with Picturesque principles would have been regarded as unnatural.

In 1838, Loudon published his *Arboretum et Fruticetum Britannicum*, his seminal work on the trees and shrubs of Britain (Loudon 1838a). This mentions numerous examples of ancient and remarkable trees, specifying their location, age and dimensions, and is the most extensive record of Britain's heritage trees to be published in the nineteenth century. In 1841, the botanist James Grigor (1811–1864) published *The Eastern Arboretum* (Grigor 1841), which was very favourably reviewed by Loudon in his *Gardener's Magazine* (Loudon 1840b). In the words of the book's subtitle, it contains a 'Register of Remarkable Trees, Seats, Gardens, etc., in the County of Norfolk'. It was illustrated with drawings of some fifty of these trees, which are also noticeable for showing no evidence of propping or bracing. As well as various books on ancient trees, horticultural periodicals also had a major role in promoting the topic. Of all the early gardening magazines, the *Gardeners' Chronicle* was particularly prominent in publishing articles and readers' letters about ancient trees. As interest in ancient and remarkable trees continued to grow among professional gardeners, garden owners, botanists and others, calls were made to provide some national record of these trees. In 1862, the *Proceedings of the Royal Horticultural Society* was one of the first journals to suggest this idea and also that the Royal Horticultural Society (RHS) should appoint an Arboricultural Committee to coordinate the gathering of this information (*PRHS* 1862, 765). While the RHS's Arboricultural Committee was eventually established, its proposed tree register never materialised in any meaningful form. However, calls continued to be made for such a register. In 1883, the editors of the magazine *Woods and Forests* decided to take the initiative themselves. In an announcement to their readers they declared:

> We shall be pleased to record in *Woods and Forests* any observations furnished by our readers with regard to measurements of the finest specimens of trees in the British

> Isles, both indigenous and exotic. Such a record will, we think, prove useful in furnishing data respecting the relative rates of growth of cultivated trees'
> (*Woods and Forests* 1883a, 32).

From time to time *Woods and Forests* did publish details of ancient and remarkable trees supplied by its readers. Unfortunately, the magazine went out of business less than eighteen months after that announcement. As well as calls for a national register of these trees, there was discussion in the horticultural press about the need to standardise how those trees were measured as different authorities on the subject had different methods. The Scottish Arboricultural Society was one of the first to take the lead and set out its own rules for measuring the girth of ancient trees. The renowned tree expert Angus Webster was quick to endorse the Society's method and declared in *The Garden* magazine:

> I wish that recorders of the dimensions of old and remarkable trees throughout the country would adhere rigidly to the rules laid down by the Scottish Arboricultural Society, that is, to take all girths of tree trunks at 2ft and 5ft from ground level, unless good reason – owing to a swelling or protuberances – for departing from these can be adduced. The reasons for so doing are obvious and need not be repeated, while a certain standard and uniformity in measurement is all important.
> Webster 1890, 385

As the Victorian era progressed, the belief that the natural character of ancient trees must not be disturbed, regardless of their dilapidated state, slowly began to be challenged. With the emergence of gardenesque principles and the focus on the individual tree rather than just the landscape, it became acceptable to give some artificial support to weak and heavy branches in an effort to retain these and prolong the life of the tree. An examination of many images of ancient trees dating from the eighteenth and nineteenth centuries confirms that some supports and restraints began to be used on a few high-profile trees from the 1840s onwards. An example is the Cowthorpe Oak, which formerly stood in a field in the village of Cowthorpe, near Wetherby, in Yorkshire. This features in Hunter's edition of Evelyn's *Silva* when it was already an ancient tree (Evelyn 1776, 500). It is also one of Strutt's tree portraits in his *Sylva Britannica*, where he describes it as 'undoubtedly the largest tree at present known in the kingdom' (Strutt 1830, 37). In these and other images, there is no sign of any tree support or restraint system. Then, a few wooden props appear under some of its low heavy branches, as evidenced in a painting of the tree by John N. Rhodes in 1840. By 1872, when a picture of the tree was published in the popular magazine *The Graphic*, it had several sturdy wooden props supporting a number of branches (*The Graphic* 1872, 253) (Figure 48). The tree was famous as 'the largest oak in England' and the picture shows a crowd gathered around it. The status of the tree as a popular tourist attraction was probably one of the reasons why its owners decided to install the wooden props, to reduce the risk of a branch collapsing and causing injury or worse.

Relevant periodicals and their arboricultural content

FIGURE 48. In Victorian times the Cowthorpe Oak became a popular tourist attraction. By 1872 the landowners had installed several sturdy wooden props to support some of the heavy branches, hoping this would reduce the risk of a branch collapsing and causing injury or worse. Note the two boys climbing up one of the props.

The development of the horticultural press from the 1820s onwards had an enormous impact on horticulture and related industries. By the 1840s there were several influential magazines and journals that focused on horticultural, landscape arboricultural and forestry topics. They had the capacity to interact with their readers, something not possible with books, as much of the material was submitted by correspondents in the form of articles, notices and letters. For the historian, the Victorian gardening press gives an invaluable insight into the experiences, ideas and concerns of those professionals working in horticulture and related fields.

Loudon's *Gardener's Magazine*, launched in 1826, was Britain's first periodical devoted solely to horticulture. Loudon not only owned the magazine, he was also its 'conductor' or editor. Over some eighteen years, until it closed following his death in 1844, it held a position of extraordinary influence and authority. Often fiercely critical and occasionally pontificating, the magazine reflected Loudon's own personality and social and professional values. Issues of social class come through in many of the contributions, highlighting important differences of view between the landowners and those employed as gardeners.

While the *Gardener's Magazine* was widely read by wealthy garden owners, it was especially valued by professional gardeners, although their access to copies was often difficult as these were initially very expensive. With regard to the magazine's focus on trees, in its first few years there is much about fruit trees, regular references to forestry but much less on ornamental trees. However, from 1828 the list of contents included a subheading entitled 'Arboriculture' that embraced all matters relating to trees, both utilitarian and ornamental. This prompted more tree-related articles, although these often focused on just two topics, moving large trees and tree pruning.

In 1833, a magazine entitled *The Gardener's and Forester's Record* was launched, conducted by Joseph Harrison, gardener to Lord Wharncliffe at Wortley Hall in Yorkshire. In his introductory remarks to the first issue, Harrison tells us much about the scope of the magazine and his use of relevant terms. He writes, 'The nature of the *Gardener's and Forester's Record* is not brought forward with a view to being a rival to any other publication, but as a *necessary auxiliary* to aid in forwarding the interests of horticulture' (Harrison 1833, 2). He then goes on to explain that 'arboriculture has been introduced to its pages' because on many estates the fruit or kitchen gardener 'has the superintendence of the woods and plantations, and is the Planter and Forester'. This remark by Joseph Harrison is enlightening. As many estate gardeners were expected to take responsibility for the woods and plantations, he believed there was a market for a periodical that covered these topics as well as general horticulture. It is also clear, not only from these remarks but from the content of the magazine, that for Harrison the term 'arboriculture' embraced both commercial forestry and amenity trees. The magazine initially included a number of articles on amenity trees, although this declined in the second and third volumes. The magazine ceased publication in May 1836, having failed to attract sufficient subscribers.

The Horticultural Society of London (later the Royal Horticultural Society) was founded in 1804 and from 1807 it published its *Transactions*, a record of its activities that also included articles of horticultural interest. These articles focused mainly on general gardening and contained little concerning amenity trees. When its title was changed to the *Proceedings of the Horticultural Society of London* in 1838, the content remained much the same. While there were many articles on fruit trees, there was little on ornamental trees and any mention of 'arboriculture' was in the context of forestry and considered beyond the scope of the Society. In 1846, the *Proceedings* became the *Journal* and from 1866 it was called *Journal of the Royal Horticultural Society*. Although the Society had an Arboricultural Committee throughout most of the 1860s, this does not appear to have any significant impact on increasing the arboricultural content of the *Journal*, although there was slightly more of this in the 1890s. While many of the Society's members were interested in ornamental trees, this focused more on their aesthetic and botanical aspects rather than the practicalities of tree care.

Launched in 1841 and initially edited by Joseph Paxton and John Lindley, *The Gardeners' Chronicle* has had a major influence on the development of professional arboriculture in Britain. Indeed, this continues today as it is published under the title *Horticulture Week*. On the front page of its first issue, an item called 'Prospectus of *The Gardeners' Chronicle*' declared, 'Another peculiar feature will be the introduction of information relating to Foresting, or Arboriculture' (*Gardeners' Chronicle* 1841, 1). However, arboriculture was understood by the editors to be mainly concerned with timber production. In 1844 the magazine introduced a regular item called 'Calendar of Operations for the Ensuing Week', which had a section entitled 'Arboriculture' (*Gardeners' Chronicle* 1844, 24). The contributor was 'W.B.', who was probably William Billington, a forester. The emphasis of this regular feature was on plantations rather than individual or small groups of trees. By the 1870s, the magazine was including forestry matters under the heading 'Forestry', although those doing the work were still referred to as 'arboriculturists' as well as 'foresters'. By 1877, the section 'Forestry' was frequently written by Christopher Young Michie, a forester, but in reference to amenity trees he often refers to the 'arboricultural profession'. The first mention of 'tree surgery' in *Gardeners' Chronicle* (1873, 947) occurs in 1873 and is a short notice reprinted from the *Journal of Chemistry* in the United States. However, the term 'tree surgery' was hardly used by British writers in the Victorian era. While the title 'forester' becomes increasingly used in *Gardeners' Chronicle* towards the end of the nineteenth century, the magazine editors and contributors seem unsure how to refer to those primarily engaged with amenity trees. Sometimes they are called 'arborist' or 'arboriculturist' and often just 'tree expert'. When the term 'planter' was used, this referred to a landowner involved in tree planting rather than a professional employed to do it.

The first issue of *The Garden*, a magazine founded by William Robinson, appeared in 1871. From the start, *The Garden* has a section entitled 'The Arboretum' that focuses on trees and shrubs and arboricultural topics. From the early 1880s, the well-known tree expert Angus Webster was making frequent contributions to the magazine, along with others, under the regular heading 'Trees and Shrubs'. However, when it came to using terms to describe the discipline of planting and caring for amenity trees, or giving a name to those who specialised in this work, *The Garden* had the same difficulty as other magazine and journals. Robinson also launched some other relevant magazines, especially *Gardening Illustrated*, but this did not have as much arboricultural content as *The Garden* and was aimed more at the middle-class garden owner.

The Scottish Arboricultural Society was founded in 1854 and the English Arboricultural Society was founded in 1882, and despite their titles they focused on forestry. Both societies will be discussed later but here we are concerned with their members' periodicals. The *Transactions of the Scottish Arboricultural Society* first appeared in 1856 and it contains a record of the Society's activities as well as articles of interest submitted by its members (*Trans. RSAS* 1905, 8). The first volume of the *Transactions*, published in 1858, has material that is pure forestry with nothing specifically related to amenity trees. Indeed, in his opening address to the Society, its President, James Brown, made it clear that the Society viewed arboriculture as synonymous with forestry (Brown 1858, 1). In the *Transactions* of 1872 there is a lengthy article entitled 'Practical Arboriculture' written by William Gilchrist, the forester at Cluny Castle (Gilchrist 1872). Gilchrist writes mainly about commercial forestry and he uses the term 'forester' seventeen times. However, he uses the term 'arboriculturist' twice and on one occasion this is in the context of ornamental trees, which suggests he might be drawing a distinction between the role of the forester and the arboriculturist (*ibid.*, 40). During the remainder of the nineteenth century the Society's *Transactions* did include a few articles focusing on ornamental trees, but these were mainly concerned with the trees themselves rather than with practical tree care. In 1884, when the English Arboricultural Society published its own *Transactions*, this followed a similar pattern to that of the Scottish Society.

There had been talk in the 1870s among Britain's foresters about the need for a new journal devoted exclusively to forestry. Eventually, this became a reality in May 1877 with the launch of *The Journal of Forestry and Estates Management.* This received some welcome advance publicity in the pages of the *Gardeners' Chronicle*, initially with a letter from Robert Hutchison, Vice-President of the Scottish Arboricultural Society. The editors of the *Gardeners' Chronicle* had expressed concern that the proposed journal might 'interfere' with the work of the Society. However, Hutchison was quick to respond and declared, 'The founders of the proposed *Journal of Forestry* are all good friends and allies of the Arboricultural Society' and that the new publication would help to strengthen the Society (Hutchison 1877, 537). As well as carrying the occasional excerpt from this new *Journal of Forestry*, the *Gardeners' Chronicle* also took advertising from it. In 1880, a large advertisement appeared that announced the commencement of a new volume of the *Journal of Forestry*, which stated, 'The editor has pleasure in announcing that he has arranged for a large number of articles on almost every conceivable branch of Arboriculture and Estates Management' (*Gardeners' Chronicle* 1880a, 633). Clearly, the confusion of terms continued with this forestry journal. The first volume was concerned largely with commercial forestry and there was very little arboriculture or even estate management. From November 1882 until November 1884 its editor was Francis George Heath, who did his best to attract a more general readership. As a result, the *Journal* became a hybrid between a technical forestry publication and something resembling *Country Life* magazine. The *Journal of Forestry* ceased publication in April 1886 due to financial difficulties.

In December 1883, an entirely new magazine was launched that focused exclusively on tree-related subjects. This was called *Woods and Forests* and from an early stage its editor was William Goldring, the much-respected landscape architect and naturalist. This modestly priced magazine was published by William Robinson. Each issue was from fourteen to sixteen pages in length and the content included a broad range of articles about trees and their cultivation. However, much of the content was not original and many items were reproduced from elsewhere, including from *The Garden*, its sister magazine. Nevertheless, there was significantly more arboricultural content in *Woods and Forests* than in the *Journal of Forestry*, although it seems to have been directed less at the professional and more at the enthusiastic amateur or landowner. Although *Woods and Forests* was well received and played an invaluable role in promoting knowledge and discussion of arboricultural subjects, it was not a financial success. In March 1885, after only sixty-eight issues, it was absorbed in *The Garden* where a section entitled 'Woods and Forests' becomes a regular feature of that magazine. As a result, there was an increase in the content of *The Garden* relating to trees and shrubs and a little more on practical arboriculture.

In reviewing the arboricultural content of British horticultural and tree-related periodicals in the nineteenth century, one feature stands out. From the 1870s there is a major increase in articles and letters on the topic of urban trees. This might be expected given the progress of the Victorian park and street tree movements, which flourished from the 1860s onwards. This coverage was to have a major impact on increasing the profile of urban trees among the public and in helping to raise the status of those professionals engaged in planting and caring for urban trees. Periodicals such as *Gardeners' Chronicle*, *The Garden* and *Woods and Forests* led the way in their coverage of this topic.

As well as being concerned with developments at home, the British horticultural press was also interested in progress elsewhere in the world and the editors reproduced articles taken from relevant overseas periodicals. Items from the *Bulletin d'Arboriculture*, the publication of the Cercle d'Arboriculture of Belgium, appeared occasionally in the *Gardeners' Chronicle*, *The Garden* and similar periodicals. Of course, those Belgian articles were often about fruit trees as the French and Belgians commonly used the term arboriculture in this context. An important American magazine was *The Horticulturist*, which ran from 1846 to 1875. It was edited by Andrew Jackson Downing, the eminent American landscape architect, until his death in 1852. While some of its articles were reproduced in the Victorian horticultural press, its overall impact in Britain is uncertain. Another American periodical whose content was sometimes reproduced in the relevant British press was *Garden and Forest: A Journal of Horticulture, Landscape Art and Forestry* (Hou 2013). Commencing publication in 1888, it existed for only nine years. However, this ground-breaking magazine was more advanced than similar British periodicals, especially in its coverage of urban horticulture, landscape architecture and planning, and by addressing these topics from an environmental perspective. The initiator and conductor of *Garden and Forest* was Charles Sprague Sargent of the Arnold Arboretum, although the practical editing was done by William Augustus Stiles. The *Gardeners' Chronicle* (1892a, 509) occasionally carried advertisements for *Garden and Forest*. While it undoubtedly had readers in Britain, its immediate impact here was probably quite limited, although indirectly this was considerable in the long term.

Important books on tree botany and dendrology

The nineteenth century could be described as the 'golden age' of horticultural and arboricultural publishing, not only because of the founding of many periodicals but also because of the huge increase in relevant books compared to the previous century. Some of those books that are concerned with practical tree care will be mentioned later. In this section we detail the more important books concerned with botany and dendrology, which also had practical value to professionals working with trees, particularly concerning their identification and planting requirements.

One of the earliest nineteenth-century works was by the English botanist, Aylmer Bourke Lambert (1761–1842), and entitled *A Description of the Genus Pinus* (Lambert 1803). This described all the conifers then known and was a sumptuously illustrated folio volume. A second edition was produced between 1828 and 1837, and a third and smaller octavo edition in 1832. Many of the new conifers discovered by David Douglas and others were described for the first time by Lambert. An early nineteenth-century work that covered all known trees was entitled *Dendrologia Britannica* and was published in 1825 by the botanist Peter Watson (1761–1830) (Watson 1825). This was issued in two volumes containing 172 beautiful hand-coloured plates. Loudon described *Dendrologia Britannica* as the 'most scientific work devoted exclusively to trees which has hitherto been published in England' and it was an inspiration for his own *Arboretum et Fruticetum Britannicum* (Loudon 1838a, 1, 188).

The many coloured plates in these books made them hugely expensive and their limited sales never remotely covered publication costs (Elliott *et al.* 2011, 71). This restricted their impact and usefulness to the landed class and ensured that professional gardeners did not get any benefit, unless lent a copy by their employer, which was unlikely. Their impact was also limited by the lack of illustrations of complete trees, instead just showing individual parts such as their leaves, twigs, flowers and seeds.

First published in 1838, Loudon's *Arboretum et Fruticetum Britannicum* is arguably the most important systematic study of the trees and shrubs growing in Britain ever produced (Loudon 1838a). In this monumental eight-volume work each species of tree and shrub is considered in various categories embracing its description, distribution, history, properties and uses, propagation and culture. Loudon also refers to notable individual trees at various locations in Britain, Ireland and continental Europe. In the four volumes of illustrations the trees and shrubs are shown in remarkable detail. They were all drawn from life, some from the grounds of Syon House at Kew but most from the Loddiges arboretum, located at their famous nursery in Hackney. There are, however, some significant omissions from the *Arboretum Britannicum* which are very revealing. The ghost of the book that Loudon had begun to compose entitled the *Encyclopaedia of Arboriculture*, which was written in manuscript but never published, haunts the pages, helping to explain the omissions (Elliott *et al.* 2011, 83). In the

Arboretum Britannicum there is a lack of information concerning the physiology and anatomy of trees and arguably more about this in his *Encyclopaedia of Agriculture* (Loudon 1825). The *Arboretum Britannicum* also has comparatively little on practical arboriculture and the tree sections of his *Encyclopaedia of Gardening* give some indication of the material that might have been included in the proposed *Encyclopaedia of Arboriculture.* According to Loudon (1838a, I, xii) the *Encyclopaedia of Arboriculture* was originally intended to be included in the *Arboretum Britannicum*, but because of the eventual length of the text he had decided to publish this separately, and this would have included all the practical aspects of the propagation, planting, maintenance and management of trees for both timber and ornament. In compiling the *Arboretum Britannicum* Loudon was assisted by some 800 people who completed a questionnaire that he circulated and which supplied a huge amount of information that was included (Loudon 1838a, I, xv). While the *Arboretum Britannicum* was beyond the pocket of most professional gardeners, it was issued in a single, if very bulky, volume as an abridged edition in 1842.

The *Arboretum Britannicum* was very well received by both professionals and amateurs in the gardening world. The eminent botanist Sir William Hooker referred to the work as 'one of the highest importance and greatest utility to the arboriculturist' (Hooker 1839). Hooker uses the term 'arboriculturist' in the context of trees for both profit and pleasure, just as Loudon does. The cost of producing the *Arboretum Britannicum* was enormous but this was not measured in financial terms alone. Loudon's health was already seriously compromised before he began the work and his wife Jane believed that her husband's physical problems were deepened by the debts that he laboured under from the 1830s, largely due to the *Arboretum Britannicum.* In Jane Loudon's view, 'in his determination to make the work as perfect as possible, he involved himself in the difficulties which hastened his death' (Loudon, J.W. 1845, xxxviii). However, the *Arboretum Britannicum* was also the source of some financial relief to his wife after he died when a special appeal to sell the remaining copies raised hundreds of pounds (*ibid.*, xlvi). Loudon had written the *Arboretum Britannicum* for the public good and it remains one of the greatest works on arboriculture.

With the introduction into Britain of many wonderful conifers from the western coast of North America, there was a market for books featuring conifers of all types. One of the first was by George Gordon (1858) entitled *The Pinetum.* Perhaps the most outstanding was by Edward Ravenscroft and entitled *The Pinetum Britannicum* (Ravenscroft 1884). This monumental three-volume work, issued in its entirety in 1884, was illustrated with William Richardson's hand-coloured lithographs of trees shown in their native habitat or under cultivation in British gardens. A less expensive book on conifers was *A Manual of the Coniferae* published in 1881 by the nursery firm of James Veitch and Son of Chelsea, the anonymous author being Adolphus H. Kent (Kent 1881).

For the professional who needed a convenient and authoritative handbook on the trees growing in Britain, these publications must have generated mixed emotions. No doubt, when they were lucky enough to glimpse them, they were full of admiration for these sumptuous productions. However, they must have also felt some frustration that not only could they not afford copies, but these books were also impractical for use in the field. What professionals really needed was a modestly priced, convenient handbook of the type produced in the twentieth century but which was unavailable in Victorian times.

Arboricultural practices

Many of the relevant books published in the nineteenth century were manuals on the practical aspects of amenity horticulture and arboriculture. Some were directed towards the professional gardener but many more were written for garden owners, ranging from those who had small suburban plots to those with large estates and extensive gardens. However, even those manuals written by and for professional gardeners often had quite patchy coverage of arboricultural practices. While there was often plenty on tree planting, especially large tree transplanting, and various aspects of pruning, especially for fruit and ornamental shrubs, there was usually only a brief mention of other aspects of practical arboriculture, such as preventative or remedial tree care. For information on practical arboriculture the pages of gardening magazines such as the *Gardeners' Chronicle* and *The Garden* would often be more helpful.

After all the rancorous debate concerning William Forsyth's plaister for restoring decayed trees (Forsyth 1802, 321), public confidence in tree experts was understandably dented. People were now more likely to doubt the opinion of 'experts' on other aspects of tree care and to question just how reliable was the science behind this. The need for scientific proof to support practical activities applied to gardening in general and the horticultural establishment was keen to provide this. One important contribution was John Lindley's *The Theory of Horticulture; or, an attempt to explain the principal operations of gardening upon physiological principles* (Lindley 1840). Lindley was an eminent botanist and horticulturist who was widely recognised for his work at the Royal Horticultural Society and elsewhere. In the preface to his book he states it was 'written in the hope of providing the intelligent gardener, and the scientific amateur, correctly, with the rationalia of the more important operations of Horticulture' (*ibid.*, v). In dedicating his book to Thomas Andrew Knight, the person who led the criticism of Forsyth's plaister, Lindley also made clear his personal stance on those earlier events. Loudon, who Lindley had worked closely with on some of his books, was quick to praise this more scientific approach, especially with regard to trees. He remarked, 'The nineteenth century has commenced with a much more scientific mode of planting and managing trees than formerly existed' (Loudon 1835a, 360).

Tree planting

With regard to establishing young trees in plantations, the often confusing recommendations of the late eighteenth century were still widely repeated. For example, Loudon (1804, 120) recommends that transplants should generally be used but then states, 'Most of the deciduous trees, and particularly the oak and ash, should be raised from the seed, where they are finally to remain', and he quotes Marshall and Miller in support of this (*ibid.*, 123). Loudon still believes that planting in an irregular pattern is acceptable, although he adds, 'In very extensive plantations, it becomes an object to plant in rows, as thus the trees may be more easily cultivated' (*ibid.*, 121). He later refines his views on this, stating, 'With respect to the disposition of trees in a plantation, where art is avowed and purposely displayed, the more regular the plants are placed, the better is the end attained; but, where nature is to be imitated, irregularity will best sustain the character' (Loudon 1822, 1096). He adds that a natural arrangement is particularly important with 'detached groups, which operate so powerful an effect in laying out the grounds of a residence'.

Another respected early nineteenth-century authority on planting was William Pontey who wrote *The Profitable Planter: A treatise on the theory and practice of planting forest trees* (Pontey 1808a). Pontey had a major nursery at Kirkheaton near Huddersfield and was renowned for his expertise on trees. His book details the techniques and equipment for forestry planting, which are quite similar to those used today. Throughout the nineteenth century the debate continued about whether the roots of young trees should be pruned back before planting, not just to remove broken or bruised roots, but to achieve root/shoot balance. The topic featured frequently in the horticultural press and a typical exchange of views appeared in the *Gardener's Magazine* in 1830. Responding to an earlier letter, James Elles (1830, 546), an English head gardener in Ireland, stated 'That the plan of not pruning transplanted trees is strictly in accordance with the principles of physiology, I think few men will hesitate to admit'. A note added by Loudon at the end of his letter endorses this view.

While most books on tree planting were concerned with forest trees in plantations, some covered ornamental trees. One example is *Practical Hints on Planting Ornamental Trees* by John Standish and Charles Noble (1852), partners in a nursery in Bagshot, Berkshire. Most of the book is taken up with descriptions of individual tree species, with a discussion of their ornamental features and their soil and climate requirements. Only a small proportion is on planting and establishment and there is almost no science to support this. However, the book was well received by the horticultural world. Of those Victorian writers on ornamental trees, most favoured planting these in the spring. For example, Ravenscroft (1883, 35) recommended planting 'just when the trees are starting into growth; for the winter is such a fearful ordeal to all vegetation [and] by planting in the spring the shock is not so great'.

One method of planting ornamental trees that was popularised by Loudon was mound planting, which was done to display the roots to good effect. Other landscape gardeners and designers adopted Loudon's technique and it became quite common in private gardens and the new public parks. Loudon explained the reasoning behind mound planting:

> In general, trees are planted so deep that this ramification [of the main surface roots] never appears above the surface, and the trunk of the tree seems fixed in the ground like a post which has been driven into it; an appearance as contrary to truth and nature, and also to the health of the tree, as the shaft of a column without a base or a capital would, if employed in a building, be to architectural taste. To prevent this monstrous and unnatural appearance from occurring in the Derby Arboretum, I have directed all the trees to be planted on little hills, the width of the base being three times the height of the hill, so that the junction of the main roots with the base of the trunk will appear above ground.
>
> Loudon 1840c, 72–73

Most experts on planting agreed on the need for some type of support for newly planted trees larger than transplants, such as a tree stake, and also some protection, such as a tree guard. While some form of support was recognised as necessary, tree stakes were generally considered unsightly and to be avoided if possible (Kemp 1850, 202). Most Victorian texts that give guidance on young tree planting recommend the use of tall stakes. Indeed, this type of staking, where the top of the stake reaches into the crown of the tree, continues to this day, despite modern research demonstrating the benefits of short stakes (Patch 1984). However, one of the first professionals to suggest short stakes was John B. Webster, a Scottish forester and gardener who worked in Armagh, Ireland, and later at Balmoral Castle. In 1888 he wrote a letter to *The Garden*, stating:

> I would merely state, keep the root firm and the head will take care of itself. In staking young trees, say from 8–10 feet high, I never use a stake more than 18–24 inches above the ground, and this I find to be neater, cheaper and more effective in keeping the tree in its proper position than in cases where the long stake is used. … Trees that are properly tied to a short stake are not only kept firm at the roots, but the stem can oscillate in a natural way with the wind, so that the pressure by the wind on the top is less, to a large extent, than in cases where the tree is tied to a long, stiff, unyielding pole. Movement by the wind, in a natural way, is also beneficial in promoting the health, strength and stamina of the tree in producing a uniform thickness of stem in proportion to its height.
>
> Webster 1888, 544

Webster's remarkable insight into the benefits of short staking was written without the advantage of modern research. Looking back on this now, it is surprising that his views did not attract more comment or have some impact on changing planting practice. However, the usual Victorian practice of using tall stakes continued into the twentieth century and can still be seen in many planting schemes across Britain.

To give individual young trees some protection against damage or destruction, the use of tree guards was often recommended. A concern of Loudon was that the colour of these guards should be appropriate and he recommended painting them brown or green to blend into the landscape (Loudon 1804, 156). As well as the wooden tree guards used in the previous century, some companies now manufactured a range of metal tree guards. As the Victorian

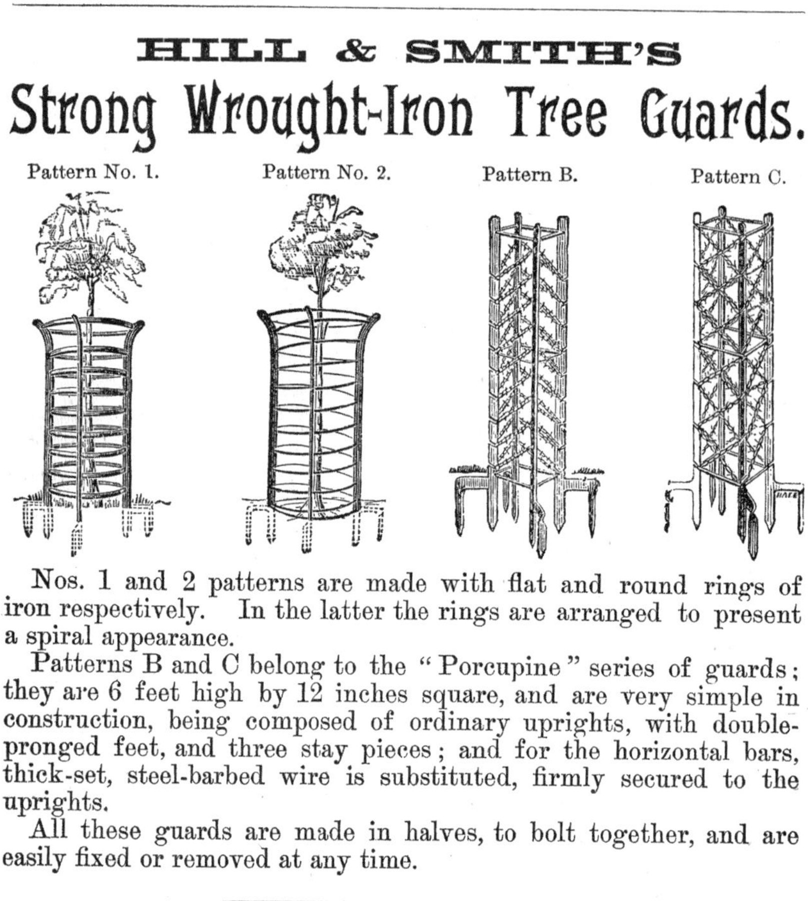

HILL & SMITH'S

Strong Wrought-Iron Tree Guards.

Pattern No. 1. Pattern No. 2. Pattern B. Pattern C.

Nos. 1 and 2 patterns are made with flat and round rings of iron respectively. In the latter the rings are arranged to present a spiral appearance.

Patterns B and C belong to the "Porcupine" series of guards; they are 6 feet high by 12 inches square, and are very simple in construction, being composed of ordinary uprights, with double-pronged feet, and three stay pieces; and for the horizontal bars, thick-set, steel-barbed wire is substituted, firmly secured to the uprights.

All these guards are made in halves, to bolt together, and are easily fixed or removed at any time.

FIGURE 49. During the Victorian era metal tree guards became increasingly popular for high-profile plantings in urban areas, especially in streets. This range of guards, advertised in 1892, came in two halves that could be bolted together for easy assembly or removal.

era progressed, these became increasingly popular for high-profile plantings in urban areas, especially in streets (Figure 49). Some manufactures offered for sale a combination of both tree support and tree guard, such as the Waverley Tree-Guard and Support (*Gardeners' Chronicle* 1894, 309). From early Victorian times commercially produced tree-ties, used to attach the young tree to the stake, were also available and made from a variety of materials. Alternatively, the professional could make their own tree-ties using rope, cloth or straw (Masters 1891, 76) (Figure 50).

As the Victorian street tree movement gathered pace from the 1860s it brought new opportunities and constraints for tree planting. Local authorities across Britain were now looking to plant up many of their streets and roads with trees. The challenge was daunting. The density of the urban fabric with its buildings, street furniture, overhead and underground services and much more, often offered little space for trees to establish and thrive (Johnston 2017, 244–271). There were also unseen problems to overcome, such as soil and air

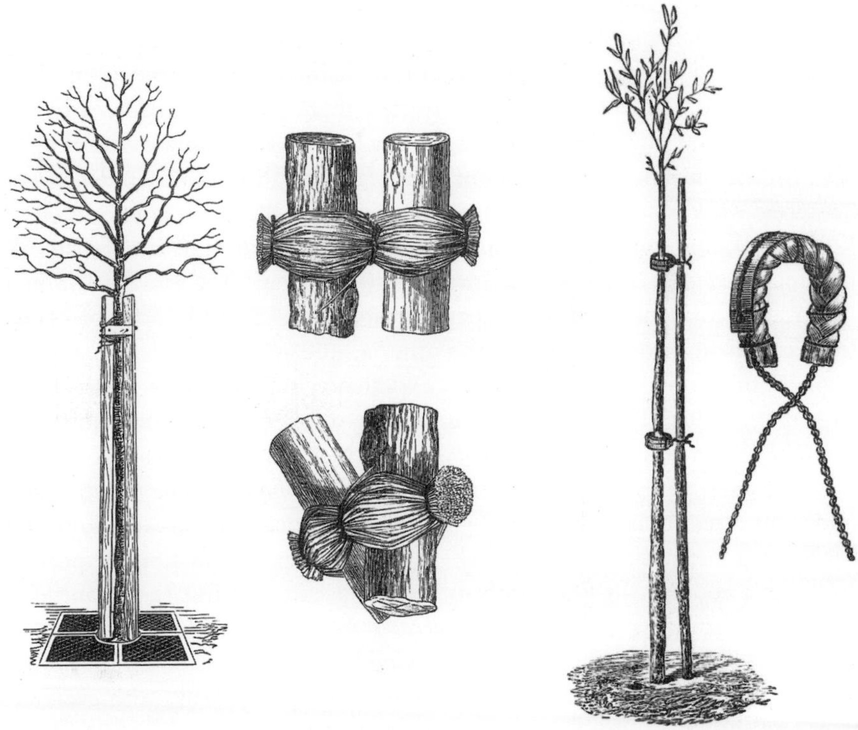

FIGURE 50. Various tree-planting accessories from 1891. Left: Thames Embankment planting with the stake acting as a guard. Centre: tree-ties made from straw. Right: A more flexible tree-tie made from wire but with a padded section to protect the bark.

pollution, and mineral deficient and compacted soil. One street tree planting project that deserves a special mention was the planting on the Thames Embankment. This generated a huge amount of debate, not just because of the prestigious location but also for the way the trees were planted, and this debate occupied many column inches in the popular and professional press. The Embankment trees were planted using at least two different methods of staking and tying, and in one method the stake also acted as a form of tree guard (Masters 1891, 76) (Figure 50).

Transplanting large trees

In the eighteenth century the transplanting of large trees was commonly undertaken by some leading landscape gardeners such as 'Capability' Brown. However, in the nineteenth century it became something of a small industry. As the century progressed an increasing number of professionals gained reputations for their expertise in moving large trees, often over considerable distances, and their services were much in demand. This was also a time of innovation in tree moving, not only with novel techniques but also with the equipment used.

One early nineteenth-century exponent of large tree moving was William McNab (1780–1848). McNab (sometimes spelt M'Nab) was born in Ayrshire and worked as a gardener for the Earl of Haddington at Tyningham in East Lothian (Fletcher and Brown 1970, 83). His character and abilities were quickly recognised and he was recommended to William Aiton for a position at the Royal Botanic Gardens at Kew. He started there in 1801, aged twenty, and became foreman in 1803. His success at Kew led to his recommendation by Sir Joseph Banks for the post of Curator at the Edinburgh Botanic Garden, where he started work in 1810. In 1820, Dr Robert Graham, the newly appointed Regis Keeper at Edinburgh, was confronted with the major task of transferring the Botanic Garden to a new site at Inverleith (*ibid.*, 104). The move was a huge undertaking that included the preparation for removal and transfer of mature trees and shrubs from one site to the other a mile away. McNab was asked to organise this as he had already gained experience in successfully transplanting large trees while at Kew (Sang 1830, 10). Some of the trees were over 40 ft (12.2 m) in height and others were 100 years old. McNab undertook this task between 1820 and 1823, with very few trees and other plants failing to be re-established under his care. To move the larger trees, he developed what became known as McNab's transplanting machine, which moved them in an upright position and with a rootball. In 1830, McNab gave an account of this transplanting in his book *Hints on the Planting and General Treatment of Hardy Evergreens in the Climate of Scotland* (McNab 1830). Following his success at Edinburgh, McNab was acknowledged as an expert in tree transplanting, especially with evergreen trees and shrubs.

The precise dimensions of McNab's transplanting machine were not widely known until 1861 when details were published in the *Gardeners' Chronicle* (1861a, 457) (Figure 51). The information was supplied by James McNab, who had succeeded his father as Curator at Edinburgh in 1849. In 1859, the gardens were extended and it was necessary to transplant a large number of trees and shrubs. This prompted James NcNab to strengthen his father's machine, enabling a larger rootball to be lifted with the aid of fewer men than were formerly required.

The next person to gain a reputation for tree transplanting was an aristocrat, not a professional. Sir Henry Steuart (1759–1836) was a wealthy Scottish landowner at Allanton House in Lanarkshire. In his desire to improve the landscape at Allanton, he experimented with moving large trees into prominent positions rather than planting young trees and waiting many years for these to mature. Steuart's methods are described in his book *The Planter's Guide*, which appeared in two editions in 1828, the second being seventy pages longer (Steuart 1828). Steuart believed that with an intelligent application of scientific methods and the proper use of his 'improved' planting machine 'an *entire Park* could thus be *wooded at once*, and forty years of life anticipated' (*ibid.*, 10). He even asserted that all this could be 'accomplished at *a very moderate expense*, and brought within the reach, not only of the great and opulent but of any person of limited fortune' (*ibid.*, 11). Not everyone endorsed his transplanting methods and some aspects would be questioned today. While he recommended

that the rootball be prepared over two seasons, he then lifted the tree and moved it bare-rooted without any protection for the roots. When the tree was planted in its new location, no form of support was given even in exposed situations. Steuart (*ibid.*, 122) states, 'Supports or props, whether composed of wood, cordage or any other material, are of little avail in giving stability'. While Steuart places great importance on his transplanting machine, it was, by his own admission, similar to that used by 'Capability' Brown (*ibid.*, 246). It was also similar in appearance to the machine used by Samuel Hayes in Ireland in the 1790s, except that there is no small trailing wheel. The tree was kept in the horizontal position during transport by two 'balancemen' who sat on the trunk and among the branches (*ibid.*, 251) (Figure 52).

Steuart's book *The Planter's Guide* gained favourable reviews by Sir Walter Scott (1828) in the *Quarterly Review* and another by James Main (1828) in the *Gardener's Magazine*. The publicity surrounding the book attracted many visitors to Allanton and its attractive and mature landscape was seen as vindicating Steuart's methods. However, he had many critics, not least among professional gardeners, not only because they could not endorse his transplanting methods but also because they objected to some of his disparaging remarks about them in his book. While most professionals maintained a diplomatic silence, one individual spoke up against Steuart, although he did this anonymously. Edward Sang was a nurseryman at Kirkcaldy in Fife who had considerable experience in arboricultural work. In 1830 he published a pamphlet entitled *Strictures on Sir Henry Steuart's Planter's Guide* using the pseudonym of 'A Planter of Some Experience' (Sang 1830). However, a review of the pamphlet in the *Edinburgh*

FIGURE 51. The precise dimensions of William McNab's tree transplanting machine were not widely known until 1861 when the details and this illustration were published in the *Gardeners' Chronicle* by his son James.

FIGURE 52. Sir Henry Steuart's transplanting machine in action in the 1820s moving a large tree in his landscape park at Allanton in Scotland. The tree was kept in the horizontal position during transport by 'balancemen' who sat on the trunk and among the branches.

Literary Journal (*ELJ* 1830, 112) names the author as Edward Sang. This was a suitably robust response to Steuart's unfounded assertions and personal slights that had appeared in his *Planter's Guide*. Sang (1830, 9) claims that Steuart had got his ideas on transplanting from James Stewart, the gardener to Sir John Hope at Pinkie, something never mentioned by Steuart himself. James Stewart described his methods in a paper read to the Caledonian Horticultural Society in 1811 and published in its *Memoirs* a few years later, and these are remarkably similar to those later used by Steuart (Stewart 1814). Sang is particularly quick to defend his friend McNab against Steuart's assertion that it was he who suggested to Graham and McNab at the Edinburgh Botanic Garden that they should use 'the ingenious method of Lord Fitzharding, in cutting round the plants [and leaving them] for two or three years after' (Steuart 1828, 63). Sang states that McNab had publicly denied getting the idea from this source, which he later confirmed in his book (McNab 1830, 6). Sang also challenges Steuart's claim that he was the first to introduce such a method of tree transplanting into Scotland by pointing out that 'John Reid … in his *Scots Gardener* in 1683 had taught his countrymen this very method' (Sang 1830, 12). Steuart had also given the credit for the tree moving at Edinburgh to Graham instead of McNab, and Sang corrected his 'misapprehension'. Loudon was very supportive of Sang and wrote a glowing review of the *Strictures* in his *Gardener's Magazine*, declaring,

'We are not sorry to see this pamphlet, in which practical gardeners are relieved of the aspersions thrown out against them by Sir Henry Steuart' (Loudon 1830, 87). He added, 'Whoever possesses the *Planter's Guide* will do well to possess also the *Strictures*' (*ibid.*, 93). In Loudon's 1835 edition of the *Encyclopaedia of Gardening* he gives just one machine for tree moving, that designed by Mr Saul, which is similar to McNab's machine (Loudon 1835a, 553).

By the 1830s, instances involving very large trees or trees in prominent locations were attracting the attention of the regional and local newspapers. One example was the moving of a large weeping ash tree, undertaken by Joseph Paxton, the head gardener at Chatsworth. This beautiful tree, which had been growing in a small Derby garden for nearly fifty years, was purchased from the owner by the Duke of Devonshire to replant in his gardens at Chatsworth (*Derby Mercury* 1830, 3). This was accomplished by Paxton with the assistance of forty labourers and several horses, using a transplanting machine similar to Steuart's. Although the removal was successful, a turnpike along one of the roads had to be dismantled to enable the enormous tree to pass, as well as the gates and part of the wall at the entrance to Chatsworth Park. This story was reported in several British newspapers and attracted much comment.

During the 1830s and 1840s, the topic of large tree moving was featuring regularly in the horticultural press with descriptions of the techniques being used and new machines being invented. One issue often debated was whether the crown of the tree should be pruned back, supposedly to compensate for root loss in the transplanting process. The arguments for and against were often a repeat of those pertaining to newly planted young trees. The prevailing view was that pruning should be avoided, unless to remove branches damaged in the move, especially when the tree was moved with a large rootball. Even Steuart, who moved his trees bare-rooted, was against the 'lopping or cutting off the Tops or side Boughs, or both, at the period of Removal', although he felt that more research was needed on this topic (Steuart 1828, 78–80). Others disagreed and maintained that 'the top should be mutilated in proportion to the damages sustained by the roots' (Howden 1835, 15). Another topic to be debated regularly was the type of physical support that should be given to the tree in its new location. While Steuart was against any kind of support, he does propose an alternative of some sort of 'retaining-bank' but the details of this are unclear (Steuart 1828, 286). The general consensus was that some support was necessary to secure the root system and prevent this from rocking about in the planting pit, which would reduce the chances of the tree establishing itself in the new location. Various arrangements of ropes and wires, together with a variety of ties and anchor points, were suggested, with some being quite imaginative. Because any form of visible support was considered by some to be unsightly, a few professionals suggested the use of underground guying systems. William Thom, the Scottish amateur tree planter, declared, 'I now prefer … the new method of supporting recently transplanted trees by horizontal props, concealed under, or level with, or immediately above,

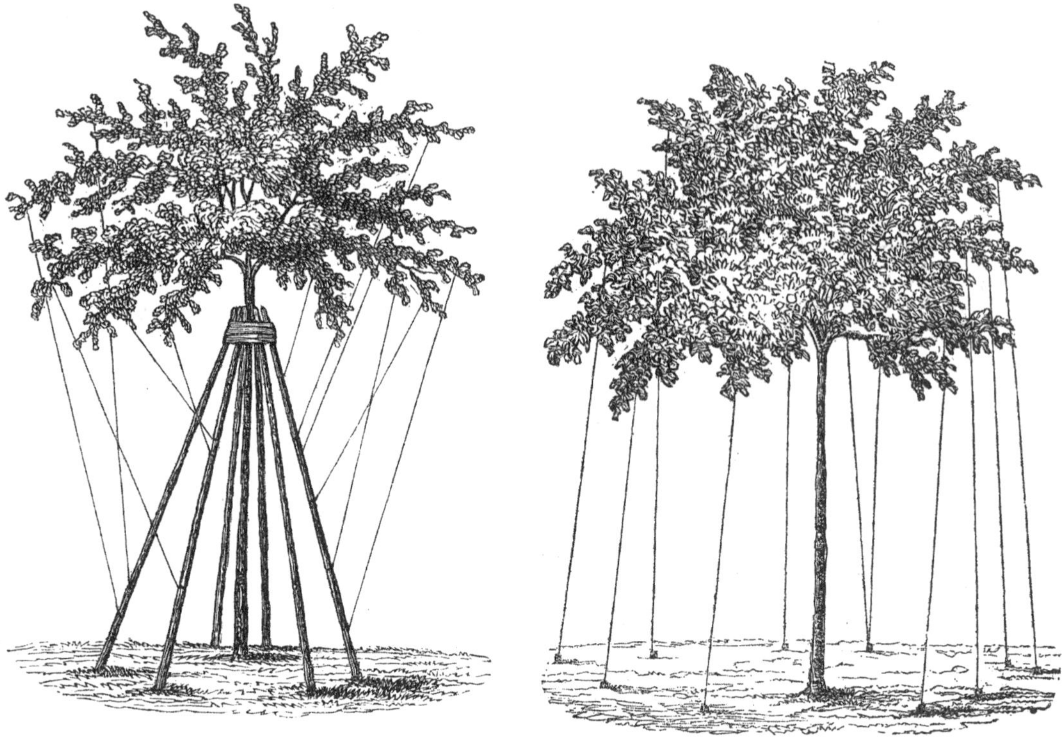

FIGURE 53. Samuel Taylor (1839), a gardener in Norfolk, devised these two novel methods of securing newly transplanted trees against the wind. He was inspired by the way Gulliver was held down by the Lilliputians.

the surface' (Thom 1831, 452). This type of tree root anchor consisted of two round poles in the form of a cross laid over the rooting area with the tree in the middle and the poles held down by stakes driven into the ground. It was similar to the method of underground guying being used at that time by Paxton at Chatsworth (Murphy 1831, 297). Some professionals who were not keen on root anchors devised their own novel methods of securing trees against the wind. One was Samuel Taylor, a gardener in Stoke Ferry, Norfolk. His proposal to use numerous cords to tie down the tree was inspired by the way Gulliver was held down by the Lilliputians (Taylor 1839). However, this does look rather bizarre by today's standards (Figure 53).

Some landscape professionals developed their own original methods for transplanting large trees. One was James Munro, the manager of Dickson and Turnbull's Nursery at Brechin, and his method was described in the *Gardener's Magazine* (Loudon 1833). Munro was dissatisfied with the accepted method of root preparation that involved digging a trench and filling it with loose soil and leaf mould. Instead, he left the trench empty, roofed it over with boards and then covered this with one inch of soil. The tree was left like this for a year. On reducing the rootball to proper dimensions for removing the tree, the old roots were found furnished with fibres and matted sufficiently to retain enough earth to protect them during removal. According to Loudon, 'The object which Mr Munro had in view was, instead of encouraging the growth of fibres at the extremities of the amputated roots, as is done when the earth is filled with loose soil, to have them formed within the ball of earth, and not on the outside of it' (*ibid.*, 217).

A few years later, Munro had the honesty to admit in the *Gardener's Magazine* that his method had not been a success. In analysing this, he suggested the trees should have been left for three years rather than only one before they were moved (Munro 1842, 104). Also, that compost or manure should have been added to the backfill when replanting.

Some professionals invented their own transplanting machines. One was James Kidd, the gardener to Lord Kinnaird of Rossie Priory, Perthshire. This could be described as a gigantic five-pronged fork, mounted on a pair of wheels and axle, the latter serving as a fulcrum for lifting the tree, after the roots had been prepared in the usual manner with a trench (Kidd 1840, 254). According to Kidd, the construction and operation of this machine involved little expense. Another interesting machine was constructed by an amateur, Colonel George Greenwood, an Old Etonian and army officer. Greenwood's interest in geology and soils led him to explore the topic of tree moving, which was featuring regularly in the popular press. In 1844, he published *The Tree Lifter* which set out his thoughts on the subject and also described his novel machine called the Tree-lifter. In 1853, an expanded second edition was published (Greenwood 1853). Greenwood claimed, 'Among the advantages of transplanting with the "Tree-lifter" may be reckoned its cheapness. Its simplicity is such that the whole may be performed, and even singled-handed, by a common labourer. One man may plant one tree per day, of from twenty-five to thirty feet in height' (*ibid.*, 1–2). Although the tree was moved immediately, this was done with a rootball. The horticultural world was not enthusiastic about Greenwood's method.

From the 1820s to the early 1840s, it was Loudon and his regular coverage of tree transplanting in his *Gardener's Magazine* that gave much encouragement to those involved. In his book *The Suburban Horticulturist* he gives a good summary of the various techniques that were commonly employed at that time and the advantages and disadvantages of each method (Loudon 1842, 312–324).

Of all the nineteenth-century individuals who got involved in transplanting large trees, the 'master of the art' was the Scotsman William Barron (1805–1891). Furthermore, while Joseph Paxton was arguably the most successful Victorian landscape gardener, Barron was perhaps the greatest tree expert of the Victorian era, considering that Loudon died in 1843. Barron was born in Eccles, Berwickshire, the son of a gardener, in 1805 and not 1800, as often cited. After serving his three-year gardening apprenticeship he worked at the Edinburgh Botanic Garden before heading south to England to work at Syon House in Middlesex. In 1830, he became head gardener to the Earl of Harrington at Elvaston in Derbyshire. It was during the development of this astonishing garden that Barron developed and refined his expertise in tree transplanting. Hundreds of trees, including very large and mature specimens, were moved across Derbyshire and adjacent counties to Elvaston. The trees were often so large that as they passed through the streets of Derby windows were frequently broken in houses on either side (Elliott 1986, 84). Barron's first tree transplanting machine was constructed in early 1831 and that autumn a 43 ft (13.1 m) tall cedar of Lebanon was moved into the gardens at Elvaston. However, unlike many of his contemporaries, Barron's transplanting method did not involve preparing the roots over two or three seasons (Barron 1852, 26). He considered the delay with this method both expensive and unnecessary. Instead, Barron created a very large rootball and took great care when cutting any roots to ensure as little disturbance as possible. Also in contrast to Steuart, Barron moved his trees in an upright position rather than horizontally. In his view, 'That system had only been applied to deciduous trees, which were removed when dormant, and could be carried with their stems in a horizontal position … but I saw the utter impossibility of thus treating the wide-spreading, horizontal, and brittle branches of Cedars of Lebanon, of the magnitude described, which if mutilated or broken, would not be worth removal' (*ibid.*, 27).To facilitate his method, Barron designed and constructed his own carriage or machine (*ibid.*, 25–48). The transplanter was assembled around the tree and the exposed ball of roots and soil was wrapped and tied before being winched up to the transporter. This whole operation normally involved several men. In a further departure from the method used by Steuart and many others, Barron did not go to the expense of digging and preparing a large pit to receive the tree. In his view:

> Much advantage would be gained in every way by placing the tree on the *surface* of the ground and by adding soil for the roots to be planted in. By this means the roots instead of being cramped in a pit, and arriving speedily at the subsoil, have the depth of natural soil to sport in, besides the free access of the air (so essential) being also secured. The trees when planted have, on this plan, the appearance of being place on raised bases, and in this way, height is gained.
>
> Barron 1852, 37

Although Barron's practice of 'hillock planting' came as a surprise to many of his fellow professionals, it was also endorsed by many, including the respected gardening author Charles McIntosh (1853, 2, 365).

After the death of the fourth Earl and the public opening of Elvaston, Barron was encouraged by the fifth Earl to develop the commercial potential of the nurseries. He was also encouraged to undertake designs for commissions for public parks, cemeteries and private gardens. In 1862, after the death of the fifth Earl, Barron resigned his position at Elvaston and purchased 40 acres in Barrowash to develop as a nursery. This not only proved profitable through the sale of nursery stock but it also gave him an excellent base for his tree-moving and landscaping operations, which he was now undertaking for various clients. The firm of William Barron and Son was always very adept at marketing their services, not only through well-produced catalogues and advertising but also through publicity in the newspapers for some of their high-profile projects. An example is the press coverage in 1855 for his removal of a large tree in Regent's Park (*ILN* 1855, 357) (Figure 54). It was not long before the firm had accumulated

FIGURE 54. William Barron and his skilled workforce using one of his machines to transplant a large tree at the Royal Botanic Society's Garden in Regent's Park. This high-profile commission in 1855 generated much public interest in tree moving.

an impressive portfolio of successfully completed national and international commissions. In 1865, the French newspaper *L'Univers Illustre* (1865, 673) was reporting on Barron's tree moving exploits in a Paris park. Another high-profile commission in Britain was undertaken for the Maharaja Duleep at Elveden Hall in Suffolk (*Gardeners' Chronicle* 1869, 365). This involved a fine purple beech, nearly 50 ft (15.2 m) tall and weighing over 20 tons. Rather than using one of his transplanting machines, Barron employed another technique which he often used with very large trees that were only being moved a short distance. The *Gardeners' Chronicle* reported this as follows:

> [The tree] had a platform of strong timber constructed underneath, and was raised upon rollers laid on planks by means of powerful jack-screws. This being done, the tree was drawn on to its new site with the aid of pulley-blocks of unusual size, and maintained throughout in an upright position. The complete success of the operation reflects the greatest credit on Mr Barron, who on its termination received three hearty cheers from those who were fortunate enough to witness it.
>
> *Gardeners' Chronicle* 1869, 365

Barron not only undertook tree-moving projects, he also offered for hire or sale a range of his transplanting machines. In 1886 he gave details of this in a letter to the *Gardeners' Chronicle*, stating, 'Our machines have been let out on hire for over thirty years. Besides having supplied machines to order of different sizes, we keep six sizes for hire, capable of removing trees from 12 tons to half a ton, which have this autumn been employed on seven estates in five different counties' (Barron 1886).

By the 1870s, machines had been constructed and supplied to many members of the aristocracy and gentry. In 1866 one was purchased by the Royal Botanic Gardens at Kew and became known as 'The Devil' by the gardening staff after several were injured by it (Desmond 2007, 419). By the end of the nineteenth century machines had been supplied to several international clients. In operating his business Barron took full advantage of the expanding rail network to transport trees from his nursery and also his transplanting machines and various parts for them. He also regularly used the railways to transport his specially trained workforce to the different projects. When a client wanted a large tree to be moved, they invariably engaged Barron and Son in a package deal that included the hire of the machine, the workforce and other incidental costs such as travel expenses.

By the 1870s, William Barron's tree-moving exploits were known nationally thanks to the publicity surrounding his high-profile projects. True to the spirit of the Victorian age, his work was viewed as heroic engineering and an arboricultural equivalent of the civil engineering triumphs of Isambard Kingdom Brunel. However, one particular commission was to prove controversial and led to many people questioning the ethics of some tree-moving projects. While working at Elvaston, Barron had taken trees from places where they had stood for centuries, often much admired by generations of local people, and removed them to a secluded garden for the private pleasure of one aristocratic couple. Interestingly, little was said about this at the time. However, in 1880 when Barron was engaged to move the famous Buckland yew, it was a different story. This venerable tree is mentioned in the Domesday Book and had been growing in the churchyard at Buckland, near Dover, for over 1,000 years (*Gardeners' Chronicle* 1880b, 305–306). Some 150 or 200 years previously, a brick vault had been built close to the trunk of the old tree on the south side, and in excavating for this vault all the roots on that side had been cut. The tree was now suffering and the plan was to move it 60 yards (55 m) away from the church where it could have another lease of life. However, this proposal was met with some vociferous opposition, not least from the editors of the *Gardeners' Chronicle* who declared, 'We would deem the removal of such a tree with such a history as an act of sacrilege' (*ibid.*, 306). Barron was shocked at this reaction and concerned about the damage to his reputation. He responded in a lengthy letter to the magazine, setting out the reasons for the tree moving (Barron 1880). At the end of Barron's letter the editors added the acerbic comment, 'Even after Mr Barron's explanation we are not disposed to alter our opinion as to that point. That Mr Barron could, given the proper appliances, move Westminster Abbey itself we do not doubt, but it would be none the less an act of sacrilege to attempt it' (*ibid.*, 557).

Despite the opposition, Barron went ahead and moved this massive tree using his system of platform, rollers and pulley-blocks that enabled the tree to

FIGURE 55. William Barron (top hat and umbrella) with his family and workforce photographed with the Buckland Yew in 1880. Moving this ancient tree, which is mentioned in the Domesday Book, created some controversy not only in professional circles but also in the popular press.
CREDIT: BOB HOLLINGSBEE.

be dragged along a specially prepared trench (Figure 55). The vicar of Buckland was delighted with the work and declared, 'The tree, by universal consent, looks far grander than it has ever done before, and when the grass grows will be a beautiful object' (*ibid.*, 556).

In the end, this incident did little to damage Barron's towering reputation and he continued to lead his very successful and much-respected firm of nurserymen and landscape gardeners. In 1885, his arboricultural expertise was recognised by Parliament when, at the age of eighty-one, he was summoned as its first expert witness by the Select Committee on Forestry (*Gardeners' Chronicle* 1891, 524). When Barron died in April 1891, the *Gardeners' Chronicle* honoured him with a lengthy obituary. His work was always motivated by deeply held moral and religious beliefs, something that is very evident from the pages of his ground-breaking book, *The British Winter Garden*. After his death, his son John continued his work and the firm of William Barron and Son achieved further success in the early decades of the twentieth century.

The success achieved by Barron and the acclaim he attracted encouraged others to try their hand at large tree transplanting. One of the most notable was Stewart McGlashen (1807–1874), a Scottish sculptor and mason, who was born in Campbeltown in western Scotland but spent his working life in Edinburgh. In the late 1840s, McGlashen (sometimes spelt M'Glashen) became fascinated with the idea of tree moving and decided to invent his own machine. Its most innovative feature was a series of 'spade-blades' that were used to lift out the rootball intact, without the need to prepare the tree's root system over two seasons (Figure 56). The basic method is similar to the Vermeer tree spade used in the 1970s. With McGlashen's machine it was claimed that the task of tree moving 'could be accomplished with ease and expedition, and at infinitely less expense, and that trees of much greater magnitude can be removed than has hitherto been supposed' (*Gardeners' Chronicle* 1852, 804). In 1852, the Caledonian Horticultural Society gave a favourable account of McGlashen's machine after members of its committee had seen it in operation (*Gardeners' Chronicle* 1853a, 6). In 1853, a demonstration of the machine was given to the Fellows of the Horticultural Society, in the presence of Prince Albert, at the Society's Gardens at Turnham Green (*Gardeners' Chronicle* 1853b, 163). The tree moved was a black Italian poplar some 55 ft (16.8 m) tall and the combined weight of the tree, earth and machinery was more than 13 tons (*ibid.*, 164). The tree was then lowered into a pre-dug hole about 40 ft (12.2 m) away. The entire operation took most of the afternoon and employed four men. The report of the demonstration in the *Gardeners' Chronicle* was largely favourable, although some reservations were expressed about aspects of the method and equipment. Some months later the editors of the *Gardeners' Chronicle* were expressing doubts about the basic 'tree spade' method. They stated:

> To wrench a tree out of the ground by main force is so contrary to the true principles of transplantation, that the consequences could not but be looked at with apprehension. We do not ourselves doubt that success would attend the trail, and the event justified our expectation; for the tree is now in good health, and has not suffered in any material degree. Is, then, the machinery employed by Mr M'Glashen to be recommended? And can violent wrenching be substituted safely for careful extraction?
>
> *Gardeners' Chronicle* 1853c, 755

In answer to their own questions, the editors gave McGlashen's machine their cautious and qualified approval. The doubts they expressed about the basic method did not impact adversely on the number of commissions for tree moving that McGlashen received. In 1854, he was invited to give a demonstration of his machine in the Parc Saint-Cloud, Paris, in the presence of the Empress of France (*L'Univers Illustre* 1854, 448). This was several years before Barron undertook a tree-moving commission in that city. Like Barron, McGlashen also saw the commercial benefits of offering versions of his machine for sale or hire. In 1853, in partnership with Messrs Young and Company of London and Edinburgh, he produced a pamphlet giving the details and prices for the various patterns and sizes of his machines (*Gardeners' Chronicle* 1853d, 807). McGlashen also invented a much smaller version of his machine, using the same tree-spade principle, for removing shrubs and young trees (McIntosh 1853, 2, 386).

FIGURE 56. Stewart McGlashen's tree transplanter, invented in the late 1840s, had an innovative feature comprising a series of 'spade-blades' (b). These were used to lift out the rootball intact, without the need to prepare the tree's root system over two seasons.

Another tree-moving machine often compared with those of Barron and McGlashen was constructed by Mr Mackay, gardener to Edward Strutt of Kingston Hall, near Derby. Charles McIntosh (1853, 2, 381) regarded this as an improvement on Barron's larger machine, although differing little in its construction or mode of operation. James Brown gives a detailed account of Mackay's machine in his second edition of *The Forester* and recommends its use (Brown 1851, 471–479). In a letter to the *Gardeners' Chronicle* in 1853, Barron confirms that 'the machine used by Strutt [Mackay's machine] was made from measurements and drawings taken from my large transplanting machine, with very slight modifications' (Barron 1853). In 1875, James Kay, the forester on the Bute Estate at Rothesay, published an article giving details of a new tree-moving machine he had invented (Kay 1875). This described the operation of his machine and included two plans showing its construction, which is similar to that developed by Barron.

The interest in tree-moving among private estate owners and public parks managers continued through the 1870s. To meet this demand there was now a tree-moving mini-industry comprising several major companies, with hundreds of machines for sale or hire, and an extensive workforce engaged in contracts throughout Britain and overseas. All this activity and interest was reflected in the articles, letters and advertisements that continued to fill the pages of the professional and popular press. The coverage in the professional press in the 1870s included two series of articles that were essentially a review of the various techniques and machines then in use. The first was in 1873 by Charles Kelly, a landscape gardener based in Tarporley, Cheshire, initially published in the *Gardener's Magazine* (not Loudon's magazine) and later issued as a small book (Kelly 1873). The second was in 1878 by James McNab, the son of William McNab, and these articles were published in *The Garden* (McNab 1878). By the mid-1880s, however, there was a significant decline in the amount of coverage on tree transplanting in both the professional and popular press. This was due to a significant drop in demand for tree moving by private clients in particular, although what caused this is a matter of speculation. Tree moving, especially done regularly, was a very expensive enterprise. When the Victorian landed estates felt the full economic shock of the agricultural depressions between 1873 and 1896, this type of expenditure would undoubtedly have been questioned. Furthermore, it seems likely that the novelty of tree moving was at last wearing off. It had often been commissioned by the landed aristocracy as a 'fashionable thing to do', to impress and amuse your friends and connections. It was really about the excitement on the day, enjoying the spectacle of moving a massive tree. Once planted, it soon became just another mature tree in the landscape. Although some projects continued to be commissioned, the 'golden age' of Victorian tree moving had come to an end.

Tree pruning and wound treatment

The subject of tree pruning was another major preoccupation of horticulturists and tree-related professionals in the nineteenth century. Like the topic of tree transplanting, it received a huge amount of coverage in the professional press and also featured quite often in newspapers and popular magazines.

At the start of the century the principles of the Picturesque movement continued to exert an influence on some aspects of landscape management and this included tree pruning. Many writers still had an aversion to pruning of any kind and recommended that this should be avoided if at all possible to allow the tree to grow naturally. Even in the 1840s, some continued to endorse this approach, such as one anonymous writer who declared, 'Ornamental trees require not the assistance of the pruner. Their branches should be secure from the unhallowed interference of his chisel and his knife' (*Gardener's Magazine* 1842, 112). Although Loudon was not against pruning in principle, in the Derby Arboretum he only allowed this in limited circumstances, such as removing dead wood, broken branches or basal suckers on some grafted species (Loudon 1840c, 79). These severe restrictions on pruning were to ensure the trees retained their natural shape.

The discussion about tree pruning focused mainly on fruit trees or pruning to improve the quality of timber in forest trees. As fruit tree pruning is beyond the scope of this book, this will not be explored. However, the debate about pruning forest trees did impact on ornamental trees as it concerned methods of branch removal. The essence of the debate was about whether snags should be left after pruning and how long these snags should be. It was generally believed that an absence of pruning would result in very knotty and poor quality timber. However, not everyone agreed and Stephen Ballard published a book with the title: *A Treatise on the Nature of Trees and the Pruning of Timber Trees; showing the impossibility of increasing the quantity or improving the quality of timber by pruning* (Ballard 1833). Some foresters only objected to pruning conifers because 'they bleed so much' (Monteath 1824, 106). Among the majority of professionals who acknowledged the value of pruning, some recommended 'close pruning', removing the branches flush with the stem. However, others thought this was damaging to timber quality because when the wound healed over decay would often be found. Some professionals favoured 'snag pruning' which involved cutting off the branch a short distance from the stem, although others thought that leaving a stump had no advantages (Gilchrist 1873). Some of the foremost authorities on forestry contributed to this debate through their articles and books, expressing their views in an often passionate and occasionally acrimonious manner. In the early years of the debate they included William Pontey (1808b), Walter Nicol (1803 and 1812), Loudon (1822), Robert Monteath (1824) and Sir Henry Steuart (1828). All these individuals were Scotsmen.

One name that stands out in this debate on pruning is William Billington, an Englishman who began as an apprentice gardener before eventually taking senior positions in forestry with the Royal Forests (*Gardeners' Chronicle* 1861b, 1047). Billington proposed an improved mode of snag pruning where the lower branches were cut back, but always preserving two or three secondary branchlets, with the aim of keeping the branch alive and avoiding the risk of it rotting back to decay the stem. His method, variously called 'foreshortening', 'cutting in' or 'terminal pruning', was described in his book *A Series of Facts, Hints, Observations, and Experiments on the Different Modes of Raising Young Plantations of Oaks* (Billington 1825). While Loudon gave the book a favourable review in his *Gardener's Magazine*, others were not so kind. Billington was clearly sensitive to this criticism and noted that it came mainly from Scotsmen. Loudon responded by saying,

> He seems to think there is a sort of tacit conspiracy on the part of Sir Henry Steuart, Mr Withers, and others, to, what he calls, 'keep him down', an idea so absurd as not to require refuting. He also seems to think that Sir Henry, and also ourselves, do not much like him, because he is not a Scotchman.
>
> Loudon 1830, 473

While Billington remained a controversial figure throughout his career, he was highly regarded by many and when he died in 1861 his anonymous obituary in the *Gardeners' Chronicle* described him as 'the well known Arboriculturist' (*Gardeners' Chronicle* 1861b, 1047). After Billington's death the debate about the pruning of forest trees continued, mostly repeating the same well-rehearsed arguments, until it eventually lost momentum in the 1880s.

In all the discussion on tree pruning what was often absent was any practical guidance on precisely how branches should be removed. Walter Nicol (1769–1811), the Scottish landscape gardener based in Edinburgh, is notable for stating the need for an undercut when pruning limbs to prevent what he called 'laceration' (tearing) (Nicol 1799, 180). While John Evelyn and Moses Cook had mentioned this in the seventeenth century, this level of practical detail was overlooked by many authors in both the eighteenth and nineteenth centuries.

There was also much debate in Victorian times about how trees healed their pruning wounds and how this could be encouraged. Some of this focused on what was the best time of the year to prune trees. James Main, in his book *The Forest Planter and Pruner's Assistant*, recommended the spring for most routine pruning. He asserted, 'The months of April and May are the best seasons for pruning forest trees, because the wounds made by the removal of such small branches will be *nearly healed over* before the end of summer' (Main 1839, 75). This was a view shared by many at that time, although as the century progressed early summer pruning for forest trees came more into favour (Webster 1887). However, the logic still seemed to be that it was preferable to prune during the most active growth, not when the tree is dormant, as mostly practised today. One early researcher into pruning was William Pontey (*c.* 1770–1831), nurseryman and gardener to the Duke of Bedford. In his book *The Forest Pruner or Timber Owner's Assistant*, he notes 'the astonishing and successful exertions of simple nature in healing the wounds of trees' (Pontey 1808b, 25). Pontey's early insight into the ability of trees to close their own wounds successfully without any human assistance was based on observation.

In the aftermath of Forsyth's widely discredited 'plaister', other tree experts offered their own formulas for the treatment of pruning wounds. In Nicol's *Planter's Kalendar* we find an early reference to the value of using coal tar (Nicol 1812, 474). In 1822, Loudon gave a summary of current practice in his *Encyclopaedia of Gardening*:

> Small wounds, such as are required to be made by judicious pruning, easily heal up themselves; large wounds, by amputations of branches, above six inches in diameter, should, if possible, never be made. Even wounds of six inches diameter, or under, will heal quicker by the application of any material that excludes the air and preserves the wood from corruption; and we agree with Sang [Nicol 1812], in recommending coal tar, or the liquor produced from coals in manufacturing gas. It is, however, less favourable to the progress of the bark over the wound than a coating of clay or cow-dung covered with moss to keep it moist. Pontey [1808b, 228] recommends putty and two coats of paint over it.
>
> Loudon 1822, 1112

Many compositions suggested as suitable for pruning wounds could best be described as 'weird and exotic' and not based on any scientific approach to the problem. Needless to say, they did not prove to be effective, and in some cases were actually harmful to the tree, and quickly forgotten. The search for an effective solution prompted professionals to investigate compositions that were being used in other countries. These were occasionally reported in the horticultural press, for example, a type of varnish that was used in Germany and said to be extremely useful (*Gardeners' Chronicle* 1858a, 190–191). This was initially used in grafting in Germany but then also applied to pruning wounds. Another interesting composition from the United States was reported in *The Garden* (1879, 219) as 'elastic waterproof cement for protecting wounds'. This comprised 'tar and beeswax, mixed together when heated, to which is added dry and sifted clay'. By the 1880s, Britain's most commonly used composition for pruning wounds was coal tar, which had been used from as early as 1810. The editors of *Woods and Forests* magazine described its advantages and also its disadvantages:

> It often happens that, either by intention, as in pruning, or by accident, trees are wounded in various ways. A common practice is to cover large wounds with coal tar, but this is objected to by some as injurious to the tree. Experiments go to show, however, that its use in covering large wounds is not injurious; but that, on the contrary, a callous readily forms under the tar on the edges of the wound, and that the wounded part is thus protected from decay. There is, nevertheless, another objection; for if the tar is applied a little too thick, the sun melts it and it runs down on the bark of the tree. This can be obviated by mixing and stirring, and thus incorporating with the tar about three or four times its weight of powdered slate, known as slate flour – the mixture being known as plastic slate and used for roofing purposes. … It is easily applied with an old knife or flat stick, and though it hardens on the surface, it remains soft and elastic underneath. The heat of the sun does not melt it, the coldest weather does not cause it to crack, neither does it peel off.
>
> *Woods and Forests* 1883b, 6

In the 1860s, while British tree experts were endlessly discussing tree pruning, often going round in circles, the French were making progress in this vital aspect of arboriculture. In 1864, Count Amédée Des Cars published his great work entitled *L'Élagage des Arbres* (Des Cars 1864). This was based on another great work by M. de Courval in 1861 entitled *Taille et Conduite des Arbres Forestiers.* Des Cars's remarkable book was the first to describe the modern approach to tree pruning and to do this in considerable practical detail. However, it remained largely unknown outside of France until 1881 when the seventh edition was translated into English by Charles Sargent, then Professor of Arboriculture at Harvard College (University) in the United States (Des Cars 1881). As a result, the book became widely read by professionals on both sides of the Atlantic. The attention it attracted in Britain was due largely to extracts being published in the professional press, especially the *Journal of Forestry*, and it was eventually published as a book in Britain in 1893.

Des Cars was very critical of the poor standard of tree pruning he observed in France and elsewhere, declaring, 'The appearance of many trees, their trunks covered with gaping wounds, protuberances, and the stumps of dead branches, clearly indicates that they have received careless or ignorant treatment' (Des Cars 1881, 5). His book sets out what he believed was 'best practice' in the hope that others would follow. With regard to removing large branches, he recommends what is essentially the three-cut method used today, with undercut, top cut and then final cut, although he refers to these as 'notches' (*ibid.*, 35–36). He then states that all wounds 'should be covered with a coat of coal tar applied with an ordinary painter's brush. This simple application then produces a sort of instantaneous cauterization, and preserves from decay wounds caused either by pruning or accident' (*ibid.*, 58). Des Cars's endorsement of coal tar gave a boost to its use in Europe and North America.

In his book Des Cars recommends the dendroscope as an aid to tree pruning. This little instrument assisted the tree worker 'to judge intelligently which branches should be removed or shortened in order to reduce it to the desired shape'. With the aid of diagrams (Figure 57) Des Cars describes its construction and operation:

> A dendroscope may be made from a piece of thin board or cardboard (a playing-card answers the purpose), in which a hole of the shape it is desired to reduce the tree to has been cut. Across the middle of the hole, from top to bottom, a piece of wire is stretched to serve as a guide to the eye. Holding the dendroscope at the level of the eye, with the wire opposite the centre of the trunk of the tree to be studied, the operator approaches the tree until the bottom of the cut falls on the trunk at the ground line. It is easy to see at a glance with the aid of this contrivance what operations should be performed in order to reduce the tree to the desired shape.
>
> Des Cars 1881, 25–27

The dendroscope was said to be very useful in pruning large numbers of street trees to the same symmetrical form and dimensions. When it was demonstrated at the 1867 World's Fair in Paris, it generated interest among some overseas tree experts, particularly from America. Although the dendroscope was mentioned in the tree-related press in Britain, the extent of its use by British professionals is unknown.

The dramatic increase in urban tree planting in the public parks and streets meant that by the 1870s many of these trees were quite large. Even in their early stages of growth some would have required formative pruning. However, many trees in close proximity to buildings and people would have increasingly required some regular pruning for safety and the avoidance of nuisance. This could involve tasks such as the removal of deadwood, cutting back basal growth, crown thinning and crown lifting. Systematic pruning was especially necessary with many street trees, such as the planes and limes, which would have outgrown their situation had they not been pollarded every few years. With this essential tree work being undertaken regularly in parks and streets, the subject of tree pruning was brought to the attention of the public in a way that had never happened before. It was almost inevitable that this tree work, or the lack of it, would generate some criticism towards those responsible, either the local authority staff or the contractors. The standard of pruning applied to the trees in London's parks was a frequent target for criticism, not only from the public but also from many professionals. Another popular target was the numerous street trees that required regular pollarding, a procedure that many deemed unsightly and which could have been avoided if smaller-growing trees had been selected. Much of the criticism claimed that the poor standards of pruning were due to the lack of a properly trained workforce. Many articles and letters in the popular press and professional periodicals highlighted the need for specially trained tree workers who were skilled and experienced in tree pruning. However, it would be some time before this need was addressed properly.

By the 1800s the tools used for tree pruning had developed little since Roman times and new developments were initially slow to arrive. Small diameter pruning was usually done with the pruning knife, while pruning shears were used for topiary and hedge cutting. Large pruning shears were powerful tools and could remove relatively large branches (Figure 58). Some professionals were still using the chisel for pruning, although Pontey was not in favour of this. Where branches are too large for the knife, Pontey (1808b, 233) believed the saw was 'the only practicable tool for pruning on a large scale'. Robert Thompson, in his popular book *The Gardener's Assistant*, summarises the use of saws for pruning:

> For large limbs the saws used by carpenters will answer, only the teeth may require to be wider set if the wood is soft and full of sap. Pruning saws are most required for ordinary pruning; they are employed for cutting off smaller branches than the preceding, and not being so broad, are better adapted for cutting close to the fork of branches, or where a broader plate could not be introduced.
>
> Thompson 1859, 131

FIGURE 57. The dendroscope assisted the tree worker in judging which branches to remove or shorten to reduce it to the desired shape. When this instrument was demonstrated at the 1867 World's Fair in Paris, tree experts recognised its value in pruning large numbers of street trees to the same symmetrical form and dimensions.

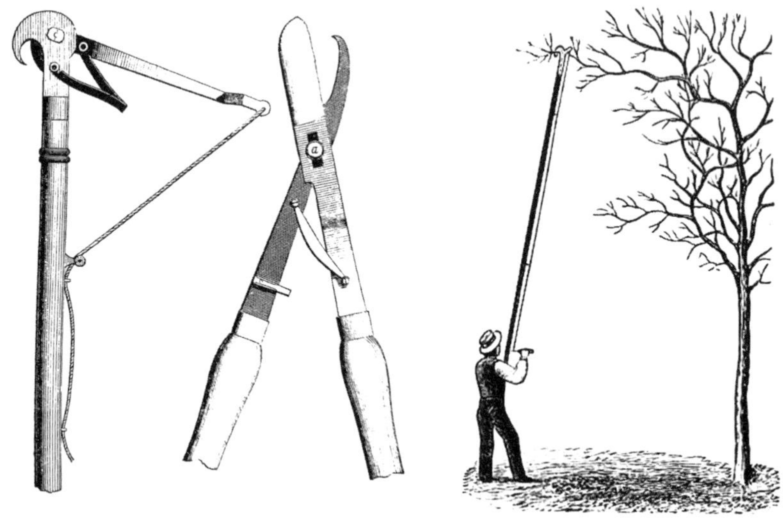

FIGURE 58. Nineteenth-century tree-pruning equipment. Left: the French invented averruncator; centre: large and heavy-duty pruning shears; right: the British-invented long reach pruner in operation.

It was not until the 1870s that there was any major development in the design of pruning saws. In a letter to the *Gardeners' Chronicle*, Jongkindt Coninck (1874), a nurseryman from the Netherlands, describes and illustrates a type of saw used at the Pomological Institute near Stuttgart and claims he had not seen anything similar in his visit to England. From Jongkindt Coninck's written description and illustration this is clearly a bow saw, with a replaceable blade secured by screws at either end of a metal frame, similar to that used today. However, as it is hardly mentioned again until the 1900s, it seems it was slow to be adopted for tree pruning in Britain.

In the opinion of Des Cars (1881, 22) the most convenient pruning tool was the straight-bladed cleaving knife, which had long been used in the Netherlands and had recently been improved by De Courval. The blade was reinforced in the middle to increase its strength and concentrate the weight. Des Cars describes how this and other pruning tools can be used when working at height in trees:

> In the north of France this tool is generally hung on an iron hook attached to a leather strap buckled round the workman's waist, who is thus perfectly free in his movements [Figure 59, left]. In pruning tall trees, or trees otherwise difficult to climb, the leather belt may with advantage be passed over the shoulder, thus bringing the pruning knife under the arm in a position from which it cannot be easily dislodged in climbing [Figure 59, right]. To ensure greater safety in climbing tall trees, a stout cord attached to the workman's waist may be fastened around the trunk in such a manner as to prevent, in case of accident, a dangerous fall. A hatchet is useful, and facilitates the operation of pruning; it may be used with one or both hands, and serves to lop off large branches, protuberances on the trunk, or the dead stumps of branches, which from their hardness would soon dull the edge of the best pruning knife. A saw, too, is very useful in cutting large branches, but

FIGURE 59. The straight-bladed cleaving knife was a popular pruning tool in the Netherlands and France in the 1870s that could be conveniently carried around the tree. This illustration from Des Cars (1881) shows the pruning knife carried at the waist (left) and carried under the arm (right).

> it requires so much practice to use this tool skilfully that it cannot be generally recommended.
>
> Des Cars 1881, 22–23

While the Romans had a range of pruning shears or scissors, it was not until the early nineteenth century that scissor-action tools emerged in a radically new design. In 1819, the aristocrat Bertrand de Molleville was listed in the French horticultural almanac *Le Bon Jardinier* as the inventor of secateurs. However, it seems that secateurs were rather slow to be adopted in Britain, while another similar French invention, the averruncator, was more readily accepted (Figure 58). Loudon describes this as having

> a compound blade attached to a handle from five to eight feet in length, and operating by means of a lever moved by a cord and pulley. Its use is to enable a person standing on the ground to prune standard trees, which it readily does when the handle is eight feet long, to the height of fifteen feet, and, by using step-ladders, any greater height may be attained.
>
> Loudon 1822, 320

The averruncator is mentioned frequently in Victorian gardening literature, which suggests it was widely used. The Standard Manufacturing Company of Derby produced a range of pruning equipment and advertised widely in the horticultural and popular press. In 1879, the company was advertising a long reach pruner, very similar to the averruncator (Figure 58). Its special merits were said to be its rigidity, lightness, simplicity of action and cheapness. With regard to its operation, it was claimed,

> It makes a very clean cut, and is worked with much less trouble than the French instrument now in use, in which the rope line attached to the knife is so liable to get out of order; an objection avoided in the new instrument by the substitution of an iron rod, worked like a lever, and less cumbersome cutting gear.
>
> *Gardeners' Chronicle* 1879, 269

This pruner was made with poles ranging in length from 4 to 10 ft (1.2 to 3.0 m). Another French invention similar to the averruncator was the *echénillior*, which had 'three ten foot and one five foot poles, with a brass socket and key fitted on one end of each, so that the whole can be readily connected together' (*Journal of Forestry* 1878, 141). With regard to secateurs, even by the late 1870s there was a continuing reluctance on the part of British gardeners to use this tool. In 1878, an anonymous letter to *The Garden* (1878a, 202) was full of praise for this tool but claimed it was rarely seen in British gardens and complained that only one retailer had it for sale in their catalogue. However, when it was known that William Robinson favoured secateurs for pruning, they got a much-needed boost.

Pruning mature trees always requires some means of access into the crown of the tree to undertake the work. Loudon (1822, 334) describes the wheel platform which is 'a flat surface of boards generally five or six feet square, placed on the summit of a frame with wheels; it can thus be moved along lawns and walks, and is used chiefly for clipping lofty hedges'. Loudon says that 'such a machine is used in shearing the magnificent hornbeam hedges in the imperial gardens at Schönbrunn [in Vienna]' (Figure 60). Loudon also described various types of ladder, and, while he mentions them in connection with orchard work, there is no indication of their use with large mature trees. Nevertheless, ladders were much used in the nineteenth century for access into mature trees, as they had done in earlier times. Des Cars is quite specific in his recommendations for ladders:

> Each labourer should be equipped with a light ladder, proportionate to the height of the tree on which he is to operate, and broader at the base than at the top. De Courval recommends that the feet of ladders intended for this purpose should be pointed, to prevent them from slipping. This is a good plan, although hardly sufficient to prevent accident, and the top of the ladder should be fastened with a strong rope to the trunk of the tree to prevent it from being thrown down by falling branches.
>
> Des Cars 1881, 23–24

Des Cars (1881, 24) mentions the use of 'hooks or spurs' as an aid to tree climbing, now often called climbing spikes. These are lengths of metal about 18 in. (46 cm) long that are attached to the inner lower legs by means of straps. The metal has a spike protruding at the bottom which is thrust into the trunk as the operator climbs the tree. However, Des Cars states, 'Except in very exceptional cases, or where very large trees are to be operated on, the climbing spurs sometimes used by professional pruners should not be allowed'. Des Cars acknowledges that the numerous small wounds made by the spikes are injurious to the tree.

In all the nineteenth-century accounts of tree pruning, there is hardly any mention of the use of ropes as an aid to tree climbing. In Des Cars's description of the use of the pruning knife, given above, he talks of 'a stout cord attached to the workman's waist' and fastened around the trunk to prevent the climber from falling. This description, however, raises more questions than answers when examined closely. There is another description of tree climbing using rope, published anonymously in *The Garden* (1874, 573), although this is in conjunction with using climbing spikes. The author states, 'A cord is carried around the waist, and if it be necessary to do any work while up the tree, the climber can pass the cord around his body and the tree, and make himself secure. The rope also answers to draw up tools or anything else that may be wanted'.

Another interesting instance of using ropes for tree climbing can be found in an 1883 account of the pruning of the Meikleour beech hedge. This arboreal wonder still exists at Meikleour in Scotland and holds the record of being the tallest and longest hedge in the world. Planted in 1745, it was left to grow in

FIGURE 60. The wheel platform was described by Loudon in 1822 after he had seen it used to prune the magnificent hornbeam palisades in the imperial gardens at Schönbrunn in Vienna. This shows a version of that wheel platform still being used for pruning palisades at Schönbrunn.

memory of soldiers who died in the Jacobite rising and now reaches nearly 100 ft (30 m) in height and 1,740 feet (530 m) in length. Every few years it is pruned to keep it in shape. Nowadays this is done by aerial platform but in Victorian times it was a daunting arboricultural challenge. A fascinating account of how this was done is given by Thomas Hunter in his book *Woods, Forests and Estates of Perthshire*:

> The operation, however, is now much facilitated by the adoption of an ingenious, but simple, method, which the Dowager Marchioness of Lansdowne observed in use upon the tall lime hedges at Versailles. Her Ladyship was so much struck by the machine in use there, that she procured a plan of it, and had one constructed by Mr Scrimgeour, Methven, under the superintendence of Mr Matheson, the land steward. The machine is simply a double ladder, 30 feet high, supported on four wheels, and by means of which the heavier branches are cut with a single stroke saw. The twigs are then cut by an averruncator to the height of 45 feet. Up to this point the work is comparatively easy, but beyond that height there is not a little difficulty and danger. One of the foresters climbs as near to the top as possible, fastening

> himself to the strongest branch he can get, by means of a belt round his waist, and holding on to another branch by his left hand. With his right hand he wields a light billhook reaching to a height of 75 feet, which is as high as can be trimmed with safety, and which is quite high enough, as the uppermost branches gradually taper towards the apex of the trees. The hedge is, altogether, the most wonderful thing of the kind in existence.
>
> Hunter 1883, 381

From the research conducted for this chapter, it is clear that climbing with ropes in tree work was done in a very basic manner in Britain in the nineteenth century, and only then when other means of access was not possible. Meanwhile, in the 1890s the French were using specially constructed safety belts and climbing ropes in what was the early development of modern tree-climbing methods. As it would be some time before this equipment reached Britain, it will be discussed in the next chapter.

Treatment of cavities in trees

The public's interest in ancient trees that gained momentum in Britain from the early nineteenth century began to be expressed in an increasing desire to preserve these trees through remedial tree work. This, in turn, was reflected in articles in the tree-related professional press. While Forsyth's plaister had proved to be a false dawn in tree care practices, nevertheless the search continued to find ways in which old and decayed trees could be preserved and their lifespan extended. As with the treatment of pruning wounds, there was a mixture of bizarre remedies and sound arboricultural practice. For example, consider the following summary of recommendations for 'renewing the health of old and decaying trees' from the highly respected forester, James Brown:

> Remove all dead branches; bole and large branches should be scrapped clean of moss and washed down with brush and cold water; then dig a trench round the trees as if you were transplanting it, cutting the roots in the same manner; then add to the trench well-rotted cow-dung, leaf mould, powdered lime in carefully specified proportions; fill up the trench to the top with a mixture of this compound and soil from the trench; remove the top 4–5 inches of the turf and soil growing in the rooting area; then spread this compost all over the rooting area from the trunk to the trench about eight inches deep; finish off with a good watering and tramping.
>
> Adapted from Brown 1851, 482–483

While Forsyth's plaister was discredited, his basic approach to the treatment of cavities in trees was still largely accepted. Donald Beaton (1853, 177–179), a professional gardener, gives an account of the renovation of old trees in *The Cottage Gardener* magazine. Of particular interest is his description of the efforts of John Caie, another professional gardener, to rejuvenate two large old oaks at Bedford Lodge, near Kensington, around 1830. These trees 'were in the last stages of decay' and 'the few limbs were stag-headed'. Beaton details the measures that Caie took: 'First of all, he had all the branches cut off as close and as smooth as possible, then he scooped out all the decayed mass from the hollow centre, and scrapped the sides of the shell down to the ground, taking off every particle of decayed matter'. Then, to exclude air and moisture penetrating the newly exposed wood, Caie had these extensive basal cavities completely filled with 'brick-bats in Roman cement'. What would be poor arboricultural practice today was, according to Beaton, instrumental in promoting the revival of the trees. He states, 'The time since it was done is now too far distant to allow any doubt as to the final result', with 'the cement in the centre being fast covered by the formation of new wood growing upwards and over it'. Caie's work has additional historical interest in being one of the earliest accounts of the use of cement in cavity treatment. In the garden of the Horticultural Society at Chiswick in the 1850s, tree cavities were filled with well-tempered clay which was then coated with Stockholm tar and pitch (*Gardeners' Chronicle* 1857, 348). While the material used to fill cavities would vary, the basic approach remained the same. As Des Cars (1881, 53) observed, 'These operations resemble, if such a comparison is admissible, the fillings performed by dentists, and with the same object, to check the progress of decay'. Beaton, Caie and other tree experts were just following the usual anthropomorphic approach to cavity work, which would continue until scientific research exposed the fallacy of this in the 1970s.

Tree felling and stump removal

When trees were being removed for land clearance and it was necessary to also remove the stump, the normal practice was to dig a trench around the trunk, cut all the roots at a distance of a few feet from the tree and then use a team of men or horses to pull the tree over (Loudon 1822, 1119). In this way much of the root was removed and the ground could be easily reinstated. Felling trees where the stump remained was another matter. In previous times the large axe was the tool of choice for this work. Now, in the early years of the nineteenth century, the crosscut saw begins to be used for felling in a manner that resembles modern techniques. In his book *The Forester's Guide*, the experienced Scottish forester, Robert Monteath, gives a detailed account of this, including how trees with a lean were dealt with:

> In cutting large trees, that stand perpendicular, in order to make the tree fall the way required, enter the cross-cut saw on that side of the tree it is intended to fall, and cut it about a third part through; then enter the saw at the other side, and when it is cut so far as to admit a wedge, place the wedge exactly opposite the way you want the tree to fall, and keep driving it slowly til the tree is nearly cut through. Be sure the saw meets the opposite cut equally on both sides; and do not imagine, as some have done, that, by making the saw meet soon through the one side than the other, the tree will fall to the side soonest met or first cut through: on the contrary, the side of the tree longest in being cut is that side to which the tree will naturally fall. By attending to the above hints, it will be found very easy to make a tree fall any way; but if the tree has a heavy top, or if it inclines very much to that side on which it is not intended to fall, it will be necessary to make use of a rope. For this

> purpose the connecting rods will be found very advantageous, by having a hook at the end of one of them, so that the rods may carry up the rope, and fix it on the top of the tree, without the very dangerous and laborious work of a person going up the tree to fix it.
>
> Monteath 1820, 191–192

In terms of modern tree-felling techniques what is noticeably absent is the use of a sink cut or 'mouth', cut into the base of the tree on the side that it should fall, to aid its fall in that direction. It is also apparent that climbing the tree to attach a pulling rope was considered dangerous, although some authors later recommended this if the operative was wearing climbing spikes (*The Garden* 1874, 573). Even though using the crosscut saw for felling was recommended by authors such as Monteath, it was slow to be adopted by many professionals. As late as 1886 an anonymous professional wrote, 'That the axe should be the favourite tool with woodmen is only natural, seeing that its use is a much more exhilarating occupation than the cramped position one is obliged to assume in working a saw in the operation of tree felling' (*The Garden* 1886a, 353). However, the writer conceded that in different districts of Britain the use of the felling axe still varied greatly and that the saw could be useful in felling large trees. The type of saw used for felling was not the old pit saw used for converting timber. According to Andrew Slater, a forester in Lancaster, 'The common cross-cut saws are those principally used for felling, and they have a wooden handle at each end, projecting at right angles to the back of the plate. They have various styles of teeth, but the "peg-tooth'd" which cuts one way, and the "lance-tooth'd" which cuts both ways, are those now mostly used' (Slater 1878).

As tree felling was especially arduous work it is not surprising that with the Victorian passion for innovation in engineering there were various attempts to mechanise this. The most notable example was a steam-powered 'tree-feller', the invention of Ransome and Company of Chelsea, which was demonstrated at Tulse Hill, London, in the presence of the famous politician William Gladstone (*Gardeners' Chronicle* 1878, 171). A young oak tree with a diameter near ground level of about 3 ft (0.9 m) was chosen. It took four men to carry the machine to the tree and two men to operate it. Within four minutes of the steam being turned on, the tree was felled. Gladstone was most impressed and said he believed the machine to be far quicker at felling trees than the usual manual methods. News of this demonstration and Gladstone's attendance was reported in the national press, including *The Graphic* (1878, 153), which carried an illustration of the event on its front page (Figure 61). Another similar tree-felling machine that got less publicity was manufactured by Worssam and Company of Chelsea (*Journal of Forestry* 1880, 409). The saw was powered by a flywheel, driven by hand, and also with steam power by a small portable engine after a little modification to the flywheel shaft.

The removal of large tree stumps was another very arduous task usually involving a small team of men and some considerable time. Various alternatives to heavy manual work were explored. Loudon (1822, 315) describes a hydrostatic press which worked on the same principle as the German Devil from Evelyn's time, although with much greater effect. The only difficulty was in finding a suitable and convenient fulcrum. The use of explosives was a popular method of stump removal in Victorian times and the professional press has several reports of its use. Dynamite was normally inserted into auger-holes bored into the stump. As might be expected with the use of explosives, fatal accidents happened. Such an incident occurred in 1882 on the estate of Joseph Gillot at Solihull and was reported in the local and professional press:

> Some large trees on the estate having been blown down by the recent gales, it was decided to remove the stumps by blasting. Several stumps had been destroyed in this way, Mr Gillott, with Mr Sidney Mitchell, a Birmingham solicitor, and his son, assisting the operation. A splinter from one of the largest trees struck Mr Mitchell on the head, dashing out his brains. The deceased gentleman was standing at the time near Mr Gillott, who had a narrow escape, even at a distance of 52 yards from the charge.
>
> *The Garden* 1882, 226

Towards the end of the nineteenth century some types of machinery and heavy equipment then being used in agriculture were introduced into tree work. This included the use of traction engines and winches instead of teams of men or horses for tasks such as pulling over trees and removing tree stumps.

Tree surveys and inspections

The huge increase in urban trees in Victorian times prompted a need for regular tree surveys and inspections by those responsible for them. With countless thousands of large mature trees growing in close proximity to people, buildings, urban infrastructure and increasing volumes of traffic, the risks and potential liabilities of these trees was becoming apparent. Public authorities and private owners realised that the way to reduce the level of risk was to monitor those trees on a regular basis. Of course, the extent to which this actually happened varied enormously, just as it does today. To facilitate these tree surveys and inspections a number of instruments and devices were used.

A brief description of a dendrometer, an instrument for measuring the dimensions of standing trees, was given in 1827 in *The Gardener's Magazine* by the inventor, Archibald Gorrie, a gardener at Annat Gardens in Perthshire (Gorrie 1827). However, Loudon adds a note to the article stating that a similar instrument had recently been made the subject of a patent by James Rogers of Marlborough. In 1840, Edward Sang published a description of a hypsometer, an instrument to measure the heights of trees, which had been invented by John Sang, possibly a relative (Sang 1840). This instrument was endorsed enthusiastically by Loudon.

In 1869, James Kay, a forester on the Bute Estate in Scotland, published details of his 'instrument for measuring the heights of trees' (Kay 1869).

THE GRAPHIC

AN ILLUSTRATED WEEKLY NEWSPAPER

VOL. XVII.—No. 429
Regd. at General Post Office as a Newspaper

SATURDAY, FEBRUARY 16, 1878

WITH EXTRA SUPPLEMENT

PRICE SIXPENCE
Or by Post Sixpence Halfpenny

TREE FELLING BY MACHINERY—MR. GLADSTONE WATCHING A TRIAL OF THE NEW PATENT STEAM FELLER NEAR TULSE HILL

FIGURE 61. This front page from *The Graphic* in 1878 shows a steam-powered 'tree-feller' being demonstrated at Tulse Hill, London, in the presence of the famous politician William Gladstone. The machine was the invention of Ransome and Company of Chelsea.

The Scottish Arboricultural Society was very impressed with this instrument and awarded Kay a first-class silver medal. Details of Kay's dendrometer were published again in 1883 in an article which stated it had the following advantages: 'First, its simplicity; second, no calculation is required; third, the height of an object can be obtained at any convenient distance; and fourth, the height of any part of a tree, such as length of bole, can be ascertained from one station' (*Journal of Forestry* 1883, 21). However, this article prompted a flurry of correspondence regarding the originality of Kay's instrument and he was accused of copying a German instrument invented by a man called Koenig. However, Kay was able to end the controversy by pointing out that Koenig's instrument had only been brought to the notice of British foresters in 1875, long after he had published details of his dendrometer in 1869.

In order to detect the presence of decay in standing trees various tools and instruments were used. Often a wooden mallet was employed to tap the trunk of the tree and if a hollow sound was heard this would indicate decay. Various types of augers were also used to bore into the trunk. The colour of the wood that was removed might indicate decay or the auger itself would find a cavity that was not apparent from an external visual assessment. Around 1855, the increment borer was developed in Germany by Max Pressler, a forester and economist (Grissino-Mayer 2003, 63). This is used to extract cores from trees for analysis of growth trends based on inspection of the tree's ring patterns. Growth rings also help determine whether trees are showing signs of poor growth. In the assessment of decay, increment cores provide direct visual means of examining zones of discoloured or decayed wood. The extent to which the increment borer was used in late Victorian Britain is impossible to determine, and, although it is mentioned in some forestry manuals, no reference to it was found in the literature relating to amenity trees.

With regard to assessing the health of amenity trees, the British literature on tree pathology was quite limited until the end of the nineteenth century. In 1892, Charles Curtis published his book *The Manifestation of Disease in Forest Trees*, the first British book devoted entirely to the subject of tree health (Curtis 1892). However, the most significant contribution on this subject came from the English translation of a German text. Just when German ideas and practice were having such a profound impact on British forestry, so it was a German scientist, who, in the words of the eminent British scientist Marshall Ward, 'succeeded in founding a plant pathology really worthy of the name' (Ward 1894, v). Professor Robert Hartig's (1894) *Text-book of the Diseases of Trees* remains one of the great works of science in any field and it established him as the 'Father of Tree Pathology'. His advances were particularly ground-breaking in our understanding of tree decay. In accordance with the principles of abiogenesis, scientists had traditionally believed that tree decay led to fungal growth. Hartig stated that the reverse was the case and developed an entirely new model where trees were wounded and fungi infected the wounds, leading to tree decay (Hubert 1931). Hartig's work is also notable for its attention to specifically urban problems. He discusses the death of urban trees caused by atmospheric pollution, the toxic effect on tree roots of coal gas escaping from underground pipes, the harmful impact on tree growth of some hard surfaces for pavements and streets, and other typically urban problems.

Arboricultural consultancy in the urban environment

The enormous volume of arboricultural work now being generated, particularly in urban areas, was not just at the manual level; it also extended into consultancy and advisory work. Those professionals who were regarded as tree experts started to carve out a role for themselves by providing the type of expertise that would eventually form a major part of the workload of the modern arboricultural consultant. Some aspects of that consultancy role are outlined here.

With many more trees in close proximity to built development and urban infrastructure, and potentially in conflict with this, it was understandable that the legal aspects of arboriculture would become far more significant. An early indication of this with respect to trees in residential gardens was highlighted in an article in *Gardeners' Chronicle* in 1842. The author believed that some account of the legal situation was necessary 'as trees become the frequent source of disputes between neighbours' (*Gardeners' Chronicle* 1842, 126). Another article in the *Gardeners' Chronicle* highlights the problems caused by the roots of trees to building foundations. This related to a row of Ontario poplar trees in a residential garden that were said to have damaged the foundations of a neighbour's house 25 ft (7.6 m) away. The soil around the trees was identified as 'a cold clay' and the roots had 'completely disintegrated the concrete' (Ferrey 1878). This is an early account of an area of work that would later preoccupy many modern arboricultural consultants. That same year in *Gardeners' Chronicle*, Christopher Young Michie, the well-known tree expert, had published an article on the problems of trees near to buildings, although he did not mention subsidence. Instead, he focused on windthrown trees and fallen branches damaging buildings and the potential for trees to cause injury or death to people. He concluded by issuing a warning against planting forest trees close to buildings (Michie 1878).

When trees cause damage to property or injury to people, or the trees themselves are the subject of damage, the question of compensation arises. This usually requires the input of an arboricultural consultant, to assess the extent to which the tree is liable and to assess the scale and cost of the damage done. There was an interesting case in 1892 when the owner of two poplar trees agreed to the railway company topping his trees because they were said to be interfering with the railway line. However, he did not want them removed because they 'deadened the noise of passing trains' and acted 'as a screen against the sulphurous emanations' (*Gardeners' Chronicle* 1892b, 115). Nevertheless, the railway company's workers felled the trees. The case went to court and after seeking expert advice regarding the value of the trees the court awarded damages of £200. A case reported in 1897 was one of the first to highlight the problems that can arise when street trees are not maintained regularly. A passenger travelling on the top of a tramcar was struck in the eye by the branch of a tree at the edge of Victoria Park in London. The local newspaper reported that the individual 'brought an action against the County Council, as the responsible arboriculturists, for not keeping the park trees under proper control. The plaintiff claimed damages and the jury assessed them at £1,250' (*Pall Mall Gazette* 1897). In coming to its judgment the court took expert advice and decided that the tree had not been properly maintained.

As many towns and cities expanded rapidly in the late Victorian era with new residential suburbs, concerns were expressed about the many trees and woodlands being lost to urban development. The property developers and their building contractors were the targets for much criticism. The *Gardeners' Chronicle* was always very keen to highlight these concerns and often carried reports on this from the national and local press. For example, in 1881 it reported:

> The *Richmond and Twickenham Times* puts in a strong plea for the preservation of large trees by builders when in the course of their work they cut up properties on which good trees exist. We fear the plea is addressed to deaf ears, for the builder is gifted with little sentiment, and cares less for trees, let them be ever so ancient or noble, if they stand in his way.
>
> *Gardeners' Chronicle* 1881a, 660

The professional press was always keen to point out that property developers and builders needed to get expert advice on how trees could be retained in urban development projects. One high-profile example of this caught the nation's imagination and did much to raise awareness of the topic. The site chosen for the Great Exhibition of 1851 was located in an area of Hyde Park containing some mature elm trees (*ILN* 1850a). Six eminent architects and engineers were selected for a Building Committee that would choose the design for the building to house the Great Exhibition. They included Isambard Kingdom Brunel, Charles Barry, Robert Stephenson and Sir William Cubitt. The competition to design the building attracted 245 entries but the committee was not satisfied with any of these. With time running out, Joseph Paxton was approached, after his work on the Great Stove and Lily House at Chatsworth. Paxton's proposal was for a glass and iron structure based on the Chatsworth Lily House but on a much larger scale. Paxton asked the committee what should be done with the trees on the site, adding that he could enclose them in the building, if required (*ILN* 1850b). Opponents of the exhibition had fought bitterly against cutting down the elm trees, especially the three large elms opposite the Prince of Wales Gate which had not yet been felled. The committee agreed to Paxton's proposal and the design was altered to enclose these trees. The result was the barrel-vaulted transept which was to be the building's most graceful and distinguishing feature. Progress in the construction of the building was reported regularly in the popular press with the widely read *Illustrated London News* taking a particular interest. It published several articles charting the progress of the work that included images and details of how the elm trees were enclosed in the overall structure of the building (*ILN* 1850c). The building was dubbed 'The Crystal Palace' by Douglas Jerrold writing for *Punch* magazine in 1850. This whole episode certainly put the spotlight on Joseph Paxton, not only as the architect of this magnificent building but also as the arboricultural consultant whose heroic feat of ingenuity saved the three beautiful elm trees. In the six months that the exhibition was open its 827,280 visitors were able to marvel at the spectacle of these towering trees now enclosed in their Crystal Palace, yet still looking in excellent health (Figure 62).

FIGURE 62. Joseph Paxton's heroic feat of ingenuity saved three beautiful elm trees by enclosing them in the main transept of the Crystal Palace. During 1851 the Great Exhibition's 827,280 visitors were able to marvel at the spectacle of these towering trees which were still looking in excellent health.

While horticultural and tree-related periodicals in the late Victorian era published an increasing number of articles on urban trees, this was not reflected in books on gardening and landscape. They generally had very little coverage relating specifically to urban trees. While the arboricultural aspects of gardening in urban residential gardens were occasionally mentioned, arboriculture in the public urban realm got little consideration. A notable exception was a contribution by Robert Walker, the tree expert and parks manager for Aberdeen Corporation. In 1890, he wrote a 24-page booklet entitled *On the Planting of Trees in Towns* that set out his approach to urban tree management and his views on suitable species for planting (Walker 1890).

Nurseries and tree production

The introduction of new tree and shrub species accelerated dramatically in the nineteenth century. In just the first thirty years of the century it was estimated that 699 new species arrived in Britain (Loudon 1838a, 1, 126). The desire to acquire these among the owners and managers of parks and gardens gave a tremendous boost to the nursery industry. With regard to nursery stock propagation, the long-established leadership of the French continued into the nineteenth century. The French superiority in grafting and budding was apparent in Charles Baltet's *The Art of Grafting and Budding*, published in Britain in 1873 as a translation of his French book *L'Art de Greffer*. The preface, which was probably penned by William Robinson, states, 'The art of Grafting and Budding has for many years been practiced in France on such a scale, and with such remarkable success, that the gardeners of that country are now far in advance of all others in this branch of horticulture' (Baltet 1873, Preface).

The nineteenth century saw the development of some major commercial nursery companies that gained a national reputation for their extensive range of high-quality trees and shrubs. Loddiges nursery in London was founded by Joachim Conrad Loddiges (1738–1826) who took over the nursery founded in Hackney by John Busch in 1771 (Solman 1995, 24). Loddiges was born in Lower Saxony and came to Britain as a young man. After 1816, Loddiges nursery began to secure a unique position in the history of the trade. Now under the leadership of Joachim's son George, it embarked on a series of prestigious developments

INDIA
INDIA

that included spectacular hothouses and an arboretum which attracted visitors from all over Europe. By 1826, the arboretum displayed most of the 2,664 hardy trees and shrubs listed in the Loddiges's catalogue. It was popularised as 'The Hackney Botanic Nursery Garden' by Loudon and it played a crucial role in the production of his *Arboretum Britannicum* (Loudon 1838a). Soon after Loudon had sketches taken of all their trees and shrubs, Loddiges had most of their forest trees removed because they were now crowding out other plants. Loudon lamented this 'incalculable loss [of] a collection of specimens such as could be found assembled together nowhere else in the world' (*ibid.*, 1, 131). The nursery closed in stages between 1852 and 1854 due to expiry of the lease from St Thomas's Hospital.

The Veitch nurseries were the largest family-run nursery business in Europe in the nineteenth century. The business was founded by John Veitch, a Scottish horticulturist, who moved to Devon and established his first nursery near Killerton by 1800 (Heriz-Smith 1988, 41). With his business steadily expanding, Veitch moved to larger premises at Mount Radford, Exeter, in 1832. He was soon succeeded in the business by his son James and grandson James junior, with James taking over the Exeter nursery, while James junior was sent to London to train for two years as a nurseryman. He later returned to Exeter, where he helped his father improve and expand the Exeter nursery, before acquiring premises in Chelsea, London. During the Victorian era the Veitch family business made a huge contribution to British horticulture and arboriculture. This also involved much more than their highly successful nursery stock operation that supplied a superb range of high-quality trees and shrubs to clients throughout Britain. The family also sponsored some eminent plant collectors of that era, who brought back new introductions from around the world that would then feature in the Veitch catalogue. They were also responsible for the publication of *A Manual of the Coniferae*, which for many years was an authoritative work on the subject (Kent 1881). A detailed history of the Veitch family business is set out in the book *Hortus Veitchii* (Veitch 1906).

William Barron's nursery at Barrowash in Derbyshire was another that was notable for offering a very wide range of trees and shrubs. Indeed, a few years after founding the nursery in 1862, Barron claimed that it had purchased 'every new plant introduced into this country' over the previous two decades (Elliott *et al.* 2007b, 137). However, the nursery was especially famous for specialising in conifers. Barron was to play a leading role in fostering the mid-Victorian fashion for evergreens and in promoting much greater use of conifers in amenity landscapes. He was critical of the deciduous trees that were popular in most parks and gardens. In his book *The British Winter Garden*, he stated that they gave a 'continued litter of decayed leaves' and displayed 'lifeless stems' throughout the winter (Barron 1852, 10–11). Furthermore, despite the best efforts of the horticultural press and botanical collectors 'to enrich our shrubberies and plantations, what is so common as to see, even at the present day, *close to our mansions*, such common-place things as elms, ashes, sycamores, poplars, or any other rubbish that the nearest provincial nursery may happen to be overstocked with'. No doubt Barron's views did much to boost his nursery's sales of conifers.

The counties of Surrey and Sussex became renowned in Victorian times as the location of some prominent nurseries. One of these was the Knap Hill Nursery of the Waterer family. In 1829, the German nurseryman Jacob Rinz described this as 'the largest and finest collection of American plants I ever saw' (Rinz 1829, 382). Another notable business was Lowfield Nurseries at Crawley, Sussex, formally established in 1871 by the Cheal family (Benton 2002, 15). Along with some other successful nurseries, Cheals offered a landscape design and construction service (Figure 63). This side of the business expanded rapidly and the firm collaborated with some leading landscape architects, such as Thomas Mawson (*ibid.*, 128). Cheals was one of the top four nurseries in England to come to prominence in the public park movement during the 1880s and 1890s. The other three were Blackhouse of York, Barron of Barrowash and Veitch of Exeter. Some leading garden designers were also involved, but in the late Victorian era it was more common for park schemes to come to the nursery trade, often as a result of winning municipal contracts with their prize-winning designs (Jordan 1994, 103).

With street tree planting schemes in the late Victorian era, nurserymen were often engaged not just to supply the trees but also undertake the planting and initial maintenance. For example, some of the first street trees to be planted in Birmingham city centre, in Broad Street in the winter of 1875–1876, were supplied and planted for Birmingham Corporation by Chad Valley Nurseries (*Gardeners' Chronicle* 1876, 355). In 1878, the local authority in Southwark, London, placed a contract with Messrs Goff and Sons, nurserymen based on Old Kent Road, to plant plane trees along St George's Road, from the Elephant and Castle to the Westminster Bridge Road (*The Garden* 1878b, 500). There are many other examples of nursery companies engaged in both the supply and planting of street trees in towns and cities across Victorian Britain (Johnston 2017). However, some contemporary commentators complained about the poor quality of the trees being supplied for street planting. The following comment from an anonymous correspondent in *The Garden* is typical:

> In street planting it is important that really good, stout and straight trees be procured. It is painful to see the miserable material under the name of trees that is sometime planted; and when the costs of the guards, stakes, etc. is considered, it will be seen that a few shillings more than is usually paid per tree would really be a great gain. It is surely false economy to put a guard costing £1 to a tree costing 2*s*.
>
> *The Garden* 1886b, 86

As in the previous century, many nineteenth-century nurserymen also diversified into the sale of tools and equipment used in horticultural and landscape work. This was a natural development since gardeners, landscapers and arborists would regularly visit commercial nurseries to buy their trees, shrubs and other plants. Some retailers' advertisements specifically mentioned

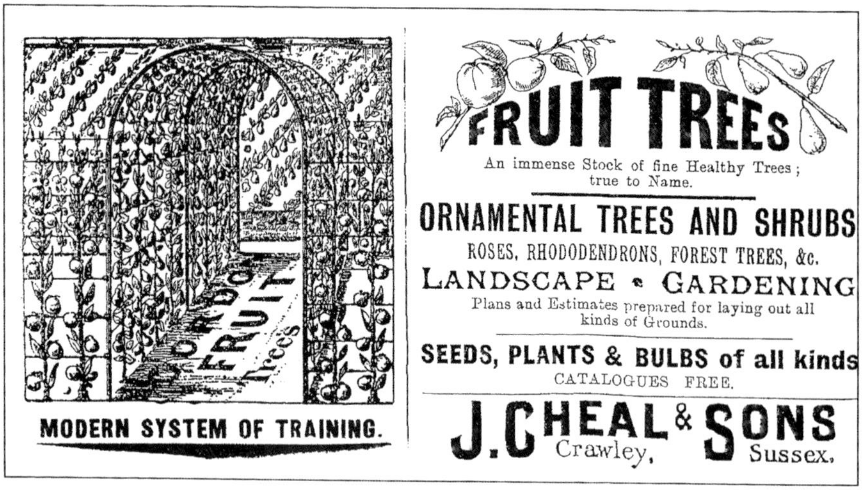

FIGURE 63. Cheals of Crawley were among a number of leading nursery companies specialising in trees and shrubs that also offered a landscape design and construction service. This advertisement is from *The Sussex and Surrey Courier* in 1899.

arboricultural tools and products, such as those of John Browning of Perth (*Perthshire Advertiser* 1899, 1).

Tree-related professionals and their representation and education

The public interest in horticulture and landscape, especially in trees and shrubs, that had grown steadily in the Georgian era continued into the new century. This was not just a fascination of the upper class on their country estates but also an increasing preoccupation of the expanding urban middle class. Loudon did much to foster this interest and he appealed directly to the suburban middle class to plant more trees and shrubs in their residential gardens. From the 1840s onwards there had also been a huge increase in publicly owned urban trees in Britain. All this tree planting led to a dramatic increase in the volume of arboricultural work required. However, who was undertaking this work?

In determining what professionals were involved in the planting and care of trees and shrubs during the nineteenth century we are faced with the difficulty of defining the relevant professional disciplines. There is a challenge because of the different terms used to describe those disciplines and the fact that those terms changed as the century progressed. That is certainly the case with the disciplines of landscape design and construction, forestry and arboriculture. In the case of landscape professionals, in the eighteenth century these individuals would commonly self-identify as landscape gardeners, landscape improvers or similar. This continued into the nineteenth century, and even in the 1890s the term landscape gardening was widely used, as in Milner's (1890) book *The Art and Practice of Landscape Gardening*. However, by that time the term was starting to be replaced by 'landscape architecture'. Loudon was instrumental in promoting the term landscape architecture among landscape professionals and in its eventual adoption by the modern profession in the twentieth century.

He took up the term from its originator Gilbert Laing Meason's (1828) work *On the Landscape Architecture of the Great Painters of Italy* and gave it publicity through his book *The Landscape Gardening and Landscape Architecture of the Late Humphry Repton* (Loudon 1840a). The first professional in the nineteenth century to call himself a landscape architect was Thomas Mawson (1861–1933), who later reached the height of his fame in the early twentieth century.

With regard to tree-specific disciplines such as forestry and arboriculture, the various terms used in these areas of expertise in the nineteenth century can be quite confusing. The term 'arboriculture' comes to the fore in the 1820s, as it was much used by Loudon. In his *Encyclopaedia of Gardening* there is a major section entitled 'Arboriculture, or planting' (Loudon 1822, 1072–1149). It is mainly about trees in plantations and what is now regarded as forestry. There are some references to ornamental trees but this is limited mainly to the landscapes of country estates. The section includes Loudon's 'Arboricultural Catalogue' where he describes the principal timber trees (*ibid.*, 1134–1198). The next major section in Loudon's *Encyclopaedia of Gardening* is entitled 'Landscape Gardening' and it is here that we find a lot of the material relating to the planting and care of amenity trees (*ibid.*, 1149–1198). The use of the term arboriculture as almost synonymous with forestry continued into the mid-Victorian era and is typified by two popular books published around this time. In 1868, John Grigor produced his first edition of *Arboriculture or a practical treatise on raising and managing forest trees* (Grigor 1881). The book is concerned almost exclusively with forestry in a rural context. In 1880, William Ablett published *Arboriculture for Amateurs* (Ablett 1880). Although predominantly about forestry, this did take a slightly broader view of arboriculture by including some chapters on trees for ornamental use. However, urban trees were not mentioned.

While the professionals undertaking commercial forestry were usually known as foresters, it is interesting that the work itself was frequently referred to as arboriculture, particularly by the estate owners. This was often the case where the work involved traditional woodland management, such as coppice woodland or coppice with standards. The term forestry was more likely to be applied to plantation forestry, especially where conifers were used. Some eminent foresters, such as Arthur Forbes, believed that 'the arboricultural aspect of English woods is one which is often pushed to the front as an excuse for bad forestry. We are told that landowners do not maintain their woods for profit so much as for their ornamental features, and as an amenity to a country estate' (Forbes 1906, 31). As a result, what was done had neither the attractiveness of amenity woodland nor the sound silviculture of proper plantation forestry.

The late nineteenth century witnessed the start of a debate about making a clear distinction between the scope of arboriculture and forestry (or 'sylviculture' as it was sometimes called). In the inaugural lecture of the School of Forestry at Edinburgh in 1889, William Somerville (1890, 403) defined 'forestry' as including formal management of land for sustained timber production and 'arboriculture' as relating to the more general management of individual trees. Somerville's views reflected those of an increasing number of professionals who were also starting to make this distinction between forestry and arboriculture, which is evident from the pages of the professional press. However, some influential figures were not persuaded and one leading voice was William Robinson, who expressed his forthright views through a book review of *The Forester* by James Brown (1894), enlarged and edited by John Nisbet. Claiming that Nisbet had 'taken such liberties with the original', Robinson declared:

> The editor tells us that the new edition is an endeavour 'to engraft the modern Continental science of sylviculture upon the old British stock of arboriculture', but [this is] a hopeless kind of task. … Dr Nisbet's attempt to make it appear that 'arboriculture', as that word has always been used in Britain, means something different from 'sylviculture', and that the two may stand in relation of stock and scion is ridiculous. … 'Forestry' and 'foresters' are two good and expressive old English words … and we should be sorry to see either dispensed with to make room for mouthfuls as 'sylviculture', sylviculturist, with 'arboriculture' and arboriculturist added to convey the rest.
>
> Robinson 1894, 468–469

With regard to the terms used to describe professionals working with amenity trees, a similar confusion emerges to that surrounding the disciplines. The term 'arborist' had been used since at least the early seventeenth century for someone who was a tree expert. This continued in the nineteenth century but to a lesser extent. It was often replaced by the term 'arboriculturist', and while this was used to describe someone engaged in both forestry and arboriculture (in their modern usage), it was also used to describe someone exclusively engaged in forestry. It is interesting that Loudon, the greatest nineteenth-century tree expert, frequently uses the term 'arboriculture' but not 'arboriculturist' or 'arborist', and those terms only appear in his *Gardener's Magazine* when used by a contributor.

While it was common for professionals to advertise their expertise in tree work, few would identify their occupation with an arboricultural title. In Loudon's *Encyclopaedia of Gardening* he lists thirty-four categories of gardener covering most of the specialist areas, such as landscape gardener, florist and market gardener. In the entire list there is only one tree-specific category, given as: 'Tree Surveyor or Timber Surveyor, limits his occupation to arboriculture' (Loudon 1822, 1202). Furthermore, Loudon makes it clear that he refers to arboricultural work in the context of commercial forestry plantations. As the nineteenth century progressed examples do emerge of arboricultural job titles in their modern usage. In 1826 we find Mr T. Callum from Abingdon, Oxfordshire, advertising himself as 'Arborist, Florist, Horticulturist, Nurseryman and Seedsman' (*Berkshire Chronicle* 1826, 3). In the *London Post Office Directory* for 1841 there is an entry for Leonard Phillipps, who is listed as 'Arborist, etc.' under the entries for Vauxhall (Post Office 1841, 269). Phillipps was a prominent London nurseryman with premises close to Vauxhall Gardens. Extensive research could not find anyone advertising their service as an arboricultural consultant, even though recognised tree experts such as William Barron and Joseph Paxton were often undertaking this work. However, the *Thetford Times* in 1891 indicates that Maxwell T. Masters, recognised as an expert on urban trees, was available for arboricultural consultancy work with local authorities. The report states:

> The Walks Committee [of Lynn Town Council] reported the result of their interview with Dr Masters, of London, the arborist, with reference to the condition of the trees upon the Central Avenue. His recommendations were that certain trees be lightened by the removal of heavy branches, and that one side of the heath be closed for a year to enable the roots to be treated in a manner he proposed, and that the other side should be dealt with in a similar manner next year. The trees known as the 'Seven Sisters' required immediate removal, and all consideration of disfigurement must give way to public safety.
>
> *Thetford Times* 1891, 6

There are many examples from Victorian newspapers where the terms arboriculturist or arborist is used in the context of amenity trees. For example, in a newspaper report of a ceremonial tree planting at Chatsworth, Prince Albert is described as a 'practiced arboriculturist' (*Derbyshire Courier* 1843, 1). Also, in a letter to the editor of the *Newcastle Journal* proposing the creation of an arboretum, a correspondent writes: 'Should the foregoing suggestion meet attention, there should be no difficulty in obtaining the advice and assistance of an experienced arborist in carrying it out' (*Newcastle Journal* 1856, 7).

The use of the term forester to signify a professional working in commercial forestry was used from a very early stage in the nineteenth century. Virtually every country estate, either large or small, maintained a forestry department in order to produce home-grown timber for building work, for fencing and gate-making, and for use in the gardens and on the home farm. This would be managed by the head forester, with the assistance of under-foresters, woodmen and labourers, and the department worked in cooperation with the head gamekeeper and the hunt establishment to ensure that forestry operations did not conflict with shooting and hunting interests.

In his book *The Planter's Guide* Sir Henry Steuart (1828, 1–2) makes some caustic remarks about the ability of gardeners and foresters in respect of tree-related work. He notes 'the vast sums, which are annually laid out' on tree planting in Scotland, yet the average Scottish landowner 'knows little' about trees. He relies on his gardener, who 'ninety-nine times in a hundred, knows nothing himself, it is "the blind leading the blind" in this important branch of rural economy'. Steuart goes on, 'Sometimes the Forester is the operating person, which is still more unfortunate; for this is generally a mere lopper and cutter of wood. In ordinary cases, he is much worse educated than the Gardener, with equal pretensions as to Arboriculture, and equal ignorance'. Steuart, not surprisingly, upset a lot of people with these remarks and he attracted much adverse comment in the gardening magazines and books, especially from Sang, McNab and Loudon.

With the massive increase in arboricultural work, a debate arose regarding which type of professional was best suited to undertake this. As there were no qualifications or formal training in arboriculture in the modern sense, the choice was mainly between the horticulturist and the forester. James McNab, Curator of the Royal Botanic Garden in Edinburgh, was clear on this:

> All trees, both old and young, under municipal control, ought to be put under the management of a properly certificated forester, instead of committing them to the mercy of anyone whom the municipal owners for the time being chose to call in to attend them, and this often after mischief has been done, not unfrequently by parties who posses but a very indifferent knowledge of pruning operations. Unless an experienced inspector is often amongst such trees, and observing from time to time the requirements necessary for their preservation, it will be difficult to get them kept in the condition in which they ought to be. … The manipulation requisite for the management of forest or ornamental trees in towns is frequently entrusted to gardeners, but they often do not possess that scientific arboricultural knowledge which has from long practice been acquired by a thorough-bred forester; the latter knows the time and method of pruning, both of old and young trees, so as to prevent that internal injury that often befalls them; and, for securing the proper development of the trees in after years, he is the proper person to employ for such work.
>
> McNab 1874, 401

This bias towards foresters was also evident in practice. In Edinburgh in the 1890s, the Superintendent of Parks and Gardens was Mr McLeod. However, the Cockburn Association, a local civic society, had doubts about his level of expertise, as a horticulturist, in the tasks of street tree planting and management. Aberdeen had employed Robert Walker, a forester, and the city was very successful in planting many thousands of trees (Johnston 2017, 111–112). The Association decided to recommended to the Edinburgh Council 'that a forester of experience should either be retained by the city, or should be requested to report annually upon the state of the trees in the gardens, parks, and streets, and to advise what should be done in regard to them' (Cockburn Association 1899, 9).

The beginnings of professional representation

The blossoming public interest in gardening was reflected in the formation of the first horticultural societies, comprising individuals who wished to exchange information and ideas about plants and gardening. In London in 1804, Sir Joseph Banks, John Wedgwood and five others formed the Horticultural Society of London (Elliott 2004, 4–5). In 1821, the Society established an experimental garden in Chiswick which included an arboretum containing a great variety of trees. To encourage the introduction of new trees, shrubs and other plants, the Society began organising and sponsoring an extensive series of plant-collecting expeditions to many different countries around the world (*ibid.*, 197–211). In 1861, the Society changed its name to the Royal Horticultural

Society (RHS) when granted its Royal Charter and it has subsequently become one of the world's foremost horticultural organisations (*ibid.*, 25). Of the seven founding members of the Society, four were gentleman amateurs and three were professional horticulturists. Nevertheless, it was not founded primarily to represent professional gardeners and within the Society the relationship between the amateur and professional worlds has not always been easy. Conflicts over authority and status have bedevilled both the RHS and the wider society within which it functions. For its first eighty years, the membership of the Society consisted mostly of nobility and landed gentry. The only significant representation for professional gardeners in the first part of the nineteenth century remained the Worshipful Company of Gardeners, one of the livery companies of City of London.

From as early as the 1820s it was proposed that an organisation should be formed to promote arboriculture in Britain and raise standards in tree planting and management. Of course, with the definition of arboriculture at that time, this would embrace both the commercial and ornamental aspects of tree cultivation. One of the first to express this view was a forester, William Billington, who declared:

> We have Linnaean, Agricultural, and Horticultural Societies, and numerous other useful associations, which have no doubt done much good; why, then, may we not have an Arborecultural Society (if I may be allowed the term) as much wanted to promote and encourage the art and knowledge of forest tree culture, when it must, I presume, be acknowledged it is far behind both Agriculture and Horticulture, to which it is so nearly allied? I as an individual think it would be of infinite importance, for the good of the country in general, if something of that nature were instituted by Government, or by patriotic gentlemen who have the means, and are willing to combine it with the Horticultural societies.
>
> Billington 1825, 329

Billington's appeal reveals two interesting points. The first is that he was not looking to his fellow professionals to take the lead in establishing the organisation. Secondly, he was looking to ally arboriculture with horticulture to help promote it, no doubt aware of the growing influence of the RHS. Sir Henry Steuart echoed Billington's call for an arboricultural society because he wanted it to be regarded as a distinct field of study (Steuart 1828, 6–7). However, he was disappointed that arboriculture was invariably treated as just a branch of horticulture rather than a subject in its own right and believed this could be overcome by the formation of 'a Society, exclusively for the improvement of Arboriculture'. Although Billington and Steuart's call for an arboricultural society had no immediate result, their views were supported by many professionals and landowners, particularly in Scotland.

In 1829, a meeting was called in Dublin, chaired by the Duke of Leinster, 'for the purpose of considering the means by which the culture of forest trees might be extended and improved in Ireland' (Sadler 1878, 457). The meeting decided that a society should be formed, 'having for its object to disseminate correct and useful knowledge relating to arboriculture'. It would be designated the 'Arboricultural Society' and would focus on improving plantations and woodland. Notwithstanding the array of influential names that supported it, the Society suffered from financial difficulties and there is no further account of it beyond October 1831 (*ibid.*, 527). Nevertheless, this Irish Arboricultural Society undoubtedly influenced further developments in Britain.

In the 1830s there were more calls for the formation of a national arboricultural society with some appearing in Loudon's *Gardener's Magazine*. In an article addressed to its 'conductor', James Munro, based in Brechin with Dickson and Turnbull's Nursery, complained, 'while communications tending to promote the interests of horticulture have poured in upon you from every quarter, arboriculture has seemed to be totally disregarded' (Munro 1833, 551). Munro thought the best way to remedy this was the formation of a 'National Society for the Promotion of Arboriculture'. In 1835, Loudon gave his own support to these calls, believing one of its first tasks should be to establish an arboretum with 'specimens of all the trees and shrubs that will endure the open air in Britain' (Loudon 1835b, 283). Meanwhile, some local horticultural societies were being formed and these were urged to take an interest in arboriculture. In an open letter to the committee of the recently launched Birmingham Horticultural Society, Ryland Bedford, rector of Sutton Coldfield, expressed the hope 'that while engaged in encouraging the science of horticulture, you will not entirely overlook that of arboriculture' (Bedford 1829). Bedford, who was keen to promote the planting of trees on wasteland and roads, published the letter at the front of his book *The Midland Forester*.

In terms of a national organisation to promote arboriculture and forestry, the Scottish Arboricultural Society was eventually founded in 1854 (*Gardeners' Chronicle* 1855, 97). Its First General Meeting was held at the Institution Rooms, Edinburgh, when office bearers and committee members were elected. The majority were titled 'Forester' and employed by members of the aristocracy and gentry. There were also a few nurserymen. In their report of this meeting in the *Gardeners' Chronicle*, the editors refer to the discipline as 'foresting' (*ibid.*, 100). In his opening address, James Brown, President of the Society, stated,

> Should the art of arboriculture be discussed in as skilful a manner as these two sister arts [agriculture and horticulture] … we have no reason to fear that a similar result will not follow in regard to forestry. … Hitherto foresters have been content to act by the uncertain light of their own experience. But now the application of science to those branches of cultivation is taking firm hold on the minds of all improving and enlightened men.
>
> *Gardeners' Chronicle* 1855, 100

The proliferation of terms is clear to see. However, to be fair to Brown, he was probably trying to be inclusive, aware that his audience had different views on what terms were appropriate. In his publications it is clear that he regards 'arboriculture' to be mostly about forestry and that the professionals involved should be called foresters.

In its early years the Society was keen to establish branches in different parts of Scotland and one of the first was formed in Brechin (*Gardeners' Chronicle* 1858b, 459). By 1869, the Society had about 350 members and the membership list shows that the majority were professionals, mostly foresters and assistant foresters but also agents, land stewards, nurserymen and seedsmen. There were hardly any members listed as gardener or landscape gardener and none was listed as arborist or arboriculturist (*Trans. SAS* 1869). In 1884, the Society was instrumental in organising Britain's first International Forestry Exhibition, held in the grounds of Donaldson's College, Edinburgh (*Journal of Forestry* 1884, 239–244). The many exhibits included a section devoted to nurserymen that contained a vast collection of trees and shrubs. While the machinery displayed was almost exclusively for timber production, a significant number of the tools and equipment could also have been used for amenity trees, especially those exhibited by the nurserymen. There was also the French stand of the Comte des Cars showing 'tools and instruments used in pruning, consisting of an averruncator, handbills, saws, pincers, chisels, and chisel with down cutting hook' (*ibid.*, 279). The exhibition received patronage from Queen Victoria and other members of the royal family. In 1887, Queen Victoria granted a Royal Charter to the Society and it became the Royal Scottish Arboricultural Society.

With the success of the Scottish Arboricultural Society pressure started to mount on the RHS to undertake some initiative in the field of arboriculture, certainly in respect of its activities in England. In the late 1850s, the RHS formed a number of new committees that specialised in different aspects of horticulture, promoted in part by the appointment of Prince Albert as its President. In 1858 it instituted its Fruit Committee and, although this was focused firmly on pomology, aspects of its work had arboricultural relevance. This was followed in 1859 by a Floral Committee on similar lines. At the RHS Council meeting in June 1859, William Paul proposed that there should be an Arboricultural Committee (Elliott pers. comm., 2019). This was subsequently announced in an editorial in the *Gardeners' Chronicle* (1860, 72). In the *Proceedings* of the RHS in 1863, there is a list of the members of its Council, Officers and Committees (*PRHS* 1863, 60–63). Its Arboricultural Committee was chaired by the Duke of Buckingham and Chandos and included forty-seven other members. Of these, the majority were from the aristocracy and gentry. There were a few nursery owners and some keen amateurs but hardly anyone who could be described as a professional. This was in contrast to the membership of the Floral Committee and the Fruit and Vegetable Committee which both had substantial numbers of professionals. The Arboricultural Committee does not appear to have been very active and it seems that no annual reports were issued. The *Gardeners' Chronicle* and other horticultural magazines would occasionally urge the committee to take action on some matter, for example, to set up trials to ascertain the suitability of trees for urban planting, but this invariably had little effect (*Gardeners' Chronicle* 1863, 483). After 1863, no further mention could be found of the activities of this

Arboricultural Committee, either in the RHS archives or in the horticultural press.

In 1882, the English Arboricultural Society was founded following an inaugural meeting at Hexham in Northumberland (James 1982, 5). The initiators were encouraged by the success of the Scottish example and keen not to repeat the limited impact of the RHS's short-lived Arboricultural Committee. The decision to use the word 'arboricultural' instead of 'forestry' in the title of the Society was probably because the Scottish Society had adopted this and it was felt that it would be more in accord with that body to do the same. Nonetheless, there was still a degree of interest in what is now regarded as arboriculture. Of the two founders of the Society, Henry Clark was a forester but the other was John W. Robson, a nurseryman and an authority on ornamental as well as forest trees (*QJF* 1929, 117). Furthermore, from an analysis of the occupations of the Society's first members taken from a list in 1884, it is probable that some 20 per cent were interested in amenity trees (James 1982, 122). During its early years, the Society appointed its presidents from among its members who were major landowners or aristocrats (*ibid.*, 23). Although they provided impressive figureheads, they took little or no part in the Society's activities.

Between the 1880s and 1914, both the English and Scottish Arboricultural Societies were involved in a long series of royal commissions and government enquiries that touched on aspects of forestry (Foot 2010, 34). While the two societies agreed on many things, they also had their differences. The Scottish were more interested in afforestation, while the English focused more on improving existing woods.

Education in arboriculture and related subjects

Throughout the nineteenth century there was still no specific education or training for those engaged in arboricultural work. It continued to be included with horticultural or landscape work, for which an apprenticeship was the usual route for employment progression. By Loudon's time, an apprenticeship lasted only three years, followed by a succession of jobs as a journeyman. In his last book, *Self-Instruction for Young Gardeners*, Loudon aimed to fill a major gap in horticultural education. The book's scope was wide-ranging 'because gardeners are frequently required to practice the business of the forester, and sometimes that of the bailiff; occasionally they are elevated to the stewardship of an estate' (Loudon, J.C. 1845, 1). However, there was no mention of arboriculture or arboriculturists. The larger nurseries and estates each provided training opportunities for gardeners who passed between them during their careers, while landowners and agents often sought recommendations for prospective gardeners from nursery companies. The botanical society gardens also provided training for those who wanted to progress in horticulture as well as specialising in arboricultural work before there were any formal educational courses at leading botanical gardens. A little formal training began at Kew Gardens in

1859 but this only took place after the normal six-day working week (Paterson 2008, 172). Following questions in Parliament, more permanent and structured educational arrangements were instituted at Kew. The School of Horticulture was established and in 1870 Kew began to issue diplomas in horticulture. From the 1850s, when local authorities had increasing responsibility for parks, gardens and street trees, they adopted their own system of apprenticeships in horticulture.

While formal education in horticulture was discussed early in the nineteenth century, it took a long time to achieve. In 1820, the Horticultural Society looked forward to the day when it could establish 'a National School for the propagation of Horticultural knowledge' (Elliott 2004, 314). With the acquisition of the Chiswick garden, the possibility of a school of horticulture suddenly came within the Society's grasp. However, training of its gardeners continued 'on the job'. From 1866, in collaboration with the Royal Society of Arts, the RHS offered a Certificate comprising subjects necessary for qualification as a gardener and more academic skills necessary to become an Associate of the RHS. Elementary education for schoolchildren had been made compulsory in 1870, and was made free of charge in 1891. Local government reorganisation in the wake of the Third Reform Act resulted in the creation of County Councils in 1888, and the following year the Technical Education Act gave them powers to set up courses with some government funding. In 1891 the RHS offered its services to any County Council which included horticulture in its programmes (*ibid.*, 317). The first to do so was Surrey and the first results, in 1892, were good. Encouraged by this success the Society agreed to supervise horticultural examinations anywhere in the country. The old RHS Certificate exams held with the Royal Society of Arts were revised to form a new RHS General Examination that was launched in 1893. Details of the syllabus published in 1897 show it had two sections, entitled 'Elementary Principles' and 'Horticultural Operations and Practice'. The latter section includes an item entitled 'Arboriculture: Trees and Shrubs and their Culture' (*JRHS* 1897/98, 310). In 1881, Edward Milner established a School of Gardening in conjunction with the education department of the Crystal Palace Company (*Gardeners' Chronicle* 1881b, 437). While arboriculture is mentioned in the syllabus of this two-year course, its extent appears to be small. One of the graduates of this school was Fanny Wilkinson, who became the first professional female landscape gardener in the 1880s (Craddock 2012, 36).

At the end of the nineteenth century the position of women in professional arboriculture was unchanged; they were not involved except occasionally for tasks such as weeding or basic nursery work. However, their prospects of obtaining positions of responsibility as gardeners were at last improving thanks to developments in horticultural education. Women first attended horticultural college in 1891, when female students were admitted to Swanley College in Kent. In 1898, Studley College in Warwickshire was established exclusively to deliver horticultural education for women. Meanwhile, in 1896 the first women were taken on as gardeners at Kew, having been trained at Swanley.

While horticultural education was making good progress, forestry education was not far behind. In 1885, the Select Committee on Forestry was formed with the main purpose of considering the establishment of a Forest School. Among its recommendations the committee called for the establishment of a Forest Board, with representation from the Scottish and English Arboricultural Societies, various public and professional bodies and some members of Parliament (*Gardeners' Chronicle* 1887, 203). The Board's functions included organising Forest Schools and determining the syllabus. At the same time, moves towards more formalised forestry education were under way. In 1885, the first forestry courses were introduced at the Royal Indian Engineering College at Coopers Hill. However, these were directed towards the colonial service and arboriculture was not included. When the college closed in 1906, the forestry section was transferred to Oxford University and the focus on timber production continued. Edinburgh University had established a School of Forestry in 1889, and this institution seems to have had a more inclusive definition of forestry that partly embraced arboriculture. Around 1890, the Surveyors' Institution (now the Royal Institution of Chartered Surveyors) began offering a Special Certificate examination in forestry.

Conclusions

The nineteenth century witnessed some huge advances in the development of professional arboriculture in Britain that laid the foundations for the practice of arboriculture as we know it today. There were several major factors that contributed to this progress.

Throughout the century there was a massive increase in the number of urban trees across Britain, driven largely by the rapid expansion of many towns and cities. The suburban housing boom led to far more garden trees and there was a tremendous increase in publicly owned trees in the new public parks and alongside streets and roads. All these urban trees presented major new challenges for those responsible for them. This was not just about ensuring their successful establishment and healthy growth in challenging conditions, but also about avoiding potential conflicts with built development and urban infrastructure, especially where this might result in harm to people or damage to property. The dramatic expansion in urban tree cover prompted a huge increase in the volume of arboricultural work, particularly as these trees matured. Many of these new trees were publicly owned, which meant that local authorities now needed specialist workforces with appropriate equipment to deal with arboricultural matters, or obtain those services from contractors. This was the start of municipal arboriculture in Britain and it dramatically increased the number of people employed in arboricultural work. The nineteenth century also heralded a more scientific approach to the planting and care of amenity trees, although research into this was still at an early stage.

The emerging specialism of urban arboriculture and the unique expertise this demanded eventually led to the recognition of arboriculture as a distinct discipline. By the late nineteenth century it was no longer viewed as synonymous with forestry or just a branch of horticulture. Those who were mainly engaged in arboricultural work were increasingly considered as part of a distinct trade or occupation, closely related but still different from horticulture and forestry. However, there was not yet any recognisable arboricultural industry and although the term 'tree expert' was often used, 'arborist' and 'arboriculturist' were not yet in general use. Nevertheless, the public were becoming aware of the value of tree experts and their specialist expertise. This was due largely to the national press coverage given to the heroic exploits of some high-profile tree experts, such as William Barron's spectacular tree moving and Joseph Paxton's amazing inclusion of the majestic elms into the Crystal Palace. However, those involved in arboricultural work were also experiencing the negative side of attracting public interest. The proliferation of newspapers and popular magazines could also provide a vehicle for public criticism of their work.

Despite the tremendous progress made in professional arboriculture, some crucial elements of a recognisable industry and profession were still absent. Even with the emergence of municipal arboriculture, there were not yet any distinct arboricultural sections within local authorities. Neither were there any specialist companies that were the equivalent of modern tree work contractors or arboricultural consultants. This work was still being undertaken by gardening, landscape, nursery and forestry businesses. There was also no professional representation for British arboriculture. The English and Scottish Arboricultural Societies, while having an interest in amenity trees, had evolved primarily as forestry societies and their subsequent change of name reflected that. Lastly, while there were now separate courses and qualifications in horticulture and forestry, this was not yet the case with arboriculture.

References

Ablett, W.H. (1880) *Arboriculture for Amateurs: Being instructions for the planting and cultivation of trees for ornament and use.* The Bazaar Office: London.

Appleby, T. (1852) Jottings by the Way. *The Cottage Gardener* 8, 358.

Ballard, P. (2003) *An Oasis of Delight: The History of Birmingham Botanic Gardens.* Revised edition. Brewin Books: Studley.

Ballard, S. (1833) *A Treatise on the Nature of Trees and the Pruning of Timber Trees; showing the impossibility of increasing the quantity or improving the quality of timber by pruning.* Ledbury: London.

Baltet, C. (1873) *The Art of Grafting and Budding.* William Robinson: London.

Barron, W. (1852) *The British Winter Garden: Being a practical treatise on evergreens.* Bradbury and Evans: London.

Barron, W. (1853) McGlashen's Tree Lifter. *The Gardeners' Chronicle,* First Series, 182.

Barron, W. (1880) The Buckland Yew. *The Gardeners' Chronicle*, New Series, 13, 556–557.
Barron, W. (1886) Transplanting Machines. *The Gardeners' Chronicle*, New Series, 26, 816.
Beaton, D. (1853) Renovation of Old Trees. *The Cottage Gardener* 11, 177–179.
Bedford, W.R. (1829) *The Midland Forester*. R. Wrightson: Birmingham.
Benton, A.M. (2002) *Cheals of Crawley: The Family Firm at Lowfield Nurseries 1860s to 1960s*. Moira Publications: Uckfield.
Berkshire Chronicle (1826) Local Intelligence. *Berkshire Chronicle*, 25 March, 3.
Billington, W. (1825) *A Series of Facts, Hints, Observations, and Experiments on the Different Modes of Raising Young Plantations of Oaks*. Baldwin, Cradock and Joy: London.
Blomfield, R. (1892) *The Formal Garden in England*. Macmillan and Co.: London.
Brooke, E.A. (1858) *The Gardens of England*. T. McLean: London.
Brooks, A. (2011) *'A Veritable Eden' – The Manchester Botanic Garden: A History*. Windgather Press: Oxford.
Brown, J. (1851) *The Forester*. Second edition. William Blackwood: Edinburgh and London.
Brown, J. (1858) Opening Address. *Transactions of the Scottish Arboricultural Society* 1, 1–7.
Brown, J. (1894) *The Forester*. Edited by J. Nisbet. Sixth edition. William Blackwood: Edinburgh and London.
Cockburn Association (1899) *Report by the Council to the Twenty-third Annual Meeting of the Association, Edinburgh, May 1899*. Cockburn Association: Edinburgh.
Coke, D. and Borg, A. (2011) *Vauxhall Gardens: A History*. Yale University Press: New Haven and London.
Conway, H. (1991) *People's Parks: The Design and Development of Victorian Parks in Britain*. Cambridge University Press: Cambridge.
Corfield, P. (2012) *Vauxhall: Sex and Entertainment, London's Pioneering Urban Pleasure Garden*. History and Social Action Publications: London.
Cox, E.H.M. (1935) *A History of Gardening in Scotland*. Chatto and Windus for Flora and Sylva: London.
Craddock, J.P. (2012) *Paxton's Protégé: The Milner White Landscape Gardening Dynasty*. John Peter Craddock: Stevenage.
Curl, J.S. (2000) *The Victorian Celebration of Death*. Sutton Publishing: Stroud.
Curtis, C.E. (1892) *The Manifestation of Disease in Forest Trees*. Horace Fox: London.
Derby Mercury (1830) Weeping Ash. *The Derby Mercury*, 14 April, 3.
Derbyshire Courier (1843) The Queen at Chatsworth. *Derbyshire Courier*, 9 December, 1.
Des Cars, A. (1864) *L'Élagage des Arbres: Traité Pratique de l'Art de Diriger les Arbres Forestiers*. J. Rothschild: Paris.
Des Cars, A. (1881) *Tree Pruning: A Treatise on Pruning Forest and Ornamental Trees*. Translated from the seventh French edition by Charles S. Sargent. A. Williams: Boston, MA.
Desmond, R. (2007) *The History of the Royal Botanic Gardens Kew*. Second edition. Kew Publishing: Kew.
ELJ (1830) A Review of 'Strictures on Sir Henry Steuart's Planter's Guide'. *The Edinburgh Literary Journal* 3, 112.
Elles, J. (1830) Observations respecting Mr Howden's Management of Forest Trees. *The Gardener's Magazine* 6, 545–548.
Elliott, B. (1986) *Victorian Gardens*. B.T. Batsford: London.

Elliott, B. (1995) *The Country House Garden: From the Archives of Country Life 1897–1939*. Mitchell Beazley: London.

Elliott, B. (2004) *The Royal Horticultural Society: A History 1804–2004*. Phillimore: Chichester.

Elliott, B. (2018) Colourful Groves: The Origins of the Woodland Garden. In: Woudstra, J. and Roth, C. (eds) *A History of Groves*. Routledge: Abingdon.

Elliott, P. (2016) *British Urban Trees: A Social and Cultural History c. 1800–1914*. The White Horse Press: Winwick.

Elliott, P., Watkins, C. and Daniels, S. (2007a) Combining Science with Recreation and Pleasure: Cultural Geographies of Nineteenth-Century Arboretums. *Garden History* 35(2), 6–27.

Elliott, P., Watkins, C. and Daniels, S. (2007b) William Barron (1805–91) and Nineteenth Century British Arboriculture: Evergreens in Victorian Industrializing Society. *Garden History* 35(2), 129–148.

Elliott, P., Watkins, C. and Daniels, S. (2011) *The British Arboretum: Trees, Science and Culture in the Nineteenth Century*. Pickering and Chatto: London.

Evelyn, J. (1776) *Silva: or a Discourse of Forest-Trees, and the propagation of timber in His Majesties Dominions*. With notes by A. Hunter. J. Dodsley: London.

Ferrey, E.B. (1878) Trees and Foundations. *The Gardeners' Chronicle*, New Series, 10, 726.

Fletcher, H.R. and Brown, W.H. (1970) *The Royal Botanic Garden Edinburgh 1670–1970*. HMSO: Edinburgh.

Foot, D. (2010) *Woods and People: Putting Forests on the Map*. The History Press: Stroud.

Forbes, A.C. (1906) *English Estate Forestry*. Second edition. Edward Arnold: London.

Forsyth, W. (1802) *A Treatise on the Culture and Management of Fruit-Trees*. T.N. Longman and O. Rees: London.

Fricker, L. (1974) John Claudius Loudon: The Plane Truth? In: Willis, P. (ed.) *Furor Hortensis: Essays on the history of the English Landscape Garden in memory of H.F. Clark*. Elysium Press: Edinburgh.

Frost, P. (1841) Rhododendrons. *Gardeners' Chronicle*, First Series, 6, 85.

Gardeners' Chronicle (1841) Prospectus of *The Gardeners' Chronicle*. *The Gardeners' Chronicle*, First Series, 1, 1.

Gardeners' Chronicle (1842) Law of Gardens; Overhanging Trees. *The Gardeners' Chronicle*, First Series, 2, 126.

Gardeners' Chronicle (1844) Calendar of Operations for the Ensuing Week. *The Gardeners' Chronicle*, First Series, 4, 23–24.

Gardeners' Chronicle (1852) Transplanting Trees. *The Gardeners' Chronicle*, First Series, 11, 804.

Gardeners' Chronicle (1853a) Caledonian Horticultural. *The Gardeners' Chronicle*, First Series, 12, 6.

Gardeners' Chronicle (1853b) Editorial. *The Gardeners' Chronicle*, First Series, 11, 163–164.

Gardeners' Chronicle (1853c) Editorial. *The Gardeners' Chronicle*, First Series, 48, 755–756.

Gardeners' Chronicle (1853d) Description of Mr M'Glashen's Patent Apparatus. *The Gardeners' Chronicle*, First Series, 51, 807.

Gardeners' Chronicle (1855) Scottish Arboricultural Society. *The Gardeners' Chronicle*, First Series, 7, 97 and 100.

Gardeners' Chronicle (1857) Wounds in Old Trees. *The Gardeners' Chronicle*, First Series, 20, 348.

Gardeners' Chronicle (1858a) A Varnish. *The Gardeners' Chronicle*, First Series, 11, 190.

Gardeners' Chronicle (1858b) Brechin Arboriculture. *The Gardeners' Chronicle*, First Series, 23, 459.

Gardeners' Chronicle (1860) Editorial. *The Gardeners' Chronicle*, First Series, 4, 71–72.

Gardeners' Chronicle (1861a) Tree Transplanting. *The Gardeners' Chronicle*, First Series, 21, 456–457.

Gardeners' Chronicle (1861b) Death of Mr Billington, the well known Arboriculturist. *The Gardeners' Chronicle*, First Series, 48, 1047. *Gardeners' Chronicle* (1863) Editorial. *The Gardeners' Chronicle*, First Series, 23, 483.

Gardeners' Chronicle (1869) Removal of a Large Tree. *The Gardeners' Chronicle*, First Series, 14, 365.

Gardeners' Chronicle (1873) Tree Surgery. *The Gardeners' Chronicle*, First Series, 28, 947.

Gardeners' Chronicle (1876) The Planting of Trees in Large Towns. *The Gardeners' Chronicle*, New Series, 6, 354–355.

Gardeners' Chronicle (1878) Mr Gladstone and Tree Felling. *The Gardeners' Chronicle*, New Series, 9, 171.

Gardeners' Chronicle (1879) A New Tree Pruner. *The Gardeners' Chronicle*, New Series, 12, 269.

Gardeners' Chronicle (1880a) Advertisement for the *Journal of Forestry*. *The Gardeners' Chronicle*, New Series, 13, 633.

Gardeners' Chronicle (1880b) Removal of an Old Yew Tree. *The Gardeners' Chronicle*, New Series, 13, 305–306.

Gardeners' Chronicle (1881a) The Preservation of Suburban Trees. *The Gardeners' Chronicle*, New Series, 16, 660.

Gardeners' Chronicle (1881b) School of Gardening. *The Gardeners' Chronicle*, New Series, 15, 437.

Gardeners' Chronicle (1885) Waddesdon. *Gardeners' Chronicle*, New Series, 23, 820–822.

Gardeners' Chronicle (1887) Forestry – Report of the Select Committee of the House of Commons. *The Gardeners' Chronicle*, Series 3, 2, 203.

Gardeners' Chronicle (1891) Obituary for William Barron. *The Gardeners' Chronicle*, Series 3, 9, 522–524.

Gardeners' Chronicle (1892a) Advertisement for *Garden and Forest*. *The Gardeners' Chronicle*, Series 3, 11, 509.

Gardeners' Chronicle (1892b) What is the Value of a Tree? *The Gardeners' Chronicle*, Series 3, 11, 115.

Gardeners' Chronicle (1894) The Waverley Tree-Guard and Support. *The Gardeners' Chronicle*, Series 3, 15, 309.

Gardener's Magazine (1842) Remarks on ornamental Forestry. *The Gardener's Magazine*, New Series, 8, 112.

Gilchrist, A. (1873) On Different Methods of Pruning Forest Trees. *The Garden* 4, 489–491.

Gilchrist, W. (1872) Practical Arboriculture. *Transactions of the Scottish Arboricultural Society* 6, 6–50.

Gordon, G. (1858) *The Pinetum: Being a Synopsis of All the Coniferous Plants at Present Known*. Henry G. Bohn: London.

Gorrie, A. (1827) Description of a Dendrometer invented by Mr Archibald Gorrie. *The Gardener's Magazine* 2, 8–13.

Greenwood, G. (1853) *The Tree-Lifter; or, A New Method of Transplanting Forest Trees.* Second edition. Longman, Brown, Green and Longmans: London.

Grigor, J. (1841) *The Eastern Arboretum*. Longman, Brown, Green and Longmans: London.

Grigor, J. (1868) *Arboriculture, or a practical treatise on raising and managing forest trees.* Edmonston and Douglas: Edinburgh.

Grissino-Mayer, H.D. (2003) A Manual and Tutorial for the Proper Use of an Increment Borer. *Tree-Ring Research* 59(2), 63–79.

Hansard (1833) *Public Health*. HC Deb, 21 February 1833, Vol. 15, 1049–1059. Available at: http://hansard.millbanksystems.com/commons/1833/feb/21/public-health (accessed 2 May 2014).

Harrison, J. (1833) Introductory Remarks. *The Gardener's and Forester's Record* 1, 1-3.

Hartig, R. (1894) *Textbook of the Diseases of Trees*. Translated by William Somerville and revised and edited by H. Marshall Ward. Country Life: London.

Heriz-Smith, S. (1988) The Veitch Nurseries of Killerton and Exeter c.1780–1863: Part I. *Garden History* 16(1), 41–47.

Hooker, W.J. (1839) Review of Loudon's *Arboretum et Fruticetum Britannicum. Annals of Natural History,* 3, 188.

Hou, S. (2013) *The City Natural:* Garden and Forest *Magazine and the Rise of American Environmentalism*. University of Pittsburgh Press: Pittsburgh, PA.

Howden, J. (1835) On the Necessity of adjusting the Tops of transplanted Forest Trees to their Roots. *The Gardener's Magazine*, New Series, 1, 15–16.

Hubert, E.E. (1931) *An Outline of Forest Pathology*. Wiley and Sons: London.

Hudson, D. and Luckhurst, K.W. (1954) *The Royal Society of Arts, 1754–1954*. John Murray: London.

Hunter, T. (1883) *Woods, Forests and Estates of Perthshire: With Sketches of the Principal Families in the County*. Henderson, Robertson and Hunter: Perth.

Hutchison, R. (1877) The Proposed 'Journal of Forestry'. *The Gardeners' Chronicle*, New Series, 7, 537.

ILN (1850a) Site for the Great Exhibition of 1851. *Illustrated London News*, 17 August, 125–126.

ILN (1850b) Mr Paxton's History of the Building of the Great Exhibition of 1851. *Illustrated London News*, 19 October, 322–323.

ILN (1850c) The Great Exhibition Building – Details of Construction. *Illustrated London News*, 14 December, 452–454.

ILN (1855) Removal of a tree by Barron's machine, in RBS's Garden, Regent's Park. *Illustrated London News*, 22 September, 35wwww7.

L'Univers Illustre (1854) Nouvelle machine à transplanter les arbres. *L'Univers Illustre*, 30 December, 447–448.

L'Univers Illustre (1865) Chariot perfectionne pour la transplantation des arbres. *L'Univers Illustre*, 25 October, 673.

James, N.D.G. (1982) *A Forest Centenary: A History of the Royal Forestry Society of England, Wales and Northern Ireland*. Basil Blackwell: Oxford.

Johnston, M. (2015) *Trees in Towns and Cities: A History of British Urban Arboriculture.* Windgather Press: Oxford.

Johnston, M. (2017) *Street Trees in Britain: A History.* Windgather Press: Oxford.

Jongkindt Coninck, A.M.C. (1874) Pruning Saws. *The Gardeners' Chronicle*, New Series, 1, 93.

Jordan, H. (1994) Public Parks, 1885–1914. *Garden History* 22(1), 85–113.

Journal of Forestry (1878) The Sécateur and Echénillior. *Journal of Forestry* 2, 140–141.

Journal of Forestry (1880) A Tree-Feller. *Journal of Forestry* 4, 409.
Journal of Forestry (1883) Kay's Dendrometer. *Journal of Forestry* 8, 21–22.
Journal of Forestry (1884) The International Forestry Exhibition, Edinburgh, 1884. *Journal of Forestry* 9, 239–244, 251–255, 269–280.
JRHS (1897/98) Examination in Horticulture. *Journal of the Royal Horticultural Society*, New Series, 21, 309–312.
Kay, J. (1869) Instrument for measuring the heights of trees. *Transactions of the Scottish Arboricultural Society* 5, 170.
Kay, J. (1875) On a new Transplanting Machine. *Transactions of the Scottish Arboricultural Society* 7, 186–189.
Kelly, C.R. (1873) *Transplanting Large Trees and Shrubs.* E.W. Allen: London.
Kemp, E. (1850) *How to Lay Out a Small Garden*. Bradbury and Evans: London.
Kennion, E. (1815) *An Essay on Trees in the Landscape.* T. Bensley: London.
Kent, A.H. (1881) *A Manual of Coniferae.* James Veitch and Sons: London.
Kidd, J. (1840) Description of a Transplanting Machine invented by James Kidd. *The Gardener's Magazine*, New Series, 6, 252–254.
Lambert, A.B. (1803) *A Description of the Genus Pinus*. George White: London.
Lindley, J. (1840) *The Theory of Horticulture; Or, An Attempt to Explain the Principal Operations upon Gardening.* Longman, Orme, Brown, Green and Longmans: London.
Loudon, J.C. (1803) Letter to the Editor dated 22 December 1803. *Literary Journal*, December 31, 1803, cols 739–742.
Loudon, J.C. (1804) *Observations on the Formation and Management of Useful and Ornamental Plantations; on the Theory and Practice of Landscape Gardening.* Archibald Constable: Edinburgh.
Loudon, J.C. (1812) *Hints on the Formation of Gardens and Pleasure Gardens.* J. Harding: London.
Loudon, J.C. (1822) *Encyclopaedia of Gardening*. Longman, Hurst, Rees, Orme and Brown: London.
Loudon, J.C. (1825) *Encyclopaedia of Agriculture*. Longman, Hurst, Rees, Orme, Brown and Green: London.
Loudon, J.C. (1830) A Review of *Strictures on Sir Henry Steuart's Planter's Guide. The Gardener's Magazine* 6, 87–93.
Loudon, J.C. (1832) Book Review: Practical Hints on Landscape-Gardening by S. Gilpin. *The Gardener's Magazine* 8, 700–702.
Loudon, J.C. (1833) Notice of a new Method of Transplanting large Trees, invented by Mr James Munro. *The Gardener's Magazine* 9, 217–218.
Loudon, J.C. (1835a) *Encyclopaedia of Gardening.* A new edition. Longman, Rees, Orme, Brown, Green and Longmans: London.
Loudon, J.C. (1835b) Editorial Note. *The Gardener's Magazine*, New Series, 1, 281–284.
Loudon, J.C. (1837) Arboricultural Notices; or, Notices of new hardy Trees and Shrubs. *The Gardener's Magazine*, New Series, 3, 18–33.
Loudon, J.C. (1838a) *Arboretum et Fruticetum Britannicum or the Trees and Shrubs of Britain.* In eight volumes. Longman, Brown, Green and Longmans: London.
Loudon, J.C. (1838b) *The Suburban Gardener and Villa Companion.* Longman, Orme, Brown, Green and Longmans: London.
Loudon, J.C. (1840a) A Summary View of the Progress of Gardening. *The Gardener's Magazine*, New Series, 6, 617–631.

Loudon, J.C. (1840b) Review of *The Eastern Arboretum* by James Grigor. *The Gardener's Magazine*, New Series, 6, 352–354.

Loudon, J.C. (1840c) *The Derby Arboretum: A catalogue of the trees and shrubs.* Longman, Orme, Brown, Green and Longmans: London.

Loudon, J.C. (1842) *The Suburban Horticulturist.* William Smith: London.

Loudon, J.C. (1843) *On the Laying Out, Planting, and Managing of Cemeteries, and on the improvement of Churchyards.* Longman, Brown, Green and Longmans: London.

Loudon, J.C. (1845) *Self-Instruction for Young Gardeners, Foresters, Bailiffs, Land-Stewards, and Farmers.* Longman, Brown, Green and Longmans: London.

Loudon, J.W. (1845) An Account of the Life and Writings of John Claudius Loudon. In: Loudon, J.C. *Self-Instruction for Young Gardeners, Foresters, Bailiffs, Land-Stewards, and Farmers.* Longman, Brown, Green and Longmans: London.

Main, J. (1828) The Planter's Guide by Sir H. Steuart. *The Gardener's Magazine* 4, 115–125.

Main, J. (1839) *The Forest Planter and Pruner's Assistant.* Ridgeway: London.

Major, J. (1852) *The Theory and Practice of Landscape Gardening.* Longman, Brown, Green and Longmans: London.

Masters, M.T. (1891) Trees and Shrubs for Large Towns. *Journal of the Royal Horticultural Society*, New Series, 13, 71–98.

McIntosh, C. (1853) *The Book of the Garden. 1: Structural, 2: Cultural.* William Blackwood and Sons: Edinburgh.

McNab, J. (1874) Tree Management in or near Towns. *The Garden* 6, 401.

McNab, J. (1878) Transplanting Trees. *The Garden* 13, 29–30, 59–62, 81–82, 97–98, 130–131.

McNab, W. (1830) *Hints on the planting and general treatment of hardy evergreens, in the climate of Scotland.* Thomas Clark: Edinburgh.

Meason, G.L. (1828) *On the Landscape Architecture of the Great Painters of Italy.* C. Hullmandel: London.

Michie, C.Y. (1878) Forestry. *The Gardeners' Chronicle*, New Series, 10, 412.

Milner, H.E. (1890) *The Art and Practice of Landscape Gardening.* Published by the Author: London.

Monteath, R. (1820) *The Forester's Guide and Profitable Planter.* James Foreman: Sterling.

Monteath, R. (1824) *The Forester's Guide and Profitable Planter.* Second edition. Stirling and Kenney: Edinburgh.

Munro, J. (1833) The Formation of a National Society for the Promotion of Arboriculture suggested. *The Gardener's Magazine* 9, 551–552.

Munro, J. (1842) On Transplanting large Trees. *The Gardener's Magazine*, New Series, 8, 102–106.

Murphy, E. (1831) Observations made during an Arboricultural Tour in Scotland and England, during the Autumn of the Year 1830. *The Gardener's Magazine* 7, 295–298.

Newcastle Journal (1856) Letter to the Editor: An Arboretum. *The Newcastle Journal*, 13 September, 7.

Nicol, W. (1799) *The Practical Planter, Or, A Treatise on Forest Planting.* Printed for the Author: Edinburgh.

Nicol, W. (1803) *The Practical Planter, Or, A Treatise on Forest Planting.* Second edition. J. Scatcherd and H.D. Symonds: London.

Nicol, W. (1812) *The Planter's Kalendar; or the Nurseryman's and Forester's Guide.* Edited and completed by Edward Sang. Archibald, Constable and Co.: Edinburgh.

O'Reilly, C. (2013) From 'the people' to 'the citizen': The Emergence of the Edwardian Municipal Park in Manchester, 1902–1912. *Urban History* 40(1), 136–155.
Pall Mall Gazette (1897) Occasional Notes. *The Pall Mall Gazette*, 20 November, 2.
Patch, D. (1984) *Tree Staking*. Arboriculture Research Note 40. Department of the Environment, Arboricultural Advisory and Information Service: Farnham.
Paterson, A. (2008) *The Gardens at Kew*. Frances Lincoln: London.
Paul, W. (1892) *Contributions to Horticultural Literature; Being a Selection of Articles Written for Gardening Periodicals From 1843 To 1892*. William Paul and Son: Waltham Cross.
Paxton, J. (1835) Some Account of the Arboretum lately commenced by His Grace the Duke of Devonshire at Chatsworth, in Derbyshire. *The Gardener's Magazine*, New Series, 1, 385–395.
Perthshire Advertiser (1899) Advertisement for John M. Browning, Seedsman and Florist. *The Perthshire Advertiser and Strathmore Journal*, 17 May, 1.
Phillips, H. (1823) *Sylva Florifera: The Shrubbery Historically and Botanically Treated*. Longman, Hurst, Rees, Orme and Brown: London.
Pontey, W. (1808a) *The Profitable Planter: A treatise on the theory and practice of planting forest trees*. Second edition. J. Harding: London.
Pontey, W. (1808b) *The Forest Pruner, or Timber Owner's Assistant*. Second edition. J. Harding: London.
Post Office (1841) *Post Office London Directory 1841*. Kelly's Directories Ltd: London.
PRHS (1862) XCV Monthly Summary. *Proceedings of the Royal Horticultural Society* 2, 757–766.
PRHS (1863) List of Council, Officers, and Committees for the year 1863. *Proceedings of the Royal Horticultural Society* 3, 60–63.
QJF (1929) Obituary Notice, J.W. Robson. *Quarterly Journal of Forestry* 23, 117.
Ravenscroft, B. C. (1883) *Town Gardening*. George Routledge: London.
Ravenscroft, E. (1884) *The Pinetum Britannicum: a descriptive account of hardy coniferous trees cultivated in Great Britain*. W. Blackwood and Sons: Edinburgh.
Ravetz, A. (2001) *Council Housing and Culture: The History of a Social Experiment*. Routledge: London.
Repton, H. (1840) *Landscape Gardening and Landscape Architecture of the Late Humphry Repton*. A new edition by J.C. Loudon. Longman and Co.: London.
Rinz, J. (1829) Remarks on various gardens about London. *The Gardener's Magazine* 5, 379–384.
Robinson, W. (1869) *The Parks, Promenades and Gardens of Paris, Described and Considered in Relation to the Wants of Our Cities*. John Murray: London.
Robinson, W. (1870) *The Wild Garden, or, Our Groves and Shrubberies Made Beautiful*. John Murray: London.
Robinson, W. (1883) *The English Flower Garden*. John Murray: London.
Robinson, W. (1892) *Garden Design and Architects' Gardens*. John Murray: London.
Robinson, W. (1894) Brown's Forester. *The Garden* 46, 468–469.
Robinson, W. (1900) *The English Flower Garden*. Eighth edition. John Murray: London.
Robinson, W. (1906) *The Garden Beautiful: Home Woods, Home Landscape*. John Murray: London.
Rutherford, S. (2013) *The Victorian Cemetery*. Shire Publications: Oxford.
Sadler, J. (1878) Arboricultural Societies. *Journal of Forestry* 2, 457–459, 572–578.
Sang, E. (1830) *Strictures on Sir Henry Steuart's Planter's Guide*. John Anderson: Edinburgh.

Sang, E. (1840) Description of the Hypsometer, an Instrument invented by John Sang, Esq. *The Gardener's Magazine*, New Series, 6, 190–193.

Scott, W. (1828) Review of *The Planter's Guide* by Sir Henry Steuart. *Quarterly Review*, March, 303–344.

Sedding, J.D. (1891) *Garden-Craft Old and New*. Kegan Paul, Trench, Trubner and Co.: London.

Simpson, J. (1900) *The New Forestry, Or the Continental System Adapted to British Woodlands and Game Preservation.* Pawson and Brailsford: Sheffield.

Slater, A. (1878) Tree-felling Saws. *Journal of Forestry* 1, 725.

Smith, C.H.J. (1852) *Parks and Pleasure Grounds; or Practical Notes on Country Residences, Villas, Public Parks, and Gardens.* Reeve and Sons: London.

Solman, D. (1995) *Loddiges of Hackney: The Largest Greenhouse in the World.* The Hackney Society: London.

Somerville, W. (1890) Inaugural lecture in the course of Forestry, Edinburgh University, 23 October 1889. *Transactions of the Royal Scottish Arboricultural Society* 12, 403–417.

Standish, J. and Noble, C. (1852) *Practical Hints on Planting Ornamental Trees.* Bradbury and Evans: London.

Steuart, H. (1828) *The Planter's Guide.* Second edition. John Murray: London.

Stewart, J. (1814) On transplanting large Fruit-Trees, whether Wall-Trees, Espalliers, or Standards. *Memoirs of the Caledonian Horticultural Society* 1, 202–206.

Strutt, J. (1830) *Sylva Britannica; or, Portraits of Forest Trees distinguished for their antiquity, magnitude or beauty*. Longman, Rees, Orme, Brown and Green; London.

Taylor, S. (1839) On different modes of securing newly planted Trees against high winds. *The Gardener's Magazine*, New Series, 5, 545–547.

The Garden (1874) Method of Climbing Trees. *The Garden* 6, 573.

The Garden (1878a) The Secateur. *The Garden* 13, 202.

The Garden (1878b) Street trees. *The Garden* 14, 500.

The Garden (1879) Elastic waterproof cement for protecting wounds on trees. *The Garden* 16, 219.

The Garden (1882) Fatal accident from blasting tree stumps. *The Garden* 21, 226.

The Garden (1886a) The axe in tree felling. *The Garden* 30, 353.

The Garden (1886b) Street Planting. *The Garden* 29, 86.

The Graphic (1872) The Cowthorpe Oak, near Wetherby. *The Graphic*, 14 September, 253.

The Graphic (1878) Tree Felling by Machinery. *The Graphic*, 16 February, 153.

Thetford Times (1891) Lynn Town Council. *Thetford and Watton Times*, 28 March, 6.

Thom, W. (1831) On the Supporting of recently removed Trees. *The Gardener's Magazine* 7, 445–454.

Thompson, R. (1859) *The Gardener's Assistant.* Blackie and Son: London.

Trans. SAS (1869) List of Members, corrected to March 1869. *Transactions of the Scottish Arboricultural Society* 5, Appendix A, 1–7.

Trans. RSAS (1905) History of Royal Scottish Arboricultural Society, 1854–1905. *Transactions of the Royal Scottish Arboricultural Society* 18, 5-19.

Veitch, J.H. (1906) *Hortus Veitchii: A History of the Rise and Progress of the Nurseries of Messrs. James Veitch and Sons.* James Veitch and Sons: London.

Walker, R. (1890) *On the Planting of Trees in Towns.* University Press: Aberdeen.

Ward, H.M. (1894) Preface to the English edition. In: Hartig, R. *Textbook of the Diseases of Trees*. Translated by William Somerville and revised and edited by H. Marshall Ward. Country Life: London.

Watson, P.W. (1825) *Dendrologia Britannica, Or Trees and Shrubs That Will Live in the Open Air of Britain Throughout the Year*. John and Arthur Arch: London.

Webster, A.D. (1887) Forestry: Pruning Forest Trees. *The Gardeners' Chronicle*, Series 3, 1, 770–771.

Webster, A.D. (1890) Knole Park and its Trees. *The Garden* 37, 385–386.

Webster, J.B. (1888) Staking Newly-Planted Trees. *The Garden* 33, 544.

Woods and Forests (1883a) Tree Record. *Woods and Forests* 1, 32.

Woods and Forests (1883b) Covering for tree wounds. *Woods and Forests* 1, 6.

CHAPTER EIGHT

The Rise of the Tree Experts, 1900–1945

When Queen Victoria died in 1901 and Britain entered the Edwardian era, most people expected life to continue in much the same way. The nation's position as the world's leading economic and political power seemed secure, and, while there were enormous inequalities at home between the super rich and the masses of urban poor, there was confidence among the ruling and middle classes that any difficulties could be overcome without much upheaval. Following the end of the Boer War in South Africa in 1902, there were several years when British soldiers were not engaged in any major overseas conflicts. However, all that confidence and security would soon disappear and be replaced by a feeling that 'the old order is doomed'.

By 1910 the rural economy was experiencing fierce competition from colonial and American food imports, resulting in 3 million acres (1.2m ha) being taken out of agricultural production (Schama 2002, 410). By 1911 just 8 per cent of the 45 million people in Britain were living off the land and the balance of political power was now shifting decisively away from the landed class to the urban and industrial elite. In 1914, the relative peace that had endured was fatally shattered with the outbreak of the First World War, a conflict that many thought would only last a few months. With over 700,000 British soldiers killed in action and many more seriously injured, hardly a family in the country was unaffected. When the horror finally ended in 1918 almost every aspect of life in Britain had changed radically or was about to be transformed. In another seismic shock to the old order, the Irish War of Independence (1919–1921) led to twenty-six of Ireland's thirty-two counties gaining their independence from Britain in the creation of the Irish Free State.

The break-up and dispersal of Britain's landed estates that had begun before the war accelerated dramatically in the 1920s. Higher taxation, death duties, the relentless decline in income from landed property and cultural pessimism all combined to encourage selling. The years from 1918 to 1921 saw the greatest shift in land ownership since the Dissolution of the Monasteries some 400 years before. After the war Britain experienced turbulent economic and political times that impacted mainly on its towns and cities. Those fortunate to live in the new industrial villages such as Port Sunlight and Bournville were hardly typical of the 40 per cent of the population who, by the turn of the twentieth century, lived in cities of more than 100,000 inhabitants (Schama 2002, 414). For many working-class families, living conditions remained squalid despite the huge expansion of new suburban districts. Popular discontent at the blatant inequalities in British society in the 1900s fuelled political and industrial unrest, leading to the rise of socialist ideas, the growth of the Labour Party, the increasing power of trade unions and the militant Suffragette movement demanding votes for women. After the war, as this political turmoil continued, Britain was plunged into the Great Depression, the global economic downturn that began in the United States in 1929. By the summer of 1932 some 3.5 million people were out of work nationally and in some areas unemployment had reached 70 per cent. Many more people had only part-time work. Eventually, unemployment began a modest fall in 1934 and fell further in 1935 and 1936. However, the rise in employment levels occurred mostly in the south of England, where lower interest rates had spurred the house building boom, which in turn helped to drive a recovery in domestic industry. Many other parts of Britain remained severely depressed for most of the 1930s.

In September 1939, Britain was once again engulfed in armed conflict following the outbreak of the Second World War. In the course of the war 384,000 British soldiers died in combat and there were 70,000 civilian deaths, due largely to German bombing raids which devastated large swathes of Britain's urban and industrial areas. When peace finally returned in September 1945, the nation faced a huge rebuilding task to repair or replace its devastated industry, infrastructure and housing stock.

Sources of information

Most of the major sources of information for this chapter were highlighted in the previous chapter, although in some cases there is an increased volume of material. This is especially the case with photographic material and other images relating to practical arboriculture and with the number of relevant advertisements appearing in horticultural and other professional periodicals. It is also notable in this period that much of the information relating to the development of professional arboriculture in Britain, in both the public and private sectors, comes from newspapers and popular periodicals in the form of news items and advertisements.

The published literature on the history of arboriculture in Britain has relatively good coverage of some of the major nineteenth-century developments, with accounts of the work of Loudon, Barron and other important contributors. By contrast, there is very little coverage of the first half of the twentieth century regarding innovations in practical arboriculture or the development of arboriculture as a trade or profession. For many people now working in the arboricultural industry there is a perception that professional arboriculture in Britain only began after the Second World War. This perception was confirmed by the series of twenty-three interviews that were conducted as part of this research with some prominent individuals in arboriculture and related disciplines to gain their insight into developments in professional arboriculture in Britain in the twentieth century. Most were unable to contribute much concerning developments in the first half of the twentieth century, in contrast to their far more extensive knowledge from 1945 to the present day. However, research for this chapter has discovered a wealth of material that indicates this was a crucial period in the early development of both public and private sector arboriculture in Britain.

Edwardian Britain and its amenity tree cover

The pattern of increased amenity tree cover in Britain that began in the late Victorian era continued in much the same way in the Edwardian period. At the beginning of the twentieth century large ornamental gardens were still to be found in the grounds of most landed estates, which included many tree-related features as well as extensive tree cover in the wider parkland. The rapidly expanding suburban districts were creating many thousands of new houses with gardens, most with some opportunity for planting trees or shrubs. In the public realm new municipal parks were still being established and this involved further tree planting and the retention of existing mature trees. In the streets and roads of urban Britain, tree planting was continuing as it had done in the late Victorian era.

The development of Britain's amenity tree cover in the first half of the twentieth century has been described in this author's two previous books on the history of urban trees and arboriculture (Johnston 2015 and 2017). For this reason only an outline is given here to provide some context for the rest of the chapter that focuses directly on the development of professional arboriculture.

Edwardian gardens and public parks

During the last years of the nineteenth century there was much emphasis on the revival of historical garden styles, especially with the attempt to identify and promote the English vernacular style. The most common vernacular style, because it was widely used in smaller urban and suburban gardens, was that known as Arts and Crafts. This artistic and design movement stood for traditional craftsmanship using simple forms and often applied medieval, romantic or folk styles of decoration. The first sign of the movement's influence in the garden was the revival of interest in rustic work, with rustic bridges and arbours becoming increasingly common in the late Victorian era. In the Edwardian era its emphasis on combining both hard and soft landscaping became particularly evident with plants becoming the softening 'flesh' around the 'bones' of the garden that was the hard landscaping, which often included terraces, paths, pools, summerhouses and pergolas (Jennings 2005, 10). Planting often came right up to the house with wisteria, roses, clematis, honeysuckle and jasmine trained up walls, around windows and over doors. Notable proponents of the Arts and Crafts style of garden design included Gertrude Jekyll and Edwin Lutyens, a partnership of self-taught gardener and professional architect that was responsible for several great gardens of the late Victorian and Edwardian period. Jekyll was also a prolific author on garden topics, and for trees and the woodland garden her book *Wood and Garden* remains a classic (Jekyll 1899). Perhaps the leading professional exponent of the Arts and Crafts garden style was Thomas Mawson, who set out his ideas on garden design in *The Art and Craft of Garden Making*, a sumptuous book that makes extensive reference to trees (Mawson 1900).

A classic feature of larger Edwardian gardens was to divide the outdoor space into smaller 'rooms'. This was usually achieved with neatly clipped yew hedges, which required high levels of skill and experience to maintain and could involve much expense where there were great lengths of this (Figure 64). The classic example of using hedges to create small outdoor spaces is Lawrence Johnston's garden at Hidcote Manor in Gloucestershire. Other dividing or defining features of the Edwardian garden that involved trees included low box hedges to delineate beds, and beech, lime and hornbeam trained into 'pleached' screens. Nut walks became increasingly popular and many garden enclosures functioned as orchards.

The revival of topiary that began in the late Victorian era with the return of the formal garden continued in the Edwardian era. This was despite the efforts of William Robinson and his well-known distaste for the unnatural and 'pretentious' appearance of this ancient art form. The use of neatly clipped hedges as garden dividers did much to promote topiary, which also now involved the popularisation of figurative topiary. Many of the books on garden design by some of the most respected authors of the day also featured examples of topiary and helped ensure its place as a typical feature of the Edwardian garden. Some books focused exclusively on this, such as *The Book of Topiary* by Charles Henry Curtis and William Gibson, which not only gave practical advice but also some historical background on this ancient art (Curtis and Gibson 1904). Many garden owners preferred not to spend much time and effort, either themselves or by their gardeners, in training up seedlings into mature specimens of topiary. For those admirers of topiary, there was now a ready-made alternative. Two firms of nurserymen, William Cutbush and Sons in London and Cheal and Sons in Sussex, specialised in the production of 'off the shelf' topiary, and contemporary photographs show their standing grounds or exhibition displays packed with birds, spirals, boats and tables shaped from yew and box (*ibid.*, 33). The prices of individual specimens were governed chiefly by their size, age and the complexity of the design. The yew tree being of slower growth than the box was, size for size, the more expensive of the two (*ibid.*,

FIGURE 64. Larger Edwardian gardens were often divided into smaller 'rooms' and this was usually achieved with neatly clipped yew hedges. This photograph from the 1910s shows a gardener using an instrument for determining the batter of the hedge and checking the accuracy of clipping.

35). These two nurseries were instrumental in both stimulating and satisfying the Edwardian fascination with all kinds of topiary.

In the cities and their suburbs, far from the world of William Robinson and Gertrude Jekyll, gardening was becoming an increasingly fashionable pastime. This had its own horticultural style that was often significantly different from the large gardens of the rich and famous. The demand for housing continued in the first decade of the twentieth century and new extensive middle-class suburban estates appeared on the outskirts of many towns and cities. These now came with back gardens as standard and significant front gardens for the larger properties. Just as the width of the road, types of street trees, street names and other features were used to draw social distinctions between localities, the status of an area was further confirmed by the extent and style of the garden space provided at the back and front of the house (Long 1993, 40). As the numbers of new amateur gardeners increased dramatically, a horticultural retail industry grew to cater for them selling gardening products, equipment, plants and literature. Of course, most middle- and working-class families could not create gardens of similar style and scale to those fashioned for wealthy clients by the leading garden designers of the day. However, thanks to the growth of horticultural literature full of imaginative garden designs, together with the greater availability of plants and equipment, limited or miniature versions of some of these garden features could be attempted. So, the flower garden, the herbaceous border, the rock garden and the water garden, along with the pergolas and trellised roses, could all find homes in Edwardian town gardens, if not all in the same garden.

At the start of the twentieth century there were still many large gardens owned by the rich and famous that employed numerous gardeners. Indeed, the Edwardian era is often regarded as the high point of the landed estate and its extensive ornamental gardens. Initially, this level of prosperity continued to be sustained by substantial financial resources inherited from the late Victorian era. However, within a few years all this began to change as the depression in the rural economy began to bite. The steady break-up and dispersal of these landed estates had begun and soon many gardeners whose families had often served their employers for generations were now seeking alternative work. Sometimes this could be found in urban areas as a gardener in one of the landscape businesses providing horticultural services to the ever-increasing number of small suburban gardens. Perhaps they might be fortunate in finding work in one of the public parks.

While some large public parks were still being created, there was increasing recognition that these did not solve the problems of access to open space for those living in the densest urban areas. There was a need for smaller, more local recreation grounds devoted to playgrounds and sports to provide people with opportunities for more active recreation (Conway 1991, 209). Concern about the health and physique of young people had been highlighted by the publication of the Report on Physical Deterioration in 1904 and medical reports of the poor fitness of army recruits (*ibid.*, 221). The resulting pressure led to the successful development of some small parks and recreation grounds, often by converting disused burial grounds and churchyards into open spaces for recreation (*ibid.*, 212). Large areas of existing public parks were also converted to recreational use. However, this undoubtedly had a negative impact on the tree cover of these public open spaces. Sports pitches required large expanses of mown grass, which conflicted directly with the provision of extensive tree cover. With the new recreation grounds and playing fields, there was a tendency to think of them as just that without any attempt to introduce ornamental horticultural features (Conway 2000, 119). As a result, these areas usually had very few trees. Where planting did occur this was likely to be limited to single rows or narrow shelterbelts of trees around the perimeter and at strategic points between sports pitches to provide players and spectators with some protection from inclement weather.

One major tree planting initiative in the 1900s concerned efforts to restore derelict land in the Midlands region of England. The work of the Midlands Reafforesting Association (MRA), formed in 1903, is an early example of using forestry to restore land devastated by the Industrial Revolution (*Gardeners' Chronicle* 1904a, 447). As a result of industrial contraction in the latter part of the nineteenth century, vast swathes of derelict land had emerged within the Black Country. While the MRA planted up sites totalling 83 acres (33 ha) before dissolving in 1925, it failed in its mission of reafforesting the Black Country (Webber 2008). Nevertheless, its main legacy was in literally 'preparing the ground' for other local organisations and government agencies that emerged some sixty years later, which had much greater success in greening those blighted landscapes. Although the activities of the MRA were reported in the horticultural press at the time, the work was viewed as forestry rather than arboriculture.

Just as many public parks and private gardens were undergoing some significant changes in the latter part of the Edwardian era, the whole process came to a shuddering halt with the outbreak of the First World War. The world of amenity horticulture and arboriculture was largely 'on hold' for the duration of the war, although many of these changes would continue again when conflict had ended.

The First World War

The First World War had a hugely negative impact on ornamental gardening and landscaping in Britain, and this also included work on amenity trees. With the nation on a war footing much of this work was not considered necessary and the men who had been employed on it were desperately needed in the armed forces or what were regarded as essential occupations. While the loss of male gardeners led to dramatically reduced workloads in most parks and gardens, it also resulted in a substantial increase in female gardeners. However, there is no evidence that women replaced men for major arboricultural work where this was still undertaken.

Throughout the war the focus on tree planting and management was very much on fruit tree growing and forestry for timber production. The economic benefits of these trees played a vital role in the war effort. Fruit production helped to feed the nation and timber was essential in construction and for pit-props in mining. Because of the shortage of men in forestry, women were trained to take their place in what traditionally was an almost exclusively male industry. The foundations of the Women's Forestry Service were laid in 1916 and this later became part of the Women's Land Army. By January 1918, 400 women were employed as forestry workers. They were not just engaged in nursery work or planting and weeding, tasks that women had previously undertaken on some estates. They were also involved in the very arduous work of tree felling and timber production (Sutherland 1918, 81). While they accomplished these tasks admirably, the clothing supplied was hardly suitable for tree felling with a crosscut saw (Figure 65). After the war, women were no longer employed in this aspect of forestry as men resumed their former roles.

During the war the chronic shortage of timber caused by the German U-boat attacks on merchant shipping was a wake-up call for the nation. When hostilities ceased the government resolved that never again would Britain find itself in such a vulnerable position. In 1919, the Forestry Commission was founded to establish a strategic reserve of timber, mainly in the upland districts of Britain where the forestry work would also provide much-needed employment. While the focus of the Forestry Commission in its early years was firmly on commercial timber production, it would later recognise the recreational potential of its forests and also play a role in the development of arboriculture in Britain.

The inter-war years and amenity tree cover

The trends in the pattern of amenity tree cover in Britain noted in the Edwardian era largely continued in the post-war years. The economic decline of the landed estate meant much less funding for their amenity landscapes and the further loss of experienced gardeners and arborists who were forced to seek alternative employment. While in the early inter-war years these unemployed but skilled workers might have moved and found work in urban parks and gardens, these prospects vanished overnight when the Great Depression descended on Britain. This severe economic situation infiltrated every corner of society, affecting people's lives between 1929 and 1939 and even beyond. In the research for this chapter it has been difficult to find any significant assessment of how this impacted on the world of amenity horticulture, much less on the development of professional arboriculture. However, there is no doubt that this difficult period witnessed a dramatic downturn in the levels of expenditure on amenity horticulture and arboriculture and consequently in the numbers of those employed on this work.

Gardens and parks in the inter-war years

Gardening style had been fundamentally affected by the war. The move from bedding-out to herbaceous borders was accelerated as people were keen to

FIGURE 65. The Women's Forestry Service was founded during the First World War due to the shortage of men in the forestry industry. They were not just engaged in nursery work or planting, but also in the particularly arduous task of tree felling.

avoid the expense of the heated greenhouses needed to raise tender bedding plants. The popularity of flowering shrubs increased greatly. According to Roger Crompton Notcutt, the famous Woodbridge nurseryman: 'Towards the end of the nineteenth century and at the beginning of the twentieth, flowering shrubs also began to gain in popularity, and since the war this has been still more marked' (Notcutt 1926, 1–2). He emphasised their labour-saving properties: 'Once shrubs have been properly planted the cost of upkeep is small compared with that of plants which must be housed in the winter in heated greenhouses. In fact, a garden of flowering shrubs goes a long way towards solving the labour difficulty'.

Immediately after the First World War there was an enormous increase in local authority house building throughout Britain and further massive expansion of the suburbs. In England, over four million new suburban homes were built between 1919 and 1939, making what had been the most urbanised country in the world at the end of the First World War the most suburbanised by the beginning of Second World War (Hollow 2011, 2). Local authorities built over one million of all suburban homes in this period. With all the new suburban gardens there was great expectation that this would usher in a new horticultural era. The provision of gardens varied enormously in both scale and quality, as did the wider landscapes on these new suburban estates. New private estates in suburbia were generally planned at a low density of twelve houses an acre, which gave room for gardens of moderate size. This housing boom in the inter-war years gave many families a garden for the first time. However, since many of these were quite small, any arboricultural element was more likely to focus on shrubs. If trees were planted, these were mainly small-growing ornamental species or fruit trees, as recommended by most of the popular gardening books and magazines of the day. Even when mature, these small trees did not present any major challenge to the home gardener in terms of pruning or eventual removal. This is underlined by the absence of advice in most of the popular literature on how to undertake any work on large mature trees or how to go about employing tree care professionals.

The renaissance of topiary in gardens of all sizes continued after the war. One connoisseur of the art that stands out is Nathaniel Lloyd, who developed his remarkable garden at Great Dixter. The garden was in the Arts and Crafts style and topiary of all shapes and sizes was a major feature. The fame of this garden undoubtedly helped to maintain the public's enthusiasm for having at least some topiary in their gardens. Lloyd's love of topiary inspired him to write *Garden Craftsmanship in Yew and Box*, a practical book based on his vast experience of establishing yew and box topiary (Lloyd 1925). One nursery that made a name for itself with topiary in the 1920s and 1930s was the Kew Topiary Nursery owned by John Klinkert (Figure 66). His advertisements and catalogues offered a huge variety of sculptured trees that would have been the envy of any ambitious clipper (*Country Life* 1922). In an interesting indication of the popularity of topiary, a London firm of advertising agents formulated a scheme by which specimens of topiary were utilised as an adjunct to advertisements for food, beverages, polishes, pocket pens and other commodities (*Gardeners' Chronicle* 1922, 106). The idea was that suitable-shaped yew trees, bearing the name of the manufacturer, their branded goods or trademark, would be planted alongside railway lines, by roads or on suitable sites in cities, to attract attention to the product being advertised. For example, to promote a brand of tinned milk they would offer the figure of a cow.

FIGURE 66. One nursery that made a name for itself with topiary in the 1920s and 1930s was the Kew Topiary Nurseries owned by John Klinkert. This advertisement from 1922 proclaims 'the finest Stock of Topiary Work in the Country'.

Public parks in the inter-war years

Public parks and open spaces also underwent significant changes in the inter-war years that impacted on their tree cover. The trend towards the provision of more active recreation facilities that was apparent in the decade before the First World War continued when the conflict ended. This not only involved the creation of new recreation grounds that were specifically designed to accommodate outdoor sports, it also impacted on the established municipal parks, where large expanses of existing landscape, often ornamental areas, were redeveloped to accommodate recreational facilities (O'Reilly 2013, 112–113). Much of this new provision was aimed at the population's growing demand for more recreation and sports facilities, fuelled partly by the increase in workers' leisure time with more holidays. The result was the emergence of a tension between competing interests in the municipal park as they struggled to redefine themselves as multi-functional open spaces. Leading figures in park administration grew alarmed at the consequences of this with William Wallace Pettigrew (1937, 4) expressing concern that ornamental horticultural features were being viewed as 'unwanted extravagances'. As with the situation before the war, establishing or maintaining areas of mature tree cover was not a priority in either these new recreation grounds or in the redesigned recreation areas of established parks. This was partly because of the cost of planting and maintenance, but also because the trees could impede the sporting and leisure activities.

Some influential landscape architects were highly critical of the standard of park design around this time and particularly the choice of trees in the limited planting that was undertaken. Thomas Mawson stated, 'In no department of municipal enterprise is there such a lamentable absence of artistic expression, or even reasonably convenient planning, as in our public parks' (Mawson 1911, 161–162). For Mawson it seemed that anyone was regarded as capable of successfully directing landscape gardening, from the Borough Engineer through to the Chairman of the Parks Committee. He believed the 'artistic gardener' with a wide knowledge of trees and shrubs should be employed for this work. However, rather than employ such a man, the work was frequently left to the practical gardener who loved rarity, variety and novelty. The outcome was 'the huddled masses of sickly, half-starved arboricultural curiosities, which so often do duty for park plantations'.

Street and roadside trees

The massive programme of street tree planting across Britain that began in the Victorian era and continued through Edwardian times resumed again in the inter-war years. This involved not only the streets in the heart of many towns and cities but also those in the rapidly expanding suburban districts that were being created to meet the huge demand for housing. Tree planting in the streets of most middle-class private developments often happened before all the houses had been built, or very soon after. Tree planting in the streets of many local authority housing estates could often take much longer or did not happen at all. As in Victorian times, the extent and type of street tree planting in residential districts often reflected the social status of the residents. In many densely developed and older urban areas, London plane and varieties of lime tree were still commonly planted, although in more suburban districts other more ornamental species were often used.

A remarkable programme of street tree planting in the inter-war years took place in the London Metropolitan Borough of Bermondsey. Led by the pioneering Labour politician Ada Salter, the achievements of her Beautification Campaign in the 1920s and 1930s have been well documented (Johnston 2017, 180–183). Further research, however, has identified an earlier contribution to street tree planting and urban greening in the borough that has received little recognition. Walter Herbert Aggett worked in private gardens and at Kew before taking the senior horticultural post at Bermondsey Council around 1895 (Anon 1927). As early as 1900 he embarked on a major programme to plant flowering cherries and other trees in many of the borough's streets. He also planted hundreds of trees as part of his restoration of three small parks that had become derelict. In his twenty-nine years at Bermondsey Aggett undoubtedly helped to prepare the ground for Salter's Beautification Campaign.

In 1925, the Roads Improvement Act was passed, which had a substantial impact on the planting of roadside trees and shrubs (Wells 1970, 295). For the first time local authorities were empowered to acquire highway marginal land for the purposes of planting and amenity. With the great programme of arterial road construction that was then underway a new chapter opened up on highway planting. Initially, local authorities with their lack of specialist knowledge found they were ill equipped to meet this new challenge and their early efforts were often unsatisfactory. However, help was soon at hand in the form of the Roads Beautifying Association (RBA). Formed in 1928 by Dr Wilfred Fox, a dermatologist with an interest in the environment, this voluntary organisation was to play a leading role in promoting and advising on roadside planting over the next three decades (*ibid.*, 296). Through a personal contact Fox was able to enlist support for his campaign for highway planting from the editor of the *Daily Express* newspaper. The editor was also a personal friend of the then Minister of Transport and a meeting with him and Fox was arranged. As a result, the Minister invited Fox to assemble a voluntary organisation of horticultural experts for the purpose of advising local authorities on the planting and preservation of roadside trees.

In the early days, expert advice was offered free of charge to local authorities by the RBA and its acceptance was encouraged by the Ministry of Transport. The organisation's first technical committee included prominent horticulturists of the day, a number of whom had aristocratic backgrounds. The RBA also gained the services of William Jackson Bean, recently retired as Curator of Kew Gardens, who became the RBA's horticultural advisor, a post he retained until shortly before his death in 1947 (Bean 1929). The types of road dealt with by the RBA in its early days were usually bypasses around towns and villages, link roads between towns and road-widening schemes, all generally urban in character. In Scotland in 1930, the Royal Scottish Forestry Society decided to follow the example of the RBA and institute a similar scheme. In a circular sent to all local authorities the Society declared:

> As far as possible, an expert arboriculturist with knowledge of the locality involved will be selected for the services offered by the [Royal Scottish Forestry] Society. The nominal fee which the Society have decided to ask local authorities to pay will be handed on in full to the arboriculturist who does the actual detailed work, and no monetary advantage will accrue to the Society through the scheme.
>
> *Aberdeen Press* 1930

In 1937, the passing of the Trunk Roads Act was a significant event in the administration of Britain's roads (Wells 1970, 299). Some 4,500 miles (7,242 km) of trunk road were transferred from local authority control to the jurisdiction of the Ministry of Transport. Dual-carriageway roads, which had previously been confined to a few urban areas, now became quite common in towns and cities, particularly as main arterial roads and around the urban fringe. The Act also signalled a new phase in the history of the RBA. In 1938, it was appointed as advisor on trunk-road planting to the Ministry of Transport. Between thirty and forty trunk-road plantings were designed by the RBA, of which twelve were carried out before the Second World War.

While the significant amount of street and roadside tree planting now taking place was welcomed by the public, some professionals warned that problems lay ahead with the maintenance of many schemes. The respected arboriculturist Denis Le Sueur stated that the people who often designed these schemes did not have the necessary expertise to consider their future maintenance requirements at the design stage. While he did not identify these individuals, it seems reasonable to assume he was referring to landscape architects and horticulturists. He believed that specialists should also be involved 'with some knowledge of road work, of entomology, plant disease, tree surgery, and manuring' (Le Sueur 1929, 417). In an editorial in 1935, the *Gardeners' Chronicle* concurred with this view that tree experts and other specialists should be engaged to design roadside planting schemes. As an example, they highlighted American practice in the State of Massachusetts where 'the man who makes the roads has nothing to do with planting them. That is the office of the Tree Warden or, in towns, of the City Forester' (*Gardeners' Chronicle* 1935, 73).

Local authorities and urban arboriculture

Since mid-Victorian times Britain's urban and suburban districts had seen a steady increase in the tree cover. This was due largely to the extensive tree planting in the new public parks, in streets and roads, and in the rapidly expanding suburban residential districts. Many of these trees were on public land owned by local authorities, which were legally responsible for their maintenance and management. Many trees that were planted in the early decades of the Victorian parks and street tree movements had now reached maturity or were sufficiently mature to require regular maintenance. With many growing in close proximity to various elements of urban infrastructure it was expected there would be potential conflicts that could only be avoided through regular maintenance. These conflicts included trees blocking out light to buildings, causing an obstruction to traffic and pedestrians or interfering with overhead or underground services. Furthermore, many street trees were large forest-type trees planted in relatively narrow streets that would always have required substantial and regular pruning, often pollarding, to keep them from outgrowing their situation. However, from an examination of the professional literature and local authority documents it seems that many authorities did not anticipate the demands that all this work would place on their finances or existing workforce.

Standards in arboricultural work

With this increasing volume of much-needed tree maintenance work, together with a lack of funding and a shortage of skilled workers, it was inevitable that there would be a public backlash about this. Of course, public criticism of local authority tree maintenance work, or the lack of it, had already been noticeable in late Victorian times. This was especially the case with the regular pollarding of street trees which many members of the public, and many professionals, regarded as unsightly (Johnston 2017, 305–311). When the work was done it often prompted adverse comments about the appearance of the trees; when it was not done this could lead to complaints about the trees 'getting out of hand'. However, the situation now was on a different scale. The vigorous debate in public and professional circles about the standards of local authority tree maintenance work began in the 1900s and continued largely unabated until the outbreak of the Second World War. While most of the comment was negative, some individuals were brave enough to take an opposite position and defend the local authorities. However, many of those on both sides of the argument agreed that much of the problem was caused by a lack of skilled arborists to undertake the work. A few examples from this protracted debate are given here.

It seems that much of the criticism relating to both individual incidents and local authority practice in general would often start in the popular press and then be picked up and commented on by the horticultural press. As might be expected, the debate also had a seasonal element, with most press coverage appearing in the spring and early summer. One of the more authoritative voices to make an early twentieth-century contribution was Ronald Munro Ferguson, a Member of Parliament and prominent Liberal politician. In 1901, Munro Ferguson published an influential paper entitled 'The Arboricultural Adornment of Towns' in the *Transactions of the Royal Scottish Arboricultural Society* that included a list of species for urban planting recommended by Professor Bayley Balfour of the Royal Botanic Garden, Edinburgh (Munro Ferguson 1901). While he unequivocally promoted the planting of many more trees in British towns and cities, he complained that local authorities were doing far too little tree planting in suburban areas, and doing this rather poorly, in comparison to private developers (*ibid.*, 389). In 1906, *The Garden* magazine republished an article from the United States that was entitled 'The Tree-Butcher' as the editor considered this entirely appropriate to Britain. It stated:

> It is generally appropriate on the approach of spring to utter a note of warning concerning the trimming of trees. More or less pruning or other care is necessary every year, either to promote health and vigour, remove dead wood, or to induce desirable changes in form or growth; but such work should never be done except by expert tree men.
>
> *The Garden* 1906, 206

The following year *The Garden* (1907, 71) published an editorial on the pruning of trees and shrubs that specifically highlighted 'The fact that many trees and shrubs are being ruined or badly crippled in their infancy, by the neglect of pruning or by the injudicious use of the pruning knife'. Careful and selective pruning in the first ten years after planting was vital to avoid problems later in the tree's life. Local authorities were not mentioned specifically and the editorial was probably responding to a neglect of this work by both the public and private sectors. Meanwhile, the regular pollarding of street trees was attracting the majority of the adverse comment. The sight of many of the nation's street trees being subjected to this regular 'mutilation' was too much for some tree experts, such as Angus Webster (1909, 400). At the same time, there were other professionals who declined to get involved in any blanket condemnation of local authorities and highlighted what they regarded as examples of good practice. In a contribution to the *Journal of the Royal Horticultural Society*, read by many professionals and influential amateurs, Mr R. Lewis Castle discusses standards of arboricultural work in a general article on urban trees. He declares, 'The excellent work performed by the London County Council in this direction merits the highest commendation, and an admirable example has been set for provincial towns to follow (Lewis Castle 1906, 84–89). In 1910, the *Gardeners' Chronicle* published an editorial on tree management that was in response to recent complaints in the popular press about the standards of planting and pruning of trees on Hampstead Heath, which was then managed by the London County Council. The editors state:

> While we disagree with the assertion made in *The Times* that 'among many gardeners and foresters there is a certain professional satisfaction in the use of the saw for its own sake', we know it to be a fact that the untrained men often employed by public bodies to carry out this kind of work are responsible for a great deal of wanton destruction. The untrained men are, as in the present case, often confounded in the public mind with the trained exponents of the art of tree culture, who are blamed for the shortcomings of the labourers.
>
> *Gardeners' Chronicle* 1910, 232

According to the *Gardeners' Chronicle* one of the causes of poor arboricultural work among local authorities was that some considered street tree pruning as little more than an opportunity for 'poor relief' and saw no problem in engaging unemployed men for this work (*Gardeners' Chronicle* 1910, 232). However, the message from the professional press that this work did not just require skilled gardeners but tree experts was mostly ignored. The *Gardeners' Chronicle* also believed there was a lack of understanding among the public of the reasons for much arboricultural work, such as pollarding:

> The whole subject of the planting and maintenance of street trees is necessarily attended with many difficulties which the layman can hardly be expected to appreciate. It is very easy to condemn – true tree lovers can hardly refrain from condemning – the annual mutilation of street trees. This method of dealing with trees is abhorrent to the instincts of the arboriculturist, and we are assured that the Parks superintendents, under whose direction the work is invariably carried out, thoroughly dislike having to do it. Unfortunately, however, there is but little choice in the matter, for the conclusion is forced upon us that trees growing in crowded thoroughfares must be kept in bounds, no matter how much they are mutilated in the process, or how much labour is spent on them each season.
>
> *Gardeners' Chronicle* 1910, 232

One very high-profile case regarding standards of arboricultural work gives an indication of the power of this topic to arouse national interest. In 1911, criticism was appearing in the popular press regarding the pruning of the recently planted plane trees along the Mall in London (*Gardeners' Chronicle* 1911, 144). The government responded by requesting Professor Balfour of the Royal Botanic Garden, Edinburgh, to produce a report about the treatment the trees had received. As this broad avenue formed the main approach to the nation's monument to Queen Victoria, public interest was enormous. In his report Balfour not only approved of the pruning and exonerated those responsible but he also described the trees as 'patterns of what street trees should be'. His report was presented to both Houses of Parliament by command of King George V and reprinted in full in some gardening magazines and professional journals (Balfour 1911, 144–146).

After enduring years of public criticism about the standard of their arboricultural work, morale among local authority arborists was understandably low. With regard to much routine tree work they were tired of being placed in the position of 'damned if I do and damned if I don't'. Angus Webster spoke for many of his arboricultural colleagues when he highlighted this problem: 'Too often, also, it is the uninitiated in the art of tree culture and management who take up the cudgels against the official placed in charge of the work; who, in many instances, is not allowed to reply through the Press to the silly and unwarranted charges levelled against his management' (Webster 1913, 207).

Throughout the 1920s and 1930s this criticism of local authority arboricultural services continued in much the same vein. Articles and letters on the topic appeared regularly in the national and local newspapers and also by some professionals writing in the horticultural press. Occasionally, someone from the much-maligned ranks of professional gardeners and arborists would respond on behalf of their colleagues, pointing out the difficulties the work presented, especially with regard to street trees. In 1926, Kew's highly respected tree expert, William Dallimore, wrote: 'Under present conditions the persons who have charge of such trees are to be congratulated on the general success of their methods, for they have a difficult problem to solve' (Dallimore 1926, 46).

The structure of local authority arboricultural services

The problem for local authority arboricultural services was not only a lack of expertise but also the often fragmented responsibility for trees within the local authority structure. While trees in parks normally came under the control of the park superintendent, this individual could be responsible for one specific park or a group of parks. Where different individuals were responsible for the daily management of different parks, standards of arboricultural work could vary greatly. That individual may have expertise in general horticultural but that was no guarantee that they had much knowledge of practical arboriculture. There were still no formal educational courses in arboriculture and the extent to which this was covered in horticultural courses varied significantly. Furthermore, there was still a belief in some quarters that a background in forestry rather than horticulture was more appropriate for someone charged with the management of urban and amenity trees. Whoever directed the work and whatever their background, much depended on the ability of those doing the practical work. Whether this was parks department staff or contractors, the level of arboricultural expertise could vary widely. Where a local authority had responsibility for trees on its housing estates, any arboricultural work might be undertaken by manual workers in its housing department, or by drafting in workers from the parks department or using outside contractors.

The situation with street trees was even more complex. Within different local authorities the overall responsibility for street trees varied between different professionals and even some individuals with no professional background. Municipal engineers were often entrusted with street trees along with many other responsibilities. By the early twentieth century, municipal engineering had become a very broad discipline embracing many aspects of local government, such as town planning, roads, parks and public housing (Buchanan 1985). Many of these areas were subsequently embraced by specialist individuals with their own professional bodies, but nevertheless still came under the overall responsibility of an individual whose job title was Borough Engineer or similar. Another group of professionals that often had managerial responsibility for street trees in the early twentieth century was the Borough Surveyors, who were qualified and experienced property professionals. Unfortunately, these officers with engineering or surveying backgrounds invariably had no specialist training or experience in the planting and management of street trees and often learnt about the topic 'on the job'. Of course, that did not preclude some of them from making a valuable contribution to the greening of the streets in their district. The practical work of planting, maintaining and removing street trees was undertaken either by local authority workers or by private contractors. In the case of the local authority staff, this was often done by its parks department workers (Pettigrew 1937, 182). With both in-house staff and contractors, the standard of work would depend entirely on the level of arboricultural expertise available.

While local authorities were obliged to undertake arboricultural work because of their legal responsibility for trees on their land, how they organised this was entirely a matter for individual authorities. As noted, on an operational level this was invariably split between different departments, such as the parks, housing, highways and other departments. At the administrative level of the Council and councillors, these committees would usually have a similar scope of responsibilities as the operational departments. However, the research for this chapter has uncovered a few local authorities at this time that regarded arboriculture as a sufficiently specialist area to warrant a committee or sub-committee to oversee this work. For example, in Worthing, West Sussex, in the late 1920s we find that Worthing Borough Council has its Trees Sub-Committee

(*West Sussex Gazette* 1927). This dealt with all matters relating to trees, including tree planting, preservation and maintenance work. In the 1930s in Cheltenham, Gloucestershire, we find that Cheltenham Corporation has a Trees Committee. The town was rightly proud of its trees, not only along its famous Promenade, but also in its many parks and open spaces. In a report in a Derby newspaper by somebody hoping their town would follow suit, it was stated:

> The [Cheltenham] Corporation is most solicitous for their [trees'] wellbeing and employs a 'tree man', skilled in pruning, as well as an arboriculturist. For many years the Corporation has had its 'Trees Committee', and the duties in which that body specialised are still conscientiously carried out by members of the Council who 'view' any tree to which drastic treatment is to be applied.
>
> *Derby Telegraph* 1935

The reference to Cheltenham Corporation employing a 'tree man, skilled in pruning, as well as an arboriculturist' promoted further research to establish whether any other local authorities at this time had specialist tree officers. While some authorities in late Victorian times had individuals who were recognised as tree experts, such as Robert Walker at Aberdeen, their posts embraced wider aspects of horticulture or parks management. In the limited published literature on the history of professional arboriculture in Britain it has been assumed that tree-specific posts in local authorities began to be created from the 1950s onwards (Bridgeman 2013, 215; Johnston 2015, 25). It can now be shown this occurred much earlier. From the early 1900s some local authorities recognised the need to have a specialist post for someone with arboricultural expertise. This tree expert could be engaged on an honorary basis, as an external consultant or as a full-time member of staff. In Alloa in Scotland in 1905 the local newspaper reported that 'Baille Calder has executed his first commission as honorary arboriculturist to the Town Council in quite a satisfactory manner' (*Alloa Advertiser* 1905). This commission involved organising and supervising the planting of a row of street trees on either side of Alexandra Drive. Also in 1905, it was reported that Dr P.W.G. Nunn, Medical Officer of Health for Bournemouth, was recommending 'that the council should appoint a skilled arboriculturist to advise and control year by year the planting of pine trees wherever possible' (*Morning Post* 1905). In this case a new post was not created and the role was filled by an existing member of staff. In 1920, a newspaper report on a commemorative tree planting by the Bishop of Bristol, organised by Bristol Corporation, refers to 'Mr Ingles, the Corporation's tree expert' who selected a tree and helped organise the planting (*Western Daily Press* 1920). Other local authorities at this time employed tree experts whose job titles were also tree-specific. In Lancashire in the 1920s, Harry Abbott was employed by Burnley Corporation as its 'head forester and tree expert' (*Burnley Express* 1928). While Abbott seems to have been highly regarded for his work with trees, he did hold some unconventional views. He believed 'that the impurities in the Burnley atmosphere have nothing to do with any difficulties there may be in the growing of trees in the town and district' (*Burnley News* 1928). He declared that the role of local air pollution in retarding the growth of trees around Burnley was 'a great fallacy' and the low rate of tree survival in the area was due to poor tree selection and planting technique.

In Essex during the 1930s, John Williams was employed as tree warden by Essex County Council with duties that related specifically to 'arboriculture and landscape work' (*Essex Chronicle* 1937). It seems likely that his job title as tree warden was influenced by reports that these posts were becoming common in the United States and the calls for such posts to be established in Britain (*Birmingham Post* 1939). Williams's duties also included surveying and inspecting trees along streets and highways. In 1938, Brentwood Council received a petition from residents in Spital Lane asking that action should be taken 'concerning dangerous trees' (*Essex Newsman* 1938). The Borough Surveyor reported the county tree warden would be making an inspection. Also in the 1930s, there is reference to another local authority officer who was employed full-time as a tree expert, this time in Yorkshire. Leeds Council employed 'Mr W.H. Evans, arboriculturist to the Leeds City Parks', whose responsibilities not only included general arboricultural matters but also the tree and shrub nursery at Roundhay Park (*Leeds Mercury* 1939). Some county councils appointed forestry officers, primarily to manage woodlands and plantations belonging to the local authority, although these individuals often advised on all tree-related matters including trees along highways.

In addition to their own staff engaged on arboricultural services, some local authorities also employed private sector assistance. Consultants might be engaged to give advice and technical support, while contractors could be used to undertake practical work such as tree planting, pruning and felling. The work of consultants and contractors is covered later in this chapter.

The emerging specialism of urban arboriculture

With the increasing focus on urban trees and the challenges of planting and managing them in the densely developed urban environment, it was not surprising that some individuals in horticulture, landscape and forestry would focus on this. Reading through the pages of the horticultural and tree-related periodicals some names appear frequently against articles on various aspects of urban arboriculture. However, one name stands out for his prolific and pioneering contribution to the literature on urban trees and arboriculture: Angus Webster.

Angus Duncan Webster (1855–1931) was born at Balmoral in Scotland where his father was head forester to Prince Albert. After gaining experience in forestry and horticulture, Webster moved south around 1880 to take up the position of forester at Penrhyn Castle, North Wales (Linnard 1985, 265). It was there and at Lord Derby's estate at Holwood Park, near Bromley, that he gained wider experience of amenity trees and gardening and began his writing career.

He was soon recognised as an able and knowledgeable all-rounder in forestry and horticulture and moved to London in 1896 to become Superintendent at Regent's Park. He remained in that position for twenty-four years until he retired in 1920 (*Gardeners' Chronicle* 1920, 14). It was during his many years based in London that Webster gained his unrivalled reputation as a specialist in the field of urban arboriculture and where he made his unique contribution to the literature on this subject.

After writing several articles on trees for professional periodicals, a book on flowering trees and shrubs (Webster 1893a) and experiencing the success of his very popular *Webster's Practical Forestry* (Webster 1893b), he embarked on his first major work on urban trees. In 1910, he produced Britain's first substantial book on this topic entitled *Town Planting* (Webster 1910). This was based on an essay for the Royal Scottish Arboricultural Society which won him their Gold Medal. *Town Planting* focuses largely on how to adapt urban arboriculture to contend with the smoke problem of large towns and cities and includes chapters on planting, pruning and the management of decaying trees. It also gives lists of the trees and shrubs suitable for urban planting. Webster had always been interested in how tree surgery could be used to prolong the life of trees and in 1916 he published *Tree Wounds and Diseases: Their Prevention and Treatment* (Webster 1916). This book, which had considerable relevance to urban trees, is discussed later. In 1918, Webster produced *Seaside Planting: For Shelter, Ornament and Profit*. This was based on his experience of creating coastal plantations in Wales and England and gives details of how to select, plant and manage trees, shrubs and other plants that would tolerate the salt-laden air around Britain's coast. His book also refers to arboricultural work in the parks and gardens of Britain's seaside resort towns that were now popular holiday destinations. Ten years after publishing *Town Planting*, Webster followed this up with an expanded study of the capital's trees entitled *London Trees* (Webster 1920). This focuses on the amenity features of the trees that would thrive in the difficult growing conditions of the capital but does not go into detail on the techniques of practical arboriculture. In all his writing Webster was keen to emphasise that trees could offer much-needed solutions to the problems of urban living, such as beautifying and softening the city landscape, purifying the atmosphere, regulating urban climate, enriching the soil and acting as potent symbols of peace, tranquillity and the natural world (Webster 1910, 47–50). Webster eventually retired from the Royal Parks in 1920, and soon after, having established excellent professional contacts in North America, he emigrated to the United States where he died in 1931. While Webster made a huge contribution to British arboriculture, especially in an urban context, he rarely used the terms 'arborist' or 'arboriculturist' and always considered himself first and foremost as a forester.

While progress was being made in advancing urban arboriculture with publicly owned trees, there was little interest in allowing local authorities to exert much influence over the privately owned trees in Britain's towns and cities.

The power of local authorities to control what happened to trees on private land was still very limited. It would be some decades before the introduction of Tree Preservation Orders (TPOs) or any rigorous system of planning permission for new developments. However, an early example of legal protection for privately owned trees similar to the TPO system has been identified. In the late 1920s and 1930s, the London County Council (LCC) had powers whereby trees on private land above a specified height and girth could be 'registered' by the LCC as suitable for preservation (Parker 1930, 253–254). The owners of registered trees must then be notified of this and, thereafter, could not cut down or wilfully destroy such trees without the consent of the LCC. If the owner felt aggrieved at a decision to refuse consent they could appeal to a Court of Summary Jurisdiction and the Court might reverse the decision and grant consent. Unfortunately, there is no information on how successful these legal powers were in preventing the destruction or serious damage to trees on private land in London around this time.

Much of the progress made in Britain in developing the specialism of urban arboriculture came from the influence of ideas and practice in the United States. A substantial amount of literature was now being produced by the American government, universities and individual professionals on the topic of urban trees and urban arboriculture. Much of this literature was being featured in the professional press in Britain, in the form of book reviews and short articles. It included the United States Department of Agriculture's series of bulletins on topics such as street tree planting and urban tree management and also other literature from American educational establishments, such as Cornell University and Syracuse University. As mentioned above, reports in the British press regarding the employment of specialist Tree Wardens and City Foresters by American local authorities were now prompting calls for similar posts in Britain.

A landmark in American literature on urban tree management came in 1911 with the publication of William Solotaroff's (1911) *Shade-Trees in Towns and Cities*. This was the first substantial book in English on urban tree management and Solotaroff set out some of the basic principles that should underlie this. The book's publication was mentioned in some British professional periodicals and Denis Le Sueur, the highly respected arboriculturist, did his best to promote what he realised was a ground-breaking work. Of particularly interest was Solotaroff's radical idea of putting a financial value on urban trees to enable officials and politicians to more than justify the expense of tree management programmes. However, the book's American perspective may have deflected many British readers from recognising its true value.

Relevant periodicals and influential books on trees

During the first half of the twentieth century many horticultural and tree-related periodicals mentioned in the previous chapter continued to feature articles and other items related to arboriculture. There were also some new titles launched that made regular contributions on the subject. At the same time some major books on tree botany and dendrology were published.

In terms of horticultural periodicals, the *Gardeners' Chronicle* and *The Garden* continued to be at the forefront in publishing material related to amenity trees and arboriculture. As in Victorian times, much of this was concerned with the trees themselves, such as their ornamental qualities and descriptions of important treed landscapes and tree collections. With both these titles there was a noticeable reduction in the early decades of the twentieth century, in comparison to the Victorian era, in the amount of material they published relating to practical arboriculture. In the 1900s and 1910s, the *Gardeners' Chronicle* published an increasing amount of material relating to forestry, which may have been an attempt to broaden its readership and reflect the debate at that time about the need for a national forestry strategy. It is interesting that the terms 'arborist', 'arboriculturist' and 'arboriculture' now appear less frequently in these magazines, even when an article is concerned with amenity trees.

In 1903, the publisher of *The Garden*, William Robinson, launched his monthly periodical *Flora and Sylva*. As well as being a giant in the world of gardening, Robinson had a life-long love of trees and implicitly understood their importance in the landscape, both urban and rural. In his introduction to the first issue he clearly states his intention to give substantial coverage to trees:

> As regards the plan of Flora and Sylva, it struck me that periodicals of this kind had always been over much devoted to flowers and plants as distinct from trees and shrubs, while every day of my life I see more and more the beauty and value of the tree. So I married Flora to Sylva – a pair not far apart in nature, only in books.
>
> Robinson 1903, 1, Introductory Letter

As well as including many of Robinson's own articles on the cultivation and uses of trees and shrubs, *Flora and Sylva* had contributions from other notable tree and shrub experts of the day, such as Augustine Henry, Maurice de Vilmorin, Ernest Wilson and William Jackson Bean. The magazine also included topical articles on current developments in horticulture, forestry, arboriculture and urban parks, including critical evaluations of government policy. Unfortunately, the substantial production costs of *Flora and Sylva* led to this sumptuous publication ceasing after only three years.

The *Journal of the Royal Horticultural Society* continued to occasionally include articles on amenity trees but, like the popular horticultural magazines, these focused mainly on the trees themselves rather than practical arboriculture. When arboriculture is mentioned this is usually in a French or Belgian context relating to fruit trees, and rarely in a British one. The term 'tree surgery' occasionally appears in the 1910s but almost exclusively in an American context, such as an article reproduced from an American magazine or the review of an American book. The *Transactions of the Royal Scottish Arboricultural Society* continued to be published, although when the Society changed its name in 1930 this was renamed as the *Scottish Forestry Journal*. Up to that point, the contents of its *Transactions* still give the impression that the Society considered 'arboriculture' to be mostly about woodlands and there was little that related to practical tree care. The Royal English Arboricultural Society began publishing a new journal in 1907 entitled the *Quarterly Journal of Forestry*. With this came new ideas and it was announced that responsibility for certain subjects would be delegated to selected members of the Publications Committee who were described as sub-editors (Fisher 1907). With regard to arboriculture these were Henry John Elwes and Augustine Henry.

Two magazines focusing on trees, arboriculture and forestry should be mentioned, although they were directed mainly at the 'engaged amateur'. In 1924, Richard St Barbe Baker founded the society of the Men of the Trees as a pioneering environmental organisation with a global focus on trees and forests. Its *Trees* magazine was first published in 1936 when it was issued six times a year, although from the 1940s this was reduced to three times a year. There is no doubt that the organisation generated considerable public interest in trees at a national and international level, and *Trees* magazine occasionally published articles of relevance to practical arboriculture. A similar publication was launched in 1932, although this was not part of any membership organisation. Lonsdale Ragg, a member of the Men of the Trees and a well-known author on natural history, was the founding editor of *The Tree Lover*, a quarterly magazine. While arboriculture was frequently mentioned, the magazine had a very broad definition of this that embraced all trees, woodlands and forests, and their management. However, there was surprisingly little about practical arboriculture or about urban trees. *The Tree Lover* ceased publication in 1945 following the death of Ragg.

When *Landscape and Garden* journal was launched in 1934 this was directed at all professionals involved in amenity horticulture and landscape architecture. It was issued under the auspices of the Institute of Landscape Architects, which had been formed in 1929, and the journal's editor was Richard Sudell (1892–1968), the well-known landscape architect and author. *Landscape and Garden* soon established an excellent reputation for its authoritative and stimulating content, which included a significant amount of material relevant to arboriculture and urban trees. It ceased publication in 1939 at the outbreak of the Second World War.

A few significant general books on trees were also published around this time. Henry John Elwes and Augustine Henry (1906) produced their classic text *The Trees of Great Britain and Ireland* that remains one of the world's great works on trees. This remarkable seven-volume study records the details of many significant specimens of over 500 species of tree then growing in England, Wales, Scotland and Ireland. The first edition of the work, which appeared between 1906 and 1913, was limited to 500 copies. While its cost placed it far beyond the reach of most professionals, it nevertheless had a huge impact on the ruling class on whose land many of the trees stood. It was followed by another classic from William Jackson Bean (1914), a member of staff at Kew, entitled *Trees and Shrubs Hardy in the British Isles*, initially published in three volumes. This remains a standard textbook after eight editions and is now in five volumes. In 1923, William Dallimore and A. Bruce Jackson published *A Handbook of Coniferae*, regarded for many years as the definitive work on conifers. As well as contributing enormously to the scientific study of trees, these monumental works also did much to stimulate public interest in Britain's ancient and remarkable trees. This undoubtedly encouraged many of those tree owners to seek professional advice regarding their care and preservation.

Overseas influences on practical arboriculture in Britain

With the dawn of the twentieth century, there were two significant overseas influences on the development of British arboriculture. One came from continental Europe and the other from the United States.

In some major French cities, especially Paris, many of the tree-lined boulevards and public open spaces now contained mature trees that required regular attention from specialist professionals and skilled manual workers. This demand for arboricultural expertise helped prompt the publication of various French works on arboriculture. Sargent's English translation of the classic work on pruning by Des Cars had been published in the United States in 1891 but was only published in Britain in 1893. It was still having an impact on British arboriculture in the early years of the twentieth century, especially as there were no comparable British texts. Another notable French contribution to arboricultural literature was a book on street trees, published in 1896 but whose impact in Britain also continued into the early twentieth century. This was Adolphe Chargueraud's classic text *Les Arbres de la Ville de Paris*, published just two years before he died. Chargueraud held the post of Professor of Arboriculture for the City of Paris and also had much influence on the management of urban trees across the city. This remarkable account of urban arboriculture went far beyond any contemporary British or American texts. Chargueraud's book was given a very favourable review in the *Gardeners' Chronicle* (1896, 400). The editors noted that it had been published under the auspices of the Prefect of the Seine, and believed that the London County Council would do a useful service if it arranged for the book to be translated into English. Unfortunately, it never was. However, further recognition in Britain for Chargueraud's book came in the 1920s and 1930s, due to the connection made between Denis Le Sueur, the eminent British arboriculturist, and the Department of Bridges and Roads for the City of Paris. Le Sueur was hugely impressed by French advances in arboriculture and did much to promote this in Britain (Johnston 2017, 75).

While the French were making great progress in practical arboriculture, especially in relation to urban trees, the Germans were also doing the same. In 1914, an excellent book by Otto Hübner entitled *Der Straßenbaum in der Stadt und auf dem Lande* (The Street Tree in the City and the Country) was published in Berlin (Hübner 1914). It is a pity that this ground-breaking text was published just before the outbreak of the First World War, ensuring that it would have almost no impact outside of the German-speaking world.

The birthplace of modern arboriculture is often thought to be the United States around the beginning of the twentieth century. That popular view in Britain fails to give recognition to slightly earlier French and other European advances around that time. Nevertheless, because of the common language and 'special relationship' that has existed between Britain and the United States since Victorian times it is understandable that advances in practical arboriculture made on the other side of the Atlantic Ocean would have considerable impact in Britain. Before exploring this it should be mentioned that the man widely regarded as the founder of the tree care industry in North America was actually an Englishman. John Davey was born in Somerset in 1846 and undertook an apprenticeship in horticulture before emigrating to the United States in 1873 (Andrews 2012). He then gained further experience of tree work before establishing the Davey Tree Expert Company, later to become one of the largest tree care companies in the world (Pfleger 1977, 7). He also wrote *The Tree Doctor*, a pioneering work on tree surgery, first published in 1901 and later as an expanded edition in 1907 (Davey 1901 and 1907). While Davey's book had a huge impact in North America, it seems to have received little attention in Britain. This is in contrast to other American books that came a few years later.

One of the first American books on arboriculture to be widely read in Britain was Liberty Hyde Bailey's *The Pruning-Book*, which went through several editions (Bailey 1898). As it was specifically about pruning it will be discussed later. German-born Bernard Fernow's book *The Care of Trees in Lawn, Street and Park* covered most aspects of arboriculture (Fernow 1910). Due to some book reviews in the professional press it was quite widely read in Britain (*Trans. RSAS* 1913). A book that seems to have had even more impact was *Practical Tree Repair* by Elbert Peets of the School of Landscape Architecture at Harvard University (Peets 1915). Its wide readership in Britain was not only due to favourable reviews but also because it was published in London as well as New York (Bartlett 1915). Peets disliked the terms 'tree surgery' and 'tree doctoring' as these 'produce in people's minds an anthropomorphic idea of a tree which is fatal to the correct understanding of a tree's nature and a tree's needs' (Peets 1915, i). However, by 1925 he seems resigned to its use when he wrote an article entitled 'Tree Surgery' for the *Gardeners' Chronicle* and asks: 'Is "tree surgery", as we term it in America, much practiced by English horticulturists and foresters?' (Peets 1925, 205). As if to answer his own question, Peets mentions only one British text, written by Angus Webster. This prompted Webster to respond with a letter to the *Gardeners' Chronicle*, stating, 'My experience is that tree repairs are but little practiced in this country, unless in the case of a few public parks and botanic gardens' (Webster 1925, 241).

During the 1920s and much of the 1930s various booklets from the US Department of Agriculture and other American institutions on subjects such as tree surgery, transplanting and pruning were mentioned in the British professional press. In the late 1930s and early 1940s there were two books that not only gained much recognition in the United States but also generated significant interest in Britain. In 1937, Millard F. Blair published his classic text *Practical Tree Surgery*, based on his own extensive knowledge and practical experience (Blair 1937). This included a short declaration entitled The Tree Surgeon's Creed, which set out Blair's personal and professional ethics. Because his book was published shortly before the outbreak of war, it was not until the late 1940s that it became more widely read in Britain. His Tree Surgeon's Creed is known to have been adopted by some British tree experts in the 1950s. In 1941, perhaps the first truly academic work on arboriculture in the twentieth century was published. This was by P.P. Pirone, the eminent plant pathologist at Rutgers University, New Jersey, and entitled *Maintenance of Shade and Ornamental Trees* (Pirone 1941). As it was published by Oxford University Press this helped to get it noticed in Britain, although its scholarship would still have ensured its wide acclaim among academics and professionals. However, because of the distractions of the war it would take British readers several years to fully appreciate this important text.

Renowned tree experts and centres of excellence

Individual gardens or establishments that are recognised for their high standards in practical arboriculture can often be instrumental in driving forward standards nationally by encouraging others to achieve similar levels. In previous eras these included establishments such as monasteries and royal gardens. Their high standards were often due to some very talented individuals who worked there and were renowned tree experts. From the 1900s onwards these 'centres of excellence' became increasingly influential though their exposure in the popular and professional media.

During the 1920s and 1930s, the Royal Botanic Gardens at Kew became recognised as a centre of excellence for arboriculture as leading figures on the staff established their reputations in this field. Perhaps its most notable tree expert was William Dallimore (1871–1959), who spent most of his career at Kew, first as a student before moving to the arboretum and rising to the position now known as Assistant Curator (Figure 67). His book *The Pruning of Trees and Shrubs* was published in 1926 and describes the methods practised at Kew (Dallimore 1926). Although William Jackson Bean (1863–1947), author of *Trees and Shrubs Hardy in the British Isles*, was regarded as another Kew tree expert, his books on practical arboriculture were directed more at the amateur. Another prominent Kew expert was Arthur Osborn (1878–1964), in charge of the arboretum in the 1930s, who wrote *Shrubs and Trees for the Garden* based on his extensive practical experience in arboriculture (Osborn 1933). While Kew was gaining much attention with its tree experts, Edinburgh Botanic Garden was not to be left out. In the first two decades of the twentieth century it had its own expert in Professor Isaac Bayley Balfour (1853–1922). Although his

FIGURE 67. William Dallimore, photographed at the request of Sir William Thiselton-Dyer, with his 'various appliances, saw, chopper, standard pruners, short pruners, wire pliers, tarpot and tarbrush, ladder, etc.' for tree surgery.
CREDIT: THE LIBRARY AND ARCHIVES, ROYAL BOTANIC GARDENS, KEW.

expertise ranged across horticulture, forestry and agriculture, he was described by Frederick Balfour (no relation) 'as the best pruning authority of his time' (Balfour 1935, 173). This was endorsed by the government in 1911 when it asked him to advise on the pruning of trees along the Mall in London, and his report was presented to Parliament.

One arboriculturist who was not a product of either Kew or Edinburgh RBG really stands out in this era. Along with Angus Webster and William Dallimore, Denis Le Sueur (1898–1969) was the outstanding British tree expert of the first half of the twentieth century. Le Sueur began his career studying agriculture at Wye College from 1908 to 1911. After serving in the Army during the First

World War and recovering from his injuries, he studied arboriculture and forestry, later obtaining the special certificate in forestry from the Institution of Chartered Surveyors (Anon 1969, 247). Le Sueur began his career as a forestry consultant in London, frequently engaged by the Corporation of the City of London where a significant portion of his work was with amenity trees (Boulton 1969, 247). In 1925 he was appointed as Lecturer in Forestry at the Royal Agricultural College, Cirencester, and held this post until 1946. His most lasting contribution to arboriculture was his book *The Care and Repair of Ornamental Trees* (Le Sueur 1934). This was the most important work on practical tree surgery to be published in Britain in the first half of the twentieth century, and, while some of the techniques he describes have now changed, it contains much that today's arboriculturist would recognise as good practice. In contrast to some of his contemporaries, Le Sueur has no problem referring to this as arboriculture and sees it as a distinct field in its own right, despite his own forestry background. He often complained about the lack of tree knowledge by the 'man in the street' and among those in government who were ultimately responsible for the nation's amenity trees and woodlands. In his view, 'The value of trees, especially in urban and suburban districts, is not sufficiently realised. Far more education in arboriculture is required, and it is the rising generation that requires that education' (Le Sueur 1934, 2).

He also thought that arboriculturists needed a far better understanding of tree science if they were to care for trees in the most effective way. He declared:

> Before a doctor or veterinary surgeon is allowed to study the advanced branches of his profession he must do a course of physiology … So it is with the arboriculturist. A tree is a living thing, and before attempting to cultivate it, to repair its wounds or improve its health, it is absolutely necessary to have some knowledge of its constitution and the way in which it feeds and grows.
>
> Le Sueur 1934, 3

Few of Le Sueur's contemporaries would have objected to the anthropomorphic reference as this was widely accepted at the time. Many arboriculturists thought that encouraging the link between tree work and 'dentistry/doctoring' would help raise the status of their trade or profession. To support his views on the importance of tree science, Le Sueur began his book with a chapter on tree biology and physiology. In a long and very distinguished career, Le Sueur also published a book on shelterbelts and wrote numerous articles on arboriculture and forestry for popular and professional periodicals.

While Kew and Edinburgh RBG were the acknowledged centres of excellence for practical arboriculture, other bodies were gaining a reputation for high standards of tree work. In the public sector, the City of London Corporation was praised for its pioneering approach to tree surgery, due largely to the influence of its consultant, Denis Le Sueur. London County Council, although sometimes criticised for its tree work in a few locations, was generally recognised as delivering a high standard of arboricultural work in most of its parks and open spaces. There is no doubt that Angus Webster had a major role in ensuring this.

Arboricultural practices

Having discussed a few of the nation's most prominent tree experts from the first half of the twentieth century, we now consider some contributions to the development of practical arboriculture made by these individuals and others. This includes a range of practices, such as tree planting, pruning, felling and cavity treatment.

Tree planting

With regard to tree planting in both commercial forestry and amenity situations, there does not seem to have been any dramatic advance from the late nineteenth century in how these trees were put in the ground. With forestry planting it was now widely accepted that planting should be in regular rows, rather than randomly, as this made weeding much easier and a more random appearance could be achieved later when thinning. According to John Simpson, author of *The New Forestry*, 'Without regular lines of some sort men get into confusion, and a very irregular plantation is the result. I saw an attempt at planting in an informal manner not long since, and a fine mess had been made of it' (Simpson 1901, 234). Occasionally, the horticultural press would feature a new tree planting technique, observed by one of their readers. For example, in the *Gardeners' Chronicle* in 1928, Victoria Slate (1928, 87) wrote about a French method for inducing the quick establishment of trees. This involved daubing the roots and branches with a mixture of clay and well-rotted manure before planting. This was done to diminish transpiration and quicken the root action of the tree. According to Slate, 'All French gardening books and papers advise readers not to forget to do this'. Most Victorian writers on ornamental trees favoured spring planting and most authorities from this era agreed with that. However, referring specifically to urban trees, Angus Webster (1909, 262) wrote, 'Autumn or early spring planting is to be recommended, the former time being in most cases preferable'.

The considerable challenges to planting trees in the urban environment in the Victorian era were even greater in the early years of the twentieth century. This not only related to air pollution and physical constraints above ground but also to the myriad difficulties facing newly planted trees below ground. Not only were there more physical constraint in terms of underground services, the soil was often very poor, highly compacted or polluted. According to Webster, as the composition of street soils could be 'positively destructive to vegetation of almost every description' with mortar, bricks, gravel, wood and other substances, pits of sufficient size to allow roots to breathe needed to be dug and filled with rich soil (Webster 1910, 14–19). Many of the techniques used at this time to overcome the difficulties of tree planting in urban areas have been described elsewhere (Johnston 2015, 78–89; 2017, 244–266).

Much of the innovation in tree planting during this era concerned the use of commercial products as aids to planting and establishment. This included various tree-ties, stakes and tree guards. Before the production of specially designed tree-ties different materials were used to tie newly planted trees to their stakes. Most commonly used were rope or wire. The likelihood of the wire cutting into the delicate bark of the young tree, especially when this was attached quite tightly, does not seem to have attracted much concern. Where rope was used some experts recommended that this should be specially prepared tar rope as this was more durable (Webster 1909, 262). In accordance with modern practice, it was often suggested that some kind of spacer should be used between the stake and the tree so that damage to the bark by chaffing did not occur. From the 1900s, it was common to see commercially produced tree ties offered for sale in the gardening magazines. One of the earliest was invented by Edward Beckett, gardener to Lord Aldenham at Elstree, which involved the use of a strap made of cloth and twine and a metal buckle. When properly affixed the strap would form a 'figure of 8' round the tree and ensure it was securely supported (*Gardeners' Chronicle* 1903a, 327). Another popular product in the 1920s was Palmer's Ever-Ready Leather Tree-Tie, with the manufacturer declaring its 'cheapness, durability and neatness the great feature, making them far superior to any known tree-tie' (*Gardening Illustrated* 1922, vii) (Figure 68).

Most newly planted amenity trees of any size required tree stakes to support them, at least for the first few years. These stakes were available from most nurseries, timber merchants and building suppliers. Those made of sweet chestnut were generally regarded as best, although other species such as larch were used. One end of the stake was pointed and all of the stake, or sometimes just the part that was driven into the ground, would be coated with creosote, which gave it a significantly longer life than tar (*Gardeners' Chronicle* 1923, 161). Commercially produced metal tree guards were often used, especially for urban tree planting in high-profile locations. From the photographic evidence it appears that, where metal types were used, wrought-iron tree guards were most popular and generally more favoured by the experts. In the words of Webster (1909, 262), 'The iron tree guard has many advantages over those of wood or expanded metal and being made in two sections it can be readily placed in position after the tree has been planted'. These guards, however, were very expensive and this could add considerably to the cost of large-scale planting schemes.

Transplanting large trees

While the Victorian 'golden age' of large tree moving was now over, there was still plenty of activity in this specialist aspect of practical arboriculture. The coverage of this topic in the professional press had declined significantly in the

TREWHELLA BROS.

MONKEY JACKS,

as used throughout Australasia, are specially designed for **saving labour** in uprooting trees and removing stumps, roots, stones, etc., as well as for general heavy lifting purposes on an estate or farm.

Call and inspect sample (and obtain full particulars) at our Works, or at the Offices of Messrs. NEWLAND, TARLTON & Co., Ltd., 166, Piccadilly, London, W.

Inquiries through your Ironmonger, or direct to

TREWHELLA BROS., 6, Alma St., Soho, Smethwick, BIRMINGHAM, will receive prompt and careful attention.

Trewhella Bros.

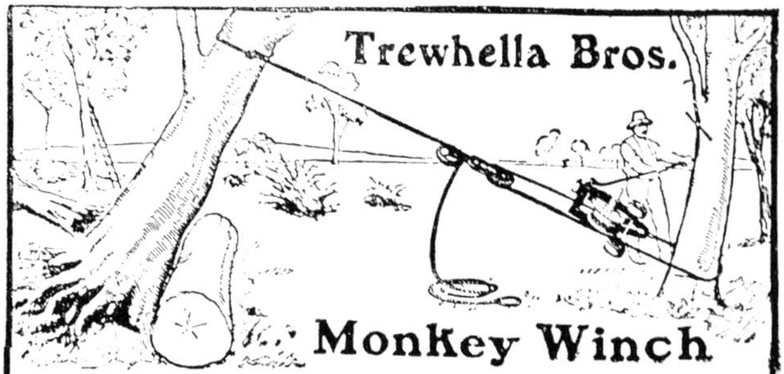

Monkey Winch

For uprooting trees and stumps there is nothing to equal this machine. It will pull down a sturdy tree in a few minutes.

May we send you a catalogue?

TREWHELLA BROS. Pty. Ltd.,
12, Island Road, Handsworth, BIRMINGHAM.

WADE PORTABLE CROSS-CUTTING SAW and CIRCULAR SAW ATTACHMENT

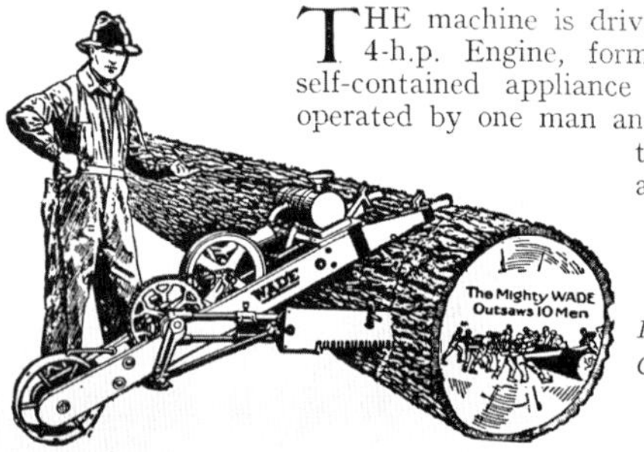

THE machine is driven by its own 4-h.p. Engine, forms a complete self-contained appliance that can be operated by one man and moved from tree to tree like a wheelbarrow.

SEND FOR ILLUSTRATED CATALOGUE

CHRISTY & PENNY, Ltd., 219, QUEEN'S RD., BAYSWATER LONDON

STIHL CHAIN SAWS

for TREE-FELLING and CROSS CUTTING LOGS

Suitable for CARPENTERS, BUILDERS, MINEWORKS, etc.

CUTTING TIME from 25 SECONDS

CLEAN STRAIGHT CUT

PETROL DRIVEN or ELECTRIC MOTOR TYPE for a.c. or d.c.

CONSIDERABLE SAVING

Ask for particulars and Leaflet B.3

KIRCHNER & Co. (London) Ltd
117-119 City Road, London, E.C.1

Telephone : Clerkenwell 6284

PALMER'S

EVER-READY LEATHER

TREE TIE

Cheapness, Durability & Neatness the great feature, making them far superior to any known Tie.

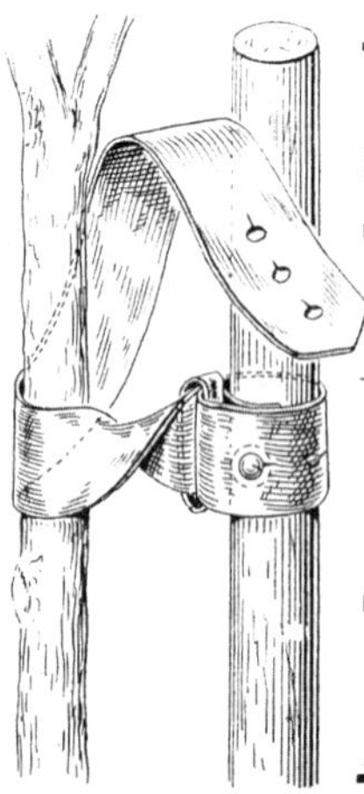

CATALOGUES apply to —

E.W.STARMER & Co., Ld., ETHEL STREET, NORTHAMPTON,

or nearest Nurseryman.

STOCK SIZES ranging from 24 to 6 inches.

PRICES

4/9 to 9d. per dozen.

AGENTS WANTED IN EVERY TOWN & COUNTY.

Tree surgery

AT pruning time a treatment with XL ALL Styptic is essential for the good health of the tree. It is useful, too, for bleeding on vines and so forth.

XL ALL STYPTIC

Of all nurserymen, seedsmen, and florists, or write for particulars to—

G. H. RICHARDS LTD.

234, Borough High Street, London, S.E.1

The SPEAR-FAST Saw

'TAKES THE IRK OUT OF WORK'

INVALUABLE FOR USE ABOUT THE FARM AND ESTATE

Suitable for cutting Firelogs, Garden and Fencing Posts, etc. The narrow blade of alloy steel is specially hardened and tempered to be suitable for these duties, and is fully tapered from the cutting edge to the back

The special teeth are designed to cut and to clear the sawdust in each direction of motion, and can be sharpened very easily without requiring great skill. They retain their keen sharpened edge for a long time much longer than most saws.

The lightness of the whole tool makes it exceptionally easy to work with and the tubular handle provides a good grip which does not tire.

The "Spear-Fast" Saw is made in three sizes, 24 in., 30 in., and 36 in., and the blades are easily renewed.

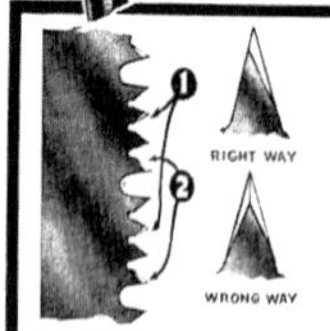

EASILY SHARPENED

The "Spear-Fast" Saw is easily sharpened. It will be seen that the teeth run in groups of four, and that in each direction of movement, the second two teeth in each group do the cutting, the clearing being done by the first two teeth in the following group.
Start sharpening all the teeth marked 1 in the illustration throughout the whole length of the blade, sharpening only one edge of each tooth to a thin knife edge (the right way is shown above).
When you have reached the end of the blade go back and sharpen all the teeth marked 2 in the illustration.
Then turn the saw over and repeat both operations.
Use a F142 Cross-Cut File 6" for Sharpening

SIZE	COMPLETE	EXTRA BLADES
24"	6/6 each	2/6 each
30"	8/6 „	2/9 „
36"	10/- „	3/3 „

OBTAINABLE FROM IRONMONGERS AND TOOL DEALERS

SPEAR & JACKSON LIMITED
ÆTNA WORKS • SHEFFIELD

FIGURE 68. Product advertisements and dates: Trewhella Monkey Jack (1910), Trewhella Monkey Winch (1923), Wade Portable Dragsaw (1925), Stihl Chain Saws (1938), Palmer's Tree Tie (1922), XL ALL Styptic (1934), Spear-Fast Saw (1938).

1890s but started to increase again in the 1900s. Once again, many professionals wanted to report their experiences of this and recommend what they regarded as the best techniques. In truth, there was not much that was new in all this, the topic having been thoroughly covered in the Victorian era. It was now generally agreed that severely pruning the branches to compensate for root loss in the removal process was a 'barbarous expedient … and perhaps the practice is not yet quite extinct' (*Gardeners' Chronicle* 1903b, 85). There was now much emphasis on having a large rootball, sufficiently prepared in advance, if at all possible. However, some professionals were still happy to move even large trees immediately.

John Nisbet's major work on forestry entitled *The Forester*, published in 1905, includes a section on transplanting large trees (Nisbet 1905, 1, 482–491). Of the two machines mentioned, he gives a detailed description and diagram of Mackay's tree transplanter (see previous chapter) but calls this the Kingston Machine (*ibid.*, 486–488). The other transplanter is called the Parisian Machine (*ibid.*, 489). The description and diagrams of this are taken from William Robinson's book *The Parks, Promenades and Gardens of Paris* (Robinson 1869, 551–556). This machine was now at least thirty-five years old but Nisbet still includes it as one of his two examples, rather than a more recent British machine.

The Royal Botanic Gardens at Kew attracted much attention in the first few decades of the twentieth century for its exploits in moving large trees. Some of Kew's tree experts gave accounts of this in the pages of the horticultural press. In 1903, William Dallimore wrote a two-part article for *The Garden* about Kew's success in moving large trees and shrubs. The first part focused mainly on the technique of using wooden planks, rollers and pulleys, while the second described their experience with one of William Barron's large transplanting machines (Dallimore 1903). When the trees were loaded onto the machine, a team of horses was used to take them to the required part of the garden (Figure 69). In 1905, William Jackson Bean of Kew gave another two-part account of their tree-moving exploits, this time in the *Gardeners' Chronicle* (Bean 1905). In the article he describes this work, commonly involving trees of five tons with rootball, as the most demanding of all arboricultural operations and something that 'lazy and incompetent workmen should never be allowed to take up, unsupervised' (*ibid.*, 353). Much of the article repeats what was recognised as good practice, including the best time for the removal being the winter months, or the middle of October to the middle of March (*ibid.*, 373). Not be outdone, in 1914 Arthur Osborn of Kew gave another account of their tree-moving work, also in *The Garden* (Osborn 1914). Of those tree experts not at Kew, Denis Le Sueur has a short section on tree moving in his classic book, which is based on the work of William Barron and Son. Also, writing in the *Journal of the Royal Horticultural Society* in 1963, he gives an interesting account of transplanting two large oaks in 1935 at Park Royal, London (Le Sueur 1963). The trees were not a balanced pair as one had been damaged by a bomb blast

FIGURE 69. One of William Barron's large transplanting machines being used at the Royal Botanic Gardens, Kew, around 1905. When the trees were loaded onto the machine, a team of horses was used to take them to the required part of the garden.

CREDIT: THE LIBRARY AND ARCHIVES, ROYAL BOTANIC GARDENS, KEW.

in 1917 with consequent retarded growth. The trees were moved by Barron and Son using one of their larger machines. Owing to imminent building operations the work had to be undertaken immediately and the trees were lifted out of the ground without any root preparation and replanted in another part of the site. Writing nearly thirty years later, Le Sueur was able to report that the trees were still very healthy.

Occasionally, a new transplanting machine would be invented, such as the one made by Faulkner and Sons, engineers at Hersham, Surrey, and constructed for use by the Royal Parks (*Gardeners' Chronicle* 1904b, 276–277). Faulkner's machine was used in the prestigious project to move numerous trees in St James's Park that were blocking the view of Queen Victoria's new memorial that was constructed outside Buckingham Palace when the Mall was redesigned.

There were still some landscape contractors specialising in moving large trees and the most famous was still William Barron and Son. After the great man died in 1891 his son, John, took over the running of the company, which included its nursery and landscaping business, and the sale and hiring out of transplanting machines, as well as tree-moving projects in Britain and overseas. In 1907, the company was engaged as aconsultants by the Old Keuchenberg Botanical Gardens in Frankfurt-on-Main, Germany, to supervise the transplanting of a

FIGURE 70. An advertisement from 1935 for William Barron and Son promoting their tree-moving services and appealing to landowners not to destroy trees but to consult them first. It names Mr E.H. Howard as the company's tree-moving expert and consultant.

massive yew tree some 300 years old (*Gardeners' Chronicle* 1907, 222). During the 1920s and 1930s the company was advertising its services regularly, often in rural lifestyle magazines such as *Country Life*. Around 1930, when the third generation of the Barrons retired from the business, the family was no longer represented in the firm and it was taken over by Mr E.H. Howard (Gossiper 1931). Howard had served his apprenticeship more than thirty years before with John Barron and had remained with the company ever since. In subsequent advertisements, Howard was named as the company's tree-moving expert and consultant (Figure 70) (*Landscape and Garden* 1935). In 1932, Barrons were involved in another high-profile project that made the pages of the national and professional press. They were asked to move a huge plane tree on a Kensington estate to a position 20 ft (6 m) away so that it would cut out the view of a block of flats from the new house that was being built (*Derby Telegraph* 1932). The tree and its rootball weighed some 40 tons and the whole 'Herculean task' involved four specially trained men, supervised by Howard.

Another company with a reputation for transplanting large trees was Cheals of Crawley in Sussex. This was also a firm with a nursery business which then diversified into landscape work, leading to it developing its expertise in tree moving. It appears that Cheals preferred to use the method of a specially prepared trench together with planks, rollers and pulleys to move their trees, although this limited the distance they could be transported (Benton 2002, 53). In one of their largest and lengthiest projects, at Hever Castle in Kent from 1904 to 1908, a large number of Scots pine were transported from Ashdown Forest, 12 miles away (*ibid.*, 155). It is recorded that each tree required a team of four horses and ten men to move it, which implies that a transplanting machine was used. Since there is no evidence that Cheals designed and constructed their own machine, this may have been purchased or hired from another company. Some of Cheals's more spectacular tree-moving projects attracted the attention of the popular and professional press. Such a project involved the removal of a large cedar (*Cedrus deodara*) at Paddockhurst, Sussex (*Gardeners' Chronicle* 1909, 398). This tree stood in the forecourt of the mansion, but additions to the building necessitated moving the tree a distance of some 20 yards (18 m). The tree and its rootball weighed up to 75 tons and at that time was perhaps the largest tree to be moved in Britain.

Tree pruning

As in the Victorian era, the subject of tree pruning continued to feature prominently in the professional press and occupy the minds of the most able tree experts. However, the pruning debate had moved on from focusing mostly on improving timber quality to concentrate more on amenity trees.

Two works from overseas authors continued to provide excellent practical and authoritative advice on this topic. The first was Sargent's English translation of *Tree Pruning* by Des Cars (1881), published in the United States in 1881 and later published in Britain in 1893. The second work was by Liberty Hyde Bailey, horticulturist and co-founder of the American Society of Horticultural Science. Bailey's *The Pruning-Book*, later known as *The Pruning Manual*, was first published in 1898 and went through numerous editions (Bailey 1898). However, it was not long before some of Britain's more prominent tree experts made their own contribution to the literature on pruning. One of the first in the twentieth century was William Jackson Bean who published an article on this topic in 1906 in the *Gardeners' Chronicle* (Bean 1906). Bean was quick to point out that successful pruning had to be based on a sound knowledge of the characteristics of each individual tree. He declared:

> The aim in pruning a large-growing tree is to make it as perfect a specimen of its kind as is possible, and, at the same time, to preserve its peculiar characteristics. Therefore, a knowledge of its size and habit is essential. Unless the operator possesses such knowledge the plants are best left alone, for bad pruning is worse than none.
>
> Bean 1906, 226

Bean was also keen to put behind him the often acrimonious debate of the previous century that had focused on pruning for timber quality. He was aware that some foresters and even some arboriculturists were still not clear that pruning for timber and pruning for amenity were separate issues. To avoid any confusion, he insisted:

> There is, moreover, a mistaken idea prevalent that pruning such as I am about to discuss is recommended for trees grown for timber. This, however, is not the case. Forestry pure and simple is a thing quite apart from ornamental arboriculture in

> park and garden. The forester aims solely at building up a trunk which will yield the maximum amount of useful timber and occupy the least possible space. The main object of the arboriculturist, however, is to so control the growth of his trees as to produce individually beautiful specimens.
>
> Bean 1906, 226

In contrast to Bean's books on pruning that came later in his career and were directed mainly at the amateur market, this *Gardeners' Chronicle* article reached many professionals engaged in horticulture and arboriculture. While it did not break any new ground on pruning amenity trees, it seems to have been very effective in making clear that this was an arboricultural topic and not to be confused with forestry. One recent innovation in tree pruning from the Continent that Bean did not endorse was the use of Stockholm tar instead of coal tar as a wound treatment. Stockholm tar is a sticky material produced by the high-temperature carbonisation of pine wood in anoxic conditions and has a long history as a wood preservative. In 1900, it was reported that the French were favouring Stockholm tar for wound treatment because it was now thought that coal tar contained substances harmful to vegetable tissue (*The Garden* 1900, 314). However, most British tree experts continued to favour coal tar.

In 1916, Angus Webster published his ground-breaking book *Tree Wounds and Diseases* that contained a chapter on pruning (Webster 1916). Unlike many of his colleagues, he did not have an aversion to pollarding when undertaken in the right circumstances. He describes the case of some elms in Kensington Gardens and Regents Park, where 'a decided change for the better has been brought about by the careful pollarding of all dangerous and unhealthy trees, though in both cases there was at the time a public outcry at the dastardly treatment to which the trees were being subjected. Time, however, has proved that the Park authorities were justified in their action' (*ibid.*, 64–65). With regard to the mechanics of branch removal, Webster reiterates the accepted practice at that time of flush cutting, and, while he recommends using an undercut to avoid tearing the bark, he does not describe the three-cut method (*ibid.*, 69–70). After branch removal, 'The best covering for tree wounds is undoubtedly coal-tar or zinc paint thinly applied, under which a callus readily forms' (*ibid.*, 19–20). He also describes another composition 'that is in frequent use, particularly on the Continent and in America' which consists of a mixture of tar, tallow and beeswax. He also states that 'grafting wax is valuable for covering tree wounds, but is somewhat expensive'.

In 1926, William Dallimore published his book entitled *The Pruning of Trees and Shrubs* that described the methods practised in the Royal Botanic Gardens, Kew (Dallimore 1926). This pioneering text has the first detailed British account of working in a tree, the tools and equipment required, and a description of various practical tree work methods, including how to remove large branches with ropes. What stands out at the start of Dallimore's book is his belief that pruning amenity trees is a necessary evil. He states:

> Pruning, like a surgical operation, is a necessary evil, for, although its importance as an aid to the proper development of certain forms of plant life cannot be overestimated, and without it some plants would not continue satisfactorily to flower and fruit, those plants would be able more effectively to resist certain diseases were pruning not required. In the same way, a surgical operation often aids in transforming a weak and deformed boy into a strong and straight-grown man, but the man who has developed normally from boyhood has had an advantage.
>
> Dallimore 1926, 9

While this is a fascinating anthropomorphic perspective on pruning, it does not disguise Dallimore's belief that it was not always necessary to routinely prune trees. In many cases he thought that pruning was undertaken needlessly when it was better to leave the tree alone. However, Sir William Thiselton-Dyer, the Director of Kew until 1905, tended to gauge the amount of productive tree work being done by his staff by the amount of branches on the ground around the tree. To give the impression that more pruning was being undertaken than was actually the case, when Thiselton-Dyer was in the vicinity Dallimore would sometimes scatter branches from previously pruned trees under the tree he was working on (Kirkham pers. comm., 2019). In all Dallimore's published work on pruning he endorses the practice of flush cutting. For example, he states:

> When a branch is removed from a tree, it must be taken off parallel with the trunk, and on no account must a snag, even though it be but an inch or two long, be left, for when a cut is made parallel with the trunk the wound usually heals well, but when even a small snag of the old branch is left the wound rarely heals.
>
> Dallimore 1926, 15

This passage in his book is accompanied by a photograph showing an example of severe flush cutting that would be regarded today as very poor practice. However, according to Tony Kirkham (pers. comm., 2019), Kew's current tree expert, in Dallimore's private journals still held at Kew he does question the practice of flush cutting. The fact that he did not express these doubts publicly was probably because this was still the accepted practice and what was expected at Kew. In terms of the mechanics of branch removal, Dallimore (1926, 30) gives a rather long-winded description of what is essentially the three-cut method. Like his contemporaries, he also favoured coal tar as a wound dressing. With regard to the tools and equipment used in tree pruning, Dallimore gives a very detailed list. He also talks about the make-up of the workforce, insisting that:

> Two men should work together when pruning trees of even moderate size. One man must do the actual work, the other guiding him and supplying the necessary tools. It is difficult when standing or sitting in the centre of a tree or when standing on a ladder to decide which branches should be removed and which left, whilst it is impossible to distinguish the leading shoot. On that account the man on the ground must take the responsibility of directing the work. Moreover, the climber may not be able to carry all the tools he wants as he ascends the tree and his mate must pass them to him by means of the weighted line. However, both men must be able to undertake either part of the work in order that they may relieve each other. … The

> point to commence shaping a tree is the top, but as the man ascends he should cut away dead branches and any inside wood that is not wanted. If such branches are required for climbing purposes, stumps a foot or so long should be left to remove as the man descends.
>
> Dallimore 1926, 31

It is clear from Dallimore's text that he has plenty of practical experience in undertaking tree work. This is in contrast to the published work of many of his contemporaries where the impression is often given that they had never actually worked at height in a tree. When Dallimore discusses the steps necessary to remove a large branch with ropes, his description is significantly more advanced and more detailed than anything previously published in Britain on practical tree surgery (Dallimore 1926, 36–38). It is not appropriate to reproduce that lengthy text here; however, modern arborists would undoubtedly recognise this as sound advice from a man with extensive personal experience of practical tree care.

After Dallimore's excellent work, the other pioneering book on tree care published in Britain in the first half of the twentieth century was Denis Le Sueur's *The Care and Repair of Ornamental Trees* (Le Sueur 1934). This broke new ground in arboriculture and tree surgery in many respects, not least in the technique of branch removal and in safety equipment for tree climbing. First, we consider his contribution to tree pruning. After almost every British text since Thomas Lawson in 1618 had extolled the virtues of flush cutting when removing branches, Le Sueur finally exposes this as poor practice:

> In cases of complete [branch] removal it is generally considered that the final cut should be made parallel and as close to the main axis of the trunk or larger branch as possible. There is, however, another method chiefly practiced abroad, in which the branch is taken off at that point nearest to the trunk which is more or less at right angles to the axis of the branch itself. This line will follow roughly the slightly raised area of bark found at the base of most branches, and referred to as the 'branch ring'. It is claimed that such wounds heal far quicker than those made in the generally accepted manner; and although the angle of the cut leaves a slight projection, it is gradually absorbed into the tree as girth increases.
>
> Le Sueur 1934, 18

The alternative to flush cutting that Le Sueur recommends, namely, making the final cut just outside the 'branch ring', makes sense immediately to those with practical experience of pruning who have also studied branch form and structure. However, this vital piece of pruning advice from Le Sueur in the 1930s seems to have been ignored or forgotten by the arboricultural world for over forty years until the existence of the 'branch collar' was presented as new information following American research in the 1970s (Shigo and Marx 1976).

Like Dallimore, in Le Sueur's writing on tree work the modern arborist can recognise immediately that he has first-hand experience of the work he is discussing. This comes out particularly clearly when he describes branch removal with ropes, which may also be the first time a British publication uses the term 'slinging' in this context (Le Sueur 1934, 20). Much of Le Sueur's book is concerned with materials for wound protection and he goes into detail about the different substances in current use (*ibid.*, 35–47). While he acknowledges the popularity of coal tar, he cannot give it his complete approval:

> [Coal tar] is perhaps the most universally used material for covering tree wounds. It is recommended by practically every writer on the subject, possibly because it has a certain number of advantages, the chief being that it is more or less waterproof, very adherent, is comparatively cheap, and is easy to obtain. It has, however, two disadvantages, one of which disqualifies it as a first-class dressing. It is definitely inelastic, and its insecticidal and sterilising properties are apt to be very overrated.
>
> Le Sueur 1934, 41

While finding various disadvantages in the materials commonly used as wound treatments, Le Sueur was always prepared to evaluate new products. He mentions a liquid dressing manufactured by the Powell Duffryn Steam Coal Company that had recently come on the market (Le Sueur 1934, 42). Known as 'Synthaprufe', this was a combination of tar with rubber. In Le Sueur's view it was too soon to evaluate this properly but early indications were encouraging. Another new product at this time that Le Sueur does not mention was XL ALL Styptic, part of a range of chemical treatments produced by G.H. Richards, a British company based in London. This was specifically developed for use on pruning wounds and was said to promote good health and stop or reduce bleeding (Figure 68). In his discussion of wound treatments Le Sueur readily admits that more research is needed, stating, 'Amongst the arboricultural problems that yet remain to be solved, the satisfactory protection of wounds by artificial means is one of the most important' (*ibid.*, 35).

Tree climbing and aerial access

So far in this history of professional arboriculture, very little evidence has been found of tree climbing as a means of aerial access. While ropes and other climbing aids were almost certainly used from early times, scant indication of this has been found in the literature, published or otherwise, or in visual images. Some references were found from the nineteenth century relating to climbing spikes or spurs, often used in conjunction with ropes, but there has been little else. By contrast, there have been numerous references down the centuries to ladders and platforms of various types. As well as using ladders in the conventional manner, there is evidence from the early twentieth century that they were sometimes used horizontally as scaffold in the tree, tied securely with rope (Blair 1937, 266). In this manner work could be undertaken on large and spreading trees such as cedars and oaks.

As with many aspects of practical arboriculture in Britain over the past 200 years, the introduction of tree-climbing techniques and equipment came largely from France and the United States. In Chargueraud's (1896, 102–105) classic work *Les Arbres de la Ville de Paris* there is a description of tree climbing using

FIGURE 71. French tree pruner photographed *c.* 1895, with approved safety belt and climbing rope. The belt is equipped on the right and left with a ring and carabiner allowing the attachment and easy extension of the ropes if necessary.

ropes and a safety belt. The book has a photograph of a tree pruner wearing the belt and shows the arrangement of the ropes attached to it (Figure 71). Also included is a drawing showing men working in a tree using this safety belt, with the rope looped over a branch near the top of the tree that acts as an anchor point. While this method of aerial working was being practised in large French cities such as Paris, it would be some years before it was introduced into Britain. Although William Dallimore (1926, 38) gives some detail on the use of ropes in tree climbing, there is no mention of any safety belt or harness.

The use of a safety belt seems to have been adopted rather later in the United States than in France. One of the earliest American aids to working in a tree was the 'bosun's chair', which has a centuries-old heritage among sailors and steeplejacks (Blair 1995, 63–64). An American book from 1896 has a photograph showing tree climbers in a large elm using ropes and one appears to be using a bosun's chair (Forbush and Fernald 1896, 247). This was made from a short plank of wood that acts as the seat and was suspended from the tree with ropes attached to the ends of the plank. It was one of the earliest forerunners to the modern seat-type safety harness and was commonly used for lengthy and tedious tasks such as cavity work. In 1917, Lem Stout was the first man at the American firm of Bartlett Tree Experts to use a climbing line, a full ten years after the company had been established (*ibid.*, 62). Another development around this time was the practice of fashioning the suspension harness from the climbing rope itself. Various knots were used but the most common was the bowline on a bight. Although it was lightweight, strong and inexpensive, any harness fashioned from the climbing line not only restricts the climber in recrotching options but can also be uncomfortable over long periods. It is claimed that the first safety harness for tree work produced commercially in the United States was the Nelson saddle, developed by Karl Kuemmerling and the Nelson Tree Company around 1932 (*ibid.*, 63). This was also known as the three-piece saddle or rope strung saddle. The Bartlett Manufacturing Company (1933, 8) of Detroit included a safety saddle in their 1933 arborist equipment catalogue and the illustration of this indicates it was probably the Kuemmerling designed saddle. It was made from heavy reinforced leather, with leg supports and two 'D' rings for attaching tools. While there was some contact between arborists in Britain and the United States at this time, no evidence could be found of these American safety belts or harnesses being used by British arborists.

The first confirmed example of a safety belt and rope being used in tree work in Britain comes from Denis Le Sueur. He had always been keen on promoting safety equipment and during a visit to France in the 1920s he was shown the City of Paris safety belt that was used by tree workers engaged in pruning operations on the Parisian boulevards. On his return he managed to obtain a number of these belts and in 1929 supplied them to local authority arborists at the City of London who then became the first in Britain to use them (Figure 72). Le Sueur (1949, 52) later recalled that the first man to be given the belt then fell out of a tree the same week and crippled himself for life. However, this was not the fault of the belt as this had been left on the ground! While Le Sueur (1934, 30) recommended the use of the City of Paris safety belt for all tree workers, there is no record of any public body or commercial company following his advice. Although using this safety belt was better than 'free climbing', it had its limitations. If an arborist fell out of a tree, the jerk at the end of a fall might result in internal injury. If provided with thigh loops, which converted the belt into a saddle, the risk of this was greatly reduced. During the 1930s, despite safety saddles being offered for sale by some American

FIGURE 72. The first confirmed use in Britain of a safety belt and rope for tree work was by arborists at the City of London Corporation in the late 1920s. This was the City of Paris safety belt that had been imported from France by Denis Le Sueur, the Corporation's arboricultural consultant.

arborist equipment suppliers, they were still difficult to obtain in Britain (Le Sueur 1946, 88). However, a satisfactory alternative was the rope 'safety sling', then standard safety apparatus for the US National Parks Department (Figure 73). In 1936, Le Sueur introduced the more advanced American safety sling into Britain, again for the use of arborists working on trees in the City of London (Le Sueur 1946, 88–90; 1949, 52–54). The safety sling was made up from a 100 ft (30 m) length of ½ in. (13 mm) manila rope with knots tied at one end to create a rope saddle going round the body and thighs of the climber. The other end of the rope was then passed over a strong branch located high in the tree and brought back to the climber and tied to the part of the rope that was ascending the tree with a locking hitch or prusik knot. A large loop or sling was thus made. The locking hitch jams under sudden strain, preventing the climber from

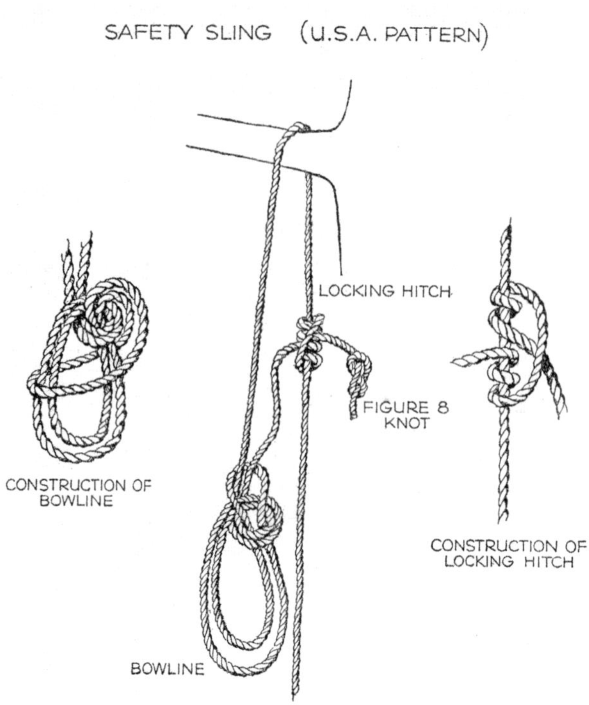

FIGURE 73. During the 1930s an alternative to safety saddles and belts was the rope 'safety sling', an American innovation introduced into Britain by Denis Le Sueur in 1936. This was made up from a 100 ft (30 m) length of rope with knots tied at one end to create a rope saddle going round the body and thighs of the climber.

falling. If the climber wishes to change their working position in the tree, they slacken the rope and slide the hitch up and down, thus making the sling smaller or larger. For those British arborists trained in the 1960s and 1970s it will be apparent that this is the same basic climbing system in terms of the use of the rope and knots that was then used in conjunction with a safety harness.

Although Le Sueur introduced the American safety sling into Britain in 1936, no record of its use by arborists outside of London could be found until after the Second World War. It does seem surprising that both the City of Paris safety belt and the American safety sling were not readily taken up in the 1930s by British arborists keen to make their working practices much safer. Le Sueur was always highly critical of tree workers who refused to use a rope and safety belt, preferring instead to 'free climb'. He states:

> There seems to be a definite streak of obstinacy in such people, who often work at a great height from which a fall may mean not only disablement but possibly a slow and ghastly death by impalement on park or other railings. Such belts need not be used at all times, but in certain cases, especially when working in dangerous trees, over railings, glasshouses, etc., their use should be encouraged. No workman should have an idea that the use of such a belt is a confession of cowardice. The use of a safety-belt is a question of common sense.
>
> Le Sueur 1934, 30

British arborists in the 1930s still preferred to work from ladders rather than use what was increasingly standard safety equipment for many arborists in France and the United States. When ladders could not get the British arborist sufficiently high in the tree, they would generally use a pole saw or simply free climb the rest of the way, sometimes using a rope to assist if the gap between branches was too great. They would then secure themselves into their working position by placing a rope around their waist and then tying this to the nearest convenient branch or the main stem.

Tree felling and stump removal

Tree felling is one of the more common tasks that the arborist is asked to undertake. Because of the risks of causing damage or injury if the operation goes wrong, it is generally something that the wise amateur leaves to the professional. While in the early years of the twentieth century it was common for home owners to undertake their own gardening work, books on gardening for the amateur mostly omit any mention of tree felling. The topic is also strangely absent from most books on arboriculture and tree surgery that are directed at the professional. In the research for this chapter a search was made of numerous books and periodicals to find some reference to the mechanics of tree felling. Most books and articles either omitted the topic or gave little coverage, excepting in terms of forest management. With regard to the practical steps that need to be taken to fell a tree in a safe manner, very little information could be found. While this was also the case in previous eras, it was expected that with the dawn of the twentieth century, with much more literature on the practical aspects of arboriculture and forestry, this vital topic would be prominent. However, in terms of arboriculture, it is almost as if tree felling was not considered to be an aspect of this, which is clearly not the case. In some respects felling could be considered the 'ultimate tree surgery' or the 'final pruning', when further treatment of the tree is pointless or it has to come down for some other reason.

While there was a dearth of information on the techniques of tree felling, there was nevertheless a fascination with innovative equipment for doing this. The use of steam-driven tree-felling machines was highlighted in the previous chapter and these continued to attract interest in the early years of the twentieth century. As late as 1916, the *Transactions of the Royal Scottish Arboricultural Society* published an article describing a new steam tree-feller (Scrimgeour 1916). This was called the Ransome Patent Steam Tree-Feller and it appears to be an updated version of the machine that was demonstrated at Tulse Hill, London, in 1878 in the presence of William Gladstone. Its development had been prompted by the significant increase in timber production over the past two years and the need to provide the 'most up-to-date labour saving machinery' during the wartime labour shortage (*ibid.*, 126). The article reported that the machine had been 'most successfully used in Scotland during the last six months'. However, as there was no further mention of this after the First World War, it is assumed it was superseded by more efficient tree-felling machines, which will be discussed shortly.

A particularly imaginative method of felling trees that is mentioned in a few articles in the 1900s is the use of electricity. In one instance, successful experiments with this were said to have been conducted in France and reported in the French periodical *Le Jardin* (*Gardeners' Chronicle* 1904c, 10). Experiments were conducted in various forests using a platinum wire heated to a white heat by an electric current that then acted as a saw to cut through the timber. It was reported that 'the tree is severed more easily and rapidly than by older methods; no sawdust is made, and the slight charring produced by the burning wire preserves the wood' (*ibid.*, 11). It was also stated that this method of felling trees with electricity was eight times as quick as using a crosscut saw. Shortly after this article there were no more reports of experiments in tree felling with electricity, which suggests that the success claimed for this may have been overstated.

Despite these efforts to develop new tree-felling equipment, using the felling axe and the crosscut saw remained the most common way of felling a tree in both arboriculture and forestry. As the twentieth century unfolded, the crosscut saw slowly replaced the axe as the tree worker's favoured tool for felling, although both were often used on the same tree. In 1949, Herbert Edlin noted, 'The custom of using the saw to aid the axe in tree felling is relatively recent, for as late as the 1920s tree fellers in Northumberland and the Welsh borders were bringing down tall trees with the axe alone' (Edlin 1949, 6).

By contrast to the practicalities of tree felling, there is plenty of information in the literature relating to the removal of tree stumps. Alternatives were always being sought to the very labour-intensive methods of removal involving grubbing and jacking the stump. The use of explosives was mentioned occasionally in late Victorian times and it continued to be an option. In 1911, at Studley Horticultural College, Warwickshire, there was a demonstration of what was called the American method of blasting (Long 1911). Holes were bored into the side of the stump, sloping downwards, the holes cleared out and Nobel's dynamite cartridges inserted. The cost of this method, including the explosives expert's fee, was 5*s* 6*d* per stump, in comparison to the old manual method costing £2 5*s*. Misfires and partial removal of the stump sometimes required fresh borings and further charges of explosives, so increasing the cost. After the First World War, the use of explosives in stump removal became increasingly common, due to the large number of people who were accustomed to handling explosive material during the conflict. Le Sueur (1934, 183–185) gives a detailed account of how this can be done safely and considers the risks to be minimal as long as the instructions are followed precisely and common sense is used.

Less dramatic methods of killing stumps and roots, although not removing them entirely, involved the use of various chemicals. It was claimed that ordinary weedkiller would stop any regrowth when inserted into small holes bored into the main roots (*Gardeners' Chronicle* 1915, 388). Le Sueur discusses the breaking down of stumps by chemical action and mentions three different approaches. The first is with a mixture of nitric and sulphuric acids poured into boreholes. However, he states that 'experiments in this direction over a period of several years show that the acid rapidly weakens, and beyond a certain definite charring in the neighbourhood of the boreholes little or no disintegration takes place' (Le Sueur 1934, 179–180). He then mentions another method that was occasionally recommended which involved placing saltpetre into holes drilled with an auger. However, he concludes, 'Widespread enquiries have failed to produce a single instance where this method has been successfully used'. Lastly, he states, 'Destruction has been attempted by means of caustic soda spread in a layer over the cut surfaces of the stump'. This worked well on softwoods but on hardwoods progress was slow and unsatisfactory. In Le Sueur's view, there were only three methods by which tree roots and stumps could be cleanly removed from the ground: by manual labour, by manual labour assisted by mechanical devices and by the use of explosives (*ibid.*, 181).

Cavity work and preventative tree surgery

The care and attention given to ancient and remarkable trees that was evident in the nineteenth century became even more apparent in the early decades of the twentieth century. As well as much more material on remedial and preventative tree surgery in books and periodicals, this was also evident from contemporary photographs of some of the nation's most famous ancient trees. These images show that considerable time and effort was being expended in trying to preserve these trees and prolong their life. Typical of this was the work undertaken on England's most famous tree, the Major Oak in Sherwood Forest. A photograph taken around 1900 shows members of the Morley family of local blacksmiths beating out sheets of lead before applying them to some of the large lower branches of the Major Oak (Figure 74). These sheets were to cover areas of

FIGURE 74. This photograph *c.* 1900 features members of the Morley family of blacksmiths beating out sheets of lead before applying them to some large branches of the Major Oak, standing behind them. Craftsmen such as blacksmiths and carpenters were also often involved in tree surgery work.
CREDIT: EDWINSTOWE HISTORICAL SOCIETY.

decay and, it was believed, to give some protection against rain penetrating the wood and exacerbating the decay. This image is also interesting because it highlights the fact that craftsmen such as blacksmiths and carpenters, and not just arborists, were often involved in tree surgery work because of the metal sheets, chains and lengths of timber that were frequently used.

An early twentieth-century contribution to the literature on remedial tree work came from William Jackson Bean in a 1906 article in the *Gardeners' Chronicle* (Bean 1906). His treatment of hollow and decayed tree trunks was quite traditional:

> Remove all, or as much as possible, of the decayed wood; wash the surfaces with a strong solution of carbolic acid; after this has become quite dry smear the surface with coal tar. This antiseptic treatment is intended to destroy the parasitic fungi and arrest the decaying process. The hole must now be filled up and made watertight. If it be small, the best 'stopping' is formed of a piece of oak made to fit; but large holes should be filled with cement. In either case a final coating of tar should be given. Except when the hole is very large, a new layer of bark will grow over the cavity when filled in the way described.
>
> Bean 1906, 245

When Angus Webster published his excellent book *Tree Wounds and Diseases* in 1916, the section on cavity treatment contained little that was new. He believed, 'The best art of the woodman has been brought to bear on the vital point of arresting decay, whether by timely filling of concrete, covering apertures so as to prevent the ingress of damp, or treating antiseptically for insect or fungoid diseases' (Webster 1916, 10). For covering apertures he recommends a plate of lead or zinc. In an effort to show he is abreast of the latest scientific developments, he adds: 'Recent experiments have proved the value of a mixture of sawdust and asphalt for filling cavities. It is particularly valuable in cases where concrete is too rigid and unyielding, such as in dealing with trunks and branches that are apt to be swayed about in stormy weather' (Webster 1916, 22–23). Some arborists preferred bricks and mortar to fill large cavities.

While the basic principles of traditional cavity work went largely unchallenged, one man was looking more closely at this. In W.E. Hiley's review of Webster's book in the *Quarterly Journal of Forestry*, he is one of the first authorities on arboriculture and forestry to question the practice of trying to remove all the decayed or infected wood from a cavity. He states:

> When the wood is thoroughly rotted it either falls away of itself or is easily removed by a chisel. But, surrounding the wood that is actually rotted, there is a layer which is merely discoloured, or may even show no indication of rot, which nevertheless contains fungal mycelium, and will ultimately become as rotted as the rest
>
> (Hiley 1917, 125).

It seems that the vast majority of arboriculturists did not pick up on the implications of this: that trying to remove sufficient material from the cavity to get back to sound wood was a rather futile exercise. If all the wood that could eventually become decayed was removed, the tree would almost inevitably become structurally unstable. However, it would be many years before research would show that trees have their own natural defences against decay and despite its good intentions traditional cavity work undermines this.

William Dallimore (1926, 39–41) in his classic work on tree pruning also includes a short section on cavity treatment, although this also repeats what was accepted practice inside the cavity itself. However, he does consider the wider context by emphasising the need for judicious pruning to lighten heavy branches that otherwise might fall due to the presence of decay or a hollow trunk. He also questions the practice of covering cavities and wounds with sheets of lead or zinc (*ibid.*, 39–41). In Dallimore's view, 'This is not good policy, for insects lodge below the covering, dirt is admitted, and very often decay goes on quite as a fast as if no covering was used'.

Denis Le Sueur (1934, 56–83) is another British tree expert to question traditional cavity work, although in the end he does repeat most of the accepted practices of the day. He asks the question, 'Should cavities be filled?' and states:

> The filling up of such holes is apparently to the layman the beginning and end of that practice, which, in default of a better word, can be termed 'tree surgery'. He has a hole in his tree, he calls in someone who fills the hole up, his tree is now quite well, and everything is all right. But is it? Fillings have been made in trees for a good many years now, and a good many trees appear to be quite healthy; but still, more trees with large cavities in them have never been filled, and they are still alive and healthy. There is no definitive ruling that a filled tree is necessarily healthier than an unfilled one, providing always that care of the unfilled one is taken in other ways. Nature can at times cope with the trouble herself, and the centre of a tree rotted out, the newer wood can go on for years and years.
>
> Le Sueur 1934, 56

Given the accepted practice and 'science' of the day, this is a very perceptive glimpse, although not yet an insight, into our modern view of cavity treatment. Having taken this bold step, however, Le Sueur does qualify his position. While clearly sceptical, he goes on: 'No attempt is being made to condemn the filling of *all cavities*', adding: 'When a definitely easily applied *elastic* filling is discovered, one that can be built up vertically like cement, the position will be different' (Le Sueur 1934, 59). He discusses at length all the different materials then used to fill cavities, highlighting what he sees as their advantages and disadvantages. Eventually, he introduces what he calls the 'open cavity method', which he claims is gradually increasing in use, not least because of its reduced cost. This follows the same basic procedure to that employed when filling a cavity, except that no filling is used and the entrance to the cavity is left open. If any cavity contains water, Le Sueur believes it is acceptable to drain this by boring a hole in the trunk below the cavity, as long as the hole is not kept open on a permanent basis (*ibid.*, 157–158). The relative effectiveness of various forms of cavity treatment remained the subject of debate right through to the early 1970s. Then, everything changed as a result of new research into tree decay processes, something that is discussed in the next chapter.

With regard to preventative tree surgery, the use of props to support large and heavy limbs on ancient and remarkable trees had become more noticeable in the late nineteenth century. However, the bracing of weak limbs to stronger parts of the tree by means of aerial cables, rods or straps only seems to have become accepted practice at the start of the twentieth century. Both propping and bracing were often undertaken along with thinning out the foliage on the branch to reduce its weight.

In the opinion of Webster (1916, 41–42), supporting heavy and diseased branches by means of wooden props was only admissible where bracing was not possible. This might be the case if there was no main trunk or strong upright limb to which the cable or connecting rod could be attached. Regardless of how well propping was done, he believed this always looked clumsy and out of place. Nevertheless, if propping was necessary, this should be done with props made of yew, oak or some other long-lasting timber. While forked props were commonly used, Webster preferred to have a slight hollow grooved out of the upper end of the prop, into which the branch would fit. Packing material could then be placed between the branch and the prop. At this time there was no mention of using metal props for this purpose. It is interesting that Le Sueur makes no mention of propping in his classic work, although he has extensive coverage of various types of bracing. This suggests he was not keen on the technique.

With regard to bracing, Webster notes in 1916 that this is now receiving more attention than in the past and he remarks 'on the number of old elms and oaks in the London district that have been operated upon of late years' (Webster 1916, 48). The method of attachment for bracing that Webster preferred involved the use of iron bands rather than chains. He states,

> Chains, though often used [for bracing], probably owing to their being readily obtainable and cheap, are for various reasons to be avoided, the flat iron band lined with leather or rubber being preferable, more easily adjusted, and less likely to cut into the bark and wood. The bands, which can be made by any blacksmith, are usually 2½ inches wide, and of the shape of the branch to be encircled, each being in two parts for ease of fixing and to allow of slackening at any future time should necessity demand.
>
> Webster 1916, 29–30

Another method of attachment that does not meet with Webster's approval is the use of rods passed through holes bored into the branch or trunk. He declares:

> Occasionally we see holes bored through the live branch or trunk in order to fix the binding rod or chain – a most reprehensible practice that, fortunately, has few supporters amongst those who have even a cursory knowledge of trees and their growth. It is, however, generally admitted that there are several disadvantages connected with bracing trees with iron rods that pass through holes bored in the trunk or branch.
>
> Webster 1916, 38

The disadvantages to which Webster refers are the risk of insect attack and fungal infection through the holes, the movement of the embedded rod causing friction with the wood, and what he believed was the inevitable weakening of the wood by boring holes into it.

Le Sueur has his own views on various methods of preventative tree surgery. When it is necessary to keep the cost of the operation to a minimum, he suggests what he calls 'slat bracing', which involves using wooden slats and wire (Le Sueur 1934, 154). The wooden slats, about 12 in. (305 mm) long by ½ in. (13 mm) square, are nailed around the outer part of the branches to be braced. Heavy wire is then passed around the branches over the slats, joined up and tightened. The wire is kept in position by staples nailed to the slats. Given Le Sueur's considerable knowledge and experience of tree surgery, it is surprising that he recommends hammering nails directly into the branches of trees, even as a measure to reduce costs. At the same time, he is critical of other destructive methods. In a substantial chapter devoted to 'bolting and bracing', he states:

> The commonest method of fastening branches together is by means of iron bands or clamps, which are passed around the limb, bolted together and connected the one to the other by means of an iron bar or chain. This type of fastening is utterly wrong, as it eventually results in the compression of the bark, and damage or death to the portion of the branch above the fastening. In theory, the bolts are occasionally loosened to allow for branch growth, but in practice this is seldom done. The only correct method for obtaining a holdfast is to pass a bolt right through the limb. There is a popular idea that such practice will have a damaging effect on the tree, but if properly done there is little danger of decay or other damage, and by letting the anchoring washer on the end of the bolt down on to the wood all danger of damage to bark or cambium is completely eliminated.
>
> Le Sueur 1934, 162

It is interesting that Le Sueur's preferred method of attachment, using bolts inserted into holes drilled right through the branch, became increasingly common after his book was published. By the 1950s it was the accepted practice, until the advent of new research into tree decay processes in the late 1970s that led to a radical revision of many tree care practices.

Tools and equipment

The first few decades of the twentieth century, perhaps even more than the late nineteenth century, were a time of great innovation with tools and equipment used in practical arboriculture. This was particularly the case with various types of mechanical equipment. Some of this new equipment is featured in advertisements shown in Figure 68. Other types of equipment, such as transplanting machines and climbing equipment, have already been covered in this chapter.

The tools used in tree pruning remained much the same as they were in late Victorian times. Pruning knives, pruning saws, secateurs and the averruncator remained standard equipment for the arborist. What was now being used increasingly often was the pole saw, effectively a regular pruning saw fixed to the end of a long pole, which could be used on the ground or in the tree. Sometimes this had sections of pole that could be screwed together to extend the length, which was useful in reaching out to those long branches when the arborist could not get close by climbing. Also becoming more common for pruning work was the bow saw. Although widely used on the Continent in late Victorian times, it was now being manufactured in Britain and advertised in the professional press (*QJF* 1938a). From the early years of the twentieth century, hedge cutting and topiary work became much easier and quicker with the invention of electric- and petrol-driven equipment for this purpose. With brand names like Little Wonder, Dextor and Tarpen, these little machines were advertised to both the amateur and professional market.

The use of steam-powered and electric-powered equipment for felling and crosscutting trees has already been mentioned, along with the traditional axe and crosscut saw. However, by the 1920s and 1930s early versions of the chainsaw were being developed. The first portable chainsaw was patented in 1918 by the Canadian millwright James Shand (Wardrop 1976, 9). After he allowed his rights to lapse in 1930 his invention was further developed by the German company Festo in 1933. Other important contributors to the modern chainsaw were Joseph Buford Cox, the inventor of the chipper-type chain, and Andreas Stihl, who developed an electric chainsaw in 1926 and a petrol-powered chainsaw in 1929, and also founded a company to mass produce them (Hall 1977, xiii–xiv). In 1927, Emil Lerp, the founder of Dolmar, developed the world's first petrol-powered chainsaw, two years before the Stihl version. By the 1930s, both types of Stilh chainsaw were being advertised in the British professional press (*QJF* 1938b). These early chainsaws required two people to operate them and the one-man chainsaw only appeared in Britain in the late 1940s.

While chainsaws were at an early stage of development, dragsaws were quite common. A dragsaw is a large reciprocating saw that uses a long steel crosscut saw to cut logs to length. Before the chainsaw became popular, the dragsaw was the most common piece of equipment for sawing through timber. However, because of their construction and weight they could only work with timber that was on the ground. The most common dragsaw used in Britain in the 1920s and 1930s was manufactured by the American company Wade, which advertised in the British popular and professional press (*Country Life* 1925a).

Two items of equipment used in tree and stump removal were developed by the Australian company Trewhella, although manufactured in Birmingham, and were known as the monkey jack and monkey winch (Le Sueur 1934, 181–182). The first to come into general use in Britain was the monkey jack, after advertisements for this appeared from 1909 (*The Field* 1909). It reduced much of the backbreaking work of stump removal and was also useful for general heavy lifting work. The monkey winch was very useful for uprooting trees and stumps and could also be used to ensure trees were felled in the right direction, being more reliable and effective than a team of men pulling on a rope. Although this was used by British foresters in India as early as 1915 (Menon 1916), it was not advertised in Britain until the 1920s but then became widely used in arboricultural operations (*Country Life* 1923).

By the 1900s, some companies were specialising in the manufacture and retailing of tools and equipment used in arboriculture. One was the Standard Manufacturing Company, based in Derby, which made and sold tree-pruning implements. They advertised in the popular and professional press as well as having stands with displays and demonstrations of their equipment at major horticultural exhibitions and shows (*Gardeners' Chronicle* 1903c, 422). The firm still exists as the Horticultural and General Standard Manufacturing Company. On a different scale there were small local retailers specialising in arboricultural products, such as W. Otton and Sons, based in Exeter, who offered a range of tree-pruning tools (*Devon Gazette* 1930). However, Britain in the early decades of the twentieth century was behind the United States. By the 1930s there were several American companies specialising in arboricultural equipment and issuing catalogues displaying a wide array of products. One of the first was the Bartlett Manufacturing Company of Detroit, Michigan, founded in 1912, which began issuing catalogues around 1915. Others included Seymour Smith of Oakville, Connecticut, and H.E. Muench of Stamford, Connecticut. However, no evidence could be found of these companies conducting any significant trade with anyone in Britain.

Tree health and tree pathology

One aspect of arboricultural work of growing interest to professionals in the early twentieth century concerned matters of tree health. While those arborists who limited themselves to basic pruning and felling work might not be expected to have much knowledge of this, for those who put themselves forward as experts in arboriculture this was a different matter. The tree expert was expected to be able to diagnose ill health in trees, identify the specific cause in terms of any pest, disease or natural agency, and prescribe appropriate treatments and precautions. While the detail of tree pathology is beyond the scope of this book, it is important to consider the implications of this work for the professional arboriculturist.

While many tree experts undoubtedly did their best for trees showing obvious signs of ill health, how would they acquire the necessary knowledge to diagnose and treat the 'patient'? As they were not tree pathologists, they invariably did not possess the right equipment to undertake scientific investigations on site and they were unlikely to have access to laboratory facilities. Furthermore, there were no formal courses in arboriculture at any university or college in Britain that would have covered this subject. Their only recourse was to learn out of books and articles about tree pathology, and most especially those that referred specifically to amenity trees. The 1894 English translation of Robert

Hartig's classic text on *The Diseases of Trees* was the standard textbook until the publication of Angus Webster's *Tree Wounds and Disease* (Webster 1916).

Webster's book was the first British text to focus on tree pathology from an arboricultural perspective. It also had an urban focus, recognising that trees under stress in an urban environment would be weakened and more susceptible to attack from pests and diseases (Webster 1916, 80). The book got a mainly warm reception from the arboricultural and forestry world, thanks to some favourable book reviews. However, as mentioned above, W.E. Hiley expressed some reservations in his review, and highlighted what he considered to be the author's lack of scientific knowledge. Hiley concluded his review by stating, 'The time for the irresponsible amateur has passed, and on the threshold of great possibilities for British arboriculture it is essential to avoid all false simplicity, and to attempt to make clear what in reality is difficult and intricate' (Hiley 1917, 125). Nevertheless, Hiley's reservations did not damage the book's popularity and it went on to be also published in the United States, where it was equally well received. America itself was a source of useful literature on tree pathology relating to amenity trees. The US Department of Agriculture's bulletins on arboricultural subjects included some on tree pathology, such as the bulletin by Jacob Kotinsky (1921) entitled *Insects Injurious to Deciduous Shade Trees and their Control.* In 1931, Ernest Everett Hubert of the University of Idaho published his major work entitled *An Outline of Forest Pathology* (Hubert 1931). While this was ground-breaking in its field, it had limited relevance to amenity trees and little impact on tree-related professionals in Britain.

Another important British contribution on tree pathology from an arboricultural perspective came with Le Sueur's classic text that included two chapters, one on fungal diseases and another on insect pests (Le Sueur 1934, 99–147). His book is also notable for giving the first detailed British account of spraying trees with chemicals as a defence and treatment against insect pests. While this had been undertaken for some time in orchards, Le Sueur believed the practice should be extended to ornamental trees. While smaller species with beautiful flowers and fruits would be the main focus of this treatment, it could also be applied to large forest-type trees. He gives details of various insect pests and the chemicals that can be used against them in what he calls 'dry' or 'wet' treatments. He also describes the types of spraying machines that were available to deliver the chemicals in a targeted manner to different sizes of trees. While Le Sueur was keen to see more spraying of amenity trees, there is no evidence that its use increased significantly after his book was published.

One aspect of tree health and pathology that featured regularly in the professional press was the threat to amenity trees from air pollution. Webster's book *Town Planting* was specifically about this and what measures could be undertaken to avoid or mitigate the problem (Webster 1910). Concern about the damage caused by pollution to vegetation of all kinds continued to preoccupy arboriculturists, horticulturists, landscape architects and other professionals right up to the outbreak of the Second World War. William Pettigrew (1928), an authority on parks management, highlighted the damage being inflicted on many urban trees in a paper presented to a conference of the Smoke Abatement League of Great Britain in Harrogate in 1928. This attracted much interest when it was reported in both the *Gardeners' Chronicle* and the *British Medical Journal*. Pettigrew's paper gave an account of the damage done to vegetation in many Manchester parks and how impossible it was to grow pines trees within 3 miles of Manchester Town Hall. On the basis of his experience he believed, 'Air pollution makes the cultivation of cedars, silver firs, spruces, pines and cypress trees an almost impossible undertaking in town gardens and parks' (*ibid.*, 3).

There does not seem to have been any arboricultural consultants making a name for themselves in tree pathology at this time, except perhaps Denis Le Sueur. However, there is an interesting item in the *Wiltshire Times* in 1931 which reports that Warminster District Council sought expert advice regarding the condition of a memorial avenue of eighty Cornish elms in its main public park (*Wiltshire Times* 1931). The Council asked Mr T.R. Peace, then assistant mycologist at the Imperial Forestry Institute of Oxford University, to act as their consultant. Peace reported that fifteen of the trees were suffering from Dutch elm disease. He recommended the removal of two trees that had much deadwood but any further action should be delayed until he had re-inspected the avenue in a year or two.

Legal aspects and tree risk management

Another area of arboricultural work that became increasingly common during the first half of the twentieth century concerned its legal aspects and the topic of tree risk management. Following the extensive urban tree planting that had begun in early Victorian times there were now vast numbers of trees growing in close proximity to people, buildings and urban infrastructure. This inevitably led to increasing instances where trees became embroiled in legal proceedings in the courts.

In November 1905, newspapers in the English Midlands were reporting on the aftermath of a tragic accident in Northfield, near Birmingham, where a falling tree had killed two women. A report of the inquest held in the Station Temperance Hotel, Northfield, stated: 'The evidence showed that, whilst they were out driving on Monday, a beech tree by the roadside in Northfield suddenly fell upon the deceased ladies and killed them instantly' (*Midland Telegraph* 1905). What is particularly interesting about this case is that it must be one of first in Britain involving a roadside tree falling and hitting a moving vehicle, which then resulted in death. What is also interesting is that the Coroner called upon a tree expert to give expert testimony, which was crucial in the outcome of the case. The report in the *Midland Telegraph* stated:

> One of the witnesses was Edward Lee, a gardener, who was called as a tree expert. He said he could not say that the tree was quite dead; it was nearly. The tree stood, he should think, about three feet from the wall. He could not say whether the roots

FIGURE 75. An emergency tree crew dealing with a large tree that has fallen across Brittains Lane, Sevenoaks, in 1935. What appears to be a dragsaw is on the ground next to the two men in the centre.

> had been injured when the wall was built. It was quite possible for the tree to grow as long as twenty years after the roots had been damaged. The roots were badly decayed, but it was impossible to discover this fact unless the roots were exposed. The Foreman of the jury asked whether the witness would have had the tree cut down if it had been in his jurisdiction. The witness was unable to say, without having seen the roots exposed, but added that he should not expect the tree to fall as it did. … The Coroner, in summing up, drew attention to the evidence of Mr Lee, which showed that it was not possible for anyone to ascertain the state of the roots unless they were exposed. Under the circumstances, the jury would have no difficulty in returning a verdict of 'Accidental Death'.
>
> *Midland Telegraph* 1905, 2

The jury duly returned a verdict of Accidental Death. It is a matter of speculation whether such a verdict would be returned today, given our understanding of 'duty of care' in relation to roadside trees and an arboricultural consultant with good legal experience acting as the expert witness.

In 1907, another interesting instance involving a roadside tree occurred in Banbury, Oxfordshire, where the tree fell across one of town's main streets. As this does not appear to have been reported in the press, it probably did not result in any death or injury. Of course, a tree falling into the road would

FIGURE 76. This council-owned tree fell across a main street in Banbury in 1907, close to the famous Banbury Cross (far left). The evidence of cavity work to the decayed and hollow base of the tree shows that council staff were aware it was structurally weak.

not have been an unusual occurrence, particularly when there were gale-force winds, when emergency tree teams would respond quickly to clear the roads (Figure 75). However, what makes this Banbury incident noteworthy from an arboricultural perspective is the evidence from a photograph taken just after the tree fell (Figure 76). It can be seen that the base of the tree is decayed and hollow, which has led to extensive cavity work involving bricks and mortar. As this street tree was almost certainly the responsibility of the local authority, this evidence of tree surgery shows that council staff were aware it was structurally weak. In today's world, once a roadside tree was identified in this condition, it would be felled. If any tree surgery, such as cavity work or bracing, was undertaken and the tree subsequently fell, the local authority would not be able to claim this as an 'act of God' and could be liable for the consequences.

Not only could legal action result when essential tree work had not been undertaken but also when damage or injury was a direct result of such work. In a case before Clerkenwell County Court in 1914, compensation was claimed for damage to an apple tree and a fowls' run caused by large branches falling on them (*Gardeners' Chronicle* 1914a, 57). This had resulted from tree work being undertaken in an adjoining garden, although the perpetrators of the damage are not identified. However, the wording of the news item suggests it was tree work contractors who were employed to dismantle the tree by the adjoining householder. The plaintiff was awarded 30*s* damages and costs after experts had assessed the cost of the damage. Another cause of legal disputes between neighbours, in which arboriculturists often acted as expert witnesses, was where tree roots or branches were said to have encroached from adjoining property. As an indication of the topicality of this issue, an editorial on encroaching tree roots and the consequences of this appeared in *Gardeners' Chronicle* in 1916. The editors gave some historical background to the topic as well as highlighting the lessons to be learned from recent court cases (*Gardeners' Chronicle* 1916, 6–7).

As arboriculturists and other professionals were increasingly acting as tree experts in court cases, some actively promoted their expertise in the legal aspects of arboricultural work. In the early twentieth century the pre-eminent figure in this field was Benaiah Whitley Adkin. He had initially trained as an architect and this work with property led him to become a Fellow of the Surveyors' Institution. Adkin was also a Life Member of the English and Scottish Arboricultural Societies and later qualified as a Barrister-at-Law. His qualifications and experience made him well suited to be a consultant in all matters relating to trees and the law and he soon gained a national reputation for this. Throughout his career Adkin was involved in many legal actions that featured trees or woodlands. His seminal book *The Law of Forestry* was published in 1914, and, while this focused on rural forests, it also included amenity trees, especially those adjacent to highways (Adkin 1914). This book and his subsequent articles in *Quarterly Journal of Forestry* represent the first significant account of the law relating to amenity trees to be published in Britain. Le Sueur's (1934, 195–210) book contained a chapter on trees and the law and this was the second substantial contribution on the legal aspects of arboriculture in Britain.

Some local authorities had their own in-house tree expert who could act as an expert witness on their behalf. For example, in October 1945 an inquest was held in Braintree, Essex, following the death of a schoolgirl when a large branch of an elm tree fell on her (*Essex Chronicle* 1945). A schoolboy was also present and, although he escaped the full force of the branch, he suffered a broken arm. This tall elm tree was standing in the school playing field and gale-force winds were blowing at the time. When the wind tore the branch off the tree, it fell, killing the schoolgirl instantly. John Williams, tree warden at Essex County Council, said there was a defect in the main trunk of the tree but that was not where the branch had blown off. He continued his evidence by stating, 'Elm trees are dangerous when they reach maturity. The branch of this tree which broke off was in the process of decay, due to old age. Elm trees are not suitable for school grounds or adjoining public highways'. On the basis of Williams's expert testimony, the Coroner recorded a verdict of Accidental Death. Again, in this modern age, with evidence of decay in the branch, the verdict may have been different. In the reports of this case there was no mention of any regular or systematic inspection of the trees in the school grounds, or who might have undertaken those inspections. Furthermore, in this author's research on arboricultural practice at this time there is almost no mention in the literature of regular tree inspections or tree risk management. The only significant contribution on this comes from Le Sueur's book. In his discussion on dangerous trees, he considered the liability of an owner of a tree that falls and causes damage. He states:

> He does not appear to be liable unless (a) the tree causing damage was known to be dangerous, or (b) he knew or ought to have known that it was dangerous, and by failing to take proper action was guilty of negligence in that he did not exercise proper care of his tree. What exactly is covered by the words 'proper care' is difficult to say, as naturally different juries will hold different opinions.
>
> Le Sueur 1934, 203

He then discusses what constitutes a dangerous tree and tries to come to a balanced view on this, stating:

> It may be considered reasonable to class a tree as 'dangerous' when in the opinion of competent persons it is liable to fall of its own accord by reason of decay, old age, disease, or similar cause, or might be brought down by an ordinary wind such as might be expected in any ordinary season or from any ordinary quarter.
>
> Le Sueur 1934, 204

He then concedes that a tree which is not apparently dangerous may fall under unusual circumstances, such as a gale of unusual severity. In such cases the owner might reasonably plead this was an 'act of God'. Le Sueur then gets to the heart of the matter. He recommends that if the tree owner is in any doubt about whether this is safe or not, 'the owner would be well advised to have his tree inspected by a competent person, and depend on his report for future action' (Le Sueur 1934, 204). He then concludes, 'The only solution to the problem seems to be frequent and careful inspection, so far as is possible'. This recommendation from Le Sueur does not appear to have attracted any published comment or endorsement from his colleagues. Furthermore, nobody seems to have asked him what constituted a 'competent person', although presumably he was referring to a tree expert like himself. However, this whole subject would become hotly debated in the late twentieth century.

Contractors and consultants

While some progressive local authorities in the early twentieth century were appointing specialist tree officers, there were also encouraging developments in the private sector. Since the seventeenth century there had been landscape contractors undertaking major tree work but now we see the emergence of a few companies specialising in tree surgery. This was prompted by the dramatic increase in residential gardens in suburban districts and by growing recognition

THE DO-IT-YOURSELF AGE

AS EASY-

AS-

FALLING OFF-

A LOG!

FIGURE 77. Most homeowners in the early twentieth century undertook their own gardening work. However, it was widely recognised that any major tree work was best left to the professional. This cartoon from *Punch* in 1919 highlights the dangers of 'do-it-yourself' tree work.
CREDIT: © TOPHAM PARTNERS LLP.

of tree surgery as a specialist activity. While most homeowners undertook their own gardening work, often as a form of recreation at weekends, it was generally considered that major tree work such as felling or heavy branch removal was best left to the professional. A cartoon from the popular *Punch* magazine in 1919 cautions the amateur gardener by humorously highlighting the dangers of 'do-it-yourself' tree work (Figure 77). Not only was the domestic garden market expanding in terms of tree work, there were also opportunities for work in the public sector. However, while many local authorities were now employing contractors for major street tree pruning and other large-scale tree work, these contracts often went to firms of landscape contractors that had the workforce and resources to operate at this scale. Some of the contractors undertaking tree work at this time and which are mentioned below are featured in advertisements shown in Figure 78.

The small tree surgery companies, often comprising just the owner and two or three operatives, usually gained work through 'word of mouth' or by advertising in the local papers. An early example of this type of enterprise was Mr A. Bartlett, a tree work contractor based in Cambridge Road, Kew, who was a former employee of Kew Botanic Gardens. In 1919, in response to a reader's enquiry in *The Field* about competent people to undertake tree work, the magazine's gardening columnist recommends Mr Bartlett, 'who is a tree expert' (*The Field* 1911). After the First World War, some ex-servicemen set up in business as tree work contractors, perhaps because they enjoyed the outdoor physical work and had acquired some basic practical skills in the army. Some advertisements for these contractors specifically mention that they are former servicemen, no doubt believing this would encourage people to employ them (*The Citizen* 1921). Around this time, an increasing number of forestry contractors saw opportunities in expanding their operations to include amenity trees, while not necessarily undertaking practical tree surgery. For example, Dr A.H. Unwin of the Woodlands Valuation Company was happy to carry out tree-planting schemes in all situations (*Country Life* 1921).

In the United States there was now a small but flourishing tree care industry. The Davey Tree Expert Company, whose origins date back to 1880, was formally incorporated in 1909 (Pfleger 1977, 35). The F.A. Bartlett Tree Expert Company was founded in 1907 and other tree surgery companies followed (Cooley 2007, 4). During the 1920s, according to Bill Matthews (1970, 21), a few British people who had been trained by these companies returned home to practice tree surgery, 'usually in a small way – very expensive treatments to very special trees'. Some British tree companies felt it was advantageous to promote a North American connection, whether this was genuine or not. In London in the late 1920s there was a firm called the Canadian Tree Felling Company, based in Hammersmith. They offered 'expert advice on trees' and would undertake a range of tree surgery work (*West London Observer* 1928). Meanwhile, some timber contractors began to see an opportunity to expand their business by taking on tree surgery work. One such firm was Higginson Brothers, who were

ADRIAN ESTLER

TREE EXPERT

PRUNING
CABLING
FEEDING
CAVITY TREATMENT
OF
LARGE TREES

INSPECTION FREE

3, Cranes Drive,
SURBITON, SURREY
ELMBRIDGE 1270

THE ENGLISH TREE EXPERT COMPANY LIMITED
COBHAM, SURREY.

Phone Cobham 392

Specialists in the treatment and preservation of ornamental and other trees.

Examinations, Reports and Estimates Submitted Free of any Charges.

THE TREE RECORD BOOK
(Newly Published)

A Book for Preserving for all time Proper Records of the Time of Planting, Position, Variety and Measurement of Individual Trees.

Full Particulars on Application.

FOR · LATE · PLANTING
OBTAIN · YOUR · TREES
& SHRUBS · FROM · THE
LAKE · DISTRICT

Telegrams "DEPENDABLE" *Windermere*

OUR NURSERY STOCK being grown on high exposed ground can be lifted with safety and removed much later than from Southern and less exposed positions.

OUR CATALOGUES of trees, shrubs, roses, and rhododendrons, and also of hardy perennials, and containing invaluable information, will be sent post free on application.

MAWSON · BROS
WINDERMERE
Robert R. Mawson Sole Surviving Partner

CANADIAN TREE FELLING Co.

Expert advice on Trees.
Felling, Fencing Estimates.
: Building Sites Cleared :

14, ANGEL RD., HAMMERSMITH.

TREE PLANTING

Thinnings undertaken, advice on woodlands given, nurseries laid down, young trees supplied, plans and estimates submitted by

Dr. A. H. UNWIN,
The Woodlands Valuation Company,
62, CHANCERY LANE, W.C. 2.

TREE-FELLING.

TREE-FELLING, TOPPING,
& ROOT GRUBBING
wanted in large or small quantities,
any district.

Apply to—

THOMAS HIGGINSON,
Timber Contractor, WARWICK

(Contractor to some of the leading City Engineers, Councils, Public Works, Land Owners, and Estates of the London, Midland, and Southern Districts)

FIGURE 78. Company advertisements and dates: Adrian Estler, Tree Expert (1936), The English Tree Expert Company (1938), Mawson Brothers (1910), Canadian Tree Felling Company (1928), Dr A.H. Unwin, Woodlands Valuation Company (1921), Thomas Higginson (1932).

in the tree felling and timber business in Warwick as early as 1926, following in the footsteps of their father. By 1931 the firm was operating as Thomas Higginson, just one of the brothers, who now called himself a 'tree felling and topping contractor' and advertising that he had 'purchased a New Machine for the pulling up of Trees and Tree Stumps' (*Warwick Advertiser* 1931). By 1933, Higginson was promoting the company as specialists in arboricultural work with a second base in south London (*Tewkesbury Register* 1933).

The 1930s saw the emergence of a number of companies that openly promoted themselves as 'tree experts'. Some of these were probably jobbing gardeners who recognised an opportunity, such as Mr A. Harley of Hoole in Cheshire. Harley called himself a 'tree expert' but in his classified advertisements offered a wide range of gardening services (*Cheshire Observer* 1930). There were also individuals who genuinely specialised in tree surgery. Adrian Estler, based in Surbiton, Surrey, appears to have been a highly skilled tree surgeon, undertaking small projects and a range of advanced tree work (*Landscape and Garden* 1936). He also seems to have been a colourful character with no head for business (ELHG 2020). Another company was E. Whatton and Sons, based in Hartwell, Northamptonshire. They were forestry and timber contractors for some years before expanding into tree surgery around 1934. In their initial advertisements for this they call themselves 'Tree-Lopping Specialists' and by the following year this had changed to 'Specialists in Tree Preservation' (*Northampton Mercury* 1935).

One of the largest firms operating at this time was the English Tree Expert Company, based in Cobham, Surrey (*QJF* 1938c). This was owned by Maurice Johannesen (Matthews pers. comm., 2012), who was working as a tree expert as early as 1933 while based in Castle Hedingham, Essex (*Essex Chronicle* 1933). Johannesen, who liked to be called Captain Johannesen, built up a successful business with a reputation for high-quality work on prestigious projects. In 1937, the English Tree Expert Company was asked to undertake work on two of Scotland's most famous trees growing in the quadrant of St Mary's College at the University of St Andrews (*The Scotsman* 1937). One was the fine holm oak (*Quercus ilex*) and the other was the ancient hawthorn (*Crataegus monogyna*) reputed to have planted by Mary, Queen of Scots. According to the report in *The Scotsman*, the latter 'was in danger of falling, but it was also treated by the experts, who skilfully grafted a composition into the trunk, which is expected to give it new strength'. That this Surrey-based company had been asked to travel to Scotland to work on these famous trees is testimony to its national reputation. In 1939, Captain Johannesen was asked to supervise an important commemorative tree planting on Magna Carta Island in the River Thames. The planting involved the mayors of London and Birmingham and was held to mark the visit by George VI to the United States (*Birmingham Post* 1939). Johannesen was in attendance because the English Tree Expert Company held the contract to maintain all the trees on Magna Carta Island. In the late 1930s the company was busy recruiting experienced 'smart men' to keep up with the demand for its services (*Surrey Advertiser* 1939).

Another company with an excellent reputation for its high standard of work, often on prestigious projects, was the Chiltern Tree Surgeons. This was formed by James H. Carey, probably in the early 1930s, who later became a prominent member of the Men of the Trees. Although he never employed more than three or four people, he was soon renowned for his advanced tree surgery, especially cavity work and bracing (Matthews pers. comm., 2012). In 1939, Chiltern Tree Surgeons were asked to report on the condition of the much-loved horse chestnut tree outside the entrance to the municipal offices of Aylesbury Borough Council (*Bucks Herald* 1939). Following the company's report, the Council's Finance and General Purposes Committee 'recommended that this firm be asked to carry out such remedial treatment to the tree as is necessary, at a cost of £14 10*s*'.

Most of the leading tree surgery companies in the 1930s were based in the south of England. However, there were companies in other parts of Britain offering a comprehensive tree surgery service. While the University of St Andrews chose to engage the Surrey-based English Tree Expert Company, there were a few tree contractors in Scotland that would have undertaken the work. One of these was James Laurie and Son, established in 1860 and based in Dundee, which had built up a substantial landscape gardening and nursery business. They were one of the first companies in Scotland to exploit the growing demand for tree surgery work. In a 1936 advertisement headlined 'Protect Your Trees', they announced they had staff 'available in any part of the country for Pruning, Wound Protection, Cabling, Cavity Filling, and General Treatment of large trees' (*The Courier* 1936).

In researching tree surgery companies in this era, much useful information was obtained from newspapers and popular magazines, in the form of news items and advertisements. However, it has been noticeable that some of the larger and more celebrated companies rarely placed advertisements in the popular press and only very occasionally in some professional periodicals. This was something that was commented on at the time by Dominic Wyndham Lewis, the well-known journalist and author. Writing in *The Bystander*, a popular weekly magazine, he declared, 'Tree surgeons in this country – they go anywhere when summoned – do not advertise to any extent, British medical etiquette naturally forbidding. In America they advertise largely' (Wyndham Lewis 1937).

The owners or managers of tree surgery companies often acted as arboricultural consultants. As noted above, those with established reputations were sometimes asked to produce written reports on the condition of trees, which their clients then considered before deciding to proceed with any practical work. Some consultants, however, were not attached to any contracting company but acted independently, giving expert advice, producing written reports and carrying out tree surveys. An early twentieth-century example of a local authority engaging an arboricultural consultant is Ramsgate Town Council, which caused a little controversy. In 1900, the council's minutes state that Alderman Wood asked for instructions for dealing with the town's trees. Several gentlemen had been written to for advice and among the responses was one from Leicester where the trees were known to be well kept (*Thanet Advertiser* 1900). This man was prepared to come to Ramsgate and advise on their trees for a charge of £10. Some councillors argued that it was preposterous to go out of the borough to get a tree expert and if their own gardener was not competent 'the sooner they got rid of him the better'. Others defended the state of the council's trees and the competency of their gardener. After much debate it was proposed that Mr Burbridge, a local tree expert from Westgate, should be approached to come and give his opinion, which the council would duly consider. This proposal was then carried with a narrow majority.

Another early twentieth-century arboricultural consultant was Abraham Stansfield of Prestwich, near Manchester, who had been Lecturer on Horticulture to the Lancashire County Council. He then became arboricultural consultant to the municipalities of Manchester and Salford and was widely regarded as an expert on urban trees (*Todmorden Advertiser* 1907). Yet another was George Harrow, who had spent many years with Veitch nurseries as manager of their Coombe Wood Nursery at Kingston-upon-Thames. He then decided to move on and set up in business as an arboricultural consultant (*Gardeners' Chronicle* 1914b).

Some prominent individuals who occasionally acted as arboricultural consultants have been mentioned. These include Isaac Bayley Balfour of Edinburgh's Royal Botanic Garden, who produced the report about tree pruning on the Mall, and Benaiah Whitley Adkin, the legal expert, who frequently dealt with cases concerning trees. William Jackson Bean, curator at the Royal Botanic Gardens at Kew, would also occasionally take on consultancy work. In 1926 he was invited by Ealing Town Council to inspect the trees on the north side of Ealing Common (*Middlesex Times* 1926). The council was undecided as to whether the plane trees or the lime trees should be removed and wanted his expert opinion. After inspecting the trees, Bean's report recommended that the planes be removed as 'they were getting too big and drastic pruning would shortly be necessary'. After the council had considered his report, it decided that the Borough Surveyor should arrange for the removal of the plane trees. Perhaps the most accomplished arboricultural consultant of this era was Denis Le Sueur, who spent much of his time in this activity alongside his teaching commitments at Cirencester. Although he was contracted to the City of London Corporation as its arboricultural consultant, he regularly undertook a range of commissions from other clients in both the private and public sectors.

Landscape architects sometimes undertook arboricultural consultancy as part of their regular work of designing and constructing landscapes. For example, they might be asked to conduct a survey of the trees on a site awaiting redevelopment, indicating their condition and recommendations for retention or removal. At the start of the twentieth century the 'bulk trade' in garden design, for both private owners and public authorities, was dominated by two firms: Thomas Mawson and Son and Milner-White and Son. Edward Prentice

Mawson had worked as an apprentice at a London architectural practice before joining his father as a partner in 1910. He became head of the landscape architecture firm when his father developed Parkinson's disease in 1923. The other big name in landscape design was Milner-White and Son. Edward Milner had developed the practice during the Victorian era and when he died in 1894 his son Henry and grandson Edward rapidly expanded the company. When they were joined by Edward White in 1902 they were soon employing a few hundred staff in the firm known as Milner-White.

In the research for this chapter into all aspects of the arboricultural industry, there was no mention of any women being involved. In view of this, it was interesting to find an article in a local newspaper from the north of England in 1924 entitled 'Tree Surgery: A Career for Women'. The article states:

> A distinctly uncrowded profession for women, and one in which the first-comers are likely to score a big success, is tree surgery. Even men tree surgeons are not by any means numerous owing in large part to their waiting for business to come to them instead of going out after it and to scales of fees that only the rich care to pay.
>
> *Penrith Observer* 1924

Nurseries and tree production

Many nineteenth-century nursery companies known for offering a wide range of high-quality trees and shrubs went on to thrive in the early years of the twentieth century. Hillier Nurseries is an example that did particularly well. In the early years of the twentieth century Edwin Hillier gradually passed the management of the business over to his sons, Edwin Lawrence and Arthur Richard, and they formally took control with their own partnership agreement in 1912 (Hillier 2014, 52). Edwin died in 1926; five years earlier, in 1921, his grandson Harold, the eldest son of Edwin Lawrence, had joined the staff. Harold was to ensure the company continued to expand in the 1930s and further enhance its reputation for trees and shrubs, including many rare and unusual varieties (*ibid.*, 92–106). The emphasis on trees was reflected in the Hillier's advertising that described the company as 'Tree Specialists' (*Country Life* 1925b). In 1935, the first Royal Warrant was granted to Hillier and Sons as Nurserymen to HRH the Prince of Wales, to be replaced on his abdication by that of HM Queen Elizabeth (Hillier 2014, 95).

In contrast to Hillier, another family dynasty of nurserymen, the Veitch family, was in decline in the early years of the twentieth century. The company's sponsoring of plant collectors, which had been at its height in the late Victorian era, was much reduced by the 1900s. In 1913, the seed business and trial site at Slough were sold to Sutton and Sons of Reading and immediately after this the Chelsea business ceased trading. In 1914, the *Gardeners' Chronicle* (1914b, 401) reported that the departments in the Veitch nurseries were 'now nearing extinction'. While the Exeter business continued to trade under Peter Veitch, the company was much diminished from its glorious Victorian days.

Many of those nursery companies focusing on trees and shrubs were also continuing to offer additional services as landscape designers and contractors. At Cheals of Crawley, Joseph Cheal played a leading role in developing the firm's landscape gardening work in the Edwardian era with designs that often had a strong emphasis on topiary (Benton 2002, 147). The Mawson family's nursery business in the Lake District continued to prosper and supply vast numbers of trees and shrubs for their own landscape schemes and many others (*Country Life* 1910) (Figure 78). Many nursery companies also had shops, often separate from the nursery grounds, as the provision of gardening supplies to both the amateur and professional market was becoming an increasingly lucrative business. An ever-wider variety of tools and equipment could be purchased, from lawnmowers to pruning knives. By the 1930s many were also offering various herbicides and pesticides, often frighteningly toxic by modern environmental standards.

Many country estates still operated their own nurseries producing trees and shrubs to meet their own requirements. However, these now concentrated mostly on forest trees for their woodlands and plantations, while ornamental trees, especially the rare and unusual species, were purchased from specialist commercial nurseries. In 1905, John Simpson, author of *The New Forestry*, published a handy reference book for owners, agents and woodmen on private estates that was entitled *The Estate Nursery* (Simpson 1905). However, estate nurseries were becoming less viable as commercial nursery companies could produce vast numbers of forest trees at very competitive prices. One such company was Howden's Hardy Highland Trees, based in Inverness, which claimed to be the most northerly commercial nursery in Britain (*Country Life* 1926).

In the public sector there continued to be a huge amount of amenity tree planting every year by local authorities and other public bodies. This demand could be met in either of two ways: by buying in nursery stock or growing their own. Many local authorities chose to purchase the bulk of their trees and shrubs from commercial nurseries but concerns were occasionally expressed in the professional press about difficulties in obtaining an adequate supply of suitable trees. These concerns related mostly to street trees where the specifications were often more exacting. For example, at a meeting of the Aberdeen Town Council in 1930, a discussion arose on the scarcity of trees for street planting (*Gardeners' Chronicle* 1930, 313). As there were 'no suitable trees to be found in the north of Scotland', the Superintendent of Parks was authorised to go south and purchase trees to the value of about £400 from nurseries in the south of Scotland or England. From the point of view of the commercial tree nurseries, they often complained that local authorities rarely gave sufficiently advanced notice of their planting requirements. In William Dallimore's words, 'They cannot be expected to stock large numbers of a wide variety of trees in the hope that they will be purchased but for which there may be little or no demand' (Dallimore 1928, 20). To avoid any difficulties in obtaining an adequate supply of suitable trees and shrubs from commercial sources, many local authorities established their own nurseries. These would also produce other plants they required, such the bedding material for displays in public parks.

The two world wars had a major impact on all nurseries and especially those producing ornamental trees and shrubs. Both demand and production fell dramatically just before, during and immediately after those conflicts. Not only was there very little amenity planting taking place but many of the men and women who had been engaged in this work were now enlisted in the armed forces or undertaking essential work. For the duration of the conflicts most tree production involved fruit trees and forest transplants.

Representation and education for tree-related professionals

After the confusion of the nineteenth century regarding definitions of tree-related occupations and the terms used for these, the twentieth century began with far more clarity. While some of the old confusion remained, especially with regard to arboriculture and forestry, the mist was clearing. In the late nineteenth century many tree-related professionals and academics were making a distinction between 'forestry', being concerned with woodlands and plantations mainly for timber production, and 'arboriculture', which focused on individual or small groups of trees for amenity purposes. However, in the 1900s there was still some disagreement on this from both horticultural and forestry sectors and it would take a while before there was consensus on the matter. Typical of the horticultural opposition was the continued intransigence of William Robinson. Regardless of his substantial contribution to arboriculture, he did not use the term himself, considering the cultivation and management of amenity trees as the province of horticulture. His frustration with the names of tree-related disciplines comes out again in his book review of *English Estate Forestry* by Arthur Forbes where he complains:

> Like so many recent books on forestry, the language used is neither simple nor clear, the free use of such new words as 'sylviculture', 'arboriculture' and the like, leading to English which is far from pure, and words about 'ornamental' woods, 'sylvicultural' woods and 'arboricultural' woods, only confuses a very simple subject.
>
> Robinson 1905, 3, 32

Within the forestry sector there was now widespread recognition that arboriculture was an area of interest quite distinct from commercial forestry but also something they should continue to embrace. This was despite many of the renowned tree experts of the era, such as those at Kew and Edinburgh Botanic Gardens, coming from a horticultural background and regarding arboriculture as an aspect of horticulture. However, given the Royal Horticultural Society's less than wholehearted embrace of arboriculture, with its short-lived Arboricultural Committee in the 1860s, foresters continued to confidently proclaim arboriculture as an aspect of forestry. Within the English and Scottish Arboricultural Societies, there were also many members from the landed class who continued to equate arboriculture with forestry. By contrast, it was their professional memberships, those working as full-time foresters or agents, who now recognised the distinction between the two. There was also a perception among professional foresters that what the landed classes called arboriculture was often just bad forestry. Typical of this was the comment made by J.G. Watson following a visit to the Woburn Estate with Augustine Henry and his students. He declared, 'The work at Woburn is on the lines of true forestry, not the casual arboriculture which is common on so many British estates' (Watson 1912).

Professional representation

Through the 1920s it was apparent that the names of the Scottish and English Arboricultural Societies were no longer appropriate in reflecting the main interests or occupations of their members. Arboriculture was now recognised as a separate discipline, although closely allied to forestry and horticulture. This prompted two closely related discussions within the Arboricultural Societies: that they should give more time to arboricultural matters (meaning amenity trees), and that they should consider changing the name of their societies. It is interesting that with the former discussion within the Royal English Arboricultural Society, two of the most prominent voices for focusing more on arboriculture were those of William Dallimore and Denis Le Sueur. Writing in the *Quarterly Journal of Forestry* in 1930, Dallimore asks the question, 'Should the Society pay more attention to Arboriculture?' (Dallimore 1930). He declares that it most definitely should, since a large proportion of its members were involved mainly with amenity trees and 'have little interest in the scientific side of silviculture' (*ibid.*, 152).

In 1930 the Royal Scottish Arboricultural Society became the first of the two societies to change its name when it became the Royal Scottish Forestry Society. In 1931 the English society followed suit and became the Royal English Forestry Society (James 1982, 49). While the changes went through relatively smoothly in both Scotland and England, there were voices of concern in the English society about proposals to balance the change of name by giving more attention to arboriculture in the Society's activities. Writing in the *Quarterly Journal of Forestry*, Mr A. Pope expressed a note of caution that was probably echoed by many members. In his view, forestry as a discipline had been around in Britain since earliest times and arboriculture had only emerged following the Reformation. He then declared, 'Forestry therefore is the original subject, and arboriculture is an offshoot from the old tree; and it should not be allowed to grow too big and overshadow forestry, which will happen if foresters do not wake up' (Pope 1931, 269). One prominent arboriculturist supporting the name change was Denis Le Sueur, although he also agreed with Dallimore that the Society should devote more of its energies to arboriculture (Le Sueur

1931, 273). Following the name change, it was announced at the next Annual General Meeting of the Royal English Forestry Society that an Arboricultural Committee had been formed 'to ensure that this side of the Society's activities were not entirely overlooked' (*QJF* 1931, 359–360). The first meeting of the Arboricultural Committee was held in November 1931 when William Dallimore was appointed chairman and approval was given to a number of proposals to increase the Society's interest in arboricultural matters (*QJF* 1932, 52). This committee continued in operation until 1937 when it was replaced by the Arboricultural Subdivision of the newly formed Technical Panel. The panel remained in existence throughout the Second World War, although it was not very active during that period and was subsequently disbanded in 1946 (James 1982, 123).

In 1925, a new forestry society had been formed entitled The Society of Foresters of Great Britain (now the Institute of Chartered Foresters) (James 1981, 220). The object of the Society was 'to advance and spread in Great Britain the knowledge of technical forestry in all its aspects'. Its members were for the most part professional foresters who were drawn from the Forestry Commission and from the teaching and research departments of those universities that were concerned with forestry. The first copy of the Society's journal appeared in 1927 under the title of *Forestry*. The Society was to show no great interest in arboriculture throughout most of the twentieth century, although that would later change.

While there was no professional organisation in Britain devoted exclusively to arboriculture, such an organisation was being formed in the United States. In 1924, the National Shade Tree Conference (NSTC) was established, comprising mainly commercial arborists and scientists in its early days, to promote higher standards and better representation for those in the tree care industry (Campana 1999, 145). This later evolved into the International Society of Arboriculture. In 1937, the National Arborist Association was formed as a trade organisation for commercial tree care companies, working in support of the NSTC (*ibid.*, 230). However, during this period in Britain, there was no obvious body for people in the arboricultural industry to join and have their interests represented. Those at the level of business owner, manager or consultant, who were likely to have qualifications in forestry or horticulture, would most likely join what were now the Forestry Societies in England and Scotland. Although the societies occasionally published membership lists in their *Journal* or *Transactions*, these do not contain sufficient information to determine the proportion of members who were primarily involved with arboriculture or amenity trees. For example, in the 1901 membership list of what was still the Royal Scottish Arboricultural Society there are a total of some 830 members (*Trans. RSAS* 1901). As might be expected, a large proportion of these, especially in the Ordinary Members category, are listed as 'Forester', 'Assistant Forester', 'Foreman Forester', 'Timber Merchant' and other forestry-related job titles. However, there are no members listed as 'Arboriculturist', 'Arborist' or similar arboriculture-related job titles.

This is despite several members, such as Isaac Bayley Balfour and John Barron, being renowned tree experts who were not involved in forestry. Furthermore, while the list includes the first or given name of almost all the members, there is only one that is female.

Another interesting insight into the world of tree-related professionals in the 1900s comes in a publication entitled *Webster's Foresters' Diary and Pocket Book* (Webster 1903). This was produced by Angus Webster and his preface notes the favourable reviews he received for the first issue of the publication the previous year. This had encouraged him to 'take up the publication of the second issue with confidence and to feel assured that he has been able to supply a long felt want to the British Forester and Arboriculturist'. At the end of the book is a 'List of Foresters and Assistant Foresters' in Britain and Ireland, which numbers some 300 individuals (*ibid.*, 37–44). From their addresses it is not possible to determine whether their work is concerned primarily with forestry or amenity trees. However, it seems likely from the overall content of this publication that most were primarily engaged in forestry. While the list includes the vast majority of their first or given names, there are no females. The *Foresters' Diary and Pocket Book* continued to appear annually into the 1930s but with little change in the composition of the list of foresters.

Some individuals who owned or managed tree surgery companies were members of voluntary organisations such as the Men of the Trees and the Royal Horticultural Society. However, the majority were probably not members of any organisations with some connection to trees. With regard to the manual workforce in arboriculture in both the public and private sectors, membership of any tree-related organisation appears to have been very unusual. Furthermore, while trades unions were gaining members throughout Britain, this had little or no impact on the arboricultural industry. Some arborists in the public sector may have been members of trades unions, but this seems very unlikely in the private sector. In 1904, the British Gardeners' Association (BGA) was formed to promote the interests of gardeners and regulate their wages and hours of working (*Gardeners' Chronicle* 1905, 364–366). William Watson, Curator at Kew Gardens, was elected as its Honorary Secretary. Two prominent tree experts at Kew were also involved, with W.J. Bean on the BGA's Committee of Selection that established the organisation and William Dallimore as a member of its first Executive Council. Sir William Thiselton-Dyer, then Director of Kew, was less than impressed with this initiative and announced his dislike of trade unions in the *Gardeners' Chronicle*. Speaking for many of his social class, he declared, 'I do not think that it would be possible to bring the scattered units of the gardening profession under the control of anything like a trades-union. If it were, I think anyone would be shy of engaging a gardener who had at his back an organisation proposing to interfere between employer and employed' (Thiselton-Dyer 1905). However, the BGA survived some early opposition from employers and the difficulties of the war years and in 1919 was issuing a new standard of wages and hours for its members (*Gardeners' Chronicle* 1919).

In the early years of the twentieth century many of the disciplines related to arboriculture were becoming increasingly organised in terms of professional representation. In the built environment sector the Royal Institute of British Architects had been formed as far back as 1834 (as the Institute of British Architects in London) and the Institute of Town Planning was founded in 1914, becoming the Royal Town Planning Institute when its Royal Charter was granted in 1959 (RTPI 1989). In terms of the landscaped environment, in 1929 the British Association of Garden Architects, which had been founded the year before, became the Institute of Landscape Architects (ILA). Thomas Mawson was the first President of ILA and also one of the first professionals in Britain to use 'landscape architect' as a professional title.

During the early 1940s an increasing number of architects and planners were elected as members of the Institute of Landscape Architects, which now sought to move away from its pre-war image as a 'domestic garden society'. While these related professions were increasingly organised and influential, there was still no professional representation for arboriculturists.

Arboricultural education

In the late Victorian era both horticultural education and forestry education had made significant advances and this continued in the early decades of the twentieth century. Horticultural education in particular was to experience a period of major growth with the development of a number of initiatives, some of these led by the Royal Horticultural Society (RHS).

The RHS General Examination had been launched in 1893 and was now proving quite popular with professional gardeners and skilled amateurs. This examination had a small element of practical arboriculture in the section on 'Horticultural Operations and Practice' (*JRHS* 1901, 262). In 1904, the RHS General was followed by more specific examinations, including in 1906 by an examination for employees in public parks (Elliott 2004, 318). The first suggestion of a National Diploma as a higher examination in horticulture was made at a meeting of the recently formed Horticultural Education Association in 1911. The examination was to be in three stages: a qualifying test, a Preliminary Examination covering horticultural operations (which included the planting, pruning and training of ornamental trees and shrubs) and a Final Examination in a specialist subject of the candidate's choice. It was hoped that the practical emphasis on the exam would circumvent the old criticisms about 'academic uselessness'. However, William Robinson still thought there were unnecessary subjects such as arithmetic and mensuration which he believed would 'preclude most of the best gardeners of today from obtaining the diploma, even if they desired to do so' (Robinson 1913). Meanwhile, horticultural education was now being developed at some universities, with Wye College and the Universities of Reading and Nottingham all offering horticultural degrees before the Second World War (Elliott 2004, 320). However, any element of arboriculture in the syllabus of these courses was quite limited, particularly in terms of tree maintenance and tree management.

Forestry education was also making progress, particularly in some universities with the development of degree-level courses. When Coopers Hill College closed in 1906, the forestry section, together with William Schlich, the senior academic, was transferred to Oxford University. However, the emphasis on timber production continued because in the view of the staff the subject of amenity trees, particularly in urban areas, was not regarded as forestry. Meanwhile, a Departmental Committee of the Board of Agriculture had been formed in 1902 to consider a range of forestry issues (Nisbet 1905, 37–42). One practical outcome of the committee's recommendations was the award of a grant of £250 per annum by the Board of Agriculture and Fisheries for the establishment of a lectureship in forestry at the University College of North Wales in Bangor. This led, in 1904, to the establishment of a School of Forestry. A landmark was achieved in 1916 when the first female graduate in Forestry Studies in the world, Mary Sutherland, graduated from Bangor. She went on to work in the New Zealand Forestry Service and played a major role in the development of that service (Roberts 2009, 29; Edwards 2020).

Outside of the university sector, vocational qualifications in forestry could be obtained through the Royal English Arboricultural Society, which had held its first examinations in 1897 (James 1982, 105). In 1913, due to a lack of candidates for its examinations, it revised these and instituted the Woodman's Certificate where the written test was discontinued. For the higher level Forester's Diploma, those candidates wishing to take this would be required to obtain the Woodman's Certificate first and then pass an examination that was entirely written and oral (*ibid.*, 106). Following these changes, there was a significant increase in the number of people taking the Society's forestry examinations.

In 1936, William Dallimore wrote an enlightening article on training in forestry and arboriculture for the *Quarterly Journal of Forestry* (Dallimore 1936). He begins by making a clear distinction between what is required in the training of foresters and that of arboriculturists. He states:

> Foresters are usually trained purely as foresters, whereas arboriculture is often regarded as a branch of horticulture and the cultivation and care of decorative trees is looked upon as a garden question. Professionally, foresters are only interested in the comparatively few kinds of trees or shrubs that can be grown as a commercial proposition in their respective countries, but arboriculturists must interest themselves in every tree and shrub that is likely to succeed.
>
> Dallimore 1936, 37

He then goes on to demonstrate how social class and educational background had such an impact on career development. He explains:

> In this country training in forestry is conducted along two separate, well defined lines. In one case boys proceed from schools to universities where they take a science or arts degree supplemented by a definite period of instruction in forest practice, with one or more vocation courses in forest areas, or they may work for a forestry degree

> only. … The other means of training is for boys to enter forests on leaving school to learn the practical side of the work. After spending at least one year, but often several years, in the forest, they may enter a forest school for two years where practical work is interspersed with courses of lectures on the theoretical side of forestry. … At the end of the training period they can obtain a certificate, by examination, from the Forestry Commission, and they may sit for a diploma examination set by either the Royal English or the Royal Scottish Forestry Society. These men take posts as foremen or foresters in State, Municipal or private forests. A few may obtain minor posts in the Colonial forest services. They usually have a good knowledge of the practical side of tree cultivation, estate work, and the management of men … but they usually lack the polish of public school and university trained men.
>
> Dallimore 1936, 39

While Dallimore is referring to forestry, the same basic framework also existed in horticulture. However, for those wishing to embark on a career in arboriculture, there was no obvious career path. There were no qualifications in arboriculture at any level and no professional body to represent those in the arboricultural industry, whether they were students or experienced professionals. For the time being, anyone wishing to train in arboriculture would have to go down the route of either forestry or horticulture and then try to specialise at a later date.

The Second World War

In 1939, the outbreak of the Second World War put a major brake on any developments in British arboriculture for the duration of the conflict. Public and private sector spending on tree work fell dramatically, as did the number of people employed on planting and maintaining urban trees, now that the nation had more urgent priories. At the same time, considerable damage was inflicted on our urban trees, and not just as a result of German bombing (Johnston 2017, 267–270). Trees that required regular maintenance were neglected and it was often only emergency work that was undertaken where there was a risk to people or property.

Despite the war bringing massive disruption to the arboricultural industry, there is evidence that some tree surgery companies continued to operate and some regular pruning work was done in streets and public parks. For example, in 1940 it was reported that the well-known yew tree in Dunsfold Churchyard, Surrey, 'has lately undergone treatment at the hands of a firm of experts in tree surgery' (*West Sussex Gazette* 1940). Later in the war a few tree surgery companies, such as Thomas Higginson of Warwick, were still promoting their services but only in a limited way in the classified advertisements of local papers (*Warwick Advertiser* 1942). These companies were able to continue because they also undertook much-needed tree felling and emergency tree work and were recognised as being in a reserved occupation. With regard to routine tree pruning by local authorities, there is some evidence of this. In Eastbourne in 1941, a disgruntled resident wrote to the local paper to ask why the Corporation's gardeners were 'lopping the trees in the Avenue when the sap is rising'

(Uppertonian 1941). However, this criticism of local authority tree work was most unusual during the war years. Although vast numbers of trees were being neglected, and any maintenance work was probably done by inexperienced men, it was probably viewed as unpatriotic to complain about this publicly. Some local authority officers with specific responsibilities for trees continued their work through the war years. For example, in December 1943 the Forestry Officer for Essex County Council was advising the removal of an avenue of poplar trees adjacent to a highway on the grounds they were unsuitable for the situation (*Essex Chronicle* 1943). Many local residents objected and the Highways Committee of Witham Urban Council decided to only remove a few trees and pollard the rest, meaning, in the words on one dissenting councillor, 'the advice of the expert had been rejected'. At least one local authority found a supplementary source of labour for arboricultural work during the war. The London Country Council used German prisoners of war to pollard an avenue of lime trees in Greenwich Park (Smith pers. comm., 2019). It seems likely that this unusual team of arborists undertook other tree work in London and other local authorities may also have used German prisoners in this capacity.

As with the previous conflict, the nation's priorities regarding trees now focused on fruit trees and forestry. Although no direct evidence could be found, it seems likely that in commercial forestry districts some arborists and gardeners who were too old to enlist were transferred into forestry work. Nevertheless there was still a shortage of forestry workers and women were again asked to fill the gap. Although it had operated from 1940, the Women's Timber Corps (WTC) was formally constituted in 1942, with the women affectionately called 'Lumberjills' (Vickers 2011, 101). From 1940 to 1945 the WTC employed just over 6,000 women who undertook various forestry tasks that had previously been done by men, including felling and timber extraction. One anonymous observer remarked, 'Some of them have even vied with the professionals in felling big timber' (Anon 1944, 14).

Although there was a dearth of literature on the techniques of tree felling before the war, a fascinating booklet on this topic appeared in 1942. Entitled *Tree Felling on Active Service* it was written by Captain N.D.G. James who would later be acknowledged as an authority on forestry and arboricultural matters. The booklet describes in detail, with numerous diagrams, how to fell trees and the equipment required to do this (James 1942). It also had a chapter on 'tactical fellings' that showed how trees can be felled to create roadblocks, provide cover for tank-hunting personnel, form footbridges across rivers and build timber roads across muddy terrain. As part of this initiative, some soldiers received practical instruction in tree felling and crosscutting timber.

Conclusions

The major increase in amenity tree cover in Britain that began in Victorian times, especially in our towns and cities, continued into the first three decades of the twentieth century. As many of these trees were now reaching maturity or were sufficiently mature to need regular pruning, this generated a huge amount of arboricultural work and created significant employment opportunities for those with the necessary skills. However, due to the lack of published accounts about the development of professional arboriculture in Britain during this period, it has been widely assumed that not much of significance occurred. However, research for this chapter has discovered a wealth of material that indicates this was a crucial period in the development of both public and private sector arboriculture.

The significant increase in arboricultural work could have been carried out under the auspices of horticulture, landscape gardening, forestry or some other discipline that had embraced this work in the past. What was different in the 1900s was that arboriculture was now widely recognised as a separate discipline, following a trend that had begun in the late Victorian era and was generally accepted by the 1930s. Along with that recognition of arboriculture as a distinct specialism was the realisation that it required specialists to undertake much of this work. While tree planting and basic pruning could be undertaken by gardeners and foresters, activities such as tree surgery, large branch removal, sectional felling, mature tree transplanting and advanced tree inspections and tree surveys all required specialist training and experience. Whenever arboricultural work of this type was required, people increasingly looked to the arborist, arboriculturist or tree expert who had the necessary skills and experience to undertake it. Without this, mistakes and bad practice would not only be harmful to the tree but could also lead to damage to adjacent property as well as injury or worse to people nearby.

From the 1900s onwards, a small but increasing number of local authorities began employing tree experts or arboriculturists, either as full-time members of staff or as outside consultants. Those that were local authority employees often had tree-specific job titles, such as Tree Warden, Tree Expert or Arboriculturist. At the same time in the private sector a small but increasing number of tree surgery companies were formed as well as arboricultural consultancy practices. These developments and others signalled the emergence of a distinct arboricultural industry in Britain. Nevertheless, there were still some major elements missing, such as a professional body to represent those in the arboricultural industry, to promote greater recognition for this and to encourage high standards of work. What was also much needed was the establishment of an educational system for the industry, to develop formal courses and qualifications in the subject at both further and higher education level. Unfortunately, the advent of the Second World War in 1939 put an enormous brake on any further developments for the duration of the conflict. It remained to be seen how much of the progress achieved over the previous three decades would continue when the conflict eventually ended and how long it would take to rebuild the momentum of the emerging arboricultural industry.

References

Aberdeen Press (1930) Planting on Roadsides. *Aberdeen Press and Journal*, 13 October, 9.

Adkin, B.W. (1914) *The Law of Forestry: In Great Britain and Ireland, Comprising the Law Relating to Trees*. The Estates Gazette: London.

Alloa Advertiser (1905) In Brief. *Alloa Advertiser*, 18 February, 2.

Andrews, B. (2012) John Davey: A Hero of Arboriculture. *The Arb Magazine* 157, 27–29.

Anon (1927) In Memoriam: Walter H. Aggett. *The Kew Guild Journal* 4, 525.

Anon (1944) *Meet the Members – A Record of the Timber Corps of the Women's Land Army*. Bennett Brothers: Bristol.

Anon (1969) A.D.C. Le Sueur. *Forestry* 42(2), 208.

Bailey, L. H. (1898) *The Pruning-Book: A monograph of the pruning and training of plants as applied to American conditions*. The Macmillan Co.: New York.

Balfour, F.R.S. (1935) The Planting and After Care of Roadside Trees. *Quarterly Journal of Forestry* 29, 163–188.

Balfour, I.B. (1911) Tree Pruning in the Mall. *The Gardeners' Chronicle*, Series 3, 50, 144–146.

Bartlett, A.C. (1915) Practical Tree Repair. *The Gardeners' Chronicle*, Series 3, 58, 225.

Bartlett Manufacturing (1933) *Bartlett Tree Trimming Equipment*. Booklet Number Nineteen. Bartlett Manufacturing Company: Detroit, MI.

Bean, W.J. (1905) Transplanting. *The Gardeners' Chronicle*, Series 3, 38, 353–355 and 373–374.

Bean, W.J. (1906) Pruning Trees and Shrubs. *The Gardeners' Chronicle*, Series 3, 39, 226–227 and 244–246.

Bean, W.J. (1914) *Trees and Shrubs Hardy in the British Isles*. In two volumes. John Murray: London.

Bean, W.J. (1925) *Ornamental Trees for Amateurs*. Country Life: London.

Bean, W.J. (1929) The Planting of Roads. *The New Flora and Silva* 1(4), 247–252.

Benton, A.M. (2002) *Cheals of Crawley: The Family Firm at Lowfield Nurseries 1860s to 1960s*. Moira Publications: Uckfield.

Birmingham Post (1939) London Letter for Women: An Idea from America. *The Birmingham Post*, 29 March, 15.

Blair, D.F. (1995) *Arborist Equipment*. International Society of Arboriculture: Savoy, IL.

Blair, M.F. (1937) *Practical Tree Surgery*. The Christopher Publishing House: Boston, MA.

Boulton, E. (1969) A.D.C. Le Sueur, OBE: An Appreciation. *The Arboricultural Association Journal* 1(9), 247–248.

Bridgeman, Peter (2013) Kewites in Arboriculture. *The Journal of the Kew Guild* 16(117), 215–221.

Buchanan, R.A. (1985) Institutional proliferation in the British Engineering Profession, 1847–1914. *The Economic History Review* 38(1), 42–60.

Bucks Herald (1939) Remedial Treatment of a Tree. *Bucks Herald*, 14 July, 15.

Burnley Express (1928) Beautiful Burnley: Tree Planting in the Streets. *Burnley Express*, 1 December, 18.

Burnley News (1928) 'A Great Fallacy': Withering of Trees Not Due to Air Impurities. *The Burnley News*, 3 October, 5.

Campana, R.J. (1999) *Arboriculture: History and Development in North America*. Michigan State University Press: East Lansing, MI.

Chargueraud, A. (1896) *Les Arbres de la Ville de Paris*. J. Rothschild: Paris.

Cheshire Observer (1930) Classified advertisement for A. Harley, Tree Expert. *The Cheshire Observer*, 8 March, 9.

Conway, H. (1991) *People's Parks: The Design and Development of Victorian Parks in Britain*. Cambridge University Press: Cambridge.

Cooley, L.B. (2007) *The Bartlett Way: 100 Years of Scientific Tree Care*. The F.A. Bartlett Tree Expert Company: Stamford, CT.

Country Life (1910) Mawson Brothers advertisement. *Country Life* 27, 693, 16 April, lxxxiii.

Country Life (1921) Dr A.H. Unwin advertisement. *Country Life* 49, 1264, 26 March, lxvi.

Country Life (1922) John Klinkert, Kew Topiary Nurseries, advertisement. *Country Life* 51, 1311, 18 February, xxxix.

Country Life (1923) Trewhella Monkey Winch advertisement. *Country Life* 53, 1362, 10 February, lxii.

Country Life (1925a) Wade Portable Crosscutting Saw advertisement. *Country Life* 58, 1489, 18 July, xl.

Country Life (1925b) Hillier and Son, 'Tree Specialists' advertisement. *Country Life* 58, 1513, 5 December, clxxix.

Country Life (1926) Howden's Hardy Highland Trees, advertisement. *Country Life* 60, 1551, 9 October, lxv.

Curtis, C.H. and Gibson, W. (1904) *The Book of Topiary*. John Lane, The Bodley Head: London and New York.

Dallimore, W. and Jackson, A.B. (1923) *A Handbook of Coniferae (Including Ginkoaceae)*. Edward Arnold: London.

Dallimore, W. (1903) Transplanting Large Trees. *The Garden* 64, 140–141 and 355–356.

Dallimore, W. (1926) *The Pruning of Trees and Shrubs: Being a Description of the Methods Practised in the Royal Botanic Gardens, Kew*. Dulau: Oxford.

Dallimore, W. (1928) Trees in Roads and Streets; Their Establishment and Maintenance. *Quarterly Journal of Forestry* 22, 19–28.

Dallimore, W. (1930) Should the Society Pay More Attention to Arboriculture? *Quarterly Journal of Forestry* 24, 150–153.

Dallimore, W. (1936) Training in Forestry and Arboriculture. *Quarterly Journal of Forestry* 30, 37–44.

Davey, J. (1901) *The Tree Doctor: A Book on Tree Culture*. Published by the Author: Akron, OH.

Davey, J. (1907) *The Tree Doctor: The Care of Plants and Trees*. The Saalfield Publishing Co.: New York.

Derby Telegraph (1932) Herculean Task for Barrowash Firm. *The Derby Daily Telegraph*, 13 January, 7.

Derby Telegraph (1935) A Derby Man's Dairy – Trees in the Town. *Derby Evening Telegraph*, 25 October, 8.

Des Cars, A. (1881) *Tree Pruning: A Treatise on Pruning Forest and Ornamental Trees*. Translated from the seventh French edition by Charles S. Sargent. A. Williams and Company: Boston, MA.

Devon Gazette (1930) Classified advertisement for pruning equipment by W. Otton and Sons, Exeter. *Devon and Exeter Daily Gazette*, 11 December, 1.

Edlin, H.L. (1949) *Woodland Crafts in Britain*. B.T. Batsford: London.

Edwards, V. (2020) *A Path Through the Trees: Mary Sutherland – Botanist and Women's Advocate*. Writes Hill Press: Wellington, New Zealand.

ELHG (2020) *Hooke Farm and Effingham Common.* An oral history from Ivor Mitchell Gillespie, recorded on 5 July 2010. Effingham Local History Group: Effingham.

Elliott, B. (2004) *The Royal Horticultural Society: A History 1804–2004.* Phillimore: Chichester.

Elwes, J.H. and Henry, A. (1906) *The Trees of Great Britain and Ireland.* Printed privately: Edinburgh.

Essex Chronicle (1933) Police News: Castle Hedingham. Motorist and Bus. *The Essex Chronicle* (Chelmsford), 10 December, 5.

Essex Chronicle (1937) Essex County Council: 'Green Murder'. *The Essex Chronicle,* 5 March, 5.

Essex Chronicle (1943) Silver End – Poplars to Survive. *The Essex Chronicle,* 3 December, 11.

Essex Chronicle (1945) Girl Crushed to Death by Tree Bough. *The Essex Chronicle,* 26 October, 12.

Essex Newsman (1938) Brentwood: Trees. *The Essex Newsman,* 8 October, 1.

Fernow, B.E. (1910) *The Care of Trees in Lawn, Street and Park.* Henry Holt and Co.: New York.

Fisher, W.R. (1907) Introductory Note. *Quarterly Journal of Forestry* 1, 3.

Forbush, E.H. and Fernald, C.H. (1896) *The Gypsy Moth.* Wright and Potter: Boston, MA.

Gardeners' Chronicle (1896) Street Planting. *The Gardeners' Chronicle,* Series 3, 20, 400.

Gardeners' Chronicle (1903a) Beckett's Tree-ties. *The Gardeners' Chronicle,* Series 3, 34, 327.

Gardeners' Chronicle (1903b) Preparing for Transplantation. *The Gardeners' Chronicle,* Series 3, 33, 85–86.

Gardeners' Chronicle (1903c) Royal Horticultural Exhibition at Holland House. *The Gardeners' Chronicle,* Series 3, 33, 419–422.

Gardeners' Chronicle (1904a) Midland Refforesting Association. *The Gardeners' Chronicle,* Series 3, 36, 447.

Gardeners' Chronicle (1904b) Tree-lifting in St James' Park. *The Gardeners' Chronicle,* Series 3, 35, 276–277.

Gardeners' Chronicle (1904c) Cutting Down Trees by Electricity. *The Gardeners' Chronicle,* Series 3, 35, 10–11.

Gardeners' Chronicle (1905) British Gardeners' Association. *The Gardeners' Chronicle,* Series 3, 37, 364–366.

Gardeners' Chronicle (1907) Transplanting Large Trees. *The Gardeners' Chronicle,* Series 3, 41, 222.

Gardeners' Chronicle (1909) Transplanting a Large Cedar at Paddockhurst. *The Gardeners' Chronicle,* Series 3, 45, 397–398.

Gardeners' Chronicle (1910) The Management of Common and Street Trees. *The Gardeners' Chronicle,* Series 3, 47, 232–233.

Gardeners' Chronicle (1911) Tree Pruning in the Mall. *The Gardeners' Chronicle,* Series 3, 50, 144.

Gardeners' Chronicle (1914a) Unskilful Tree-Felling. *The Gardeners' Chronicle,* Series 3, 55, 57.

Gardeners' Chronicle (1914b) The Veitchian Managers. *The Gardeners' Chronicle,* Series 3, 56, 401.

Gardeners' Chronicle (1915) Destroying Tree Stumps. *The Gardeners' Chronicle,* Series 3, 58, 388.

Gardeners' Chronicle (1916) The Law as to Encroaching Tree-roots. *The Gardeners' Chronicle,* Series 3, 59, 6–7.

Gardeners' Chronicle (1919) Classified advertisement for the British Gardeners' Association. *The Gardeners' Chronicle*, Series 3, 65, 1 February, iii.
Gardeners' Chronicle (1920) Retirement of Mr A.D. Webster. *The Gardeners' Chronicle*, Series 3, 68, 14.
Gardeners' Chronicle (1922) Topiary as an Aid to Advertising. *The Gardeners' Chronicle*, Series 3, 71, 106.
Gardeners' Chronicle (1923) Staking Young Trees. *The Gardeners' Chronicle*, Series 3, 74, 161.
Gardeners' Chronicle (1930) Shortage of Street Trees. *The Gardeners' Chronicle*, Series 3, 88, 313.
Gardeners' Chronicle (1935) Roadside Trees. *The Gardeners' Chronicle*, Series 3, 98, 73.
Gardening Illustrated (1922) Palmer's Tree Tie advertisement. *Gardening Illustrated* 44, 21 January, vii.
Gossiper (1931) Gossip of the Day: A Famous Firm. *The Derby Daily Telegraph*, 13 August, 4.
Hall, W. (1977) *Barnacle Parp's Chain Saw Guide*. Rodale Press: Emmaus, PA.
Hiley, W.E. (1917) Tree Wounds and Disease – book review. *Quarterly Journal of Forestry* 11, 121–125.
Hillier, J. (2014) *Hillier: The Plants, the People, the Passion.* Hilliers Nurseries: Winchester.
Hollow, M. (2011) Suburban Ideals on England's Interwar Council Estates. *Garden History* 39(2), 203–217.
Hubert, E.E. (1931) *An Outline of Forest Pathology*. Wiley and Sons: London.
Hübner, O. (1914) *Der Straßenbaum in der Stadt und auf dem Lande.* Paul Parey: Berlin.
James, N.D.G. (1942) *Tree Felling on Active Service.* A. Wheaton: Exeter.
James, N.D.G. (1981) *A History of English Forestry*. Basil Blackwell: Oxford.
James, N.D.G. (1982) *A Forest Centenary: A History of the Royal Forestry Society of England, Wales and Northern Ireland.* Basil Blackwell: Oxford.
Jekyll, G. (1899) *Wood and Garden*. Longman, Green and Co.: London.
Jennings, A. (2005) *Edwardian Gardens*. English Heritage: London.
Johnston, M. (2015) *Trees in Towns and Cities: A History of British Urban Arboriculture.* Windgather Press: Oxford.
Johnston, M. (2017) *Street Trees in Britain: A History.* Windgather Press: Oxford.
JRHS (1901) Examination in Horticulture. *Journal of the Royal Horticultural Society*, New Series, 26, 261–263.
Kotinsky, J. (1921) *Insects Injurious to Deciduous Shade Trees and their Control.* Farmers' Bulletin 1169, United State Department of Agriculture: Washington, DC.
Landscape and Garden (1935) Barron Tree Moving advertisement. *Landscape and Garden* 2(2), iii.
Landscape and Garden (1936) Advertisement for Adrian Estler, Tree Expert. *Landscape and Garden* 3(1), 4.
Leeds Mercury (1939) News from the Parks – This Will Be a Lovely Year for Roses. *Leeds Mercury*, 13 April, 4.
Le Sueur, A.D.C. (1929) Roadside Planting. *Journal of the Royal Horticultural Society* 54, 412–422.
Le Sueur, A.D.C. (1931) The Name of our Society. *Quarterly Journal of Forestry* 25, 271–273.
Le Sueur, A.D.C. (1934) *The Care and Repair of Ornamental Trees.* Country Life: London.
Le Sueur, A.D.C. (1946) Safety for Tree Workers. *Quarterly Journal of Forestry* 40, 88–90.

Le Sueur, A.D.C. (1949) *The Care and Repair of Ornamental Trees*. Second edition. Country Life: London.

Le Sueur, A.D.C. (1963) The Park Royal Tree Plantings. *Journal of the Royal Horticultural Society* 88(6), 263–269.

Lewis Castle, R. (1906) Trees for Towns. *Journal of the Royal Horticultural Society*, New Series, 31, 84–89.

Linnard, W. (1985) Angus Duncan Webster: A Scottish Forester at Penrhyn Castle, North Wales. *Scottish Forestry* 39, 265–274.

Lloyd, N. (1925) *Garden Craftsmanship in Yew and Box*. Ernest Benn: London.

Long, A.P. (1911) The Removal of Tree Stumps. *The Gardeners' Chronicle*, Series 3, 50, 411.

Long, H. (1993) *The Edwardian House: The Middleclass Home in Britain 1880–1914*. Manchester University Press: Manchester and New York.

Matthews, W.E. (1970) The Tree Scene in Britain. *Weeds, Trees and Turf*, August, 20–24.

Mawson, T.H. (1900) *The Art and Craft of Garden Making*. B.T. Batsford: London.

Mawson, T.H. (1911) *Civic Art: Studies in Town Planning, Parks, Boulevards and Open Spaces*. B.T. Batsford: London.

Menon, K.G. (1916) Trewhella Monkey Winch in Forest Works. *The Indian Forester* 42(4), 224–225.

Middlesex Times (1926) Ealing Common Trees: Expert Decides Against Planes. *The Middlesex County Times*, 13 February, 13.

Midland Telegraph (1905) Ladies Killed by a Tree – The Inquest. *The Midland Daily Telegraph* (Coventry), 15 November, 2.

Morning Post (1905) Pine Trees at Bournemouth. *Morning Post*, 21 April, 7.

Munro Ferguson, R.C. (1901) The Arboricultural Adornment of Towns. *Transactions of the Royal Scottish Arboricultural Society* 16, 388–397.

Nisbet, J. (1905) *The Forester: A Practical Treatise on British Forestry and Arboriculture for Landowners, Land Agents, and Foresters*. Volumes 1 and 2. William Blackwood: Edinburgh.

Northampton Mercury (1935) Advertisement for E. Whatton and Sons, Hartwell. *Northampton Mercury and Herald*, 31 May, 8.

Notcutt, R.C. (1926) *A Handbook of Flowering Trees and Shrubs for Gardeners*. Martin Hopkinson: London.

O'Reilly, C. (2013) 'We Have Gone Recreation Mad': The Consumption of Leisure and Popular Entertainment in Municipal Public Parks in Early Twentieth Century Britain. *International Journal of Regional and Local History* 8(2), 112–128.

Osborn, A. (1914) Transplanting Large Trees and Shrubs. *The Garden* 78, 68–69.

Osborn, A. (1933) *Shrubs and Trees for the Garden*. Ward, Lock: London.

Parker, F.W. (1930) Trees in London. *Quarterly Journal of Forestry* 24, 253–276.

Peets, A. (1915) *Practical Tree Repair*. The Field and Queen: London.

Peets, E. (1925) Tree Surgery. *The Gardeners' Chronicle*, Series 3, 77, 205.

Penrith Observer (1924) Tree Surgery: A Career for Women. *The Penrith Observer*, 2 September, 6.

Pettigrew, W.W. (1928). *The Influence of Air Pollution on Vegetation*. Paper read at the conference in Harrogate of the Smoke Abatement League of Great Britain, 29 September, 1928. Published by the Author: Manchester.

Pettigrew, W.W. (1937) *Municipal Parks: Layout, Management and Administration*. The Journal of Park Administration: London.

Pfleger, R.E. (1977) *Green Leaves: A History of The Davey Tree Expert Company*. The Pequot Press: Connecticut.

Pirone, P.P. (1941) *Maintenance of Shade and Ornamental Trees*. Oxford University Press: New York.

Pope, A. (1931) The Name of our Society. *Quarterly Journal of Forestry* 25, 268–271.

QJF (1931) Report of AGM, 1 September 1931. *Quarterly Journal of Forestry* 25, 359–360.

QJF (1932) Report of Council Meeting, 18 November 1931. *Quarterly Journal of Forestry* 26, 52.

QJF (1938a) Spear-Fast Saw advertisement. *Quarterly Journal of Forestry* 32, vii.

QJF (1938b) Stihl Chainsaw advertisement. *Quarterly Journal of Forestry* 32, ix.

QJF (1938c) English Tree Expert Company advertisement. *Quarterly Journal of Forestry* 32, January, xxv.

Ragg, L. (ed.) (1935) *The Tree Lover: A Quarterly Magazine*. Alexander Moring Ltd: London.

Roberts, D. (2009) *Bangor University 1884–2009*. University of Wales Press: Cardiff.

Robinson, W. (1869) *The Parks, Promenades and Gardens of Paris*. John Murray: London.

Robinson, W. (ed.) (1905) *Flora and Sylva*. Complete set published in three volumes, 1903–1905. Gardening Illustrated: London.

Robinson, W. (1913) National Diploma in Horticulture. *The Garden* 77, 443.

RTPI (1989) *Planning for Town and Country: Context and Achievement, 1914–1989*. 75th Anniversary brochure. Royal Town Planning Institute: London.

Schama, S. (2002) *A History of Britain. Volume 3: The Fate of the Empire 1776–2000*. BBC Worldwide: London.

Scrimgeour, J. (1916) A New Steam Tree-Feller. *Transactions of the Royal Scottish Arboricultural Society* 30, 126–128.

Shigo, A. and Marx, H. (1976) *RX for Wounded Trees*. AIB-387. United States Department of Agriculture Forest Service: Washington, DC.

Simpson, J. (1901) Forestry: Mistakes in Pruning and Planting. *The Gardeners' Chronicle*, Series 3, 29, 233–234.

Simpson, J. (1905) *The Estate Nursery: A Handy Book for Owners, Agents, and Woodmen on the Propagation and Rearing of Forest Trees*. The Country Gentlemen's Association: London.

Slate, V. (1928) A French Method for Inducing the Quick Establishment of Trees and Shrubs. *The Gardeners' Chronicle*, Series 3, 83, 87.

Solotaroff, W. (1911) *Shade-Trees in Towns and Cities*. John Wiley and Sons: New York.

Surrey Advertiser (1939) Classified advertisement for staff by the English Tree Expert Company. *Surrey Advertiser and County Times*, 22 July, 22.

Sutherland, M. (1918) Women's Labour in Forestry. *Transactions of the Royal Scottish Arboricultural Society* 32, 81–84.

Tewkesbury Register (1933) Classified advertisement for Thomas Higginson, Warwick and London. *The Tewkesbury Register and Gazette*, 19 August, 1.

Thanet Advertiser (1900) The Town Trees: Expert Advice to Be Had. *Thanet Advertiser*, 6 January, 8.

The Citizen (1921) Classified advertisement for tree felling and lopping. *The Citizen* (Gloucester edition), 2 February, 1.

The Courier (1936) 'Preserve Your Trees', advertisement by James Laurie and Son. *The Courier and Advertiser* (Dundee), 6 February, 1.

The Field (1909) Trewhella Monkey Jack advertisement. *The Field*, 23 January, xv.

The Field (1911) Answers to Correspondents: Garden. *The Field*, 16 December, 1295.

The Garden (1900) Dressing and Stoppings for Tree Wounds and Decaying Hollows. *The Garden* 58, 314.

The Garden (1906) The Tree-Butcher. *The Garden* 69, 206.

The Garden (1907) Pruning of Trees and Shrubs. *The Garden* 71, 305–306.

The Scotsman (1937) Work of Tree 'Doctors' at St Andrews. *The Scotsman*, 25 January, 7.

Thiselton-Dyer, W.T. (1905) The Kew Guild. *The Gardeners' Chronicle*, Series 3, 37, 362–363.

Todmorden Advertiser (1907) Lancashire Naturalists of Note: Mr Abm. Stansfield. *Todmorden Advertiser and Hebden Bridge Newsletter*, 27 September, 3.

Trans. RSAS (1901) List of Members and Past Presidents. *Transactions of the Royal Scottish Arboricultural Society* 16, Supplement 1–22.

Trans. RSAS (1913) *The Care of Trees in Lawn, Street and Park* by Bernhard E. Fernow. *Transactions of the Royal Scottish Arboricultural Society* 27, 119–120.

Uppertonian (1941) Avenue Trees: Letter to the Editor. *Eastbourne Gazette*, 7 May, 4.

Vickers, E. (2011) 'The Forgotten Army of the Woods': The Women's Timber Corps during the Second World War. *Agricultural History Review* 59(1), 101–112.

Wardrop, J. (1976) British Columbia's Early Experience with Chainsaws. British Columbia Provincial Museum: Victoria, BC.

Warwick Advertiser (1931) Advertisement for Thomas Higginson, Tree Felling and Topping Contractor. *Warwick and Warwickshire Advertiser*, 21 February, 1.

Warwick Advertiser (1942) Classified advertisement for Thomas Higginson. *Warwick and Warwickshire Advertiser*, 18 December, 2.

Watson, J.G. (1912) The Woburn Forests. *The Gardeners' Chronicle*, Series 3, 52, 422.

Webber, J. (2008) Greening the Black Country: The Work of the Midland Re-afforesting Association in the Early Twentieth Century. *Arboricultural Journal* 31(1), 45–62.

Webster, A.D. (1893a) *Hardy Ornamental Flowering Trees and Shrubs*. Gardening World Office: London.

Webster, A.D. (1893b) *Webster's Practical Forestry*. William Rider and Son: London.

Webster, A.D. (1903) *Webster's Forester's Pocket Diary and Note Book*. W. Rider and Son: London.

Webster, A.D. (1909) Town Planting. *The Gardeners' Chronicle*, Series 3, 45; 220, 262 and 400.

Webster, A.D. (1910) *Town Planting: And the Trees, Shrubs, Herbaceous and Other Plants that are Best Adapted for Resisting Smoke*. George Routledge and Sons: London.

Webster, A.D. (1913) Public Parks and Gardens: Tree Felling and Pruning. *The Gardeners' Chronicle*, Series 3, 53, 207–208.

Webster, A.D. (1916) *Tree Wounds and Diseases: Their Prevention and Treatment*. Williams and Northgate: London.

Webster, A.D. (1918) *Seaside Planting: For Shelter, Ornament and Profit*. T. Fisher Unwin: London.

Webster, A.D. (1920) *London Trees*. The Swarthmore Press: London.

Webster, AD. (1925) Tree Surgery. *The Gardeners' Chronicle*, Series 3, 77, 241.

Wells, D.V. (1970) History of the Roads Beautifying Association. *Arboricultural Association Journal* 1(11), 295–306.

Western Daily Press (1920) Honouring Those who Serve: Bishop of Bristol Plants a Tree. *Western Daily Press*, 2 November, 6.

West London Observer (1928) Advertisement for Canadian Tree Felling Company. *The West London Observer*, 14 December, 15.

West Sussex Gazette (1927) Week at Worthing: Tree Preservation. *The West Sussex Gazette*, 6 October, 11.

West Sussex Gazette (1940) One of Surrey's Well-Known Yews. *The West Sussex Gazette*, 11 April, 4.

Wiltshire Times (1931) The Park Trees: Expert Advice on the Memorial Avenue. *Wiltshire Times and Trowbridge Advertiser*, 10 October, 2.
Wyndham Lewis, D.B. (1937) Standing By. *The Bystander*, 10 February, 217.

CHAPTER NINE

Professional Arboriculture 'Comes of Age', 1946–Present

When the Second World War ended in Europe in May 1945 the physical destruction across Britain was enormous, with many shattered urban centres, industrial areas and housing districts. While the task of rebuilding Britain was daunting, the mood of the nation was one of determination and a recognition that now was the time not only for renewal but also for fundamental change. That mood was reflected in the political situation when the wartime leader Winston Churchill and his Conservative Party lost the General Election of July 1945 to Clement Attlee and the Labour Party.

The Labour government of 1945–1951 introduced radical policies that transformed many aspects of life in Britain. This included much-needed social reforms and a national programme of house building. Britain's post-war recovery was remarkable and by 1957 the Conservative Prime Minister Harold Macmillan confidently declared that most of the British people 'have never had it so good'. The 1960s and 1970s were a period of economic and social turmoil, marked by some protracted industrial disputes and the rise of youth culture. With Margaret Thatcher's Conservative government from 1979–1990 the country moved to the right politically with radical economic policies characterised by deregulation and privatisation. While the immediate post-war government in Britain had involved much state intervention and socialist ideals, Thatcher's government would dramatically reverse many of these policies. One consequence was a reduction in local government expenditure leading to cutbacks in local services in many districts. The return of a Labour government under Tony Blair in 1997 did not signal any major swing to the political left. While there were significant increases in spending on health care and education, there were also some controversial market-based initiatives. In 2008, the global financial crisis broke and caused the Great Recession in the economy of Britain and much of the western world. In response to this the Conservative government introduced a programme of sustained reductions in public expenditure that between 2010 and 2019 cut by more than £30 billion public spending on welfare payments, housing subsidies and social services. Local government services were especially affected with a decrease of almost

60 per cent in central government funding for local authorities in England and Wales between 2010 and 2020 (LGA 2019, 2).

Post-war Britain witnessed a major period of urban development, much of which focused on the urgent need for new housing. The government's urban renewal efforts were led by the New Towns Programme of 1946–1970. This was initially motivated by the need to relocate vast numbers of people in dilapidated and bombed-out housing. It led to the eventual creation of some thirty-two new towns across Britain and Northern Ireland, although not all were completed as originally intended. Whether the results of individual examples of new towns are admired or criticised, the sheer scale of the achievement was remarkable and Britain received international recognition for this pioneering work. It led to millions of working-class people having a better quality of life in their new homes, most of which had small gardens. Under Thatcher's Conservative government the Housing Act 1980 was passed which allowed council tenants to buy their houses and flats. By 1997 this 'Right to Buy' initiative had resulted in the sale of over 1,700,000 council dwellings, leading to a 15 per cent rise in home ownership but also a drastic reduction in the nation's social housing stock (Disney and Luo 2015, 22).

Sources of information

As this chapter covers a period of rapid progress in the development of professional arboriculture in Britain, it is not surprising that a huge amount of research material was obtained. Not only was there a vast increase in material from published literature in the form of books and professional periodicals, this also applied to popular newspapers and magazines. Towards the end of the research process the author made a request through the Arboricultural Association's *Arb Magazine* for contributions of relevant research material relating to the development of British arboriculture in the twentieth century (Johnston 2019). This yielded some very useful written and visual material. During the research process the author undertook twenty-three interviews with some prominent individuals in arboriculture and related disciplines to gain their insight into developments in professional arboriculture in Britain in the twentieth century. While this yielded little material relating to 1900–1945, it generated a wealth of material from 1946 to the present. As a huge amount of material for this period had already been gathered, one of the aims of the interviews was to get guidance on what material was essential to include and how to present this.

Trends in Britain's post-war landscapes

Immediately after the Second World War there was widespread recognition that the reconstruction of Britain's ravaged and neglected towns and cities would involve much landscaping and tree planting. Not only were the nation's publicly owned trees in poor condition but this also extended to privately owned amenity trees. The development of Britain's amenity tree cover in the second half of the twentieth century has already been described in detail (Johnston 2015 and 2017). This also included significant coverage of the constraints and threats to amenity trees. For this reason only an outline is given here to provide some context for the rest of the chapter that focuses directly on the development of professional arboriculture. Also, in previous chapters we have examined garden styles that were popular down the centuries and explored their tree-related features. However, this will not be done for this chapter. The amenity landscapes of Britain in the second half of the twentieth century can be viewed as a glorious collection of almost all those past styles and we are not concerned here with current styles of landscape design. Instead, we will outline two modern trends that have impacted on a wide range of amenity landscapes.

Following the major cutbacks in public sector expenditure that began in the early 1980s, local authorities and other public bodies have continually looked at ways to save money on their parks, gardens and other amenity landscapes. This has impacted on many aspects of the design, maintenance and management of those landscapes. For example, intricate horticultural features such as topiary, stilted hedges or pleached trees were generally avoided as they were expensive to create and maintain. Efforts to save money could also influence the selection of trees where small ornamental species requiring little maintenance were often preferred over large forest-type species that might require major tree work in the future. Arboricultural work has traditionally been very labour-intensive and labour costs have usually been the most expensive element of this. From the 1970s, the introduction of specialist heavy machinery for various types of arboricultural work greatly reduced costs in the long term, provided the initial investment in the machinery was justified.

The second trend emerged with the rising 'green' awareness of the 1980s and 1990s. With growing concerns about threats to the natural environment and the need to protect and promote wildlife, the neatly managed nature of the formal gardens of previous eras seemed out of place to many people. This led to the emergence of an ecological landscape movement that advocated the creation of semi-natural areas, particularly wildflower meadows and semi-natural woodland, in many urban parks, housing estates and other open spaces. While some of these 'ecological landscapes' were very successful, others were less so. Furthermore, this vital message about creating more environmentally friendly and sustainable landscapes has also been used to promote other agendas. During times of severe cuts in public expenditure, the creation of 'semi-natural' landscapes was sometimes viewed as a convenient way to save money by letting these sites 'manage' themselves (Johnston 1983, 279). Public concern about biodiversity loss has also been exploited to promote a 'natives only' policy for urban planting (Johnston *et al.* 2012).

Private gardens and landscapes

The break-up of Britain's landed estates that accelerated dramatically in the inter-war years had left many of their ornamental gardens in a dilapidated state. Now, after several years of wartime emergency many of these once great gardens were in a dire state. The grand houses had often been used as hospitals or accommodation for the armed forces and the upkeep of their gardens was a low priority. For a large number of these great houses and gardens, the National Trust was to be their saviour. From the late 1940s, the National Trust began acquiring major gardens under the Gardens Fund, established by the Trust and the Royal Horticultural Society (Thomas 1979, 20). The National Trust for Scotland also began performing a similar role north of the border. To renovate, maintain and manage these gardens to a high standard, the Trusts needed to employ a small army of skilled gardeners, with many of the head gardeners trained at leading horticultural establishments such as Wisley, Kew and Edinburgh. This has also involved much high-quality arboricultural work. Over the years the National Trusts have employed many practical arborists and arboricultural consultants, not so much as full-time members of staff but on short-term contracts. Other similar sources of arboricultural work have included the parks and gardens managed by English Heritage, the Crown Estate, the Royal Palaces and the Royal Parks in London. These not only have many fine trees and shrubs but also some nationally important tree collections, providing substantial employment opportunities for practical arborists and arboricultural consultants. Lastly, Britain is blessed with a remarkable heritage of several internationally renowned botanic gardens and arboretums as well as many lesser-known examples, all providing work for skilled arboricultural staff.

At the other end of the scale to those magnificent gardens were the small residential gardens of the nation's ordinary home owners and council tenants. The house-building boom of the post-war years created millions of new little residential gardens in the rapidly expanding suburbs and in new towns. However, these were not a major source of work for the arboricultural industry. Popular gardening books and magazines of the 1950s onwards would recommend planting small ornamental trees that required little pruning and could be removed by the home owner when the time came. Most gardens were too small to contain large mature trees that might require the attention of a tree surgery company. Besides, large trees had a bad reputation for causing problems such as blocking out light, encroaching roots and branches, leaves blocking gutters and even subsidence to foundations. When new local authority housing estates were developed, some existing trees were retained in the gardens of the new houses, to create a mature and attractive landscape. If these trees required any major work, this would be organised by the council's Housing Department, often using arborists from the Parks Department. Following the 'Right to Buy' initiative from the early 1980s, those former council tenants with large trees in their gardens now found themselves responsible for them. While no detailed study has been undertaken into this, anecdotal evidence suggests that many of these trees were removed soon after the 'handover', providing some short-term work for local tree surgery companies.

Expansion of municipal arboriculture

The poor condition of many local authority trees at the end of war required some immediate and large-scale responses. A number of authorities had appointed specialist tree officers before the war and some of these had continued working throughout the conflict. Now that the wartime emergency was over those officers had to cope with the public's increased expectations regarding levels of tree planting and standards of tree care. John Williams, the Tree Warden for Essex County Council, who had been appointed in the 1930s, was not only facing an increasing workload in his responsibilities for publicly owned trees, he was still expected to give advice on some high-profile, privately owned trees. For example, in 1949 he was asked to submit a report on what measures could be taken to preserve the famous oak at Great Yeldham in Essex (*Suffolk Free Press* 1949). This was one of Williams's last major assignments as Tree Warden as he retired a few months later. It is interesting that, in seeking a suitable replacement, Essex County Council advertised the post as far away as Scotland (*The Scotsman* 1949).

While existing local authority officers and manual staff did their best to restore the neglected treescapes, it really needed a greatly enlarged workforce of skilled arborists and tree officers. In the early post-war years most new arboricultural jobs were for manual workers. For example, in 1946 the Royal Borough of Leamington Spa advertised for a Forester in its Parks and Gardens Department who was 'fully experienced in arboriculture' (*Leamington Courier* 1946). As the advertisement stated that 'any applicant should not exceed 40 years', this was probably mainly manual work. In 1948, the Borough of Harrogate advertised for a Forester in its Parks Department who 'must possess a thorough knowledge of tree surgery and arboriculture' (*Yorkshire Post* 1948). Again, this was a manual post as the applicant was expected to be 'able to take down dangerous trees on Public Highways and Parks'. In these and similar appointments the applicants were required to have experience of the work, but formal qualifications in arboriculture were not mentioned as these did not exist. Many of these posts were titled 'Forester' and most successful applicants probably had a forestry background rather than a horticultural one. This tendency to view arboricultural manual work as something for a 'forester' continued for some years and even included posts in London. For example, in 1955 the Borough of Sutton and Cheam was advertising for a Forestry and Street Tree Foreman (*West Sussex Gazette* 1955) and in 1968 the London Borough of Ealing was advertising for a Working Forester (*Acton Gazette* 1968a).

While many local authorities were busy appointing arboricultural manual staff, initiatives were also being taken at officer level. The late 1940s and 1950s witnessed a flowering of municipal arboriculture in Britain with London as the initial centre and the London County Council (LCC) as the first 'arboriculturally conscious' post-war local authority. In 1946, the LCC Parks Department created a new division to look after all its trees (Leathart 1973, 103). The person appointed to organise and run this division was Hubert Taylor, a Kew-trained horticulturist who had studied arboriculture there under William Dallimore. With a team of eighty-five tree pruners, Taylor set about rectifying the lopping and topping that had disfigured so many of London's trees in the war years and before. In 1953, he appointed Ted Storey as Arboricultural Officer to take charge of these teams, the first full-time tree post for any local authority in Britain to be titled 'arboricultural' (Storey 1963; Bridgeman 2013, 215). Another team was established to take responsibility for the new Tree Preservation Orders that were being made in London (Leathart 1973, 104). In 1965, Peter Bridgeman was appointed Arboricultural Technical Officer to the now Greater London Council (GLC) under Hubert Taylor and Ted Storey (Bridgeman 1991, 17). Both Storey and Bridgeman were also Kew-trained horticulturists who then decided to specialise in arboriculture. The increasing focus on municipal arboriculture in the 1950s saw the publication of the first British book on urban tree management since Angus Webster's *Town Planting* in 1910. Ronald Morling's *Trees in Towns*, published in 1954, gave some much-needed guidance on this topic, directed mainly at local authorities (Morling 1954a).

With the GLC leading the way in municipal arboriculture, the creation of the London boroughs in 1965 provided opportunities for these new local authorities to also deliver a modern tree management service. Ealing Borough, later the London Borough of Ealing, gained prominence in the 1950s and 1960s with Fred Boddy as Director of Parks, and then under Trevor Preston in the 1970s and 1980s. Preston had trained in horticulture at Kew and joined Ealing's Parks Department in 1963 as Technical Assistant, eventually rising to Chief Leisure Services officer in 1986 (Anon 1991, 306). At the London Borough of Harrow, Don Wells, another Kewite, was appointed as Landscape Architect in 1965 with responsibility for landscaping works and tree preservation orders (Honour 1970, 272). Wells had considerable expertise in arboriculture and was also one of the founders of the newly formed Arboricultural Association. In the late 1960s, Michael Cudd, yet another Kewite, was appointed as Arboricultural Officer to the London Borough of Hounslow where he did much to raise standards of tree care and training (Storey 1984, 319). Cudd later joined the horticultural team at the GLC and remained there until his life and promising career was tragically cut short by a fatal stroke in 1984. In 1972, the City of Westminster created a tree officer post and appointed Wisley graduate Paul Akers who served in this prestigious role with distinction for many years (Ryan 2009a).

Outside of London other local authorities were recognising the importance of having specialist arboricultural staff to deal with their trees and woodlands. At Aberdeen, the authority's long-standing reputation for excellence in arboriculture continued in the early 1960s under Arroll Winning, its Director of Parks. In 1966, Winning was appointed by Sheffield City Council as its Director of Recreation. Apart from raising the standard of arboricultural work in the city, Winning's time at Sheffield was notable for the development of an integrated and specialist arboricultural service across the whole authority (Winning 1973, 158). This was in contrast to almost all local authorities at that time where responsibility for trees and arboricultural matters was split between several different departments (Johnston 2017, 57). This made coordination of the authority's arboricultural services quite difficult and encouraged an inefficient and ineffective service (Britt and Johnston 2008, 545).

The success of the GLC and Sheffield in establishing their own specialist arboricultural services encouraged other local authorities to follow. Generally, it was the larger authorities that were appointing specialist tree officers, often located within the parks department and sometimes with a brief to advise other departments. However, the smaller authorities still tended to rely on a range of existing staff. In some cases this continued to be the Borough Surveyor or Highway Engineer, or often one of the horticultural or parks officers. This reliance on horticultural expertise by many local authorities was criticised by some in the arboricultural industry who felt that these arboricultural operations were being approached from a horticultural viewpoint which they considered was less proficient than this specialist work demanded (Winning 1973, 159).

At the same time as many Borough Councils were creating new tree officer posts, often with the title Arboricultural Officer, the County Councils were continuing to appoint forestry officers (Winning 1973, 152–153). These forestry officers were usually located within the planning department where they assisted in planning control, and often with advisory services in connection with highway trees and tree preservation orders. Some forestry officers also played an important part in land reclamation schemes, especially in the industrialised northern and midland counties.

The reorganisation of local government for England and Wales in 1974 and for Scotland in 1975 had a huge impact on the structure and function of local authorities across Britain. The changes were viewed by some arboriculturists as a significant opportunity for the advancement of municipal arboriculture (Winning 1973, 153). It was hoped that the radical reduction in the number of authorities would give opportunities for the adequate staffing and funding for arboricultural services. The number of tree officers across Britain undoubtedly increased in the wake of the reorganisation. However, the reorganisation did little to halt the growing fragmentation of local authority arboricultural services and in many cases exacerbated it. Many more local authorities now sought to create specialist tree officer posts in their Planning Department, to manage increasing responsibilities for trees through the implementation of planning legislation. Previously, the Planning Department might have sought this specialist arboricultural advice from a tree officer located in the Parks Department.

Initially, many tree officer posts in local authorities were titled Arboricultural Officer and in 1971 the Arboricultural Association produced a leaflet with that title in the hope of encouraging more local authorities to create these posts (Arboricultural Association 1971). However, it rapidly became apparent that the word 'arboriculture' meant little or nothing to most members of the public (Johnston and Rushton 1999, 3). Soon, an increasing number of local authorities were using job titles such as Trees and Woodlands Officer, Tree Protection Officer or just Tree Officer. This was especially the case with new appointments in Planning Departments where there may already have been an Arboricultural Officer in the Parks Department. It is now common practice to refer to all these tree-related posts as 'tree officers'.

During the 1960s and early 1970s there was a trend in local authorities to place their direct labour force on bonus schemes, including those in arboricultural services. These schemes aimed to promote a predictable level of output, while also encouraging workers to increase their output by rewarding this with bonus payments. Some local authorities had simple schemes for arboricultural work where the time taken to complete the operation would vary very little. This might include the planting of a standard-size tree in a park, the annual pollarding of street trees in different size categories or the removal of basal suckers from lime trees. However, other authorities went to the extreme of trying to include every possible arboricultural operation in ridiculously complex schemes (LB Harrow 1977). In many cases, the more complex the scheme, the more generous it was. With some tree teams in London in the 1970s, if they worked at a reasonable pace they could achieve maximum bonus by early afternoon and then have no financial incentive to continue working that day. However, that was hardly the fault of the tree teams. Unfortunately, these anomalous situations encouraged negative attitudes towards local authority arboriculture by the private sector of the industry. Peter Honey of Honey Brothers, a tree care company in Surrey, declared:

> Local authority tree gangs will have to have a far more commercial approach otherwise they will find that carrying out tree work using their own labour may prove to be extremely expensive, and I have had practical experience of this and have seen where local authority teams are abysmally behind commercial teams as far as productivity is concerned.
>
> Honey 1972, 27

This view in the arboricultural industry of 'private sector good, public sector bad' seems to have become more widespread in the 1980s when the privatisation of local authority arboricultural services was in full swing. However, in this author's experience it was an unjustified generalisation.

In 1979, the election of a Conservative government led by Margaret Thatcher marked a radical change in the relationship between central and local government in Britain. From the start there were cuts in government funding for local authorities. The Local Government, Planning and Land Act

1980 introduced a new block grant and compulsory competitive tendering (CCT) for local authority direct labour organisations. The block grant allowed the government to impose grant penalties or 'rate-capping' on councils which exceeded expenditure targets, prompting substantial cuts in their services in order to reduce their rates. Tree planting and management was an obvious target for cuts in many authorities, being a relatively low priority in comparison to some other services. The CCT system exposed many local authority services, including tree work operations, to external competition and privatisation (Ball 1995). CCT further splintered already fragmented local authority tree services into 'the client' and 'the contractor' split and raised concerns in the industry about maintaining high standards in arboricultural work. Unfortunately, the CCT system meant that the cheapest quotation won the contract and there was often little regard to the quality of work. The more established companies generally had higher costs and there were widespread concerns that the system 'let the cowboys back in' (Ball pers. comm., 2019). It also discriminated against tree surgery companies because the all-embracing contracts often included landscape and grounds maintenance work that was beyond the scope of these specialist companies. Furthermore, the large size of the contracts made even medium-sized tree surgery companies reluctant to bid as they were not confident they could muster the resources to deliver them. Roy Finch, a tree contractor from Worcestershire, recalls attending an industry event around this time and being asked by a government minister whether he was pleased now that he was being regularly invited to tender on these large contracts (Finch pers. comm., 2019). The minister was surprised when Finch replied that the paperwork 'went straight in the bin' because it was too complicated and he had little confidence in the tendering process. In 1999, the Labour government replaced CCT in England and Wales with a system known as Best Value.

In 1978, the Standing Committee on Arboricultural Education (SCAE) attempted to give the first authoritative estimate of the number of people working in arboriculture in Britain. It estimated there was a total of 5,000 people in the whole of the arboricultural industry (SCAE 1978, 6). Of these, 9 per cent or 450 were at officer level in local authorities, which included tree officers and technical assistants. These figures were collated from estimates provided by the members of SCAE that included representatives from all sectors of the industry. In 1979, a survey was conducted by the London Borough of Harrow of the arboricultural sections of the London Boroughs (LB Harrow 1979). Of the 32 London Boroughs, excluding the City of London, responses were received from 23, providing the first detailed picture of how arboricultural services were delivered in the capital. This included work relating to planning matters. The number of officers and technical staff for each borough ranged from 1 to 7 with an average of 2.9. The number of manual staff for each borough ranged from 4 to 20 with an average of 12. Two of the boroughs, Kensington and Chelsea and City of Westminster, had no manual staff and used contractors. This survey was conducted just before the privatisation of council services which greatly reduced the extent of their direct labour engaged on arboricultural work, not only in London but nationally. The survey also shows that these tree sections, usually based in a parks or recreation department, often dealt with all arboricultural matters covering both public and privately owned trees. The situation in London where the larger boroughs had distinct tree sections comprising small teams of tree officers was also now being replicated in the larger urban local authorities outside of the capital.

The 1980s and 1990s were a period of great change and uncertainty in municipal arboriculture, with many local authority tree management services under severe pressure from lack of funding and frequent restructuring. Nevertheless, as the twentieth century ended, major advances had been made in municipal arboriculture. For example, 69 per cent of local authorities in Britain now had a least one specialist tree officer who also had a tree-related job title, such as Arboricultural Officer, Tree and Woodlands Officer or Urban Forester (Johnston and Rushton 1999, 3). In the other 31 per cent of local authorities the officer primarily responsible for trees had a general job title such as Parks, Horticultural or Landscape Officer. Furthermore, of all those officers primarily responsible for the tree matters within the local authority, some 62 per cent described themselves as coming from an arboricultural background, some 27 per cent had a professional background in horticulture, landscape or parks management and a further 8 per cent had a background in forestry (*ibid.*, 10). The remaining 3 per cent had various other professional backgrounds. What these figures show is that by the late 1990s arboricultural professionals were now the dominant force in local authority tree management in Britain.

Trees in streets, highways and motorways

Of all the trees in a neglected state at the end of the war, those requiring the most urgent attention were in streets and alongside highways. If they were to fall or lose major branches the consequences could be very serious. In London, the LCC and the boroughs made a priority of these trees. The pre-war pruning regimes for pollarded planes and limes were now re-established to ensure public safety and avoid problems such as blocking out of light to nearby buildings. However, the performance of local authorities varied considerably and a poor standard of arboricultural work was not uncommon. Some leading arboriculturists repeated the old complaint that street trees often came within the jurisdiction of those responsible for roads maintenance, who in most cases had little arboricultural knowledge (Boddy 1947, 72). There was also a growing mood among the public that large forest-types trees were inappropriate for residential streets, especially if this required 'brutal' pruning on a regular basis to keep them in check. By the 1950s, the planting of forest-type street trees had declined dramatically in many local authority districts in favour of using small ornamental trees (Hornsey 1953, 170). As well as being popular with many members of the public, this policy of replacing the 'Victorian relics' had the support of some prominent tree experts such as Fred Boddy (1947, 72) and Wilfred Fox (1953, 204) of the Roads Beautifying Association. Nevertheless, forest-type species such as plane and lime continued to be the dominant species in the older residential streets of many towns and cities. In the late 1960s, however, following the appointment of specialist tree officers in many local authorities, pollarding was again under severe scrutiny as a management practice for street trees (Boddy 1968, 16). This new generation of tree experts sought to find more imaginative and 'aesthetically pleasing' solutions to this traditional method of management. In the early 1970s, many urban local authorities tried to regrow the crowns of their pollarded street trees in an attempt to give them a more attractive and natural appearance (Brown 1972, 36). Unfortunately, the resulting crowns were often unstable due to extensive decay in the pollard points and when high winds occurred branches frequently split out, causing a danger to people, property and vehicles (Johnston 2017, 322).

While the newly appointed trees officers were taking responsibility for street trees, major highways and trunk roads often remained beyond their control. In many cases the highway authority was the county council, which had its own staff to deal with tree matters in relation to these roads. In terms of designing and implementing tree planting schemes for major roads, landscape architects and forestry officers played a major role. In the late 1940s, the Roads Beautifying Association (RBA) was continuing to offer advice to central and local government on roadside tree planting, which now focused on a new network of roads across the country. Some professionals, particularly landscape architects, were not happy with this situation. Richard Sudell, a prominent landscape architect, wrote to Geoffrey Jellicoe, then President of the Landscape Institute, insisting, 'The RBA are not in a position to implement their offer of free advice since they employ no landscape architect' (Gilson 2019, 175). Sudell and many of his colleagues were concerned that the RBA engaged horticulturists, rather than landscape architects, to give advice on tree planting and management to local authorities. In Scotland, the Royal Scottish Forestry Society also continued to offer advice on roadside planting to local authorities north of the border, although most of the RSFS's advisors were described as arboriculturists (*Falkirk Herald* 1947). By the early 1950s, both these roadside-planting schemes had been wound down and landscape architects became increasingly dominant in this field.

From the late 1950s, the government embarked on an ambitious programme to create a network of motorways across Britain. This not only resulted in the loss of many existing trees and woodlands, it also involved a massive amount of tree planting and landscaping once the motorways had been constructed. The policy for this was formulated by the government's Landscape Advisory Committee, now chaired by Sir George Taylor, formerly the Director of the Royal Botanic Gardens, Kew (Dunball 1972, 42). Work on the ground was often directed by landscape architects, with other professionals with a more horticultural approach also playing a role. Sir Geoffrey Jellicoe, possibly the most well-known landscape architect of his generation, had an early and special interest in the landscape design of motorways (Jellicoe 1958). Another influential figure who specialized in tree planting on both motorways and trunk roads was Anthony Dunball who for many years in the 1960s and 1970s was Horticultural Advisor to the Ministry of Transport. Dunball, like Wells, was another former Kew student and in 1964 he would also be one of the founders of the Arboricultural Association.

Another major post-war initiative involving extensive tree planting and landscaping concerned the restoration of derelict land. However, this employed mostly foresters and landscape architects along with specialists in ecology and soils and hardly any arboriculturists were engaged in this work.

Trees and planning

The introduction of new planning legislation in the post-war years opened up opportunities for arboricultural expertise in relation to some planning matters, such as tree preservation orders and trees on development sites. This work not only involved local authority tree officers but also private sector arboricultural consultants.

Tree preservation orders (TPOs) were introduced in the Town and Country Planning Act 1947, and although the legislation has been revised several times the same basic principles apply (Mynors 2002, 410). A TPO can be made by a local planning authority (usually a local council) to protect specific trees or an area of woodland from deliberate damage or destruction if those trees are important to the amenity of an area. TPOs make the felling, lopping, topping, uprooting or otherwise wilful damaging of trees without the permission of the local planning authority a legal offence. There are exemptions to this in exceptional circumstances such as work on a dead or dangerous tree, or work to prevent or remove a nuisance (in a legal sense). The penalties for contravening a TPO are not insignificant and anyone found guilty of this is liable, if convicted in the magistrates' court, of a fine up to £20,000. Serious offences may be committed to trial in the Crown Court where, if found guilty, an unlimited fine can be given. In determining the amount of fine, the court will take into account any financial benefit which has resulted, or is likely to result, from the offence.

A TPO is usually made by the local planning authority if the tree or trees are vulnerable and are worth saving, or this may be requested by a member of the public for the same reason. Much work can be involved in making, administering and monitoring TPOs, especially if the local planning authority has a large number. This work should also involve someone with specialist arboricultural knowledge. As already noted, when many tree officer posts were created in the 1960s and 1970s, these officers often acted as consultants to their planning departments on matters such as TPOs. Then, many planning departments appointed their own tree experts with job titles such as Tree

Preservation Officer. In recent decades, private sector arboricultural consultants have played an increasing role in the TPO system, which has generated a considerable volume of work for them (Davis pers. comm., 2019). Authorities often suggest that owners of trees with TPOs seek the advice of an arboricultural consultant in these matters. In addition to TPOs, similar legislation governs the protection of trees growing in designated conservation areas. With the amount of time and money that goes into administering TPOs, especially where there are a great many, some arboriculturists now question whether this system offers the best use of public money to protect privately owned trees.

Another area of increasing work for arboriculturists relates to planning applications and trees on development sites. When a planning authority receives an application for development it has the opportunity to scrutinise this and assess the impact of the proposals on any trees on the site, along with a wide range of other considerations. In the light of all this, it may reject the application or alternatively grant planning permission, but subject to a number of conditions. In the case of trees, any conditions may include the retention and protection of some existing trees or the planting of new trees. As with TPOs and trees in conservation areas, the responsibility for administering planning applications normally rests with the planning department of the relevant local authority. To ensure that applications are scrutinised effectively and appropriate provisions are made for trees, both now and in the future, the authority needs to employ arboricultural expertise. In 1980, the British Standards Institution published *BS 5837: 1980 Trees in Relation to Construction*, which gives recommendations on how trees on development sites can be protected (BSI 1980). While this has been updated a few times, the basic principles still apply. In the 2012 edition of this British Standard it states that the developer must commission a tree survey and constraints plan, an arboricultural impact assessment and an arboricultural method statement, to provide guidance in drawing up the application. The details of these documents are not relevant here. However, to produce them does require the developer to employ an arboriculturist with experience of development sites to spend a considerable amount of time not only surveying the site but also preparing the documents. Once planning permission has been granted and the development goes ahead, both the local authority and developer need to spend time regularly monitoring the site to ensure that all the necessary tree protection measures are in place, that nothing is being done to put the trees at further risk, and that all tree surgery and tree planting is undertaken as required. For the modern arboricultural consultant, their involvement in planning applications and development sites can account for a large proportion of their income.

One aspect of planning legislation has meant that when trees are destroyed it is often necessary to assess their monetary value. For example, in 1967 the penalty for felling a TPO tree without permission was 'a fine not exceeding £250, or twice the value of the tree, whichever is the greater' (James 1972, 204). Estimating the value of an amenity tree, rather than a timber tree, is not straightforward. Anticipating the need for an agreed method to determine the monetary value of an amenity tree, Rodney Helliwell, an arboriculturist, devised what became known as the Helliwell System (Helliwell 1967). The value is calculated by multiplying together a range of factors: size, useful life expectancy, presence of other trees, relation to setting, form, and very occasionally any 'special factors' such as historic association or very rare species. The final number is converted to a monetary value using an index-linked figure. While some professionals have questioned the basis of the system, it was nevertheless a major step forward at the time and has subsequently featured in numerous court cases, insurance claims and public inquiries.

The introduction of tree preservation orders signalled a more proactive approach towards amenity trees by central government in the post-war era. The focus on tree planting and management in the new towns movement and the many new housing developments prompted the Ministry of Housing and Local Government to publish guidance on this. In 1958, its 83-page book entitled *Trees in Town and City* set out how good tree selection, planting, maintenance, protection and management can contribute much to the beauty of urban areas and to the quality of life of the people who live there (MHLG 1958). However, while the book recognises that this work requires considerable knowledge of trees and the built environment, it avoids giving guidance about the professionals who might have this. It mentions that some local authorities and developers employ arboriculturists and landscape architects, and that some are engaged as outside consultants, but no further details are given. The book was well received by the public and most professionals, but it was also seen by some as a substitute for employing suitably qualified professionals. For example, an editorial in the *West Sussex County Times* (1958) declared, 'No local authority can afford to employ or expect to find by accident on its staff a qualified arboriculturist. But with advice on the choice of trees, their planting and care … this publication could be usefully studied in any council office'.

In the early 1970s, the increasing number of TPO appeals being made to the Department of the Environment (DoE) highlighted the fact that the department had no arboricultural technical input into the appeals process (Davis pers. comm., 2019). To rectify this situation the DoE appointed Colin Bashford as its Arboriculturist in 1975. In this post Bashford was able to influence a range of DoE initiatives and ensure they had some arboricultural input. This included the publication of the DoE Circular 39/78 entitled *Trees and Forestry* that gave guidance on tree-related matters to local authorities in England and Wales (DoE 1978). As the work of the DoE on TPO appeals was continuing to increase, Caroline Davis was appointed in 1986 to support Bashford's work on arboricultural matters. Davis had been Tree Preservation Officer with the London Borough of Hillingdon. When Bashford left the DoE in 1989 to establish his own consultancy firm, Davis replaced him as the DoE's Senior Arboriculturist (Anon 1997, 301). In her time at the DoE, Davis made some important contributions to British arboriculture, including her pivotal role in the commissioning of *Trees in Towns*, a ground-breaking research study on urban trees in England, published in 1993 (DoE 1993). This was also the first in the DoE's series of arboricultural publications entitled Research for Amenity Trees. When Davis retired early due to ill health in the mid-1990s, she was replaced by Peter Annett who had many years' experience as Arboriculturist at the Royal Borough of Kensington and Chelsea. While at the DoE, Annett issued guidance on the TPO system known as the Blue Book (DoE 2000) and played a vital role in the commissioning of *Trees in Towns II*, another ground-breaking survey of urban trees and their management in England (Britt and Johnston 2008). Both of these *Trees in Towns* surveys will be discussed later.

Advances in urban tree management

The creation of many tree officer posts dramatically broadened the scope of professional arboriculture in Britain. Previously, tree experts had focused on individual or small groups of trees or perhaps the treescape of a large landscape park or garden. Now, local authority tree officers were required to focus on all the trees and woodlands within their district. This demanded a far more planned and strategic approach to arboricultural operations, even where responsibility for trees was split between separate tree officers in the parks and planning departments. The DoE's *Trees in Towns* report was the first central government publication to give a strategic overview of urban trees in England. While aspects of tree management were included in the report, the DoE wanted to do more to encourage mainly urban local authorities to develop a strategic approach to this. Therefore it published *Urban Tree Strategies* to give guidance to authorities on the production of tree strategy documents (DoE 1994). These would be based on accurate surveys of all the trees and woodlands in the district, what is now referred to as the urban forest, and set out the necessary policies and plans to protect and enhance this, usually over a five-year period. While many tree officers already recognised that this was how the urban forest should be managed in the late twentieth century, the reality on the ground was often very different. A combination of lack of funding, shortage of staff and a huge workload meant that many tree sections were operating in 'crisis management' conditions. Only high-priority work was being undertaken, mostly to ensure public safety or prevent damage to property. Although the local authority may have produced a tree strategy, it was often impossible to deliver this, even in normal circumstances.

While the late 1980s and 1990s were challenging times for tree officers, this period also saw some significant advances. The introduction of computer-based technology into tree management in Britain in the early 1980s increased the efficiency and effectiveness of many local authority tree programmes (Hope 1982; Bickmore and Hall 1983; Thurman 1983). In particular, it enabled authorities to keep up-to-date, accurate and easily accessible records of their trees, as well as TPO records, and to use these to formulate maintenance programmes and management plans. Over the past few decades, tree management has also become more environmentally conscious. For example, there is much greater awareness of the need to recycle tree debris generated by pruning and felling operations (Britt and Johnston 2008, 208–210). Previously, much brushwood and timber would be dumped in landfill sites or burnt outside the smokeless zones. This change was facilitated by the introduction of woodchippers and other machinery that could process the tree debris into useful products.

Local authorities also became more aware of the need to involve local communities in what happens to the urban forest in their district. By the 1980s it was increasingly recognised that the provision of publicly owned trees should not be regarded as just another public service to be operated with the same degree of detachment as the provision of street lighting (Johnston 1985). There were two major reasons for this shift in approach. The first was a fundamental change in public attitudes that demanded it. The rapid growth of the voluntary sector in the post-war years has resulted in many locally based environmental and conservation organisations, amenity societies and other community groups all wanting to be consulted and involved in the management of trees in their district. The second factor was a shift in central and local government policy to reflect the public's growing interest in the environment and its desire to get involved with the decision-making process. From the 1980s, the environment had considerable political importance and the Department of the Environment was assuming far-reaching responsibilities for the quality of life in which trees could play a major part. In response to all this, tree officers found themselves developing tree programmes that focused on people as well as trees, which included various community-based projects involving elements of education, consultation and practical participation (Britt and Johnston 2008, 515–530). Many of these developments were influenced by the emerging international urban forestry movement, which will be discussed later.

Natural and manmade disasters create challenging times

While the 1960s and early 1970s were times of great opportunities for professional arboriculture in Britain, especially with regard to local authorities, the late 1970s, 1980s and early 1990s were challenging times for reasons not directly connected to politics or the economic situation. This was due to some unprecedented threats to amenity trees and local authority arboriculturists were invariably in the front line. This topic has been well documented elsewhere so only the impact on professional arboriculture is considered here.

Dutch elm disease (DED) was the first major threat to trees to emerge in the post-war era. A new and far more virulent strain of the disease arrived in Britain in the late 1960s and spread rapidly northwards, eventually reaching Scotland. By the early 1990s, DED had decimated Britain's elm tree population and radically changed the character of many rural landscapes. Elm had also been a notable tree in many towns and cities and the loss there was equally devastating. For the arboricultural industry, the death of so many trees meant a huge and sustained increase in work. With regard to those elms growing in streets or on council land, the local authorities' own tree teams could usually cope with the workload, supported by contractors when necessary. However, the vast numbers of trees growing on private land, especially in urban and suburban districts, were a matter for the property owners. It soon became apparent that in many districts the existing tree surgery companies could not cope, resulting in many new companies being established. While some contractors delivered a high standard of work, others did not. Trees may have been removed but often nearby property was damaged, the site was left in a mess and high prices were charged. This lack of professionalism by some contractors damaged the reputation of the arboricultural industry.

While DED was still raging, another disaster befell trees of all species, mainly in the south of England. The Great Drought of 1976 began the year before with an unusually dry summer, but the lack of rain continued into the following year, culminating in the driest sixteen-month period in more than 250 years. Young trees were most vulnerable in this drought, especially those in urban areas where growing conditions were often far from ideal. With young trees wilting everywhere, many local authorities did not have sufficient staff or equipment to get round to watering every vulnerable tree. In an article in the *Acton Gazette* headed 'Trees are Dying of Thirst', the London Borough of Hammersmith issued an 'appeal for residents to help out by watering any young trees outside their homes' (*Acton Gazette* 1976). According to Parks Manager Bert Hart, 'A bucket of water every other day could help them survive until we get a good rainfall'. Many local authorities across the south of England were now issuing similar appeals. When water usage was restricted, residents were asked to use bathwater as long as it was not too soapy. The public responded magnificently, not only learning the important lesson that young trees are fragile and need post-planting care, but also, when the tree survived, establishing a lasting bond with it.

The most visually dramatic disaster to befall Britain's trees was the Hurricane of October 1987. Sometimes also called the Great Storm, these hurricane-force winds resulted in 15 million trees being blown down in one night across the south and west of England. While countless homeowners lost trees in their gardens, it was the urban local authorities that were worst affected. The huge 'clear-up' operation lasted for months and involved an army of council workers, contractors and volunteers. It completely disrupted the authorities' normal tree management programmes, leading to a suspension of planting and routine pruning. The worst affected authorities took a few years to get back to anything resembling their regular tree management service.

While these natural disasters destroyed millions of trees, they also reminded the public about the enormous contribution trees and woodlands make to our urban and rural landscapes. Furthermore, these events significantly increased the public's awareness of the arboricultural industry. After each event, more people would be aware that there was somebody in their local authority who was their tree officer. Furthermore, there were teams of skilled arborists, either council workers or contractors, who had the difficult and often dangerous job of looking after the trees in their streets, parks and open spaces.

As well as these natural disasters, there were manmade calamities that destroyed many hundreds of thousands of trees. This invariably resulted from conflicts between built development and urban trees which became a regular feature of this era. The problems of tree roots causing damage to building foundations had been highlighted in the nineteenth century but from the 1950s the issue became increasingly prominent. An article by Ronald Morling in the *Gardeners' Chronicle* was one of the first to generate post-war interest (Morling 1954b). To give guidance to developers and others on the placement of trees near buildings the National House Building Council published its Practice Note 3, which gave recommended distances between buildings and trees of different species (NHBC 1969). This was revised in 1974 to include recommendations on the depth of foundations and has been further revised since. The Great Drought of 1976 really put the spotlight on those trees that have a high water demand, often a significant factor in building subsidence. Subsidence damage can be a major issue with street trees where these are growing on shrinkable clay soils. After 1976, local authorities in London and other districts with these soils faced an increasing number of insurance claims. This situation provided plenty of work for arboricultural consultants working on behalf of the property owners. However, for many tree officers the task of investigating these claims was very time-consuming, especially when they already had a heavy workload. In these circumstances their departmental head could recommend that the claim be settled through the council's insurance, which did not impact on their budget, rather than going to the expense of contesting the claim in a legal case that might later be lost. The tree was then felled, although a thorough investigation may have saved it.

Another manmade problem that killed many trees was the national programme of cable installation during the 1990s. The introduction of cable television in Britain required a vast programme of civil engineering that involved excavating service trenches along the length of almost every street in the country. Because of a lack of supervision of the works and inadequate guidance, particularly in the early stages of the programme, little consideration was given to tree roots and this led to the loss of many mature street trees. Unfortunately, the hard-pressed tree officers were invariably not in a position to monitor the work of all the different installation gangs that may have been operating in their district. In 1995, the National Joint Utilities Group published *NJUG 10*, a guidance document prepared following several years of consultation with the arboricultural profession and the DoE (NJUG 1995). While this code of practice was widely regarded as providing adequate protection for street trees, compliance with it on the part of the utility companies was purely voluntary and many trees continued to be lost.

Ancient and remarkable trees

During these natural and manmade disasters Britain lost some of its famous and much-loved trees. However, these events helped to put the spotlight on the nation's wonderful heritage of ancient and remarkable trees. Interest in this topic was also stimulated by some popular books. These include *Trees of the British Isles in History and Legend* by Jim Wilks (1972), *Tree Heritage of Britain and Ireland* by Andrew Morton (1988), *Meetings with Remarkable Trees* by Thomas Pakenham (1996) and *The Heritage Trees of Britain and Northern Ireland* by Jon Stokes and Donald Rodger (2004).

Nobody did more to promote Britain's tree heritage than Alan Mitchell, the Forestry Commission dendrologist. In 1972, his book *Conifers in the British Isles* was published by the Forestry Commission (Mitchell 1972). As well as giving authoritative descriptions of all conifers growing in Britain and Ireland, it also provided a huge amount of data on their age and dimensions. In his pioneering work over many years Mitchell travelled widely throughout Britain and Ireland measuring literally thousands of notable trees. After decades of debate there was now an agreed methodology for taking and recording these measurements. In an early compilation of Mitchell's data the Forestry Commission published *Champion Trees in the British Isles* in 1985, which was then regularly updated (Mitchell and Hallett 1985). Mitchell retired from the Commission in 1986 and with his collaborator Victoria Hallett founded The Tree Register in 1988, a charitable organisation that continues his work today. Mitchell measured and recorded over 100,000 trees on his handwritten card index before his death in 1995. While this remarkable man is remembered for his work with the Tree Register, many arboriculturists will also mention his book *A Field Guide to the Trees of Britain and Northern Europe*, which for many years was their indispensable guide to tree identification (Mitchell 1974).

This upsurge of interest in ancient and remarkable trees has had a major impact on professional arboriculture. In 1993, the Ancient Tree Forum (ATF) was founded by a group of multidisciplinary experts to promote further discussion about ancient trees and share information and expertise regarding their management. Individual members of the ATF have made significant contributions to this field, such as Ted Green, Neville Fay and Jill Butler. The work of the ATF has also prompted the publication of much-needed guidance on how to maintain and manage these 'living green monuments'. One of the first was Helen Read's *Veteran Trees: A Good Management Guide* (Read 2000) and several others have followed. Consultants and contractors who have specialised in this work have found their services much in demand from the owners and managers of historic parks, gardens, woodlands and other landscapes with ancient trees.

Arboricultural practices and equipment

The second half of the twentieth century was a period of dramatic developments in practical arboriculture. Huge advances were made in the effectiveness and efficiency of many practices thanks to some pioneering scientific research and the introduction of a wide range of new equipment and machinery. However, for the first twenty years after the war, the pace of change was rather slow.

In the immediate post-war years, Denis Le Sueur's ground-breaking book *The Care and Repair of Ornamental Trees* remained the standard British textbook on the subject (Le Sueur 1934). When this came out in a second edition in 1949 there was only a limited amount of new material, including a short chapter on war and bomb damage to trees. In the 1950s, this book was recommended reading for the forestry degree courses at Bangor University (Harris pers. comm., 2019). When Sir David Eccles, Minister of Works, appointed a standing committee in 1954 to advise him on appropriate policies for tree planting and management in the Royal Parks, of the six committee members appointed only Le Sueur had specialist arboricultural expertise (*QJF* 1954). Of Britain's three eminent pre-war tree experts of Webster, Dallimore and Le Sueur, only Le Sueur remained active in the arboricultural industry in the late 1940s and 1950s. Webster had died in 1931 and Dallimore was heavily involved in the development of the National Pinetum at Bedgebury until his death in 1959 (*Gardeners' Chronicle* 1959, 285). When Le Sueur died in 1969 professional arboriculture in Britain lost one of its greatest ever exponents (Boulton 1969).

In the 1960s there was much debate among leading tree surgery companies about standards of work. Although Le Sueur and others had set out what was considered good practice at that time, it was often not followed by many who purported to be tree surgeons. A few well-respected practitioners blamed the poor standards of some tree surgery companies on their ignorance of other aspects of arboriculture and especially their lack of knowledge of the trees themselves. Henry Girling, who had trained at Kent Horticultural Institute, complained that many tree surgeons were not arborists and consequently did not have the necessary plant knowledge (Girling 1969, 220). This view was shared by Jim Wilks, a contractor in Bristol, who stated: 'Mr Girling is very correct in assuming that some tree surgeons know little about plant material. I recall one particular well-known character who thought that Ailanthus was a Gilbert and Sullivan opera!' (Wilks 1969, 251).

Some tree surgery companies in the 1950s and 1960s were significantly influenced by developments in North America. Information was coming across the Atlantic in the form of books and magazines that promoted new ideas and practices. Personal contacts were also made, one of the first being when Tom Wilson, co-founder of Southern Tree Surgeons in Sussex, attended the International Shade Tree Conference in Toronto in 1963 (Matthews and Draper 1971, 337). The classic American text *Maintenance of Shade and Ornamental Trees* by P.P. Pirone (1941) was still the most authoritative book on the subject. It held an unrivalled position until the publication in 1983 of *Arboriculture: Care of Trees, Shrubs and Vines in the Landscape* by Richard Harris of the University of California (Harris 1983).

In the 1970s, a few British books on practical arboriculture were influential. The first was Herbert Edlin's *Collin's Guide to Tree Planting and Cultivation* (Edlin 1970). Edlin had already published a number of books on forestry topics and at that time was the Forestry Commission's Publications Officer. Despite his forestry background, the book contains a substantial amount of advice on practical arboriculture. This was followed by a book entitled *The Arboriculturalist's Companion* by N.D.G. (Jimmy) James, a companion volume to his similarly titled work for foresters (James 1972). James had been a Lecturer in Forestry at the Royal Agricultural College, Cirencester, before becoming the land agent at the Clinton Devon Estates (Harris 1993, 444). This book covers much the same ground as Edlin's work and its readable text presented in a convenient pocket-sized volume ensured its popularity with many arboriculturists. However, it will also be remembered for its use of the term 'arboriculturalist'. This spelling with the added 'al' is virtually absent in British literature before James's book but subsequently appears increasingly often. Now the public not only have to contend with the word 'arboriculture', which means nothing to most of them, but the professionals engaged in this cannot agree on what they are called. To clarify, the Oxford English Dictionary has the noun of arboriculture as 'arboriculturist'.

The most ground-breaking British book of the 1970s was Peter Bridgeman's *Tree Surgery: A Complete Guide* (Bridgeman 1976a). After training at Kew and working with the GLC, Bridgeman went on to be Head of Arboriculture at Merrist Wood Agricultural College in Surrey (Ryan 2009b). His book gives an excellent insight into the progress made by professional arboriculture in Britain since the first edition of Le Sueur's work in 1934. It is particularly enlightening on the machinery, tools and equipment used for tree surgery during the early 1970s. However, in terms of tree surgery practices, the timing was unfortunate as new research in tree care in the United States made some aspects of it rapidly out of date.

While Robert Hartig had revolutionised our understanding of tree decay processes in the late nineteenth century, many aspects of arboricultural practice had remained much the same over the past few centuries. Tree care was based largely on an anthropomorphic approach where treatments were administered to 'the patient' in much the same way as a doctor or dentist would work. Limbs were amputated, wounds were dressed and cavities were filled. Indeed, the popular name of 'tree surgery' for this work seems to encapsulate the approach. That all changed in the late 1970s as a result of new research in the United States. Alex Shigo (1930–2006) of the USDA Forest Service developed a new model for tree decay that has become known by its acronym CODIT (Compartmentalization of Decay in Trees). According to the CODIT model, when a tree is wounded its cells undergo changes to

FIGURE 79. Product advertisements and dates: The Tornado One-Man Power Saw (1950), Arbrex Pruning Compound (1969), Toms' Tree-Ties (1958), Atlas Stone Company Tree Grilles (1968), Stihl 020 AV Chainsaw (1972), Growth-Master Electric Pruner (1968), Tirfor Winches (1952), Stanton Hope, Equipment for the Tree Gang (1968).

form 'walls' around the wound, slowing or preventing the spread of disease and decay to the rest of the tree (Shigo 1977). The implications for tree care practice were enormous. Flush cutting in pruning, traditional wound sealants, cavity draining and filling and other traditional practices were no longer regarded as good practice (Shigo 1991). Although it took a while for Shigo's ideas to become accepted in Britain, once he had made a few personal appearances and his work was widely read, all this radically changed the arboricultural industry (Bridgeman 2007). Some arboriculturists still believe that everything Shigo said was 'gospel' and entirely original. While he was undoubtedly a 'giant' of modern arboriculture, this unquestioning approach does not do him justice.

In the following sections we explore how arboricultural practices evolved in the second half of the twentieth century, some of which resulted from Shigo's research. As this era witnessed a huge number of developments in practical arboriculture, only an outline of the most significant can be given here. Some of the machinery, tools and equipment used to undertake these practices are also described, with a few examples illustrated in Figure 79. What was especially notable in this era was the dramatic increase in heavy machinery used in British arboriculture from the 1960s onwards. Much of this machinery had been developed in the United States in the 1950s (Rattigan 2013), although Britain lagged some years behind in utilising this (Matthews 1970, 24), and in some cases it has never featured to the same extent as in the United States.

Tree planting

Much of the innovation in tree planting during this period concerned the use of commercial products as aids to planting and establishment. This included various tree-ties, tree guards and tree shelters (Pepper *et al.* 1985). For example, some new and more effective tree-ties appeared in the 1950s. The Rainbow tree-tie came in different sizes and was made from imperishable plastic with a rustless metal buckle that could be adjusted to allow for stem expansion (*Gardeners' Chronicle* 1951, 128). Another new brand was Toms tree-ties, made with a slotted rubber pad as a spacer between the stem and the stake and a reinforced rubber belt for attachment (*Gardeners' Chronicle* 1958, 268).

An important innovation in the successful establishment of transplants and whips in woodland-type planting was the introduction of tree shelters in the 1980s. The tree shelter was invented by Graham Tuley in Scotland in 1979 and is a translucent plastic shelter or tube used to nurture the young tree in the early stages of its growth (Tuley 1985). As well as providing a mini-greenhouse environment for the young tree, it also forms a physical barrier protecting it against browsing and chemical spray applications. Tree shelters can provide up to 200 per cent additional growth in comparison to unprotected trees. They have become widely used in woodland planting schemes in rural areas and in some projects in the urban fringe.

A major topic that preoccupied the arboricultural world in the 1980s concerned the method of staking used when planting standard size trees. It had been commonly believed that physical damage and vandalism to newly planted trees could be reduced by supporting the stem high into the crown with a tall stake. However, research involving mathematical modelling, initially conducted in the United States, demonstrated that this method of staking increased the ease with which a stem can be broken compared with an unsupported stem (Patch 1984). Short stakes are sufficient to provide good anchorage for the root system, while at the same time allowing some movement of the stem above the top tie which encourages stem diameter growth and crown development. Short stakes for standard trees are now recommended and the use of taller stakes is associated with outmoded 'Victorian' practice. However, despite the research evidence, tall stakes continue to be widely used in tree-planting schemes.

Much of the debate surrounding tree planting in past few decades has focused on the size and type of planting stock best suited for different situations. There was a tendency for landscape architects to use larger sizes of trees in planting schemes where whips or transplants would be cheaper, establish more quickly and often overtake any standard trees within a few years. The argument for using larger sizes was that the trees had more immediate impact and were less prone to vandalism. However, it was suggested that many landscape architects were keen to use larger and therefore more expensive sizes of nursery stock because their fees were often based on a percentage of the total cost of the scheme. There is no doubt that forestry-type planting schemes using transplants and whips have a role in amenity planting and are now generally used for schemes alongside motorways and trunk roads.

A major concern for all those involved in urban tree planting in the 1980s, and something that remains deeply concerning, was the high mortality rate of newly planted trees in many schemes. This was especially noticeable with street tree plantings, which are often subject to the most demanding planting conditions. The problem was highlighted in research by Tony Bradshaw, a pioneer in restoration ecology, and Peter Gilbertson, both working at the University of Liverpool (Gilbertson and Bradshaw 1985). It was argued that the poor survival rates must be attributed to the way techniques of planting and maintenance were applied in practice, rather than to any inherent problems of sites or trees in general. Later research has confirmed the high levels of mortality for newly planted trees in urban areas with losses averaging around 25 per cent (Johnston and Rushton 1998, 38; Britt and Johnston 2008, 218). As well as bad planting techniques, other factors were shown to contribute to the high levels of mortality, including poor quality nursery stock and lack of post-planting maintenance such as weeding and watering. The scale of the financial loss to local authorities every year is enormous. In an effort to promote the survival of newly planted trees, arboriculturists and landscape professionals were always keen to hear about commercial products that might assist with successful establishment. Since the 1960s it has been common practice to add various commercially produced organic and inorganic amendments to the backfill soil at the time of planting. These have included slow-release fertilisers, water-retaining substances and mycorrhizal inoculation. Nevertheless, some professionals are sceptical about many of these, suggesting that their promotion often has more to do with product marketing than with aiding tree establishment.

Transplanting large trees

After the tremendous interest in moving large trees that was evident in the nineteenth century, there had been a relative lull in this activity in the inter-war years. However, from the 1950s the topic increasingly featured in the horticultural and landscape press. For the Festival of Britain in 1951, held in London on the South Bank of the River Thames, many large trees were transplanted from various parts of the country (Woodland and Dallimore 1958, 364). This major project did much to stimulate the post-war interest in the topic. In another high-profile project in 1965, the GLC planted an avenue of

FIGURE 80. In 1965 the GLC planted an avenue of semi-mature plane trees along London's South Bank between County Hall and the Festival Hall. This photograph was taken in September 1966. The trees now make an impressive feature much admired by Londoners and tourists.
CREDIT: PETER BRIDGEMAN.

FIGURE 81. The Newman Tree Mover in action, designed by Chris Newman of Civic Trees Ltd. This shows the company involved in a planting project on the GLC Elmington Estate, Camberwell, in 1966.
CREDIT: DERIC NEWMAN.

semi-mature plane trees along the South Bank between County Hall and the Festival Hall (Ryan 2009b, 22–23) (Figure 80).

The Civic Trust, a charitable organisation formed in 1957 to improve the built environment, launched its 'Plant More Trees' campaign in 1961. This aimed to promote the planting of semi-mature trees to create an instant impact in new and existing urban developments (Mabbett 2011, 24). In 1963, Civic Trees was formed by Chris Newman as a private company specialising in tree moving, following his involvement with the Trust's 'Plant More Trees' campaign (Newman 2014, 27). He designed the Newman Frame, an apparatus that sits on top of the excavated rootball and connects to a girdle chain at the base. The tree can then be levered or lifted out of the ground and transported on a specially designed trailer pulled by a Land Rover or similar vehicle (Figure 81). The complete apparatus, known as the Newman Tree Mover and still in use today, is an inexpensive way of moving trees up to 40 ft (12 m) in height and weighing over two tons. Civic Trees also established a large nursery to supply its clients with a wide range of semi-mature trees. In 1966, the British Standards Institution published *BS 4043: 1966*

FIGURE 82. The Michigan tree transplanter, using the power scoop technique, exhibited at a Civic Trust demonstration of large tree-moving machinery at Jarrow in 1966.
CREDIT: ARBORICULTURAL ASSOCIATION.

Transplanting Semi-Mature Trees, which drew heavily on the recent experience of the Civic Trust and Civic Trees (BSI 1966a). Also in 1966, a conference on moving semi-mature trees was held in Basingstoke in October of that year, organised by the Civic Trust and attended by around 400 architects, planning officers, landscape architects, arboriculturists, foresters and other interested professionals (*QJF* 1967, 1). This event undoubtedly gave a further boost to interest in the topic.

Meanwhile, in the United States the transplanting of large trees had become common practice and new machines and techniques were being developed (Everett 1961, 249). One of the American firms at the forefront in this field was the Vermeer Manufacturing Company. In 1965 it introduced its TM-700 tree mover, a trailer-mounted machine specially designed for moving trees. Within a few years Vermeer had manufactured a range of 'tree spades' that could lift rootballs up to 2.3 m in diameter. These were often self-propelled machines consisting of a number of blades that encircled the tree, digging into the soil, and then lifting the entire tree with its roots and soil out of the ground. It could then be replanted in a similar hole that had previously been dug out by the tree spade. In 1968, Civic Trees imported the first Vermeer tree spade into Britain and Newman quickly organised a demonstration of this in Richmond Park, London, attended by individuals from the arboricultural and landscape industries (Newman 1969). Another American tree-moving machine imported into Britain at this time was the Michigan tree transplanter (Clark International 1966) (Figure 82). This used the power scoop technique where the tree was literally scooped up; a quick but rather crude method used widely in the rapid clearance of opencast mining sites (Wells 1967).

Other companies apart from Civic Trees specialised in large tree moving. Hillier Nurseries was involved in the 1951 Festival of Britain plantings and continued to attract attention with some high-profile projects. In 1969, at the request of the Royal Horticultural Society (RHS), Hillier's tree specialists gave a demonstration at the Chelsea Flower Show planting a large maple about 8 m high and weighing 6.25 tons (Hillier 1969). English Woodlands, a well-known forestry company, also got involved in tree moving. In the early 1970s, it had an Instant Tree Department with Mr P. Williams as the manager (Williams 1972). When Chris Newman died suddenly in 1994, professional arboriculture in Britain lost a major figure who had made a unique contribution to its development. In 2006, Civic Trees was acquired by the Glendale Group and continues to operate with Chris Newman's son Deric at the helm.

The transplanting of semi-mature trees often occurs today on major landscape projects but the moving of large mature trees only happens occasionally. While the level of expertise and availability of suitable technology for this have never been better, hard choices must still be made about whether this can be justified financially, even when the prospects for success are good. For the cost of moving a single large tree, many hundreds of young trees could be planted. Unless the tree will be a major feature of a prestigious development or is a rare or historically important specimen that is threatened in its current situation, the considerable expense of transplanting a large mature tree cannot usually be justified.

Tree pruning

A major contribution to the literature on tree pruning was made in the early 1970s by George Brown, an assistant curator at Kew. In 1972, he published *The Pruning of Trees, Shrubs and Conifers* and this remains one of the most authoritative works in arboricultural literature (Brown 1972). His recommended treatment of specific genera is still essential reading for arboriculturists and this excellent book was updated in 2004 by Kew's Tony Kirkham (Brown 2004).

Although Le Sueur had recommended in 1934 that flush cutting should be avoided and the final pruning cut should be made just outside the 'branch ring', his advice was not followed. In the post-war literature on pruning right up to the early 1980s flush cutting was still standard practice and leaving any sort of a protrusion or snag was the mark of the unskilled 'lopper'. Around 1979, Alex Shigo and his co-workers were recommending that the final pruning cut should be just outside the 'branch collar', the swollen area often found around the base of the branch (Shigo *et al.* 1979). The branch collar consists primarily of tissues connected directly to the tree's main trunk or the parent branch and not to the branch being pruned. Injuring or removing this collar destroys the tree's ability to create 'walls' around the wound, which slow or prevent the spread of disease and decay to the rest of the tree. However, Shigo's recommendations on pruning seem to have had little impact on British professionals until his research was endorsed by their own scientists. In 1983, the DoE's Arboricultural Advisory and Information Service published a research note entitled *A Definition of the Best Pruning Position* that included a paper by Shigo on this subject (Lonsdale 1983). Only then did most arboricultural professionals start to accept that flush cutting was no longer good practice. This method of using the branch collar to identify the proper location to remove a branch has become known as natural target pruning.

There was a similar situation with the treatment of pruning wounds. For many years professional arboriculturists had recommended that pruning cuts over 2 in. (5.1 cm) in diameter should be painted with some form of coating to protect against disease and decay (BSI 1966b). In the immediate post-war years some commercial products came on the market, such as Arbrex, endorsed by Le Sueur (1963, 266), and later the improved Arbrex 805 that contained a copper fungicide. A paste called Santar was also commonly used and this contained the fungicide mercuric oxide. In the early 1970s, a German product called Lac Balsam became popular and this is still used today as a wound sealant. Also known as artificial bark, it has an olive-green colour that blends with the bark and it was also said to include a callus-inducing hormone (Bridgeman 1976a, 54). However, following the research of Shigo and others, the use of wound sealants began to be questioned. While they were supposed to prevent decay and encourage callus growth, where was the evidence? Taking up this challenge, a small research team at the Forestry Commission was formed in 1976 and after six years of research the scientists detailed their findings (Mercer 1982). They found that bitumen and latex preparations were effective promoters of callus growth but were not effective in controlling decay fungi. Many of the products tested were both ineffective against decay and toxic to the tree. While Santar seemed to be the most suitable product there were doubts about its future use because of its active ingredient mercuric oxide.

This home-grown research signalled the end for professional arboriculturists of the time-honoured practice of painting their pruning cuts. Further research by the Forestry Commission only confirmed the earlier findings (Lonsdale 1984). What had previously been synonymous with good practice was now regarded as outdated, which was welcome news for many tree surgeons. Having a pot of tar-like substance attached to your belt while tree climbing often resulted in spillage over your working clothes, which washing machines found difficult to remove. It also meant some vigorous hand-washing at the end of every day. Some arboriculturists believe that painting pruning cuts will again become standard practice when a really effective product is developed – but that remains to be seen.

The range of hand tools used in pruning operations that were available before the war remained largely similar in the second half of the twentieth century. Among these, the pole saw was increasingly used for light pruning

FIGURE 83. In the post-war years the pole saw was increasingly used by arborists for light pruning and deadwooding. This photograph from the mid-1960s also shows the arborist wearing an early safety harness.
CREDIT: MICHAEL P. NEALE.

and deadwooding (Figure 83). However, there have been some notable advances in the design of these tools and the material used in their manufacture. Many experienced arborists comment on how much pruning saws have improved in recent decades, not least with the introduction in 1995 of Silky saws from Japan (Smith pers. comm., 2019). The most significant development in pruning equipment has been the use of chainsaws. While one-man chainsaws were introduced into Britain after the war, they were too heavy and cumbersome to be used in the tree, except for occasionally removing large limbs. In 1968, however, one the first lightweight chainsaws, manufactured by the American company McCulloch, was attracting the attention of arborists (Matthews 1968a, 3). This could be used one-handed while working in trees. This was later overtaken in popularity by the German-made Stihl 020 AV, which became standard equipment for many tree teams. Unfortunately, these lightweight chainsaws were often used for all routine tree pruning, even when the pruning saw was a safer, cheaper and more environmentally friendly alternative.

The disposal of brushwood and cordwood from pruning operations has

FIGURE 84. The Gibbs Wood Chuck was the first woodchipper to be manufactured commercially in Britain and the first machines were delivered in 1972. This shows one of the early Wood Chucks in use with the London Borough of Ealing in 1973.

CREDIT: J. GIBBS OF BEDFONT.

always been a major part of tree work. Even when cut up and stacked, the work on a large tree can generate a full lorry-load. When smokeless zones were introduced following the Clean Air Act of 1956 the option of burning it on site was often not available. The eventual answer was the woodchipper, a machine used for reducing small tree trunks, branches or brushwood into smaller woodchips, resulting in a reduction of as much as 10 per cent of the original bulk. Invented by Peter Jensen in Germany in 1884, modern versions of these machines were being manufactured in the United States in the 1950s and they became very popular with tree care companies. The first woodchipper to be manufactured commercially in Britain was the Wood Chuck, produced J. Gibbs of Bedfont in Middlesex. The basic chipper mechanism was purchased from the United States and the rest of the machine built around it. The first Wood Chucks to be used in Britain were delivered to the London Boroughs of Richmond, Croydon and Harrow in 1972, with the boroughs of Ealing and Redbridge the following year (Gibbs 1988) (Figure 84). As woodchippers became increasingly popular with arborists, other companies started to manufacture their own versions. The Arboreater, sold by arboricultural equipment supplier Michael Dalrymple, was a low-budget machine favoured by smaller tree care companies. The American companies of Asplundh and Vermeer introduced their own range of woodchippers into Britain in the mid-1970s. Not only did the woodchipper solve the arborists' brushwood disposal problem but the chips themselves were a marketable commodity with a variety of uses such as organic mulch in gardening.

Tree climbing and aerial access

In the 1930s, Denis Le Sueur introduced both the City of Paris safety belt and the American safety sling into Britain for use by the City of London Corporation tree teams. However, these do not seem to have been used elsewhere at that time. Immediately after the war, Le Sueur again promoted these techniques as well as the improved 'safety saddle' in an article in the *Quarterly Journal of Forestry* (Le Sueur 1946). However, when the second edition of Le Sueur's book appeared in 1949 he expressed his disappointment that British arborists were still reluctant to follow their American counterparts and adopt these safer working practices. He states:

> In Great Britain such precautions are seldom taken. Apart from the characteristic objection of the average British workman to anything 'new', the chief argument against the safety rope seems to be that 'the tree will be more difficult to get round'. This is probably because most high lopping is done by piecework, and if the operator has to stop cutting in order to move and adjust ropes he will not make so much money. The argument is not convincing, as a man relieved of the necessity of retaining his balance by his own effort will certainly work not only better, but faster. Insurance is no solution. Money is generally little consolation to the widow or the cripple.
>
> Le Sueur 1949, 51

For most arborists in the public and private sectors in the late 1960s and early 1970s, the common method of access into the tree continued to be the ladder. Triple extension ladders would enable the arborist to access the crown of most tall trees and from there they could free climb higher into the required working position or pull themselves up with the aid of a rope thrown over a convenient branch above. In the absence of safety belts or harnesses specifically for tree work, some arborists used the safety belts made for Post Office staff to climb telegraph poles and would attached a strop to this to tie themselves into the working position. Others experimented with modified parachute harnesses. One piece of tree-climbing apparatus that attracted some interest in the 1950s was the Swiss-made tree bicycle (Usher 1953, 120). This was used by the Forestry Commission for collecting conifer cones and by the RHS at Wisley Garden for pruning deadwood on Scots pines (Honey 1964a, 280). However, it was impractical for most tree surgery work because it could only be used on a straight trunk without branches (Figure 85).

It was not until the 1960s that the use of the safety harness and climbing rope began to be adopted by some British arborists and it took a long time

FIGURE 85. In the 1950s the Swiss-made tree bicycle attracted some interest but it proved impractical for most tree work. This shows student Colin Bashford using the equipment at Merrist Wood College in 1970–1971. Bashford would later reach the heights of becoming ISA President in 2011.

before this was widely accepted. Even at Kew Gardens in 1962, ropes and harnesses were not used for tree climbing (Bridgeman pers. comm., 2019). However, around this time a few forward-thinking tree surgery companies were providing their arborists with some of this equipment (Honey 1964a, 279–280). One of these companies was BCR Tree Surgeons, founded by Cyril Reddick and based in Surrey. In 1966, when Reddick issued his tree teams with safety harnesses and climbing ropes not everyone was keen to use them (*Evening Post* 1966). When two of his arborists were asked to work on a large tree in Sunningdale they refused to use the equipment when asked to do so by the foreman. Unfortunately, while one of the men was pruning deadwood from the top of the tree he suddenly fell. He died ten days later in hospital and an inquest was then held into his death. At the inquest the foreman said, 'The tree fellers hated wearing the harnesses because they were an encumbrance when climbing about a tree'. He then added that he had now received orders to sack anyone who refused to wear one. In his evidence Reddick said that he had pioneered the construction of these safety harnesses in England when he had persuaded a firm to manufacture a limited number of American-style harnesses for him. The manufacturer had been reluctant to agree due to the limited demand for the product but was persuaded when Reddick provided sufficient funding. Reddick told the jury, 'After spending all this money on the equipment, I plug it all I can'.

It seems likely that the American-style harness introduced by Reddick was the type called the 'sit harness' or 'saddle' which consists of a waist belt and two leg support loops. As well as Reddick, some other leading tree surgery companies were trying to source their own supply of harnesses. Some companies imported a small number of American harnesses at considerable expense (Matthews 1968b, 177). Efforts were also made to get the Forestry Commission to apply pressure on British manufacturers and this eventually resulted in the production of two harnesses, one a parachute-style harness and the other resembling the American safety saddle, the latter being preferred by most tree surgeons. With regard to the climbing rope and the knots used, this was the same basic system with the bowline and prusik knot that Le Sueur had introduced from the United States in 1936 (Le Sueur 1946, 90). Some climbers substituted the prusik knot with a mechanic device known as a prusik ascending clip, used in mountaineering (Bridgeman 1976a, 27). This gave the same action as the knot and was much faster to fit and remove, especially if the safety catch was taken off. However, this modification was potentially dangerous as the rope could jump out of the clip if held at a certain angle and then there was nothing to stop the climber falling.

At an arboricultural conference held in 1968 at Merrist Wood College, Surrey, a demonstration of tree climbing using rope and harness was given by Honey Brothers, a local tree surgery firm (Figure 86). The demonstration was supervised by Peter Bridgeman, newly appointed to the college staff, and the harness shown was manufactured by the London firm of Barrow Hepburn. After the event there was such a big demand for the harness from the arboricultural industry that the manufacturers temporarily ran out of stock (Honey 1969). Although the use of climbing rope and harness became increasingly common in the 1970s, not all sectors of the industry immediately adopted it. For example, Alistair Smith (pers. comm., 2019) recalls that, although he had been trained to use a rope and harness at Merrist Wood in 1974, when he joined the GLC afterwards this method was frowned upon by his colleagues who worked mainly from extension ladders (Figure 87). In 1977, the prusik loop system of climbing was introduced by Jack Kenyon at Merrist Wood, which was good for teaching students and inexperienced people as there was virtually no knot tying involved while climbing (Whitehead 1981).

FIGURE 86. This demonstration of tree climbing using rope and harness, organised by Honey Brothers, was held at the ABTSA conference at Merrist Wood College in 1968. Peter Bridgeman is next to the climber with Tony Honey to his right (left of photo). Behind the climber, partly obscured by the rope, is Peter Honey.

In the mid-1980s, the Forestry Safety Council tried to impose the use of the full-body harness on the arboricultural industry through its safety guide *FSC18: Tree Climbing and Pruning* (FSC 1981; Kenyon 1985). While sit harnesses at that time were not entirely satisfactory from a safety point of view, most arborists felt that the full-body harness did little, if anything, to reduce the risks inherent in tree surgery work. After much heated debate, the majority view in the industry prevailed. In the early 1990s some improved safety harnesses were manufactured and the Willans T22 with its padded seat became widely used (Bridge 2010, 52). In the past two decades increasingly advanced harnesses have become available, including the first harness specifically designed for female arborists (George 2014, 25). The quality of climbing ropes has also improved out of all recognition from the early nylon ropes used in the late 1960s.

Arborists have always been keen to explore mechanised alternatives to the very arduous work of climbing around in trees all day long. What are now called mobile elevating work platforms (MEWPs) are commonly used today in tree work operations where the nature and the volume of work can justify purchasing or hiring them. However, experimenting with some form of MEWP for tree work can be traced back to at least the early 1960s. Around 1962, Jim Wilks, the Bristol tree surgeon, purchased a fire engine and used its turntable extension ladder for access into tall trees (Girling pers. comm., 2019). After a demonstration of this at an industry event at Westonbirt Arboretum in 1965 his fire engine became renowned in the arboricultural world and a few tree contractors followed his example (Figure 88).

FIGURE 87. (above) Greater London Council arborists in action in the late 1960s. They are using an extension ladder rather than rope and harness for aerial access into a tall tree. This was still the GLC's preferred method in the mid-1970s.

CREDIT: PETER BRIDGEMAN.

FIGURE 88. (right) In 1962, the Bristol tree surgeon Jim Wilks purchased a fire engine and used its turntable extension ladder for access into tall trees. This shows Wilks giving a demonstration of this at an ABTSA event at Westonbirt Arboretum in 1965.

CREDIT: HENRY GIRLING.

Tree felling and stump removal

The basic techniques of tree felling that were used in the arboricultural industry in the 1930s and 1940s have not changed dramatically since then. The various felling cuts made, the use of some type of saw, felling wedges and pulling rope when required, are all still in evidence today. However, there has been one development that has impacted massively on so many aspects of practical arboriculture: the chainsaw.

Until the early 1950s, tree felling was undertaken almost entirely by manual methods. The everyday tools that had been used for decades were the crosscut saw and the axe, with the bow saw also employed for the smallest trees. Although companies such as Stihl and Teles had developed two-man chainsaws pre-war, these were used for crosscutting timber and considered impractical for tree felling. After the war, improvements in engine design and aluminium enabled lighter chainsaws to be made which could be used by one person. In 1949, the British company

DANGER
TREE WORK

Danarm introduced the Tornado, claiming it to be 'the world's fastest one-man power saw (*Country Life* 1950). Other early British one-man chainsaws were made by Sankey-Aspin, with its specially designed Cutesa cutting chain (*Country Life* 1956), and the Teles Little Tiger, claiming to be 'the fastest and finest of all one-man chainsaws' (*QJF* 1958). In recent decades modern chainsaws have increased dramatically the efficiency and effectiveness of tree work. However, their overuse in some situations has also had negative impacts on standards. When all pruning was done with hand tools, which took longer and demanded more physical effort, perhaps more thought went into this work and over-pruning was generally avoided.

Early chainsaws were notorious for the amount of vibration they generated, inducing a medical syndrome in the operator known as 'white finger', which appeared after using a fixed-handled chainsaw for two to five years (*QJF* 1971). The constant vibration created a tingling 'whiteness' or numbness in the fingers as the blood vessels, joints and nerves were affected. Once this problem had been recognised, chainsaws were designed with anti-vibration buffers mounted between the handle and main body of the saw, which greatly reduced the amount of vibration reaching the operator's hands.

From the 1940s onwards winches continued to be widely used in tree-felling and stump-removal operations. The Trewhella monkey winch remained popular until it was slowly replaced in the 1960s by a range of Tirfor winches that then became standard tree work equipment (Bridgeman 1976a, 44–45). The French-made Tirfor winches were much lighter, more portable and capable of being carried by one person. The Tirfor T35 with a 5-ton pull was especially popular with arborists. Various types of tractor-mounted winches developed for the forestry industry were also becoming common in tree-felling and stump-removal operations. When dismantling large trees in confined spaces, tree work contractors now often hire mobile cranes with skilled operators to assist in this difficult work.

One machine that has revolutionised stump removal is the stump grinder. This is a machine that removes the top of stumps to below ground level by means of a rotating cutting disc that chips away at the wood. The size of these machines can vary from that of a lawnmower to a tractor. Once the stump has been ground down, the surface can be reinstated, whether this is grass, paving slabs or some other material. In 1956, the Vermeer Manufacturing Company founder, Gary Vermeer, invented the first modern stump grinder and after major modifications this only required one operator. By the mid-1970s, the machines were being used by some local authorities and tree surgery companies in Britain (Figure 89). However, because of the high cost of purchasing these machines, stump grinding was often contracted out to tree companies that specialised in this work (Matthews 1970, 24).

In addition to the mechanical methods of removing tree stumps, various chemical treatments were widely used in the 1970s and 1980s. Some of these were later found to be toxic to humans, such as the popular brand called

FIGURE 89. By the mid-1970s, stump-grinding machines were starting to be used by some local authorities and tree surgery companies in Britain. This early machine, the Vermeer 2460, was exhibited at the Arboricultural Association conference at Merrist Wood College in 1974.
CREDIT: PETER BRIDGEMAN.

Brushkiller that could be sprayed onto stumps or poured into pre-drilled holes (Wilson 1981, 15).

Other tree care practices

Tree surgery companies before the war frequently undertook cavity work in decayed trunks and major branches and this trend continued after the war. Similarly, the bracing and propping of major limbs continued to be requested by garden owners and managers keen to prolong the life of their old and much-loved trees. The leading companies took a pride in their expertise in this work and considered that it marked them out as tree surgeons as opposed to 'loppers and toppers'. However, two separate developments that fundamentally changed arboricultural practice would impact greatly on this type of work. The first was the research conducted by Alex Shigo and his co-workers in the United States. Cavity work as it had been practised for centuries suddenly became bad practice and was actively discouraged. For example, it was shown that drilling a hole up into a cavity to drain out water simply punched a hole through the CODIT wall and spread decay into healthy wood. Furthermore, the idea that it was

FIGURE 90. This arborist is cable bracing a weak stem and using a draw vice to tension the cable. The 1976 photograph features Barry Phillips of Merrist Wood College. Cable bracing became less common from the 1980s with a realisation of its legal implications. CREDIT: PETER BRIDGEMAN.

feasible to remove all the infected wood and get back to health tissue was finally exposed. Although this fallacy had been highlighted back in 1917 by W.E. Hiley, the British forester, scientific research in America now confirmed this.

The other development that discouraged cavity work was mentioned in the previous chapter. Where extensive cavity work has been undertaken or a major limb has been braced or propped, this indicates that the tree has been identified as potentially unstable and possibly dangerous. Then, if that tree subsequently fails and this results in damage or injury, the owner of the tree is in a difficult legal position. The realisation of this dilemma in the 1980s, following some relevant court cases, resulted in a dramatic reduction in this type of tree surgery. Local authorities in particular were not prepared to contemplate this work, except on important trees located in a quiet part of a park or garden. For nationally important gardens with significant numbers of heritage trees, this type of work was still undertaken but only after a thorough risk assessment had been made (Figure 90).

Tree spraying has remained an unusual practice in British arboriculture, although it is widely done in North America. Part of the reason for that might be the reluctance of many British gardeners to spray any plants with chemicals following the disturbing revelations in Rachel Carson's book *Silent Spring*. Tree feeding of mature trees also remains uncommon in Britain. However, one practice related to this has received much attention, particularly in the 1990s and 2000s. When the Hurricane of 1987 struck Kew Gardens it lifted the whole root plate of the famous Turner's oak (*Quercus x turneri*) that had been planted in the 1770s (Kirkham 2008, 77). Fortunately, the tree settled back into the soil and did not fall over. However, the movement appeared to stimulate a previously sluggish tree into vigorous growth. This recovery led to a programme of decompaction of the roots of other large trees, led by Kew's tree expert, Tony Kirkham. A Terravent machine, invented in Germany, was used to inject nitrogen gas under high pressure near the roots, breaking up compacted soil to admit the passage of air and moisture (*ibid.*, 78). In a second blast, mycorrhizal fungi and a bio-stimulant were introduced to promote the rejuvenation of roots. Combined with mulching this treatment had a remarkably positive impact on the health and growth of the trees. In the late 2000s, as Terravent spare parts were difficult to obtain, the programme switched to the Air-Spade, an American-invented machine already being used by some arborists in Britain. This employs a continuous jet of compressed air which penetrates and dislodges most types of soil, but is harmless to non-porous objects such as plant roots, buried pipes, or cables. This highly effective decompaction tool has revolutionised this aspect of practical tree care and is widely used in root investigations.

Suppliers of arboricultural equipment

During the 1960s and 1970s, tree surgery companies and local authorities were introduced to a wide range of new specialist equipment. Along with some heavy machinery this included lots of smaller equipment, such as improved types of chainsaws, safety harnesses, winches and a wide range of tree surgery tools and fittings (Bridgeman 1976a, 31–53). In the mid-1960s some tree surgery companies were obtaining highly specialised items such a such as screw-eyes and eye-bolts for bracing work from local blacksmiths as they were not yet being mass produced (Bridgeman pers. comm., 2019). However, with the sheer volume of tools, equipment and associated materials now being used on a regular basis, it was only a matter of time before specialist companies would be established to supply all this to the arboricultural industry. One of the first companies to recognise this business opportunity was Stanton Hope, already a well-known supplier of forestry equipment. In the mid-1960s they began advertising that they could now offer 'equipment for the tree gang', ranging from pruning tools to measuring instruments. In 1972, the tree surgery firm of Honey Brothers established a separate company, Honey Brothers (Sales) Ltd, to retail arboricultural equipment and supplies to the industry, issuing their first catalogue in 1974 (Honey Brothers 1974). As a contracting company it had already been supplying some equipment to other contractors and when

Merrist Wood College began arboricultural courses from 1968 onwards much of the equipment was purchased from Honey Brothers (Bridgeman pers. comm., 2019). In 1970, Michael Richmond and his wife Shirley founded the company called Michael Richmond, also based in Surrey, which rivalled Honey Brothers as suppliers of equipment to the arboricultural industry.

Codes of practice and safety standards

Before guidance on arboricultural practice was available from official sources, the arboricultural industry had to rely on published guidance from horticulture, forestry and agriculture, some of which was relevant to arboriculture. However, it was clear that arboriculture needed its own guidance documents, government approved but compiled with the cooperation of the arboricultural industry. In the 1960s, two organisations were formed to represent the arboricultural industry, the Association of British Tree Surgeons and Arborists (ABTSA) and the Arboricultural Association. These will be discussed later but their role in promoting codes of practice and safety standards is mentioned here. Both organisations recognised that if standards of practice in the industry were to be improved, one of the first steps was to decide what those standards should be and to draw up specifications for how the work should be undertaken (Beeching 1965, 148).

In 1964, the Institute of Landscape Architects asked the British Standards Institution (BSI) to draw up standards for landscape operations and one of these was a standard for tree work (Beeching 1965, 148). The BSI is the UK national standards body and it publishes technical standards on a wide range of services and products (Boddy 2009). To draw up the standard for tree work the BSI formed a steering committee comprising representatives of twenty-six relevant organisations, including ABTSA and the Arboricultural Association. In 1966, the BSI published the new standard entitled *BS 3998: 1966 Recommendations for Tree Work* (BSI 1966b). While it was a tremendous step forward for the industry, this was published before the revolution in tree care prompted by Alex Shigo. Consequently, the standard became outdated at that time, although it was several years before it was eventually revised. Another early arboricultural standard published by the BSI on semi-mature trees has already been mentioned. The BSI continues to publish standards that are relevant to arboricultural work. The most recent is *BS 8545: 2014 Trees: From Nursery to Independence in the Landscape* (BSI 2014), which contains some excellent recommendations based on sound arboricultural research. However, the inclusion of the phrase 'independence in the landscape' in the title seems strange given that arboriculturists always stress that trees in the urban landscape need regular and careful attention once planted.

Arboriculture has always been a potentially dangerous occupation requiring specialist skills, not least because of the risks of working with large trees and heavy branches, often at height and in close proximity to people and buildings.

In the 1950s and 1960s, with the introduction of chainsaws and heavy machinery it steadily became a lot more dangerous. At the same time, the world of industry and commerce was becoming far more safety-conscious and there were calls to introduce some all-embracing health and safety legislation.

In 1974, the Health and Safety at Work Act came into force. The Act imposed general duties on all employers and the self-employed to ensure the health and safety of those who might be affected by their business activities, and on employees to look after their own safety. It had a huge impact on the nation's working population, including those in the arboricultural industry. One immediate effect was an increase in arborists' use of personal protective equipment, such as hearing and eye protection attached to safety helmets, and chainsaw trousers, gloves and boots (Bridgeman 1976, 22). In the decade after the Act came into force the Forestry Safety Council (FSC) was regarded as having a remit that embraced both forestry and arboriculture. However, this was not a satisfactory arrangement for the Arboricultural Association, which sought for several years to gain representation on this but it was not accepted (Preston 1991, 38). However, the FSC did accept a draft safety guide on tree climbing prepared by the Association, but when this was published as FSC18 it was in a modified form that was not acceptable to the arboricultural industry. In view of this, the Arboricultural Association set up a Working Party in 1986 to review all aspects of arboricultural practice which led to the formation of the Arboricultural Safety Council (ASC) in 1989. This was chaired by Trevor Preston and comprised representatives from the Arboricultural Association together with some forestry, trade and professional bodies and local authority organisations. In 1991, the ASC published its first safety guide entitled *ASC1: Tree Climbing Operations* which was well received by the arboricultural industry (ASC 1991). More safety guides from the ASC followed soon after.

There is no doubt that the ASC filled a much-needed role but a preoccupation with safety issues has continued to dominate many aspects of arboriculture over the past few decades. This is a reflection of an increasingly risk-averse society and it has encouraged many local authority tree officers to take a very wary approach to urban tree management.

Tree surveys and tree inspections

As the issue of health and safety began to dominate arboriculture, the safety of the trees themselves came increasingly under the spotlight. As Le Sueur had stated in 1934, to ensure as far as possible that trees were in a safe condition, what was required were frequent and careful inspections by a competent person (Le Sueur 1934, 204). Although these inspections should embrace all trees where the public or property was at risk, most attention was given to highway trees.

In 1973, the DoE issued guidance to local authorities and others regarding inspection regimes for roadside trees in its Circular 90/73 (DoE 1973). This was quite vague in specifying these inspection regimes and left this to the discretion of the tree officer by recommending 'inspections by a competent person at reasonably frequent intervals' dependent upon the 'age, kind, condition and circumstances of each tree'. In 1975, Circular 52/75 gave more realistic guidance but was still regarded as too vague by some tree officers (DoE 1975). This issue then rumbled on for many years without any clear national standard on the frequency and method for tree inspections (Eden 2007). Tree officers did their best to interpret the latest guidance but often had difficulty implementing this due to ongoing financial constraints. However, some authorities were able to create specialist tree inspector posts where the officer's primary role was to conduct tree surveys and inspections. Many local authorities have now produced Tree Risk Management Plans or similarly titled documents that set out their approach to this issue. A major step forward was the establishment of the National Tree Safety Group (NTSG) in 2007, composed of a wide range of organisations with an interest in tree risk management. In 2011, the NTSG published a document entitled *Common Sense Risk Management of Trees* (NTSG 2011). This was welcomed by the arboricultural industry for at last providing some common sense and practical advice on managing risks from trees (*Tree News* 2012).

Fatalities caused by trees invariable get wide coverage in the media, especially if there is the death of a child. When these incidents occur there is often a major reaction by local authorities and other tree owners to inspect or re-inspect their trees and undertake felling and heavy pruning if there is the slightest doubt about tree safety. One such incident occurred in 2003 when a young girl was crushed to death in her school playground by a falling tree that was uprooted in windy conditions. This tragedy at Ashtead, Surrey, attracted national media coverage, including the front page of the *Daily Telegraph* (Paterson 2003). Thankfully, this type of incident is very rare. As the NTSG's report points out, on average only six people each year in Britain are killed by falling or fallen trees or branches (NTSG 2011, 22).

While concerns about tree safety have escalated, arboriculturists have also gained a much better understanding of the nature of tree failure and how to identify risks from trees. An important contribution was made in 1994 by Claus Mattheck, a German scientist, who developed the visual tree assessment method for examining trees based on their external shape. This was set out in his book *The Body Language of Trees* (Mattheck and Breloer 1994). Another seminal book by David Lonsdale, then working at the Forestry Commission, was published in 1999 entitled *Principles of Tree Hazard Assessment and Management* (Lonsdale 1999). As our understanding of tree failure has increased, new instruments have been developed to assess the extent of decay in standing trees. One of the first was the Shigometer, introduced by Alex Shigo, and more recent equipment includes the PiCUS Sonic Tomograph, Arborsonic Decay Detector and the Resistograph (Monk 2011).

Arboricultural research

In the second half of the twentieth century arboricultural research has become the subject of much debate within the arboricultural industry. Even before the fundamental changes brought about by Shigo and his co-workers there was widespread recognition among British arboriculturists that the roots of good practice are based firmly in good research.

Throughout the twentieth century, British arboricultural research has always been bound up with the topic of forest research. Furthermore, forest research itself has been dominated by the activities, capabilities and requirements of the Forestry Commission. While much research work at the Forestry Commission in the 1950s and 1960s was focused firmly on forests and commercial forestry, there were still some significant contributions to advancing practical arboriculture. In 1956, a major British work on tree pathology was published entitled *Tree Injuries: Their Cause and Prevention* by Herbert Edlin and Maurice Nimmo (1956). This was directly relevant to arboriculture and both men worked for the Forestry Commission. Possibly the most significant British book on tree pathology appeared in the 1960s. Written by Tom Peace (1962) and entitled *Pathology of Trees and Shrubs*, it was the first comprehensive account of the pests, diseases and other agencies that were damaging to woody plants in Britain. The scope, detail and academic rigour of this 750-page book is remarkable. At the time of publication Peace was Chief Research Officer for the Forestry Commission and the project was only made possible by the Commission, which allowed him to use official time to deliver it. Although there have been some notable advances in tree pathology since Peace's book and some nomenclature has changed, it is still recommended as a major source of further information by current texts (Strouts and Winter 2000).

Even with these contributions the Forestry Commission was still slow to embrace arboricultural research directly (Johnston 2015, 94). However, in recent decades the Commission has become far more active in this field and has supported a wide range of initiatives. Much of the driving force behind that has come from the arboricultural industry. Following the ravages of Dutch elm disease there was a growth of interest in wide-ranging research on trees that were not just of silvicultural importance (Winning 1973, 160). The arboricultural industry increasingly believed it should be serviced by a research initiative that understood its particular needs and circumstances.

In September 1972, the Sixth Annual Conference of ABTSA unanimously passed a motion calling for the formation of an Arboricultural Research Working Party (ARWP) (Insley 1976, 54). This was formed under the chairmanship of Fred Last of Edinburgh University with representatives from arboricultural and horticultural organisations, other relevant bodies and the Forestry Commission. Its report was issued in January 1974 (ARWP 1974) and later published in the *Quarterly Journal of Forestry* (*QJF* 1975). Among several recommendations the report stated that 'the highest priority should be given to the creation of an integrated information and advisory service' (ARWP 1974, 12). This would be concerned with gathering relevant international literature, preparing advisory publications, delivering an advisory service and assessing the research needs of the arboricultural industry. By October 1974, the Forestry Commission had appointed an arboriculturist, Derek Patch, to begin to develop work on implementing these recommendations and further staff followed with funds from the DoE. Some of the early research objectives were to improve tree selection for amenity trees, develop enhanced techniques of tree production and establishment, and promote knowledge and control of decay in amenity trees.

The Arboricultural Advisory and Information Service (AAIS) began operating in 1976 on an initial five-year contract, based at the Forestry Commission's research station at Alice Holt in Surrey. At last, arboricultural research was beginning to get the recognition it deserved. From 1984, the AAIS produced a series of Arboricultural Practice Notes and Arboriculture Research Notes covering a wide range of topics. These quickly gained a reputation as key reference documents and played an invaluable role in keeping British arboriculturists informed of the latest research into practical arboriculture. Following a review of the AAIS in 1991 by the DoE's Arboricultural Advisory Board (AAB 1991), in 1993 it was placed under the management of the Tree Advice Trust, an independent charity, and charged with being largely self-financing (Last 1993). However, by the 2000s the Trust seems to have lost its early momentum as the AAIS and the requirement that it should generate its own income was placing constraints on its effectiveness. Some elements in the industry were not inclined to pay the required charges for its services and the increased availability of electronic information sources further reduced demands for this with a consequent decline in its income. The Trust ceased trading in 2013, coinciding with Patch's retirement after his many years of dedicated service to this initiative. The archive of its published research and practice notes was passed on to the Arboricultural Association to maintain its availability to the arboricultural industry and other professionals.

From the 1980s, the Forestry Commission was involved in 'showcasing' the latest arboricultural research through a series of Arboricultural Research Conferences, organised jointly with the Arboricultural Association. These were held in 1980, 1985 and 1990. A similar conference was held in 1995 with the support of the DoE. The papers presented at all four research conferences were published in individual volumes (Forestry Commission 1980; Patch 1987; Hodge 1991; Claridge 1997). From the early 1990s through to 2008, the DoE (succeeded by the Department for Communities and Local Government) produced a series of major publications on urban tree research. The series was called Research for Amenity Trees and nine titles were published on a range of subjects, including tree pathology, tree risk assessment and management, urban woodlands and air quality, and tree roots in the built environment.

Inspired by the Arboricultural Research Conferences of the 1980s and 1990s, a group of arboricultural and urban forestry professionals got together in early

2010 to propose an international research conference, specifically on urban trees (O'Callaghan 2012, 235). A major aim of the initiative was to build partnerships between arboriculturists and other professionals involved with trees and the urban environment and to put tree experts at the heart of the debate on this topic. In April 2011, the Trees, People and the Built Environment research conference was held in Birmingham, attracting some 400 delegates from Britain and overseas. In April 2014, a second conference was held at the University of Birmingham, attracting even more delegates. Both conferences were partnership initiatives involving a range of professional, government and academic bodies, each represented on a Steering Group which coordinated the event and drew up the programme. The conferences were hosted on behalf of the Steering Group by the Institute of Chartered Foresters (ICF) as its annual conference. Further conferences with the same name and branding have been held by the ICF, although without most of the arboricultural members of the original Steering Group. The proceedings of the first two conferences were published, making an important contribution to the literature on urban trees and urban forestry (Johnston and Percival 2012 and 2015).

In recent years arboricultural research has come under sustained pressure due to constraints on public expenditure. In response to this a small group of arboriculturists came together in an initiative to fund arboricultural research in Britain. In 2011, Fund4Trees was founded by Russell Ball, formerly the Executive Officer of the London Tree Officers Association, who organised the first 'Ride for Research' event. He took his inspiration from the 'Tour de Trees' cycling events that feature at the annual conferences of the International Society of Arboriculture to raise money for research. While the sums raised by Fund4Trees have been relatively modest, the energy and enthusiasm of the group has done much to raise the profile of research projects among arboriculturists and related professionals. They have also funded some important research undertaken by arboriculturists. One of the features of arboricultural research in Britain is that it has invariably been done by people who are not from an arboricultural background. In recent years some talented British researchers have emerged with backgrounds in arboriculture, especially Glynn Percival at the Bartlett Tree Research Laboratory in Reading and also Andrew Hirons and Duncan Slater at Myerscough College in Lancashire.

Contractors and consultants

After the Second World War there was a tremendous increase in tree work contracting companies and also in arboricultural consultancy practices. As before the war, there was often some overlap between the two where the owner or manager of the contracting company would also act as a consultant providing not only advice but also tree surveys and written reports. While many new contacting and consulting firms were being formed, there were also some that had been established before the war that continued when the emergency had

THE
CHILTERN TREE SURGEONS
(OF AMERSHAM)

Experts in General Tree Preservation

LOPPING AND FELLING OF DANGEROUS TREES

18 Queen Court, Queen Square, W.C.1
TERminus 8394

THE TREE PRESERVATION COMPANY
FERRING, SUSSEX

FULLY INSURED **ATTENTION TO ALL TREES** G. H. Goodings, H. S. Lisle

IF YOU ARE CONCERNED OVER THE CONDITION OF YOUR TREES THEN CONSULT US A TEAM OF HIGHLY EXPERIENCED MEN WHO WILL DEAL WITH DANGEROUS TREES AND ALL FORMS OF TREE SURGERY.

WE ALSO SPECIALISE IN LANDSCAPE GARDENING, PRUNING OF FRUIT AND ORNAMENTAL TREES AND SHRUBS

GORING 42027

TREE PRESERVATION
AND
TREE SURGERY

BEECHING OF ASH LTD
ASH, ALDERSHOT, HANTS
TEL. ALDERSHOT 21415

To Estate Owners—YOUR TREES!

May be seriously affected by dead branches which spread disease and are dangerous. Converted into fire logs, decaying wood will last the winter as fuel. We specialise in this work, and felling and **TREE PRESERVATION.**

THE TREE SPECIALIST CO.
A. RUMSEY, F.R.H.S.
LITTLEHAMPTON
TEL. 714 TEL. 714

THE
SOUTHERN TREE SURGERY
COMPANY

Arboricultural Consultants
Byways, Woodham Lock, West Byfleet, Surrey (Byfleet 42739)

and Tree Surgeons
25 St. Thomas Street, Winchester, Hants (Tel. 2531)

Advise on, estimate for, and carry out all tree work

Country wide, up to date service covering the safety, preservation and maintenance of all amenity trees

offered by a leading company in this specialised field

BCR
TREE SURGEONS
LIMITED

examinations
surveys
reports

all practical field-work carried out in accordance with BSI standards

approved by the Association of British Tree Surgeons and Arborists

BCR Tree Surgeons Ltd
5 Church Road
Gt. Bookham Surrey
tel: Bookham 2124

also at
Buntingford, Herts
tel: Puckeridge 681

J. H. WILKS
TREE SURGEON

Maintenance, Repair, Cleaning, Bracing, Tree Surgery to backward, suspect, and neglected trees.

THE DEANS
2, HAMBROOK LANE
STOKE GIFFORD, BRISTOL
Tel. FILTON 3301

ESTS. FREE **TREES** W. 8135

FELLING, LOPPING and Stump Removal. Tree Surgery, Tree Preservation.

PLANT FOR HIRE: — Crane, Tractors, Winches, Power Saws, etc.

CONTRACTORS TO H.M. GOVERNMENT, COUNTY AND BOROUGH COUNCILS, ETC.

VICKERS & CO.

WORTHING 8135 **77 Tarring Rd. WORTHING**

FIGURE 91. Company advertisements and dates: The Chiltern Tree Surgeons (1950), The Tree Preservation Company (1957), Beeching of Ash (1970), The Tree Specialist Company (1949), The Southern Tree Surgery Company (1963), BCR Tree Surgeons (1968), J.H. (Jim) Wilks, Tree Surgeon (1960), Vickers and Company (1958).

ended. Advertisements for some of these early post-war companies are featured in Figure 91.

Contractors

One small pre-war tree surgery contractor with a reputation for a high-quality work was Adrian Estler. He resumed operating after the war, continuing to do most of the skilled work himself and was based at his home in Effingham, Surrey. Estler was often asked to act as a consultant and this usually resulted in him undertaking any work that he recommended. For example, in 1951 he was asked by Evesham Town Council's Parks Committee to examine the avenue of lime trees in the Workman Gardens in Evesham, Worcestershire (*Tewkesbury Register* 1951). The council had become concerned that the trees were dangerous. However, Estler was able to report that they were generally in good condition and only a small amount of pruning and other work was required. Following the submission of his report, the Parks Committee 'adopted the recommendation of their Trees Sub-Committee that the Borough Surveyor should arrange for Mr Estler's report to be implemented'. Estler continued working as a tree surgery contractor until at least 1959. Another contractor operating before the war who continued afterwards was Thomas Higginson. In 1946, he placed advertisements for tree workers and a foreman, suggesting that he was now expanding his operations (*Coventry Telegraph* 1946). Higginson continued running the business until his death in 1973, when his son, also called Thomas, took over and it ran until 1992 (Higginson pers. comm., 2019). His son, Brian, now works as an arboricultural consultant, continuing the Higginson family line of private sector tree experts going back to the 1920s.

One of the largest pre-war tree work contractors was the English Tree Expert Company, owned by Maurice Johannesen (Matthews pers. comm., 2012). The firm had a national reputation and worked all over Britain. After the war the company continued its excellent standard of tree work and Johannesen frequently worked as a consultant, often on high-profile projects. In 1947, the Dean and Chapter of Chichester Cathedral engaged Johannesen to report on the trees along the Cathedral promenade in West Street, following which the English Tree Expert Company undertook some deadwooding and crown-lifting (*Chichester Observer* 1947). In 1949, the company was still operating as the English Tree Expert Company and was based at Sunbury-on-Thames. From 1947 it had been managed by Tom Wilson, a Canadian, who had served during the war as a lieutenant in General Patton's 'Hell on Wheels' Armoured Division (Matthews and Draper 1971, 337). Around 1952 the company was reconstituted as The Southern Tree Surgery Company (*West Sussex County Times* 1952), now owned and managed by Mr H.C.B. Ingram, who was a relation of Johannesen (Matthews pers. comm., 2012). The Southern Tree Surgery Company, now based at West Byfleet in Surrey, continued operating until at least 1963 (*JRHS* 1963).

FIGURE 92. Chiltern Tree Surgeons working in 1956 to preserve an old ash tree at the Royal Masonic Girls' School, Rickmansworth. The arborists are cleaning out and filling a decayed fork and dressing two open wounds.

CREDIT: CHILTERN TREE SURGEONS.

Another company with a national reputation before the war was The Chiltern Tree Surgeons, owned by James H. Carey (Matthews pers. comm., 2012). While this had operated on a relatively small scale, after the war it seems to have expanded but still specialising in high-quality tree surgery, especially cavity work and bracing (Figure 92). Although the company had its main base in Amersham, Buckinghamshire, in the late 1940s and early 1950s it also had an office in London (*JRHS* 1950) and it was 'operating for a limited period in Bristol' (*Western Daily Press* 1949). In 1954, Carey gave advice on the famous 250-year-old black poplar known as the Arbor Tree at Aston-on-Clun (*Birmingham Post* 1954). This had recently been given only four more years to live and there was concern about its safety. Carey, who a newspaper described as 'one of Britain's eminent tree surgeons', declared that there was no reason why the tree should not live for another 250 years if it was pruned. While no mention of Chiltern Tree Surgeons could be found after the mid-1950s, Carey continued to act as an independent consultant into the late 1950s.

The growth of new tree surgery companies in the late 1940s and 1950s reflected the opportunities in amenity tree care that came with the nation's post-war economic recovery. Not only did local and central government have more money to spend on amenity trees but the increase in disposable income for many families that was reflected in the gardening boom also included spending money on garden trees. Several of these new companies were formed by ex-servicemen who, in trying to adjust to civilian life, were looking for an outdoor occupation that involved a degree of excitement and possibly even risk. This was an era of pioneering tree surgery companies run by determined, hard-working and colourful characters.

One of the first new firms was The Tree Specialist Company, founded by a former Wing Commander called A.A. Rumsey and based at Littlehampton, West Sussex (Matthews pers. comm., 2012). Rumsey began operating on a small scale but by the late 1940s was advertising his services regularly in local newspapers. He initially described himself as 'Tree Specialist' and sometimes headed his advertisements with an appropriate poetry quotation, on one occasion from 'The Garden' by Andrew Marvell, one of the most famous English poems of the seventeenth century (*West Sussex Gazette* 1949a). He was also describing his company as 'an organisation devoted to the care and preservation of trees' and with 'teams of experienced men and qualified tree climbers, fully insured and working under expert supervision operating in all Southern Counties'. What is notable about Rumsey and his Tree Specialist Company is the volume and style of his advertising. Until the 1970s, most companies did very little advertising and relied mainly on personal contacts and recommendations to gain work. Rumsey not only advertised a lot but did this imaginatively. Apart from his use of poetry he would employ dramatic copywriting (*West Sussex County Times* 1949) (Figure 91). He always included the letters FRHS after his name, indicating he was a Fellow of the RHS, to add to his credibility. During the late 1940s he advertised regularly to employ 'Experienced Tree Climbers and Tree Fellers; also Pruners', suggesting his company may have been expanding (*West Sussex Gazette* 1949b). The Tree Specialist Company operated successfully throughout the 1950s and by 1958 the company was proud to declare it was acting as 'Contractors to H.M. Government, Horticultural Societies, Municipal Councils, etc.' (*West Sussex Gazette* 1958a). The company also had its own tree and shrub nursery at Roundstone, West Sussex, catering to the retail and wholesale market. Although Rumsey was still advertising for experienced staff in 1959, no mention of his company could be found after that.

Around 1950, The Tree Specialist Company had a base in Kingston-upon-Thames, Surrey, as well as its main headquarters in Littlehampton. The Kingston base was managed by Tom Wilson, the Canadian former army officer, who had joined the firm from the English Tree Expert Company (Matthews 1990). Working under Wilson was Bill Matthews, a young arborist who had spent three years of the war in the Royal Navy (Ryan 2010, 18). When the Kingston operation was unexpectedly wound up, Wilson and Matthews decided to continue in tree surgery by forming their own company. In 1952 they founded The Southern Tree Surgeons, equipped with just a length of rope, some saws, a bag of wedges and a motorbike. The founding date is now sometimes given as 1951 but in earlier accounts, including by Matthews himself, this is stated as 1952 (Matthews 1990). The name of their new company was similar to the existing Southern Tree Surgery Company and its owner, Mr Ingram, felt obliged to place an advertisement stating that his company 'has no connection with the Southern Tree Surgeons' (*West Sussex County Times* 1952).

Despite any confusion about company names, Southern Tree Surgeons founded by Wilson and Matthews soon achieved an excellent name for themselves with steady growth throughout the 1950s. Based at Crawley Down in West Sussex, the company faced some stiff local competition for work. Nevertheless, Matthew's early association with the landscape architect Lanning Roper enabled the company to work on many prestigious projects and to make some influential contacts (Ryan 2010, 19). In the early 1960s, with plenty of regular work in Gloucestershire, they decided to open a branch there at Fossebridge and also another at Redhill in Surrey. The company continued to expand through the 1970s and soon had six branches in England, making it the largest tree care company in Britain. It also opened an office in Dublin to offer tree care services throughout Ireland. Both Matthews and Wilson were now playing a key role in the development of professional arboriculture in Britain, not just through Southern Tree Surgeons but also through their involvement in various industry bodies. In the 1980s, the high standing of the company was emphatically underlined when it became the first arboricultural company in Britain to be granted a Royal Warrant. When Tom Wilson died in 1990, Matthews continued running the company but was now looking at options to sell this as he had no natural successor. In 1992, he sent his usual Christmas card to Robert Bartlett of the Bartlett Tree Expert Company in the United States with the words: 'Why doesn't the best of companies in America get together with the best company in England?' (Cooley 2007, 85). In 1993, Bartlett Tree Experts acquired Southern Tree Surgeons and the transaction was finalised the following year. Matthews still had some involvement with his former company but now concentrated on his work in the wider arboricultural industry (Anon 2018). When Bill Matthews died in 2018 he left behind a unique and lasting contribution to professional arboriculture in Britain.

Another company to emerge immediately after the war was the wonderfully named Beeching of Ash, formed in 1946 by Jim Beeching and based in the Surrey village of Ash. Beeching concentrated on the finer arts of tree surgery, such as sensitive pruning work, with cable bracing his particular speciality. He frequently expressed his concern about the standard of work carried out by some tree surgery companies and had no qualms in highlighting what he perceived as poor practice by some of his well-known colleagues (Beeching 1969, 249). At the same time, he actively promoted improved standards within the industry and was one of the leading figures in the formation of ABTSA in 1964. Like many early companies, Beeching did not advertise much but occasionally placed advertisements where he thought they may be read by more discerning clients, such as in the *Journal* of the RHS (*JRHS* 1970). Beeching was also adept at attracting free publicity through newspaper and magazine articles. In 1960, *Farm and Country* magazine ran a three-page article about how chainsaws were replacing crosscut saws, which featured the work of Beeching of Ash (Fairchild 1960). By 1970, it was a highly successful company with an established reputation for quality work. In the early 1970s, Beeching of Ash was bought by the Economic Forestry Group (EFG), and Ian Keen, an experienced EFG forester with knowledge of arboriculture, was appointed to manage the company (Keen pers. comm., 2019).

In the early 1950s a tree surgeon called Jim Wilks established a company based in the Bristol area. Due to the high quality of his work and his obvious knowledge of trees and tree surgery, he soon developed a thriving business. By the mid-1950s he was working for local authorities across the West Country as well as the owners of some large private gardens. In 1956, he became involved in a controversial incident concerning the proposed felling of some trees surrounding a new housing estate in Wells, Somerset (*Wells Journal* 1956). Members of the Wells Preservation Society (WPS) who opposed the felling had secured a reprieve for all the trees except a large beech tree. However, at the request of Wells City Council, an assistant forester at the County Planning Department had inspected all the trees, in the presence of several city councillors, and he had recommended the felling of this beech tree. Determined to save the tree, the WPS members called in Jim Wilks for a second opinion and his report stated: 'A perfect specimen, well-balanced and in good condition'. However, the majority of the Council's general purposes committee still voted in favour of removing the tree. After further objections, Wilks invited the councillors to make another inspection of the tree, this time accompanied by Richard St Barbe Baker, the founder of Men of the Trees. Baker said the tree was a beautiful amenity and dismissed as ridiculous the claim that it was dangerous. Eventually, after further debate the tree was saved. The incident did much to promote Wilks's tree surgery credentials in the West Country and after the publication of his book on heritage trees he had a national reputation as a tree expert (Wilks 1972).

Other companies established in the 1950s included Honey Brothers at Peasmarsh in Surrey, founded by Peter and Tony Honey in 1955. This company has already been mentioned in connection with the promotion of safety harnesses and its separate sales business for arboricultural equipment. By the 1990s its tree surgery contracting work was much reduced, although the sales business was flourishing. When Peter Honey died in 1995 his two sons continued for some years (Bridgeman 1996), however, only the sales business now operates and the family are not involved. Also emerging in the 1950s was The Tree Preservation Company, founded by G.H. Goodings and H.S. Lisle. It began operating around 1957 from a base in Ferring, West Sussex, and claimed to have 'a team of highly experienced men who will deal with dangerous trees and all forms of tree surgery' (*West Sussex Gazette* 1957a). The firm initially competed with Rumsey's Tree Specialist Company but ceased trading in 1968 (*Acton Gazette* 1968b). Another tree surgery firm formed in the late 1950s was Vickers and Company based in Worthing, West Sussex. It was successful in securing contracts from both central and local government bodies (*West Sussex Gazette* 1958b).

While many companies formed soon after the war were based in southern England, other companies were established elsewhere in Britain. In the mid-1950s, Welsh Tree Services began operating out of Raglan, Monmouthshire, and undertook tree surgery work as well as forestry consulting and contracting services (*Western Mail* 1955). Also in Wales, in the late 1950s, a company called Landscapes were promoting themselves as 'specialists in the preservation of mature and ornamental trees' as well as being experts in the 'lopping and felling of dangerous trees' (*Western Mail* 1959). In the north of England one of the earliest and largest companies was Northern Tree Surgeons (*Evening Chronicle* 1969). This was formed in the mid-1950s by Conrad Jorgensen, who would later play a leading role in arboricultural industry. In Scotland, specialist tree surgery companies were formed later than in England and until the 1960s this work was undertaken mainly by existing landscape and forestry contractors. The growing public interest in tree surgery was viewed as an opportunity to diversify their businesses and they already had the resources, contacts and some of the expertise to do this.

In the Midlands in the early 1960s, a tree surgery company emerged called A.J. Mackenzie Clay Limited, based in Northampton. In 1965, one of its few advertisements states: 'We are the Tree Expert Company' and 'the largest organisation in the British Isles exclusively engaged on surgery, transplanting, maintenance' (*Coventry Telegraph* 1965). No further information could be found regarding this company, except that in 1962 A.J. Mackenzie Clay was a forestry and landscape consultant based in Rugby. However, considering the other tree surgery companies around at that time, his claim to be the biggest in Britain was probably untrue.

Some other early tree surgery companies should be mentioned, not least because their owners made important contributions to British arboriculture. In 1964, Cyril Reddick founded BCR Tree Surgeons at Bookham, Surrey (Matthews 1995). Reddick has been mentioned as a pioneer of the safety harness in Britain and he also played a leading role in ABTSA. By 1968, BCR opened another branch in Buntingford, Hertfordshire, and became one of the largest companies of this era. In the mid-1960s, Roy Finch was a forestry contractor in his home county of Worcestershire (Finch pers. comm., 2019). He began learning about tree surgery through his membership of the Royal Forestry Society and in 1967 he formed Roy Finch Tree Care, specialising in tree surgery (Anon 1994, 356). In the early days of the business he displayed great ingenuity in equipping himself to undertake a wide range of practical work. For example, he made his own safety harness out of horse straps and rivets which he used until Bill Matthews introduced him to commercially manufactured harnesses (Finch pers. comm., 2019). As a member of ABTSA and then the Arboricultural Association, Finch played a leading role in the development of the arboricultural industry in the 1980s and 1990s and his considerable knowledge of practical arboriculture and safe working practices was widely recognised.

The reluctance of leading tree surgery companies in the 1950s and 1960s to advertise in popular newspapers and magazines has been noticeable in this research and made the task of identifying companies more difficult. There was also recognition that 'door-knocking' was inappropriate for 'reputable' companies (Jorgensen 1973, 3). However, it was accepted that advertising in upmarket magazines and horticultural periodicals, such as *Country Life*, *The Field* and the *RHS Journal*, not only reached many potential clients but also gave the right image for the company and the industry. When ABTSA was formed in 1964, one of its first initiatives was to establish a List of Approved Contractors. Companies that were members of ABTSA and could demonstrate a high standard of work were eligible to be included on the list, which was made available to the public, local authorities and other organisations on request. In 1971, the Arboricultural Association established its own Directory of Approved Contractors and when ABTSA and the Association merged in 1974 this became the industry's only approved list. Considering the cost of being examined and the likely benefits to the contactor, the list's value has always been a topic of debate. Approved companies previously competed with non-approved firms that relied mainly on their adverts in the local Yellow Pages. Today, the public can consult several directories of general traders on the internet and these include tree surgeons. While the public increasingly use these internet sources to find a local tree surgeon, those companies are not subject to the level of scrutiny given to those approved by the Arboricultural Association.

In the 1970s, the arboricultural industry was still dominated by small companies with only a few operatives. However, there was also a growing number of landscape, grounds maintenance and forestry companies diversifying into arboricultural work. When local authority services were privatised in the 1980s, tree work was often included in tender documents concerned with landscape and horticultural work and it was those larger broad-based companies that often got the contracts. The dominance of these companies has continued. In a recent survey of the top thirty arboricultural businesses in Britain, many that had substantial turnovers of arboricultural work also had major landscape and grounds maintenance operations (*Horticulture Week* 2020). Furthermore, several top companies were mainly involved in utility arboriculture, an area of work that had grown enormously since the 1990s. The utility arboriculture industry is concerned with tree and vegetation management adjacent to electricity supplies, railways, waterways and telecommunication networks.

Consultants

Some arboricultural consultants who were active before the war also continued in the late 1940s and beyond. Denis Le Sueur continued to build on his formidable reputation not only working for long-standing clients such as the Corporation of the City of London but also taking on other public and private sector commissions. In 1951, he published an article in the *Quarterly Journal of Forestry* on the subject of forestry and arboricultural consultancy, which he regarded as a profession on the same level as landscape architecture and urban planning (Le Sueur 1951). While much of the article is concerned with forestry, he makes it clear that 'a forester is not necessarily an expert arboriculturist' (*ibid.*, 183). He adds, 'While many arboriculturists do not pretend to be expert foresters, too many foresters consider they know all that is to be learnt about arboriculture'.

Another high-profile arboricultural consultant who continued after the war was James H. Carey, the owner of The Chiltern Tree Surgeons. In the late 1950s he found himself involved in a public inquiry where he and Le Sueur were representing opposite sides. The inquiry concerned the application of East Sussex County Council for confirmation of a tree preservation order (TPO) on 102 trees, almost all ancient pollarded beeches, on the Paddockhurst Estate (*West Sussex Gazette* 1957b). The owners had objected to the inclusion of forty-one of the trees on the grounds that they were dangerous. Carey was called by the County Council as an expert witness and, having inspected the trees, he claimed they were some of the finest pollards he had seen. However, he recommended the felling of sixty-three other nearby beeches that were not included in the order. Le Sueur was called as an expert witness for the Paddockhurst Estate. While he agreed with Carey that the trees covered by the order appeared healthy, he believed the branch structure of these ancient pollards was likely to be unstable. He also said that felling a significant number of nearby trees would expose those that remained and put them at risk of windthrow. Le Sueur added that he would not trust any pollarded beech and would state his reputation that 'after the felling of the trees recommended by Carey, many of the remaining trees would be down in two years'. The inquiry later gave its decision in favour of the Paddockhurst Estate and the TPO was not confirmed. No mention could be found of either Le Sueur or Carey working as consultants after the late 1950s.

One of Kew's tree experts in the 1930s was Arthur Osborn, and after the war he established his own consultancy practice. While his excellent reputation ensured plenty of work, this did not always go smoothly. In 1947 he was engaged by Bath City Council to report on its street trees and make recommendations for arboricultural work (*Bath Chronicle* 1947a). One controversial recommendation from Osborn was that the famous avenue of forest-type trees in Pulteney Road be removed and replaced with birch or flowering cherries. When his report was submitted, the Parks Committee initially adopted this recommendation. However, following further consultation and debate, many members of the Parks Committee had changed their mind and were opposed to this, even though 'Mr Osborn, MBE, who advised them was a very noted arboriculturist' (*Bath Chronicle* 1947b). The Committee decided to retract their initial recommendation to remove the trees and consider the matter again. Eventually, some tree surgery and pruning was carried out but the majority of trees were not felled.

Another early post-war tree consultant who undertook both arboricultural and forestry work was Robert Cowell Smith of Banchory, Aberdeenshire. In 1948 he was involved in a fascinating legal action brought by the Ben Nevis Distillery that was seeking an interdict against the North British Aluminium Company to cease manufacturing at its Fort William plant as the noxious fumes were harming the animals and vegetation of their Inverlochy estate (*The Scotsman* 1948). Appearing as an expert witness for the Ben Nevis Distillery, Smith reported that the trees and forestry plantations on the estate were 'exceedingly unhealthy'. He also disagreed entirely with the suggestion that the estate was simply unsuitable for growing timber, and also with the view that the damage to the trees might have been caused by something other than the fumes. The court eventually decided that the manufacture of aluminium could continue but with conditions attached to this.

Much of the work of these arboricultural consultants involved appearing as an expert witness in court cases and inquiries. In 1958, Le Sueur published a paper entitled 'The Forestry Expert Witness' drawing on his own extensive experience (Le Sueur 1958). Despite its title, this is probably the first article concerned with the role of the arboriculturist as an expert witness. Le Sueur states that the increasing number of legal cases concerning trees then reaching the courts was promoted by the wider availability of legal aid (*ibid*., 187). To take advantage of this growing workload, some consultants made a speciality of this work, where they were able to combine a good knowledge of arboriculture with an in-depth understanding of legal matters. One of these was Edward Brooke Boulton, who had held the post of Lecturer in Forestry at the University of Cambridge, where he was recognised for his work on timber technology. After acting frequently as an expert witness, often on cases concerned with tree failure, he became very experienced in this work and by the early 1970s was regarded as the main authority on trees and the law (Matthews 1968c, 4). Another individual with a background in timber technology who was highly regarded as an expert witness in tree failure cases was Harold Ernest Desch (Matthews pers. comm., 2012).

From the early 1970s a new generation of arboricultural consultants appeared on the scene and one of the first was Giles Biddle. After studying forestry at Oxford University he was awarded his doctorate with a thesis on poplar mosaic virus (Biddle pers. comm., 2019). Biddle had always been interested in amenity trees and after spending a year in California he contacted Bill Matthews of Southern Tree Surgeons to seek his help in starting a consultancy business. An agreement was reached to form Tree Conservation Limited whereby Biddle, Matthews and Tom Wilson would each own one-third of the company. However, Biddle undertook the majority of the work, much of which initially involved the fungicide treatment of elm trees suffering from DED. After two years Biddle bought out the other two partners and continued with Tree Conservation Limited on his own. Although the DED treatment was not very effective, the Great Drought of 1976 had put the spotlight on the role of tree roots in building subsidence and Biddle quickly established an excellent reputation for his expertise in this area. While the problem was now generating plenty of work for arboricultural consultants, particularly in the south-east of England, Biddle was the recognised authority and he published extensively on this topic, most notably with his two-volume book (Biddle 1998). Biddle played an important role in the development of British arboriculture, especially as Chairman of the Arboricultural Association, where he significantly enhanced the professional image of arboriculturists.

In 1971, the Arboricultural Association launched its Register of Consultants in Arboriculture, to ensure high standards of practice among those included on the register and to regulate this according to an agreed code of ethics (Anon 1971). Other bodies now maintain registers or directories of consultants that feature arboriculturists, such as the Institute of Chartered Foresters and Consulting Arborist Society. It is unfortunate that the small profession of arboriculture has not managed to operate a single register or directory for its private sector consultants, as some much larger professions have done.

A new type of arboricultural consultancy practice emerged in the early 1980s involving teams of consultants working together. One of the first was Peter Bridgeman and Associates, formed in 1976 when Bridgeman left Merrist Wood College. In the early days he worked mainly on his own, supported by others when necessary, but in the mid-1980s he began promoting his consultancy team through advertisements in the horticultural press (*Horticulture Week* 1986). His team included Dick Grainger and Keith Lever, also former staff at Merrist Wood, and Linda Cross, a biologist. As environmental and conservation issues attracted more public interest in the late 1980s and 1990s, some arboricultural consultants sought to broaden their client base and reflect their wider expertise by embracing an environmental and ecological perspective. One of the first was Treework Environmental Practice, formed by Neville Fay in 1990, and other similar companies have followed. At the same time, consultancy practices focusing on urban planning, landscape architecture, ecology and related disciplines sought to strengthen their consultancy teams by appointing an arboricultural specialist. Since the 1990s, several arboricultural consultants have gained reputations for their expertise in legal and planning matters, such as Dealga O'Callaghan, Les Round, Richard Nicholson and Jeremy Barrell. Although not an arboricultural consultant, Charles Mynors, an eminent barrister, established an unrivalled reputation for his knowledge of the legal aspects of arboriculture. His book *The Law of Trees, Forests and Hedgerows* has become an essential reference for all arboricultural professionals (Mynors 2002).

A major driver for arboricultural consultancy was the Town and Country Planning Act 1990 which not only had a significant role in regulating the development of land but also stimulated the demand for arboricultural expertise (O'Callaghan pers. comm., 2019). The Civil Procedure Rules, introduced following the Woolf Report of 1996, also stimulated demand for consultants as expert witnesses. The requirement for an arboricultural impact assessment in BS 5837, the written report that determines the site-specific effect of a planned development on the existing trees, has also boosted the workload of many consultants (Durdant-Hollamby pers. comm., 2019). Another source of work has been the requirement of mortgage providers for a tree survey and arboricultural report when someone takes out a mortgage, to assess the likely risk that trees might pose to people and property. All this increasing demand for arboricultural consultants has encouraged many arboriculturists to choose this career path. The numbers of consultants has also grown considerably in the past decade following the government's austerity agenda when many local authority tree officers were made redundant. Many of them subsequently went into consultancy work, prompting concern from some established firms that this was 'flooding the market' and undercutting fees. Established consultants have also complained at the loss of their local tree officer as an 'understanding' point of contact, especially when their role is taken by an officer with little knowledge of arboriculture (Adams pers. comm., 2019).

Nurseries and tree production

After centuries of excellent French literature on tree propagation, two outstanding British books on this subject were published in the late 1940s. The first was Robert Garner's classic work *The Grafter's Handbook*, based on his extensive experience as chief propagator at the East Malling Research Station (Garner 1947). The second was the excellent *Propagation of Trees, Shrubs and Conifers* by Wilfred Sheat, horticulturist at the Ministry of Transport (Sheat 1948).

Shortly after the war many local authorities set up their own nurseries producing not only herbaceous material but also trees and shrubs to meet their own requirements. Some authorities also set up 'tree banks' where young trees were grown on to large standard and semi-mature sizes before being transplanted to various planting sites. The GLC tree banks at Hersham in Surrey and Hainault in Essex were pioneering ventures in this regard (Ryan 2009b, 22) and were replicated by other authorities such as Ealing with its tree bank at Glade Lane in Southall (Thurman pers. comm., 2019). In the

1980s, however, with the escalating privatisation of public services, many local authorities closed their nurseries and relied solely on stock purchased from commercial growers.

Of the commercial tree and shrub nurseries that were prominent before the war, Hillier Nurseries is the most notable in the post-war era. Not only did the company remain at the forefront of tree production in terms of its product range and scale of operation, but Harold Hillier became recognised internationally as a leading authority on trees and shrubs. In 1957 he was awarded the Victoria Medal of Honour from the RHS and in 1983 he received a knighthood for services to horticulture (Hillier 2014, 178). Hillier's major written work was *Hillier's Manual of Trees and Shrubs*, first published in 1972 and compiled with assistance from Roy Lancaster, the eminent plantsman and tree expert (Hillier 1972). In 1980, the Arboricultural Association enhanced an already close connection with the Hillier family when it moved its headquarters to Ampfield House, home of Hillier Nurseries. When Sir Harold Hillier died in 1985, the Association continued to work closely with his son Robert who had been its Honorary Treasurer since 1976 (Anon 1994, 353).

In recent decades Barcham Trees has emerged as a major nursery for amenity trees. Based near Ely in Cambridgeshire, it was founded in 1978 by Peter Wells who had previously worked at Hillier and EFG nurseries (Appleby 2011). Wells heard about the use of white bags for container-grown trees, developed in Australia, which reduced the temperature in the container and produced a superior root system. Using this method of tree production Barcham Trees has expanded rapidly, especially through sales in the local authority sector. In contrast to its competitors, Barcham has taken an individual approach to marketing. Rather than spending money on conventional advertising, it has organised a series of free seminars and occasional one-day conferences on tree-related subjects (Wells pers. comm., 2019).

While a detailed account of trends in tree selection is beyond the scope of this book, some post-war developments should be mentioned. The trend away from forest-type trees in streets and confined urban spaces that was evident before the war became even more apparent in the 1950s and 1960s. However, instead of the usual small ornamental species of trees, other choices became available. Following tree-breeding programmes begun by Edward Scanlon in the United States, nurseries could now offer a selection of 'tailor-made' street trees. These had been specially developed with qualities that made them ideally suited as street trees and they soon became popular in local authority planting schemes. Also, from the 1970s onwards, tree officers began to experiment with planting a much wider range of species in parks, open spaces and streets. Many had been encouraged to try more unusual species during their arboricultural education by college lecturers such as the inspiring John Whitehead at Merrist Wood. To meet this growing demand, many commercial tree nurseries started to stock a much wider range.

Professional representation

By the 1940s the larger professions allied to arboriculture were now represented by a professional body. Some of these bodies had been formed in the nineteenth century and others in the first few decades of the twentieth century. Some were trade associations and others were constituted as professional institutes. However, for professional arboriculturists there was still no organisation dedicated to representing their interests. For owners of tree surgery companies and the local authority tree officers who wished to belong to a relevant body, there were alternatives. Some of them joined the two Royal Forestry Societies, the Men of the Trees or the RHS.

The Royal Forestry Society (RFS), based in England, continued to have many members whose primary interest was arboriculture rather than forestry. However, for the first decade after the war arboriculture received little attention from the RFS until N.D.G. James proposed to resurrect its Arboricultural Committee (James 1957). This was duly reformed in 1957 and it prompted an increase in the RFS's arboricultural activities, which included organising meetings at locations of special interest to arboriculturists (James 1982, 107). In 1965, the RFS organised a visit to Westonbirt Arboretum, which included demonstrations of arboricultural practice from members of the recently formed ABTSA (Leathart 1966, 61). Around 200 people attended and they watched demonstrations of tree pruning, cavity work, cable bracing, aerial access with Jim Wilks's fire engine, as well as enjoying a guided tour of the arboretum with Alan Mitchell. In the 1970s, the RFS noticed a decline in its members' interest in arboriculture and its Arboricultural Committee was dissolved in 1976 (James 1982, 124). This diminished interest was due largely to some major developments that will be described next.

Professional representation finally arrives

In the early 1960s, the arboricultural industry in Britain finally got an organisation of their own to represent them – in fact, they got two organisations. The first to organise themselves were a number of tree surgery companies. Part of the inspiration for this was the International Shade Tree Conference (later the International Society of Arboriculture), formed in the United States in 1924, which now represented professional arboriculturists in several countries (Campana 1999). In 1963, Tom Wilson of Southern Tree Surgeons had attended the International Shade Tree Conference, held that year in Toronto, Canada, and on his return several preliminary meetings were held with tree surgery colleagues. Such was the enthusiasm shown by all who attended that it was decided to go ahead and form their own organisation (*Gardeners' Chronicle* 1964a).

The Association of British Tree Surgeons and Arborists (ABTSA) was formed in 1963 with the inaugural meeting taking place on 29 April at Farnham,

Surrey (Honey 1964b). Tom Wilson was elected as the first Chairman and other founding members included Peter and Tony Honey, Bill Matthews, Jim Beeching and Cyril Reddick. Eve Deller was appointed as the paid Secretary of ABTSA, a position she held until her retirement in 1974 (Deller 1991). Much of the motivation of the early members was to establish recognised standards for arboricultural work, particularly tree surgery, and to make the public aware that this was a specialist occupation that needed skilled and experienced professionals (Beeching 1965). This was reflected in the early establishment of ABTSA's List of Approved Contractors. ABTSA membership comprised fellows, probationary fellows, associate members and student members. The fellows were its governing body and were also its approved contractors, having demonstrated the required standards of work, supplied good references, and obtained a minimum of £50,000 of insurance cover.

In 1964, another British arboricultural organisation was formed, the Arboricultural Association, this time by mainly public sector horticultural and landscape officers (Wells 1974, 218). Initial interest had been generated by an informal meeting held in Middlesex Guildhall in May 1963 and the Arboricultural Association was formally launched at an inaugural general meeting, also at Middlesex Guildhall, in May 1964. Don Wells was appointed as its first Chairman and other founding members included Hubert Taylor, Derek Honour and Anthony Dunball. It is interesting that the announcement of the formation of the Arboricultural Association in the *Gardeners' Chronicle* (1964b) makes no mention of ABTSA or any possible overlap between the two organisations, although they had similar aims. Indeed, many arboriculturists and tree surgeons were members of both organisations. However, the Arboricultural Association also had some strategic aims in terms of influencing government policy and legislation where this was relevant to arboriculture, particularly in areas such as planning and the environment. That was a reflection of the priorities of its members, many of whom were employed in local government. To ensure the Association had some political 'leverage' it appointed Arthur Skeffington as its Founding President. Skeffington was Labour MP for Hayes and Harlington who in 1967 became Parliamentary Private Secretary to the Minister for Housing and Local Government. In 1967, Lord Kennet and Duncan Sandys MP were appointed as Vice-Presidents of the Association, followed in 1968 by Earl Jellicoe as a third Vice-President. With its president and vice-presidents, the Association now had representation on both sides of both Houses of Parliament (Anon 1968). From its outset the Arboricultural Association has also had a strong educational role. This began with the publication in 1965 with its first advisory leaflet entitled *Tree Preservation Orders* and has continued to this day with a wealth of advisory leaflets, codes of practice, safety guides and other documents.

In 1969, Don Wells stepped down as Chairman of the Arboricultural Association and was succeeded by Tom Wilson, who had been the founding Chairman of ABTSA (Matthews and Draper 1971). Don Wells was a Kew graduate, as were many of the prominent early members of the Arboricultural Association such as Hubert Taylor, Sidney Pearce, Ted Storey, Frank Knight, Anthony Dunball and Peter Bridgeman. Of the eleven members of the Association's initial Executive committee, five were Kewites (Bridgeman 2013, 125). In 1966, seven of the thirteen members of its Executive Committee were former staff or students from Kew.

In its early years ABTSA's membership grew steadily and a list dated June 1970 shows that it had 137 members (*ABTSA Newsletter* 1970). Of these, two were from Wales, one from Scotland, four from Ireland (both north and south) and two were overseas. The rest were from England with the vast majority from the south. There were eight women on the list. An analysis of the publications for members produced by both ABTSA and the Arboricultural Association has highlighted some interesting attitudes in relation to professional representation and competence. It was not uncommon for members to express negative attitudes towards foresters who often were viewed as knowing little or nothing about arboriculture. As professional foresters usually had degree-level qualifications, in contrast to most arboriculturists, and many County Council forestry officers had responsibilities for arboricultural matters, there were fears that the forestry profession might try to 'take over' arboriculture. There were also tensions between ABTSA and Arboricultural Association members about their respective levels of expertise in certain areas. Typical of these exchanges is a letter from Jim Beeching of ABTSA to the editor of the *Arboricultural Association Journal*. Beeching writes, 'It is interesting to see, in the same issue, that Mr Girling is getting agitated because all tree surgeons are not arborists but he might note that there is an increasing amount of written evidence to prove that all arborists are far from being tree surgeons' (Beeching 1969).

In the early days of the two organisations, the membership of ABTSA seem to have had a more international perspective, evidenced by the number of articles and news items about overseas developments, particularly in North America. Its early links with the International Shade Tree Conference (later the ISA) were enhanced with some leading ABSTA members regularly attending its annual conference and reporting on this in the *Newsletter* (Honey 1970a, 6). Some members also combined those conference visits with short tours of North America to study arboricultural practice on the other side of the Atlantic.

One topic where the members of both organisations were in complete agreement was the need to increase public awareness of good arboricultural practice and to warn of the dangers of employing 'loppers and toppers' with little knowledge of trees. A significant achievement towards that end was made by ABTSA in 1970. One of its approved companies had been working for Roger Musgrave, a director of a leading advertising agency, and he became interested in ABTSA's aims to promote good tree surgery (Honey 1970b). At this time, *The Sun* was running a competition for charities to win a free full-page advertisement in its newspaper for the most imaginatively produced advertisement about their services. Musgrave and his team offered to work free of charge on an advert for

Trees don't like amateur surgeons...

Imagine you're a proud, handsome tree. But then someone decides you're too tall. Or blocking out too much light. So the local lopper-and-topper treats you to some amateur surgery. 'Pruning' you into a mutilated stump. Ouch.

Trouble is, it happens all too often. You can see Jack the Lopper's victims everywhere. All they're ever likely to grow again is a mass of whiskery twigs. Or the shock may kill them completely.

But it needn't happen. If you've got a problem tree, it's easy to call in a qualified Tree Surgeon. He knows how to prune it without spoiling the shape. He can thin its branches to let the light through. If it's damaged or diseased, he knows the right remedy. And his charge can easily be less than the local buzz-saw botcher's.

The Association of British Tree Surgeons and Arborists is pledged to keep Britain's trees beautiful. Ask us for a free list of our members. Their services cover the entire country. Each one has proved his skill, ability and knowledge to an examining panel, and guarantees to work to official tree surgery standards. Each one is comprehensively insured.

Your beautiful tree is a precious possession. It deserves the best possible care. We're ready to help.

The Association of British Tree Surgeons and Arborists, [illegible] Wings Road, Upper Hale, Farnham, Surrey. Tel: 5924.

Keep trees beautiful. Consult a qualified Tree Surgeon.

FIGURE 93. This ABTSA advertisement appeared in *The Sun* newspaper in May 1970. Its important message about the need to consult a qualified tree surgeon is dramatically conveyed by the stunning visual image.
CREDIT: DAVE FORD.

ABTSA. From a total of 240 entries, the ABTSA advert was picked as one of the twenty winners (Musgrave 1970). The advertisement, which urges tree owners to 'Keep Trees Beautiful' and 'Consult a Qualified Tree Surgeon', appeared in *The Sun* newspaper on Tuesday 5 May 1970 (Figure 93). It is impossible to accurately assess the impact of this advert on the general public but it must have been considerable. Certainly, the feedback from ABTSA members on this nationwide publicity was very favourable, although the image of the saw cutting into the finger did generate some controversy.

The Arboricultural Association-ABTSA merger

By 1971, ABTSA and the Arboricultural Association had coexisted alongside each other for several years and many arboriculturists were members of both. However, in such a small industry it was inevitable that questions would be raised about the need for two separate organisations, given their similar aims and activities. In early 1971, at the invitation of ABTSA, a meeting was held between representatives of the two organisations (*ABTSA Newsletter* 1971, 4). Its purpose was to consider whether sufficient common ground existed to justify establishing formal discussions on amalgamation. Following a positive meeting it was agreed that each organisation would nominate five members to a joint working party which had the following terms of reference:

> To consider proposals for the amalgamation of the two Associations, to devise plans for a single organisation which will further the development and standing of arboriculture and to make recommendations jointly to the two Associations by December 1971.
>
> *ABTSA Newsletter* 1971, 4

The Joint Working Party was chaired by Richard Haigh, an Arboricultural Association ordinary member and former Associate Member of ABTSA (Jorgensen 1972). By May 1972 it had produced its final report and it was announced that this would soon be sent to all members. Its formal recommendation was that the merger of the two organisations should take place. However, it was not until September 1974 that this actually happened. It seems that, once agreement was reached in principle, the detailed negotiation of the merger raised some difficulties that took time to overcome. Eventually, at an Extraordinary General Meeting of the Arboricultural Association held at Merrist Wood Agricultural College on the 3 September 1974, the formal merging of the Arboricultural Association with ABTSA was put into effect by the adoption of several resolutions (Lewington 1974). The new Association would be known as The Arboricultural Association (incorporating the Association of British Tree Surgeons and Arborists), later abbreviated to just 'incorporating ABTSA'. The members of an Acting Council for the newly merged Association were duly elected (Anon 1974) (Figure 94). This was an historic occasion for British arboriculture and the culmination of over three years' work, not just

1
2
3
4
5
6
7
8
9
10
11
12
13
14
15
16

FIGURE 94. Arboricultural Association (Incorporating ABTSA) Acting Council, 1974: 1. Norman Waumsley, 2. Keith Ross, 3. Arthur King, 4. Bernard Haverson, 5. Dr Giles Biddle, 6. Bill Matthews, 7. Colin Bashford, 8. Bill Lewington (Chairman), 9. Peter Bridgeman (Secretary), 10. John Chaplin, 11. Conrad Jorgensen, 12. Trevor Preston, 13. Richard Haigh, 14. Don Wells, 15. Stan Oldham, 16. Miss G. Morris. Some members not identified and some not present.

by the Joint Working Party but also many members of both organisations. Just a small number of Arboricultural Association members voted against the merger, including Derek Honour, Henry Girling and Alan Smith, a tree surgery contractor (Girling pers. comm., 2019). With very little dissent, it was hoped there would now be a unified and influential voice promoting British arboriculture.

The 1980s was a period of sustained growth for the Arboricultural Association. Now that it was the unified voice of British arboriculture it was making a real impact at central government level and having a significant influence over what happened to the arboricultural industry. While there was much work to be done on safety issues and privatisation had caused major disruption to many local authority tree sections, the prospects were nevertheless bright. The Association was also maintaining those early international connections, particularly with the ISA. In 1980, in conjunction with the ISA, it organised the First International Arboricultural Conference at the University of Sussex in Brighton (Anon 1983). The programme included many eminent speakers, including Alex Shigo, Gordon King and Larry Biles from the United States and Aloys Bernatzky from Germany.

Other developments in professional representation

With the large number of tree officers now employed by local authorities it was logical that those working in the same region would get together periodically to discuss matters of mutual interest and explore areas where they could cooperate. After some early informal gatherings of tree officers in the London area, the London Tree Officers Association (LTOA) was established to put these contacts on a more formal basis. Following the Hurricane in 1987, a consultants' report to Task Force Trees recommended the funding of a full-time executive officer for the LTOA (Johnston 1999, 248–249). When Russell Ball was appointed to this position the organisation increased its activities and developed a much higher profile. Since 1999, Becky Porter has continued to make similar progress in this role with the support of some dynamic LTOA committee members. Taking their lead from the LTOA, tree officers in other regions formed similar groups, such as the Midland Tree Officers Association formed in 1994. In 1996 there were ten regional tree officer groups, all of them in England (Hinton and Smith 1996, 197).

The Society of Foresters of Great Britain was founded in 1925 but had shown little interest in arboriculture. In 1982 it was granted a Royal Charter and became the Institute of Chartered Foresters (ICF). This transpired to be a hugely significant moment for British arboriculture but it attracted little comment at the time. The Charter granted to the ICF included a definition of forestry that embraced arboriculture. Although the leadership of the Arboricultural Association at that time largely failed to recognise the implications of this, the Association's own efforts to become the professional body for British arboriculture were now effectively over. However, as the ICF was still showing little interest in arboriculture there was no immediate concern that foresters might try to 'take over' arboriculture. While the award of chartered status to ICF professional members meant they could now designate as 'chartered forester', the Arboricultural Association did not view this as an attractive option for arboriculturists who did not see themselves as foresters. In 1997, the Landscape Institute was granted a Royal Charter and its professional members were able to designate as 'chartered landscape architect'.

While professional horticulturists had always been members of bodies such as the RHS and various botanical societies, in the early 1980s they did not have their own professional institute. This was in contrast to related disciplines such as forestry and landscape architecture that had achieved this and would later be granted a Royal Charter. In 1980, the Council of the RHS proposed the creation of an institute that could act as a campaigning body for the status of horticulture as a profession (Elliott 2004, 329). In drawing up its proposals the RHS worked closely with the Horticultural Education Association (HEA) whose members were based mainly in the colleges and universities teaching horticulture. In 1984 the proposal for an Institute of Horticulture was formally launched and the HEA agreed to be absorbed into this. By August 1984, over 600 applications had been received for membership, and the Institute of Horticulture (IoH) was officially inaugurated the next month. Some suitably qualified arboriculturists now joined the new IoH, although most retained their membership of organisations such as the Arboricultural Association. The IoH grew steadily and was soon recognised as the professional voice of horticulture. In 2014, it was granted a Royal Charter and became the Chartered Institute of Horticulture (CIH), enabling CIH professional members to designate as 'chartered horticulturist'.

The creation of two other bodies should be mentioned, although not closely connected with professional arboriculture in Britain. The first is the Tree Council, founded in 1973 with government backing as an umbrella body for organisations involved in tree planting, care and conservation. In 1975, one of the first actions of the Tree Council was to organise National Tree Week which is now an annual event. In 1990, the Tree Council launched a national network of volunteer tree wardens to support tree-related community involvement activities (Anon 1990). Tree warden schemes now operated in many towns and cities as well as rural districts, with assistance from their local authority or local conservation organisation. Tree wardens and their network on local contacts can provide valuable support to local authority tree officers, particularly in the organisation of community-based tree planting and educational activities. Some arboriculturists, such as Bill Matthews and Caroline Davis, have played a leading role in the development of the Tree Council. The second organisation is the European Arboricultural Council (EAC), formed in 1993 (Anon 1993). Originally sponsored and funded by the Arboricultural Association, the EAC is intended as a forum where delegates from arboricultural organisations throughout Europe can liaise on matters such as research, education and the improvement of working practices.

Magazines, journals and the internet

As the arboricultural industry and profession became more organised, this was reflected in a wide range of magazines, journals and other publications. Some were of a more general nature and covered a wide range of related disciplines, while others were specifically focused on arboriculture.

After the war, the *Gardeners' Chronicle* continued to publish a substantial amount of material relevant to arboriculture. From the 1960s, this also included job advertisements for local authority tree officer posts. The magazine has played a significant role in providing coverage of some important or controversial episodes in the development of professional arboriculture. *Gardeners' Chronicle* is widely read by professionals in horticulture, landscape, parks management and related fields and it keeps these individuals informed of developments in arboriculture. In 1986, it changed its name to *HorticultureWeek* and its internet output is now more significant than its hard copy.

The first issue of the *ABTSA Newsletter* appeared soon after that organisation was formed and it played a vital role in the exchange of information and ideas among its members. The *Arboricultural Association Journal* was launched in 1965 and it played a more scholarly role. When the merger came in 1974, its title was changed to the *Arboricultural Journal* and Tom Hall was appointed as editor. Hall had graduated in forestry at Edinburgh University, held a doctorate in tissue culture from Liverpool University and had been employed as Arboriculturist with the Metropolitan Borough of Calderdale. In his twenty-four years as editor he transformed the *Arboricultural Journal* into an internationally recognised academic publication and in the process did much to enhance the standing of British arboriculture (Johnston 2013). From 1981 the journal was subtitled *The International Journal of Urban Forestry*, to reflect the worldwide interest in this field (Anon 1980). When Hall retired in 1997 the journal continued under various editors but has faced increasing competition from rival publications such as the ISA's journal *Arboriculture and Urban Forestry* and the more recent *Urban Forestry and Urban Greening*. The journals of the two Royal Forestry Societies, the *Quarterly Journal of Forestry* and *Scottish Forestry*, continue to publish papers of arboricultural interest. The Arboricultural Association members' magazine, now entitled *The Arb Magazine* and published quarterly, is a high-quality publication with a readership that extends beyond just Association members. Other independent arboricultural magazines have appeared and these include *Essential Arb* and *Pro Arb*.

Mention should be given to an online bookshop specialising in literature on trees, arboriculture and forestry. Called Treesource, this was founded in 2000 by Jo Ryan, an arboriculturist from Yorkshire and graduate of Askham Bryan College and Aberdeen University. Treesource has been a popular source of books for tree-related professionals and, although not always the cheapest copies available, people preferred to buy them from Ryan who was widely respected in the arboricultural industry. In 2014, she sold Treesource to Summerfield Books and it continues to disseminate tree-related literature nationally. Ryan was also involved in a publication venture. In 1996, Chris Hinton and Lesley Smith (1996) compiled the first edition of *The Tree Resource Yearbook*, essentially a directory for the arboricultural industry that contained a wealth of relevant information. A second edition was published in 1998. In 2002 it was re-titled *Tree Sourcebook* and produced by Ryan's company Treesource, which also produced the last edition in 2006. These were very useful publications which also gave an overall picture of Britain's arboricultural industry at that time.

In the past twenty years the development of the internet has had a huge impact on every industry and profession, including arboriculture. It has facilitated the rapid and extensive dissemination of information and enabled arboricultural organisations to increase their profile. It has also seen the launch of some discussion forums relevant to arboriculture. The first was UKTC, established by Chris Hastie in 1999 and popular with tree officers and consultants. Over the past twenty years UKTC has contributed significantly to British arboriculture by enabling the sharing of a vast amount of information and ideas. Another forum called Arb Talk, founded by Steve Bullman in 2007, performs a similar role for practical tree care professionals.

Arboricultural education

In the late 1940s, there were still no separate qualifications in arboriculture, the subject being included in some horticultural and forestry examinations. However, this eventually happened through the vision of the Royal Forestry Society of England, Wales and Northern Ireland (RFS). Post-war lobbying for the RFS to develop specific arboricultural qualifications had begun with some of its own members as early as 1950 (Watkins 1950). They included William Dallimore, who not only perceived the need for arboricultural qualifications but also felt that such a move would help to enhance the standing of arboriculture as a subject quite distinct from forestry. He stated, 'There is a good deal to be said for the Society taking a greater interest in arboriculture than is done at present, and endeavouring to erase from the minds of silviculturists that arboriculture is a subject to be tolerated but not encouraged' (Dallimore 1951, 60). The RFS also received encouragement to offer arboricultural qualifications from Hubert Taylor, the prominent GLC arboriculturist (Anon 1988, 363). Finally, in 1957 the recently reformed Arboricultural Committee of the RFS proposed the introduction of two examinations in arboriculture, leading to a certificate and a diploma respectively (James 1982, 107). The first examination for the Certificate in Arboriculture was held in 1958 and the first examination for the Diploma in Arboriculture in 1959. Candidates had to pass the Certificate before they could sit the Diploma. A three-week course to prepare candidates for the Certificate was offered at Cumberland and Westmoreland College of Agriculture and Horticulture at Newton Rigg (Bridgeman 1976b, 452).

When the Certificate examination was held for the first time in 1958, twenty-nine candidates presented and five obtained the qualification, having passed both the written and practical parts of the examination. When the first Diploma examination was held in 1959, five candidates presented but only two passed both the written and practical examinations. The two successful candidates were Arroll Winning, then Director of Parks at Aberdeen, and Mr Fos of the Royal Botanic Garden, Edinburgh (*QJF* 1960, 92). The examiners were George C. Taylor, Hubert Taylor and Ray Evison. In 1961, Hubert Taylor took over as senior examiner on the death of George Taylor, a position he held for ten years (Anon 1988, 363). In 1966, the RFS Council announced that its Diploma in Arboriculture would be known in future as the National Diploma in Arboriculture and 'that a holder of this diploma shall be entitled to use the letters *N.D.Arbor.* after his name' (*QJF* 1966). Having made this assumption about gender, in 1967 Muriel Blore became the first woman to pass the Diploma in Arboriculture (*QJF* 1968, 75). She was a librarian at Pershore College who ran evening classes on arboricultural topics and organised visits to tree collections.

Now that the RFS was offering qualifications in arboriculture there was a clamour in the arboricultural industry for the establishment of some formal courses in the subject. ABTSA led the way with this, although it was focused mainly on basic practical courses. Peter Honey of Honey Brothers declared:

> Education and training need urgent consideration. Arboricultural examinations embrace too wide a field; very few tree firms have but scant knowledge of shrubs, and will have little to do with them. There is room for a different examination, limited to tree work, tree surgery and the planting of new trees. This would encourage youngsters to take an interest in tree surgery. At the moment tree surgery firms are really the only schools available.
>
> Honey 1964b, 363

His views were echoed by Jim Beeching, another leading member of ABTSA. In discussing the prospects for British tree surgery companies he wrote, 'The trouble from which all of these firms suffer is not a shortage of work but a shortage of skilled staff to carry it out' (Beeching 1965, 149). If professional arboriculture in Britain was to continue to thrive, there was an urgent need for a formal educational system that would develop courses in the subject.

In 1967, ABTSA approached Merrist Wood Agricultural College in Surrey to help organise and provide a venue for a two-day arboricultural course (Bridgeman 2010a, 25). The college was conveniently located close to some of ABTSA's leading members and the new College Principal, Tony Harris, was very supportive. The course was effectively the first annual conference for ABTSA and comprised talks and demonstrations of tree surgery techniques by several of its members. What was envisaged to be a low-key one-off gathering signalled the start of a remarkable decade of growth for both arboricultural education and the industry. The high numbers attending convinced the college that there was a demand for education and training in arboriculture. The initial step towards this was to appoint the country's first full-time Lecturer in Arboriculture, Peter Bridgeman, the Kew graduate and former GLC arboriculturist. In September 1968, the first ten-week tree surgery course commenced, followed in 1970 by a three-week tree surgery course for foremen and a much broader and detailed one-year Certificate in Arboriculture course. In 1972, the first students embarked on a three-year course for the Ordinary National Diploma in Arboriculture (Bridgeman 2010b, 23). Full-time arboricultural education was now firmly established in Britain and Merrist Wood College was its national centre. To cope with the demand for arboricultural courses, a talented team of lecturers was assembled, including Derek Patch, John Whitehead, Don Robinson and Dick Grainger. Arboricultural technicians with substantial industry experience were also appointed to provide support for practical classes. As the college's Arboriculture Department grew it gained international recognition and began to attract some overseas students. Much of this success was due to the drive and ability of Peter Bridgeman, now Head of Department, who was ably assisted by his College Principal, his staff and industry organisations.

Arboricultural courses were also being established at other education and training centres, often linked with the requirements of the Local Government Training Board that was emphasising the importance of arboriculture in its recommendations for parks and playing-field staff (Winning 1973, 154). The National Proficiency Tests Council (NPTC), an awarding body for land-based training and education, later began offering certificates of competency for specific practical skills in arboriculture. NPTC certificates have become a valuable indicator that operatives can carry out arboricultural work in a safe and proficient manner.

As well as ABTSA's efforts, the Arboricultural Association was also taking a keen interest in arboricultural education. In September 1969, the Association held a two-day symposium at Merrist Wood College to discuss arboricultural education and training and the development of appropriate examinations (Arboricultural Association 1969). Some twenty people attended including representatives from ABTSA, Forestry Commission, Institute of Landscape Architects, Department of Education and Science, Askham Bryan College and other interested parties. The symposium appointed a Joint Working Party to consider the report of its deliberations and make recommendations for the future development of arboricultural education and training. This reported in early 1970 and its findings were then put to an open meeting arranged by the Association in conjunction with ABTSA at a three-day conference in September 1970 (JWPAE 1970). In 1972, the Standing Committee on Arboricultural Education was formed to implement the Joint Working Party's recommendations, with Frank Knight as its Chairman and Bridgeman as its Secretary (Bridgeman 1991, 18). Its main role was to promote and monitor developments in arboricultural education to ensure a consistently high standard of provision across the country.

The success of arboricultural courses at Merrist Wood College led to other institutions starting full-time courses. By the mid-1970s, Myerscough (Lancashire), Cannington (Somerset), Reaseheath (Cheshire), Capel Manor (Enfield), Houghall (Durham) and Askham Bryan (Yorkshire) were among the colleges offering full-time courses in arboriculture (Bridgeman 2010c, 43). Despite this progress, all these courses were still at further education (FE) level and arboricultural education remained 'grounded' at this level for many years. The significant delay in the development of degree-level courses damaged the professional standing of arboriculture. While related professions had produced graduate-level professionals for many decades, some with chartered status, arboriculture had remained something of a craft-level occupation. As many tree officers were only qualified to FE level, this had limited their influence in the strategic aspects of urban tree management. Attitudes within the industry were partly responsible for this situation. Many senior industry figures were self-taught and had learnt 'on the job' and did not recognise the need for management level courses. Furthermore, some leading contractors were stressing the shortage of skilled manual workers and believed any new courses should concentrate on teaching practical skills (Matthews 1982, 176). It was not until 1992 that the breakthrough came when Myerscough College offered the first full-time higher education (HE) course with its Higher National Diploma in Arboriculture (Johnston and Hirons 2012, 105). From 1993, the University of Aberdeen offered a BSc (Hons) in Forestry with a specialism in Arboriculture and Amenity Forestry (Miller pers. comm., 2019). Eventually, in 1998, Myerscough College became the first British educational establishment to offer a degree course in arboriculture with its BSc (Hons) in Arboriculture. In 2007, it offered the first masters degree with its MSc in Arboriculture and Urban Forestry. Although an MSc course had been run by Middlesex University from 1998, this was actually an MSc in Resource Management with an option in arboriculture. Some other colleges soon followed Myerscough's lead and developed their own HE courses in arboriculture.

With these developments at Myerscough College, this institution now replaced Merrist Wood as the national centre for arboricultural education. That position was consolidated further in 2003 when Myerscough became the first educational institution in the world to offer degree-level arboricultural courses through e-learning (Johnston and Hirons 2012). Students from as far away as Japan, Brazil, Australia and South Africa could now gain a degree in arboriculture at Myerscough via the internet. While these developments were major breakthroughs, the future of arboricultural education at HE level remained uncertain as it was still being taught through FE colleges. In contrast to related subjects such as forestry, horticulture and landscape architecture that had been firmly established in the university sector for many years, these FE colleges generally had a weak academic and research culture, and low salaries created difficulties in the recruitment of suitable academic staff.

In the late 1980s, the RFS National Diploma in Arboriculture underwent some change. In 1987 it was renamed as Master in Arboriculture to parallel the recent change of the similar RHS qualification to Master of Horticulture (*QJF* 1987). However, following changes in the government's Department of Education regulations the RFS was asked not to use this title as it was not a postgraduate qualification. In 1990, the RFS renamed the qualification as the Professional Diploma in Arboriculture (*QJF* 1990). Interestingly, the RHS was allowed to continue with the name of its Master of Horticulture, as the RHS had a long pedigree in the field of education (Harris pers. comm., 2019).

With a steady stream of graduates now entering the arboricultural industry, a private sector educational enterprise was also making a major contribution to this. Tree Life was founded in 1997 by Dave Dowson, previously Lecturer in Arboriculture at Brooksby College, and provides arboricultural education courses at various levels through day release and distance learning. It has been especially successful in enabling people in the industry to gain the Professional Diploma in Arboriculture.

Women in arboriculture

Women had always been virtually absent in professional arboriculture but the post-war years witnessed a long-overdue breakthrough. Some women who came to prominence during this period have already been mentioned, such as Eve Deller, Muriel Blore, Caroline Davis and Jo Ryan. Others include Jean Leslie (née Berry), Executive Secretary of the Arboricultural Association for many years (Anon 1995), Ruth Chapman, the pioneering Trees and Woodlands Manager at Dacorum Borough Council (Britt and Johnston 2008, 545–554), and Lesley Adams who in 1999 was the first woman to become an Arboricultural Association Registered Consultant. These women were exceptional in many ways and not just for being among a small number of their gender working in the arboricultural industry.

The first reliable data on the gender imbalance in the arboricultural industry came from research in 1997 at the University of Ulster (Johnston and Rushton 1999, 3). This found that with the main tree-related officer post in British local authorities, only 7 per cent of the postholders were female and only 3 per cent were female with an arboricultural background. The results raised questions about whether the arboricultural profession was unattractive to women or whether they may be actively or passively discriminated against when applying for these posts. Another survey conducted in 2004 of local authorities in England found that nearly 20 per cent of all tree-related officer staff were female (Britt and Johnston 2008, 135). Progress on this issue has been achieved due to more women obtaining qualifications in arboriculture since the mid-1990s, especially at higher education level. The advent of degree-level courses that prepared students for managerial or consultancy posts rather than manual work has made a career in arboriculture more attractive to women. This issue has gained more attention recently with the formation of the Women in Arboriculture Group (McEwan 2018), an encouraging development that might also prompt action on other aspects of diversity within the arboricultural industry.

Urban tree management and the impact of urban forestry

Urban tree management in Britain underwent a radical transformation in the late 1980s and 1990s due largely to new ideas and practices emanating from the urban forestry movement in North America. While it has been stated that urban forestry is 'nothing new' in Britain and dates back to the model villages and garden cities over 100 years ago, this misunderstands the essential concept of urban forestry and confuses it with the imaginative planting and management of urban trees. However, urban forestry is a quantum leap forward in urban tree management that is rooted firmly in the late twentieth century.

While the concept has its origins in Canada in the mid-1960s, urban forestry ideas and practice have now radically transformed urban tree management in many other countries, including Britain (Jorgensen 1986; Johnston 1996). Urban forestry emphasises the importance of a planned, systematic and integrated approach to the management of the urban forest, the collective term for all the trees in and around a town or city. It also brings together in a multidisciplinary approach all the relevant professionals from both the natural and social sciences, including arboriculturists, horticulturists, foresters, landscape architects, planners, sociologists and ecologists. With the development of urban forestry has come a growing recognition of the many environmental, economic and social benefits of urban trees. This has come out of exciting new research on the importance of our urban forests and what is also referred to as green infrastructure (Forest Research 2010). However, those benefits will only be realised if the right tree is planted in the right place. Otherwise, trees can become part of the problem instead of part of the solution, and ensuring that this does not happen often requires the skills and experience of the tree expert.

Urban forestry is as much about people as it is about trees. As one early advocate put it, 'The aim of urban forestry is to improve the welfare of urban residents; the planting and care of trees is a means to that end, not an end in itself' (Johnston 1985, 126). Much of the social and psychological value of the urban forest depends on residents' involvement in the planting and care of trees and community involvement in local authority tree programmes can have a major role in that. Urban forestry promotes the use of computer-based technology and the development of long-term tree management strategies, systematic tree maintenance programmes and partnership working. This is in contrast to the mainly unplanned, unsystematic and fragmented approach that has characterised urban tree 'management' since trees were first planted in our towns and cities. Sadly, that situation continues in many local authorities in Britain, due largely to lack of funding and staff.

By the early 1980s, the advances made by the urban forestry movement in the United States were attracting much attention from relevant British professionals. However, when the British urban forestry movement emerged in the late 1980s, there was resistance to the concept among many arboriculturists (Johnston 1997a, 109). While the reasons behind this are complex, they relate to a mistaken belief that it was a threat to the professional standing of arboriculturists and re-ignited those long-standing fears about foresters trying to 'take over' arboriculture. Many British arboriculturists saw themselves as the experts in amenity tree management and regarded it as their domain. In reality, urban forestry is a multidisciplinary approach that involves not only arboriculturists but many other professionals. Some arboriculturists failed to understand the basic concept and others dismissed the term as an Americanism, even though it was being endorsed by relevant government departments and agencies in Britain. While arboriculturists in North America had advanced their professional standing by playing a leading role in the development of urban forestry, the leadership of the Arboricultural Association in the late 1980s and 1990s disassociated itself from the movement in Britain (Chaplin 1991, 3). This resulted in an historic missed opportunity to promote the influence and status of arboriculturists.

The British urban forestry movement had its origins in three separate initiatives (Johnston 1997a and 1997b). The first was the work of Northern Planners, a planning and landscape consultancy based in Scotland that embedded the underlining principles into some of their projects in the early 1980s. The second was the Forest of London Project, a tree-planting and education initiative that was Britain's first city-wide urban forestry project. Work on this began in 1984 and when it was launched in 1987 it involved all of London's local authorities, relevant voluntary organisations and hundreds of local community groups. It later inspired several similar 'Forest of …' projects in other cities. The third initiative arose in the Black Country in 1986 among a group of planners and other professionals who were interested in restoring the derelict land in the region. Of these initiatives, only the Forest of London Project was led mainly by arboriculturists. All three urban forestry initiatives came together to help organise the First UK Conference on Urban Forestry, held in Wolverhampton in 1988 and appropriately called 'A Seed in Time' (Johnston 1997b, 320).

Following the First UK Conference, many new urban forestry initiatives were launched and further national conferences were held in 1991 and 1993 (Johnston 1999 and 2000). The new initiatives included the launch of a quarterly magazine called *Urban Forests* and the establishment of the Black Country Urban Forestry Unit, which then became the National Urban Forestry Unit until it closed in 2005. The most high-profile initiative was the government's Forest for the Community programme launched in 1989 by the Forestry Commission and the Countryside Commission. This aimed to create twelve new 'community forests' on the edge of major cities in England, each with the involvement of local authorities in the region, major voluntary organisations and private sector sponsorship (Johnston 1989; Rollins 2001).

The growing focus on urban forestry in the early 1990s prompted a groundbreaking government-commissioned research study on the urban forest in England. Entitled *Trees in Towns*, it was the first such survey to give a picture of the urban forest across England and it also provided evidence to government to develop some strategic policies (DoE 1993). In 1999, the University of Ulster published *A Survey of Urban Forestry in Britain* giving the result of the first survey of tree management programmes of British local authorities (Johnston and Rushton 1999). In 2008, a more detailed survey of England was published entitled *Trees in Towns II* (Britt and Johnston 2008). This survey was commissioned in February 2004 by the Office of the Deputy Prime Minister (later the Department for Communities and Local Government) and undertaken by ADAS, a consultancy company, and Myerscough College. The findings not only gave a detailed picture of the nature and extent of the urban forest but also provided a wealth of information about how those trees were being managed. The survey report was also significant for recommending ten targets for performance in tree management that all local authorities could achieve within the next five years (*ibid.*, 104–109). These included having at least one specialist tree officer, developing a comprehensive tree strategy and installing a computerised tree management system. Also included were twelve case studies giving examples of good practice in various aspects of tree management (*ibid.*, 487–642). What was unusual about the *Trees in Towns II* research study was that the majority of those involved were arboriculturists, where previously such a study would have been done by other professionals. After a two-year delay, the government published the report in 2008 and the response from most professional organisations was overwhelmingly positive. However, instead of taking ownership of *Trees in Towns II* and promoting the ten targets, the leadership of the Arboricultural Association was strangely silent (Johnston 2010, 39–40). While the report recommended that a third *Trees in Towns* survey should be conducted ten years later, this has not happened.

By the start of the twenty-first century the urban forestry movement in Britain had dissipated as its ideas and practices became accepted into the mainstream of urban tree management. Urban forestry was no longer a 'buzzword' but now just another name for modern urban tree management. It was accepted without question that the urban forest in every local authority district needed a long-term strategy and management plan, regularly updated and based on an accurate survey of all the trees and woodlands. New methods of surveying and valuing trees and the urban forest were also being developed. An innovative management tool known as the Capital Asset Value for Amenity Trees (CAVAT), developed by arboriculturist Chris Neilan, is now widely used to express the public amenity value of urban trees in monetary terms (Neilan 2017). Also widely used is a collection of urban forestry analysis and benefits assessment tools known as i-Tree (Hall *et al.* 2018). This was developed by the United States Forest Service to quantify and value the ecosystem services provided by trees. As the arboricultural industry has moved forward, the leadership of the Arboricultural Association has also embraced the concept of urban forestry and now view this as an opportunity to promote the professional standing of arboriculturists. In recent years this has been particularly evident from the content of its publications and the programme of its annual conferences.

Difficult times for the arboricultural industry

The unified voice for British arboriculture that was achieved with the 1974 merger was sadly short-lived. By the early 1990s, many Arboricultural Association members were unhappy with their leadership on several major issues. These included a lack of progress in promoting the professional status of arboriculturists and a perceived reluctance to cultivate overseas contacts (Johnston 2001, 154). The debate over various contentious issues was frequently played out in the professional press, often in an acrimonious manner. As these issues remained unresolved, disillusioned groups of members began discussing whether to break away from the Association and form separate organisations which they hoped would better represent their interests. The first new group to be formed was the United Kingdom and Ireland Chapter of the International Society of Arboriculture (ISA), which was formally recognised by the main ISA Board in August 1992. The ISA was, and still remains, the world's largest organisation for professional arborists with chapters in many different countries. Some leading members of the Arboricultural Association publicly stated their opposition to the move, which they regarded as damaging to the future prospects for professional arboriculture in Britain (Hall 1992).

This major split in British arboriculture caused much disquiet among Arboricultural Association members and this was evident at its annual conference in Swansea in 1993 (Bassirian 1993). Following the conference, a Review Group was established with a brief to review the existing structure of the Association, to consult with the membership and make recommendations for improvements. The Review Group was chaired by Peter Bridgeman and its members mainly comprised of the chairs of the Association's committees. When the report of the Review Group was sent to all members in August 1994 it was hoped that its recommendations would address the underlining issues promoting discontent among large sections of the membership (Arboricultural Association 1994). However, the report was unable to prevent further fragmentation. In 1996, a group of contractors announced the formation of a British branch of the National Arborist Association, a trade association based in the United States (Johnston 2001, 159). In 1997, the National Association of Tree Officers was formed by several prominent local authority members of the Arboricultural Association (*ibid.*, 160). This aimed to represent tree officers nationally and work closely with the regional tree officer groups. In 2003, the Consulting Arborist Society was formed by a group of arboricultural consultants who immediately established their own directory of consultants.

Some efforts were made to reverse this fragmentation, focusing mainly on securing a merger between the Arboricultural Association and the ISA UK Chapter. The first attempt was made in 1995 but after months of talks the negotiations broke down (Wolftree 1995). Another attempt was made in 2003 when some groundwork was undertaken in advance. In June 2003 the members of the Arboricultural Association and the ISA UK Chapter were balloted to see if they would 'support the principle' of a merger of the two organisations (*Essential Arb* 2003). After the members of each voted overwhelmingly in favour of this, a joint discussion group was formed. However, after months of negotiations the talks collapsed. While the Arboricultural Association remained Britain's largest arboricultural organisation, the ISA Chapter still represented a significant number of arboriculturists.

At the same time as this fragmentation, the Institute of Chartered Foresters (ICF) was at last developing a significant interest in arboriculture and acting on the inclusion of amenity trees in its Royal Charter. From the early 2000s, suitably qualified and experienced arboriculturists were steadily joining the ICF as professional members, now facilitated by the development of higher education qualifications in arboriculture. A small number of arboriculturists in the ICF began to lobby its leadership to broaden the designation of chartered professional to include arboriculturists. A watershed moment in the ICF's championing of arboriculture came in 2008 when it was granted a variation in its Royal Charter to allow its professional members who were arboriculturists to designate as 'chartered arboriculturist' (Ibrahim 2008). After many years of arboriculturists trying to gain equal recognition with similar professionals with chartered status, this had finally arrived. It prompted a steady stream of suitably qualified arboriculturists to become ICF members and gain chartered arboriculturist status. Also, those colleges running degree-level courses were now encouraging their students to join the ICF and apply for chartered status after they graduated. However, the Arboricultural Association did not encourage its members to gain chartered arboriculturist status. Instead, it obtained permission from the Society for the Environment to confer the status of chartered environmentalist on its professional members. Since then it has urged its members to get chartered – but as chartered environmentalists, a designation that says little about their expertise in arboriculture. Clearly, the Arboricultural Association was not content to act as a trade association and still had ambitions to be a professional body.

In recent years there have been some developments with tree officer organisations. In 2009, the Midland Tree Officers Association signed an agreement with the Society of Municipal Arborists, based in the United States, and renamed itself the Municipal Tree Officers Association. It would now have a national remit and claimed to be the 'voice of municipal arboriculture'. In 2018, the National Association of Tree Officers and the LTOA decided to merge to build a stronger national organisation, although the LTOA would remain the regional tree officer group for London. Consequently, in 2020 the Association of Tree Officers was launched.

There has recently been some success in reversing this fragmentation in the representation of British arboriculture. In 2015, following further talks with the Arboricultural Association and the ISA headquarters, the ISA UK Chapter was disbanded. In mid-2016 the Arboricultural Association and the ISA Board signed an agreement whereby the Association became an Associate Organisation of the ISA. While this damaging split had been healed, more work was still required on building cooperation between relevant organisations and uniting the industry. Nevertheless, the membership of the Arboricultural Association and the ICF has grown significantly in recent years, which is a positive sign for the future of professional arboriculture.

The changing scope of professional arboriculture

It seems appropriate to conclude this history of professional arboriculture in Britain with a few words about its changing scope in recent times. In the previous chapters we have charted how arboriculture evolved over the centuries until it emerged in the nineteenth century to focus on the planting and care of amenity trees and shrubs. The second half of the twentieth century has seen some major changes in its scope.

Arboriculture has traditionally embraced all woody plants, including trees, shrubs and climbers. However, in recent decades many arboriculturists have just focused on trees and left the shrubs and climbers to amenity horticulturists and other professionals. At the same time, arboriculturists have become less involved in the design of treed landscapes. While the tree officer is still usually responsible for street tree planting schemes, the design of schemes in parks and open spaces is largely the province of horticulturists and landscape architects. This is reflected in arboricultural education courses where little time is now spent on design matters compared to courses in horticulture and landscape. Also, nursery practices are no longer regarded as an integral part of arboriculture. Anyone pursuing a career in commercial tree and shrub production would take a course in commercial horticulture not arboriculture. Many arboriculturists regard their declining role with shrubs, landscape design, nursery practices and other 'horticultural' aspects of arboriculture as a backward step that has diminished their close relationship with woody plants.

Some aspects of practical arboriculture have become less common in recent decades. The preoccupation with public safety has greatly reduced the demand for preventative tree surgery such as cable bracing and propping. Cavity work has also become unusual, not just for safety reasons but as a result of revised arboricultural practice. With regard to tree pruning, most arborists rarely get the opportunity to undertake delicate and aesthetically pleasing work specifically adapted to the individual tree and its circumstances. Most pruning is formulaic and requires very similar treatment to groups of trees, such as crown lifting or crown thinning to a set specification. Many arborists lament the reduced scope of their work and complain that they are now primarily the 'undertakers' of trees, called in to remove them when they are dead, dying or no longer wanted. While some arborists working in historic or high-profile gardens still get to undertake a wide range of practical work, for most this is no longer the case. Some arboricultural consultants have also observed that they are now increasingly perceived as simply the experts on 'problem trees', mostly called upon to investigate some conflict or difficulty involving trees.

While the scope of arboricultural work has significantly reduced in many respects, in other ways it has expanded enormously. The appointment of tree officer posts in local authorities has demanded a far more strategic and long-term perspective for arboriculture. Arboriculturists as tree officers now find themselves responsible for the overall management of the urban forest, or a large part of it, rather than just being concerned with individual or small groups of trees. The close proximity of thousands of trees to roads, buildings, utility services and other urban infrastructure has created many new challenges and new areas of work for arboriculturists in both the public and private sectors. It has also created new opportunities for work with an environmental perspective. Never has there been greater recognition of the contribution of urban trees to creating liveable and sustainable towns and cities. Together with the growing threat of climate change and biodiversity loss, this should only increase the need for tree experts. However, the reporting in the national media of recent street tree felling in Sheffield has undermined the environmental credentials of tree officers (Drury 2016). Arboricultural organisations must do more to promote a positive image for the vital work that tree officers do. They also need to create a 'greener' image for the arboricultural industry. For a small industry, it has a large carbon footprint. This is due to the huge amount of vehicle and machinery use that is common in arboricultural operations (Johnston 2010, 29–30). Not only must arboriculturists reduce their carbon footprint, they also need to proclaim their environmental credentials, such as their commitment to recycling tree debris and their role in mitigating climate change. Furthermore, as the guardians of trees, they must stress how vital these are to people's physical and mental health.

One aspect of the tree officer's role that has increased dramatically in the past few decades has been the extent of interaction with people. With the growth of the voluntary sector and often working in 'crisis management' conditions, dealing with the public has become a very time-consuming and sometimes very challenging part of the job. Arboricultural education has had to reflect that. Until the 1990s, arboricultural courses taught the prospective tree officer a lot about trees but almost nothing about the 'people' side of the work. Now, degree courses in arboriculture include a substantial amount of social science, such as how to organise and deliver effective environmental education programmes, public consultation exercises and practical participation in community tree-planting schemes.

In the last ten years, the government's austerity agenda has led to cuts in local authority arboricultural services and the loss of many tree officer posts

(Parker 2019). Without another *Trees in Towns* or similar survey the precise scale of those spending cuts or job losses is unknown. However, anecdotal evidence suggests this is substantial. Where tree officers have been made redundant what remains of their workload is often being done by professionals who are not tree experts. Moreover, the loss of many tree officer posts could lead to a narrowing in the scope of arboriculture away from a strategic and long-term perspective and back towards individual and small groups of trees. That would greatly reduce arboriculturists' ability to use their unique expertise to ensure that our urban forests continue to deliver all those crucial benefits at a time when they are desperately needed.

Conclusions

The second half of the twentieth century could be described as the era when professional arboriculture in Britain at last 'came of age'. Furthermore, the period from 1945 to the late 1970s witnessed some major developments that transformed the industry and the profession. After centuries of living under the shadow of horticulture and forestry, arboriculture finally emerged as an industry and profession in its own right. While the subject of arboriculture had been recognised as a separate discipline since late Victorian times, it was now supported by specialist arboricultural education courses. The establishment of these courses from the late 1950s, first by the Royal Forestry Society in England and then by Merrist Wood and other colleges, was crucial in enhancing the professional status of British arboriculture. When Myerscough College developed degree-level courses in arboriculture in the 1990s and 2000s, the academic standing of the subject was at last approaching that of related subjects which had been long established in the university sector.

With the founding of the Association of British Tree Surgeons and Arborists in 1963, followed by the Arboricultural Association in 1964, professional arboriculture in Britain finally had some distinct representation. When those two organisations merged in 1974 to provide a unified voice for British arboriculture, the future looked very bright. By 1980 the public sector of the industry was well developed, with at least one dedicated tree officer in most local authorities, and in the larger authorities there were often several officers and large teams of manual workers. The private sector of the industry was also thriving, with many specialist tree work contractors and a growing number of arboricultural consultants. High standards of professional practice throughout the private sector were also being promoted and regulated by the Arboricultural Association with its approved directory of contractors and register of consultants.

From the early 1980s, however, professional arboriculture in Britain has been on a 'rollercoaster ride'. Some major advances have been made that have helped consolidate and build on the remarkable progress achieved in the previous few decades. The arboricultural industry now has its own safety guides and codes of practice that are as robust as those for similar industries. At long last, arboriculturists who are sufficiently qualified and experienced can gain chartered arboriculturist status, placing them on a par with other professionals, such as architects, landscape architects and engineers. This was facilitated by the development of degree-level courses in arboriculture, something that has also encouraged more women to embark on a career in arboriculture. Some arboriculturists have also made a significant contribution to wider debates about trees, green infrastructure and the urban environment, for example, by promoting urban forestry ideas and practice. This has involved working in partnership with other professionals and engaging with local communities, the people who should ultimately benefit from this work.

While there have been some very positive developments since the early 1980s, there have also been some difficult times for British arboriculture. It has suffered periodically from major political and economic factors that have had a particularly disruptive and damaging impact on the public sector of the industry. Although these have been largely beyond its control, its professional organisations could have been more active in defending their members' jobs and interests. At the same time, British arboriculture has suffered from some internal difficulties of its own making. The industry's leadership has been found wanting at crucial times, leading to major delays in arboriculture's development and some significant missed opportunities. The fragmentation of professional representation in such a small industry has been quite damaging, with much dissipation of energy and duplication of effort. In these challenging environmental times it is regrettable to see an increasingly important industry failing to make its crucial contribution to government policy because its competing voices are either ignored or drowning each other out.

At this time of writing, Britain is engulfed in the coronavirus pandemic and this will have enormous long-term consequences. It will undoubtedly severely damage the economy, which will include an initially detrimental impact on arboriculture. However, this crisis has served to further highlight the value of nature and natural landscapes in our lives. The growing awareness that trees are vital to our health and welfare, not least in an urban environment, should mean that arboriculturists and their unique expertise will be in even greater demand in the future. However, that will only happen if arboriculturists address the current difficulties facing their industry and profession, to ensure they can respond positively and collectively to the challenges and opportunities that lie ahead.

References

AAB (1991) *Review of the Arboricultural Advisory and Information Service (AAIS)*. Report by the Arboricultural Advisory Board. Department of the Environment: London.

ABTSA Newsletter (1970) Full Membership List. *Newsletter of the Association of British Tree Surgeons and Arborists*. June, 2–5.

ABTSA Newsletter (1971) Discussions with the Arboricultural Association. *Newsletter of the Association of British Tree Surgeons and Arborists*, June, 4.

Acton Gazette (1968a) Advertisement for Working Forester, London Borough of Ealing. *Acton Gazette*, 2 May, 14.

Acton Gazette (1968b) In the High Court of Justice: In Bankruptcy. *Acton Gazette*, 3 October, 16.

Acton Gazette (1976) Trees are Dying of Thirst. *The Acton Gazette*, 29 April, 12.

Anon (1968) The Earl Jellicoe, DSO, MC. *Arboricultural Association Journal* 1(6), 138.

Anon (1971) Register of Consultants in Arboriculture. *Arboricultural Association Journal* 1(12), 319.

Anon (1974) The Arboricultural Association (incorporating ABTSA) Acting Council: Elected Members. *Arboricultural Association Journal* 2(7), 216.

Anon (1980) Special Announcement: Arboricultural Journal 1981. *Arboricultural Journal* 4(2), i.

Anon (1983) *Trees in the 21st Century: Proceedings of the First International Conference Sponsored by the Arboricultural Association in Conjunction with the International Society of Arboriculture.* A.B. Academic Publishers: Berkhamsted.

Anon (1988) Arboricultural Award 1988: Hubert Taylor MBE. *Arboricultural Journal* 12(4), 263–264.

Anon (1990) National Tree Warden Network is Launched. *Urban Forests*, Autumn 1990, 5, 1. A Seed in Time Ltd: Hoddesdon, Hertfordshire.

Anon (1991) Arboricultural Award 1991: Trevor Preston. *Arboricultural Journal* 15(4), 305–306.

Anon (1993) European Arboricultural Council. *Arboricultural Journal* 17(2), 221–222.

Anon (1994) Arboricultural Award 1994: Robert T. Hillier and Roy R. Finch. *Arboricultural Journal* 18(4), 353–356.

Anon (1995) Arboricultural Award 1995: Jean Berry. *Arboricultural Journal* 19(4), 337–338.

Anon (1997) Arboricultural Association Award 1997: Caroline Davis. *Arboricultural Journal* 21(2), 301–304.

Anon (2018) Remembering Bill Matthews. *The Arb Magazine*, Winter, 183, 14–15.

Appleby, M. (2011) Interview: Pete Wells, Founder, Barcham Trees. *Horticulture Week*, 25 March, 12.

Arboricultural Association (1969) *Symposium to Discuss Arboricultural Education, Training and Examinations.* Symposium Report, October 1969. Arboricultural Association: Worplesdon.

Arboricultural Association (1971) *The Arboricultural Officer.* Leaflet produced by the Arboricultural Association: Stansted.

Arboricultural Association (1994) *The Arboricultural Association Review: Report to Members.* Arboricultural Association: Ampfield.

ARWP (1974) *Arboricultural Research and Advisory Needs.* Report prepared and issued by the Arboricultural Research Working Party: Edinburgh.

ASC (1991) ASC1: Tree Climbing Operations Safety Guide. Arboricultural Safety Council: Romsey.

Ball, R. (1995) White Collar Compulsory Competitive Tendering (CCT): The Death Knell for Arboricultural Officers and the End of Professional Street Tree Management. *Arboricultural Journal* 19(3), 285–293.

Bassirian, H. (1993) Time to Root Out the Problems. *Horticulture Week*, 1 October, 25–28.

Bath Chronicle (1947a) Tree-lined Roads. *Bath Weekly Chronicle and Herald*, 4 January, 1.

Bath Chronicle (1947b) Pulteney Road Trees Saved. *Bath Weekly Chronicle and Herald*, 8 February, 5.

Beeching, A.G. (1965) The Association of British Tree Surgeons and Arborists. *Quarterly Journal of Forestry* 59, 144–152.

Beeching, A.G. (1969) Letters to the Editor. *Arboricultural Association Journal* 1(9), 249–250.

Bickmore, C.J. and Hall, T.H.R. (1983) *Computerisation of Tree Inventories*. A.B. Academic Publishers: Berkhamsted.

Biddle, P.G. (1998) *Tree Root Damage to Buildings: A Practical Guide to the Problems of Subsidence*. Willowmead Publishing: Wantage.

Birmingham Post (1954) Pruning Can Save ArborTree at Aston-on-Clun. *The Birmingham Post*, 22 October, 5.

Boddy, F.A. (1947) The Preservation of Trees in Towns. *The Gardeners' Chronicle*, Series 3, 121, 156.

Boddy, F.A. (1968) *Highway Trees*. Clarke and Hunter: London.

Boddy, M. (2009) British Standards and Arboriculture. *Arboricultural Journal* 32(1), 5–17.

Boulton, E. (1969) A.D.C. Le Sueur, OBE: An Appreciation. *Arboricultural Association Journal* 1(9), 247–248.

Bridge, M. (2010) The Development of Tree Climbing Techniques. *Arborist News*, December, 50–54.

Bridgeman, P.H. (1976a) *Tree Surgery: A Complete Guide*. David and Charles: Newton Abbot.

Bridgeman, P.H. (1976b) Arboricultural Education. *Arboricultural Journal* 2(10), 452–457.

Bridgeman, P. (1991) Twenty-five Years of Arboriculture: A Personal View. In: Hall, T.H.R. (ed.) *The First Twenty-five Years*. Arboricultural Association: Ampfield.

Bridgeman, P. (1996) Peter Martyn Honey 1930–1995. *Arboricultural Journal* 20(1), 123–124.

Bridgeman, P. (2007) Alex Lloyd Shigo: 1930–2006, A Personal Tribute. *Arboricultural Journal* 30(1), 5–6.

Bridgeman, P. (2010a) Merrist Wood 1967–1977. *Essential Arb* 31, 24–26.

Bridgeman, P. (2010b) Merrist Wood 1970–1973. *Essential Arb* 32, 22–23.

Bridgeman, P. (2010c) Merrist Wood 1974–1977. *Essential Arb* 33, 42–43.

Bridgeman, P. (2013) Kewites in Arboriculture. *The Journal of the Kew Guild* 16(117), 215–221.

Britt, C. and Johnston, M. (2008) *Trees in Towns II: A New Survey of Urban Trees in England and their Condition and Management*. Department for Communities and Local Government: London.

Brown, G.E. (1972) *The Pruning of Trees, Shrubs and Conifers*. Faber and Faber: London.

Brown, G.E. (2004) *The Pruning of Trees, Shrubs and Conifers*. Second edition revised and enlarged by Tony Kirkham. Timber Press: Cambridge.

BSI (1966a) *BS 4043: 1966 Transplanting Semi-Mature Trees*. British Standards Institution: London.

BSI (1966b) *BS 3998: 1966 Recommendations for Tree Work*. British Standards Institution: London.

BSI (1980) *BS 5837: 1980 Trees in Relation to Construction*. British Standards Institution: London.

BSI (2014) *BS 8545: 2014 Trees: From Nursery to Independence in the Landscape*. British Standards Institution: London.

Campana, R.J. (1999) *Arboriculture: History and Development in North America*. Michigan State University Press: East Lansing, MI.

Chaplin, J. (1991) A Quarter of a Century Endeavour: The Story of the Arboricultural

Association. In: Hall, T.H.R. (ed.) *The First Twenty-five Years*. Arboricultural Association: Ampfield.

Chichester Observer (1947) City Council Brevities. *Chichester and Southdown Observer*, 5 April, 1.

Claridge, J. (ed.) (1997) *Arboricultural Practice: Present and Future*. Research for Amenity Trees No. 6. Department of the Environment, Transport and the Regions: London.

Clark International (1966) *Michigan Model 85III Tree Transplanter*. Clark International Marketing: Camberley.

Cooley, L.B. (2007) *The Bartlett Way: 100 Years of Scientific Tree Care*. The F.A. Bartlett Tree Expert Company: Stamford, CT.

Country Life (1950) Danarm Tornado Chainsaw advertisement. *Country Life*, 24 February, 500.

Country Life (1956) Sankey-Aspin Chainsaw advertisement. *Country Life*, 29 March, 652.

Coventry Telegraph (1946) Thomas Higginson, classified advertisement. *The Coventry Evening Telegraph*, 22 October, 10.

Coventry Telegraph (1965) A.J. Mackenzie Clay Ltd., advertisement. *The Coventry Evening Telegraph*, 13 July, 2.

Dallimore, W. (1951) Examination in Arboriculture. *Quarterly Journal of Forestry* 45, 59–60.

Deller, E. (1991) Early Years. In: Hall, T.H.R. (ed.) *The First Twenty-five Years*. Arboricultural Association: Ampfield.

Disney, R. and Luo, G. (2015) *The Right to Buy Public Housing in Britain: A Welfare Analysis*. IFS Working Paper W15/05. Institute of Fiscal Studies: London.

DoE (1973) *Circular 90/73: Inspection, Maintenance and Planting of Roadside Trees*. Department of the Environment: London.

DoE (1975) *Circular 52/75: Inspection of Highway Trees*. Department of the Environment: London.

DoE (1978) *Trees and Forestry*. Circular 36/78. Joint Circular from the Department of the Environment and Welsh Office. HMSO: London.

DoE (1993) *Trees in Towns: A Survey of Trees in 66 Towns and Villages in England*. Research for Amenity Trees No. 1. Prepared for the Department of the Environment by Land Use Consultants. HMSO: London.

DoE (1994) *Urban Tree Strategies*. Research for Amenity Trees No. 3. Report to the Department of the Environment prepared by Land Use Consultants. HMSO: London.

DoE (2000) *Tree Preservation Orders: A Guide to the Law and Good Practice*. Department of the Environment: London.

Drury, C. (2016) Sheffield's Tree Massacre. *The Independent*, 12 April, 12.

Dunball, A.P. (1972) Landscape Treatment of Trunk Roads and Motorways. *Arboricultural Association Journal* 2(2), 38–43.

Eden, N. (2007) Towards a National Standard for Tree Risk Inspections. *Arboricultural Journal* 30(2), 127–136.

Edlin, H.L. (1970) *Collin's Guide to Tree Planting and Cultivation*. Collins: London.

Edlin, H.L. and Nimmo, M. (1956) *Tree Injuries: Their Causes and Prevention*. Thames and Hudson: London.

Elliott, B. (2004) *The Royal Horticultural Society: A History 1804–2004*. Phillimore: Chichester.

Essential Arb (2003) Merger Talks. *Essential Arb* 10, 6.

Evening Chronicle (1969) Winner May Even be a Poplar! *Evening Chronicle* (Newcastle), 11 April, 9.

Evening Post (1966) Tree Feller Was Not Wearing a Harness. *Evening Post* (Reading), 11 February, 6.

Everett, T.H. (1961) Moving Large Trees the American Way. *The Gardeners' Chronicle* 149, 248–250.

Fairchild, A. (1960) The Usefulness of Power Saws: Replacing Cross-cuts and Axes. *Farm and Country*, 26 October, 412–414.

Falkirk Herald (1947) Prospect of Tree-lined Roads: Arboriculturist to Give Advice. *The Falkirk Herald*, 5 February, 1.

Forestry Commission (1980) *Research for Practical Arboriculture*. Occasional Paper No. 10. Forestry Commission: Edinburgh.

Forest Research (2010) *Benefits of Green Infrastructure*. Report to Defra and CLG. Forest Research: Farnham.

Fox, W. (1953) Street Trees. *The Gardeners' Chronicle*, Series 3, 134, 204.

FSC (1981) *FSC18: Tree Climbing and Pruning.* Forestry Industry Safety Guide. Revised October 1981. Forestry Safety Council: Edinburgh.

Gardeners' Chronicle (1951) The Rainbow Tree-Ties. *The Gardeners' Chronicle*, Series 3, 129, 128.

Gardeners' Chronicle (1958) Toms Tree-Ties advertisement. *The Gardeners' Chronicle*, Series 3, 144, 268.

Gardeners' Chronicle (1959) Obituary: Mr W. Dallimore. *The Gardeners' Chronicle* 146, 285.

Gardeners' Chronicle (1964a) The Arboricultural Society. *The Gardeners' Chronicle* 156, 439.

Gardeners' Chronicle (1964b) A New Association. *The Gardeners' Chronicle* 156, 30.

Garner, R.J. (1947) *The Grafter's Handbook.* Faber and Faber: London.

George, P. (2014) The Evolution of the Arborist's Climbing Harness. *Pro-Arb Magazine*, October, 24–25.

Gibbs (1988) Chronological list of Wood Chuck sales and customers, 1972 to 1988. Internal document produced by J. Gibbs of Belfont Ltd.

Gilbertson, P. and Bradshaw, A.D. (1985) Tree Survival in Cities: The Extent and Nature of the Problem. *Arboricultural Journal* 9(2), 131–142.

Gilson, M. (2019) Confluence of the Times: Richard Sudell (1892–1968). *Garden History* 47(2), 163–180.

Girling, H.A. (1969) Arboricultural Training. *Arboricultural Association Journal* 1(8), 219–220.

Hall, C., O'Brien, L., Hand, K. and Raum, S. (2018) *Evaluation of i-Tree Eco Surveys in Great Britain. Impacts and Key Lessons: The Views of Stakeholders*. Forest Research: Farnham.

Hall, T.H.R. (1992) Editorial: Formation of ISA Chapter, UK and Ireland. *Arboricultural Journal* 16(1), 1–3.

Harris, R.W. (1983) *Arboriculture: Care of Trees, Shrubs and Vines in the Landscape.* Prentice-Hall: Englewood Cliffs, NJ.

Helliwell, D.R. (1967) The Amenity Value of Trees and Woodlands. *Arboricultural Association Journal* 1(5), 128–131.

Hillier, H.G. (1969) Planting of a Large Maple at Chelsea. *Journal of the Royal Horticultural Society* 94, 342–344.

Hillier, H.G. (1972) *Hillier's Manual of Trees and Shrubs*. David and Charles: Newton Abbot.

Hillier, J. (2014) *Hillier: The Plants, the People, the Passion.* Hilliers Nurseries: Winchester.

Hinton, C. and Smith, L. (1996) *The Tree Resource Yearbook 1996.* The Tree Resource: Byfleet.

Hodge, S.J. (ed.) (1991) *Research for Practical Arboriculture.* Forestry Commission Bulletin 97. HMSO: London.

Honey Brothers (1974) *The 1974/5 Honey Brothers Catalogue.* Honey Brothers (Sales) Ltd: Peasmarsh.

Honey, A.J. (1970a) Tour of the United States of America. *Newsletter of the Association of British Tree Surgeons and Arborists*, June, 6–7.

Honey, P.M. (1964a) The Tree Surgeon in England Today. *Journal of the Royal Horticultural Society* 89, 279–285.

Honey, P.M. (1964b) Notes from a Tree Surgeon. *The Gardeners' Chronicle* 153, 363.

Honey, P.M. (1969) Safety Note No. 2. *Newsletter of the Association of British Tree Surgeons and Arborists*, February, 2.

Honey, P.M. (1970b) Chairman's Comments. *Newsletter of the Association of British Tree Surgeons and Arborists*, June, 1–2.

Honey, P.M. (1972) The Future of Arboriculture in Great Britain. *Newsletter of the Association of British Tree Surgeons and Arborists*, May, 26–28.

Honour, D.R. (1970) Profile – Donald V. Wells. *Arboricultural Association Journal* 1(10), 271–273.

Hope, F. (1982) Well Filed Trees. *The Gardeners' Chronicle & Horticultural Trade Journal* 191, 18–19.

Hornsey, W.M. (1953) Street Trees. *The Gardeners' Chronicle*, Series 3, 134, 170–171.

Horticulture Week (1986) Peter Bridgeman and Associates advertisement. *Horticulture Week*, 21 March, back page.

Horticulture Week (2020) Top 30 Arboriculture Businesses 2020. *Horticulture Week*, 1 October, 22–23.

Ibrahim, M. (2008) Institute Creates Chartered Status for Arboriculturists. *Horticulture Week*, 3 July, 2.

Insley, H. (1976) Amenity Tree Research: A New Move in the Forestry Commission to Extend Research on Arboriculture. *Arboricultural Journal* 3(1), 54–57.

James, N.D.G. (1957) Proposal to Establish an Arboricultural Section in the Society. *Quarterly Journal of Forestry* 51, 244–246.

James, N.D.G. (1972) *The Arboriculturalist's Companion.* Basil Blackwell: Oxford.

James, N.D.G. (1982) *A Forest Centenary: A History of the Royal Forestry Society of England, Wales and Northern Ireland.* Basil Blackwell: Oxford.

Jellicoe, G.A. (1958) Motorways: Their Landscaping, Design and Appearance. *Journal of the Town Planning Institute* 44, 274–283.

Johnston, M. (1983) Urban Trees and an Ecological Approach to Urban Landscape Design. *Arboricultural Journal* 7(4), 275–282.

Johnston, M. (1985) Community Forestry: A Sociological Approach to Urban Forestry. *Arboricultural Journal* 9(2), 121–126.

Johnston, M. (1989) *Forests for the Community.* CCP No. 270. Forestry Commission and Countryside Commission: Cheltenham.

Johnston, M. (1996) A Brief History of Urban Forestry in the United States. *Arboricultural Journal* 20(3), 257–278.

Johnston, M. (1997a) The Early Development of Urban Forestry in Britain: Part I. *Arboricultural Journal* 21(2), 107–126.

Johnston, M. (1997b) The Early Development of Urban Forestry in Britain: Part II. *Arboricultural Journal* 21(4), 317–330.

Johnston, M. (1999) The Springtime of Urban Forestry in Britain – Developments Between the 1st and 3rd Conferences, 1988–1993, Part 1. *Arboricultural Journal* 23(3), 233–260.

Johnston, M. (2000) The Springtime of Urban Forestry in Britain – Developments Between the 1st and 3rd Conferences, 1988–1993, Part II. *Arboricultural Journal* 23(4), 313–341.

Johnston, M. (2001) British Urban Forestry in Transition – Developments Between 1993–1998, Part II. *Arboricultural Journal* 25(2), 153–178.

Johnston, M. (2010) Trees in Towns II and the Contribution of Arboriculture. *Arboricultural Journal* 33(1), 27–41.

Johnston, M. (2013) Dr Tom Hall: Obituary. *Arboricultural Journal* 35(1), 4–5.

Johnston, M. (2015) *Trees in Towns and Cities: A History of British Urban Arboriculture.* Windgather Press: Oxford.

Johnston, M. (2017) *Street Trees in Britain: A History.* Windgather Press: Oxford.

Johnston, M. (2019) The History of Professional Arboriculture in Britain. *The Arb Magazine* 185, 42–43.

Johnston, M. and Hirons, A. (2012) Going Online with Arboricultural Education. *Arboriculture and Urban Forestry* 38(3), 105–111.

Johnston, M. and Percival, G. (eds) (2012) *Trees, People and the Built Environment.* Conference Proceedings of 2011. Forestry Commission Research Report. Forestry Commission: Edinburgh.

Johnston, M. and Percival, G. (eds) (2015) *Trees, People and the Built Environment II.* Conference Proceedings of 2014. Institute of Chartered Foresters: Edinburgh.

Johnston, M. and Rushton, B.S. (1999) *A Survey of Urban Forestry in Britain.* University of Ulster: Coleraine.

Johnston, M., Nail, S. and James, S. (2012) 'Natives versus Aliens': The Relevance of the Debate to Urban Forest Management in Britain. In: Johnston, M. and Percival, G. (eds) *Trees, People and the Built Environment.* Conference Proceedings. Forestry Commission: Edinburgh.

Jorgensen, C.W. (1972) Chairman's Comments. *Newsletter of the Association of British Tree Surgeons and Arborists*, May, 2.

Jorgensen, C.W. (1973) Chairman's Comments. *Newsletter of the Association of British Tree Surgeons and Arborists*, January, 3.

Jorgensen, E. (1986) Urban Forestry in the Rearview Mirror. *Arboricultural Journal* 10(3), 177–190.

JRHS (1950) The Chiltern Tree Surgeons advertisement. *Journal of the Royal Horticultural Society* 75, February, xiv.

JRHS (1963) The Southern Tree Surgery Company advertisement. *Journal of the Royal Horticultural Society* 88, June, xviii.

JRHS (1970) Beeching of Ash advertisement. *Journal of the Royal Horticultural Society* 95, December, v.

JWPAE (1970) *Report on Arboricultural Training, Education and Examinations.* Joint Working Party on Arboricultural Education: Worplesdon.

Kenyon, J. (1985) Harnessing Facts. *Horticulture Week*, 12 April, 27.

Kirkham, T. (2008) The Decompaction Programme on Trees at Kew. *IDS Yearbook for 2008*, 77–80. International Dendrology Society: London.

Last, F.T. (1993) The Tree Advice Trust. *Arboricultural Journal* 17(2), 113–116.

LB Harrow (1977) *Works Specification: Part 11C, Parks Section – Forestry.* Department of Technical Services, London Borough of Harrow: Harrow.

LB Harrow (1979) *Survey of London Boroughs Arboricultural Sections.* Parks Section, Department of Engineering, London Borough of Harrow: Harrow.

Leamington Courier (1946) Advertisement by Borough of Royal Leamington Spa, appointment of Forester. *Royal Leamington Spa Courier*, 18 January, 5.

Leathart, P.S. (1966) Arboricultural Meetings 1965. *Quarterly Journal of Forestry* 60, 59–63.
Leathart, P.S. (1973) Profile: Hubert Taylor. *The Arboricultural Association Journal* 2(4), 101–105.
Le Sueur, A.D.C. (1934) *The Care and Repair of Ornamental Trees*. Country Life: London.
Le Sueur, A.D.C.. (1946) Safety for Tree Workers. *Quarterly Journal of Forestry* 40, 88–90.
Le Sueur, A.D.C. (1949) *The Care and Repair of Ornamental Trees*. Second edition. Country Life: London.
Le Sueur, A.D.C. (1951) Consultant Forestry as a Profession. *Quarterly Journal of Forestry* 45, 182–4.
Le Sueur, A.D.C. (1958) The Forestry Expert Witness. *Quarterly Journal of Forestry* 52, 187–191.
Le Sueur, A.D.C. (1963) The Park Royal Tree Plantings. *Journal of the Royal Horticultural Society* 88, June, 263–269.
Lewington, W.R. (1974) The Merger with the Association of British Tree Surgeons and Arborists. *Arboricultural Journal* 2(7), 227.
LGA (2019) *Local Government Association briefing – Debate on Local Government funding*. Issued 15 January 2019. Local Government Association: London.
Lonsdale, D. (1983) *A Definition of the Best Pruning Position*. Arboricultural Research Note 48. Arboricultural Advisory and Information Service: Farnham.
Lonsdale, D. (1984) Available Treatments for Tree Wounds: An Assessment of their Value. *Arboricultural Journal* 8(2), 99–107.
Lonsdale, D. (1999) *The Principles of Tree Hazard Assessment and Management*. Research for Amenity Trees No. 7. TSO: London.
Mabbett, T. (2011) Civic Trees on the Move. *Essential Arb* 38, 24–26.
Mattheck, C. and Breloer, H. (1994) *The Body Language of Trees: A Handbook for Failure Analysis*. Research for Amenity Trees No. 4. TSO: London.
Matthews, W.E. (1968a) General News. *Newsletter of the Association of British Tree Surgeons and Arborists*, November, 3–4.
Matthews, W.E. (1968b) Safety in Tree Work: No. 1 Safety Saddles. *Arboricultural Association Journal* 1(7), 176–179.
Matthews, W.E. (1968c) Trees and the Law. *Newsletter of the Association of British Tree Surgeons and Arborists*, November, 4.
Matthews, W.E. (1970) The Tree Scene in Britain. *Weeds, Trees and Turf*, August, 20–24.
Matthews, W.E. (1982) The Tree Scene in Great Britain. *Journal of Arboriculture* 8(7), 176–177.
Matthews, W.E. (1990) Obituary: Tom Wilson. *Arboricultural Journal* 14(2), 128.
Matthews, W.E. (1995) Obituary: Cyril Reddick. *Arboricultural Journal* 19(3), 294.
Matthews, W.E. and Draper, A.G. (1971) Profile: F.T. Wilson. *Arboricultural Association Journal* 1(12), 335–337.
McEwan, G. (2018) Arb Association Launches Bid to get More Women into Industry. *Horticulture Week*, 9 March, 21.
Mercer, P.C. (1982) Tree Wounds and their Treatment. *Arboricultural Journal* 6(2), 131–137.
MHLG (1958) *Trees in Town and City*. Ministry of Housing and Local Government. HMSO: London.
Mitchell, A.F. (1972) *Conifers in the British Isles – A Descriptive Handbook*. Forestry Commission Booklet 33. HMSO: London.

Mitchell, A.F. (1974) *A Field Guide to the Trees of Britain and Northern Europe.* Collins: London.

Mitchell, A.F. and Hallett, V.E. (1985) *Champion Trees in the British Isles*. Research and Development Paper 138. Forestry Commission: Edinburgh.

Monk, B. (2011) *Evaluation of Decay Detection Equipment in Standing Trees.* United States Forest Service. http://www.fs.fed.us (accessed 2 November 2020).

Morling, R.J. (1954a) *Trees in Towns*. The Estates Gazette: London.

Morling, R.J. (1954b) Damage by Tree Roots. *The Gardeners' Chronicle*, Series 3, 135, 118.

Morton, A. (1998) *Tree Heritage of Britain and Ireland.* Swan Hill Press: Shrewsbury.

Musgrave, R. (1970) Contest-winning Advertisement Gives Unexpected Boost to Britain's Tree Surgeons. *Newsletter of the Association of British Tree Surgeons and Arborists*, June, 14–15.

Mynors, C. (2002) *The Law of Trees, Forests and Hedgerows*. Sweet and Maxwell: London.

Neilan, C. (2017) *Capital Asset Value for Amenity Trees (CAVAT), Full Method: Users' Guide*. London Tree Officers Association: Camden.

Newman, C.J. (1969) Introducing the Tree Spade. *Arboricultural Association Journal* 1(8), 215–217.

Newman, D. (2014) Interview with Deric Newman of Civic Trees. *Pro-Arb Magazine*, November, 27–29.

NHBC (1969) *Root Damage by Trees: Siting of Dwellings and Special Precautions*. Practice Note 3. National House Building Council: London.

NJUG (1995) *NJUG 10: Guidelines for the Planning, Installation and Maintenance of Utility Services in Proximity to Trees.* Publication No. 10. National Joint Utilities Group: London.

NTSG (2011) *Common Sense Risk Management of Trees.* Report by the National Tree Safety Group. Forestry Commission: Edinburgh.

O'Callaghan, D. (2012) Nomination for the Arboricultural Association Award 2011. *Arboricultural Journal* 34(4), 234–236.

Pakenham, T. (1996) *Meetings with Remarkable Trees*. Weidenfeld and Nicolson: London.

Parker, J. (2019) DEFRA's 'Duty to Consult'. *Chartered Forester*, Spring, 26–27.

Patch, D. (1984) *Tree Staking*. Arboriculture Research Note 40. Department of the Environment, Arboricultural Advisory and Information Service: Farnham.

Patch, D. (ed.) (1987) *Advances in Practical Arboriculture*. Forestry Commission Bulletin 65. HMSO: London.

Paterson, M. (2003) Girl, 8, Killed as Tree Falls on Schoolchildren. *The Daily Telegraph*, 31 January, 1.

Peace, T.R. (1962) *Pathology of Trees and Shrubs with Special Reference to Britain*. Oxford University Press: Oxford.

Pepper, H.W., Rowe, J.J. and Tee, L.A. (1985) *Individual Tree Protection.* Arboricultural Leaflet 10, Department of the Environment. HMSO: London.

Pirone, P.P. (1941) *Maintenance of Shade and Ornamental Trees*. Oxford University Press: New York.

Preston, T. (1991) Arboricultural Safety Council. In: Hall, T.H.R. (ed.) *The First Twenty-five Years*. Arboricultural Association: Ampfield.

QJF (1954) Trees in the Royal Parks. *Quarterly Journal of Forestry* 48, 219.

QJF (1958) Teles Little Tiger advertisement. *Quarterly Journal of Forestry* 52, xxii.

QJF (1960) Arboricultural Section: Arboricultural Examinations, 1959. *Quarterly Journal of Forestry* 54, 92–93.

QJF (1966) Society Notices: National Diploma in Arboriculture. *Quarterly Journal of Forestry* 60, 6.

QJF (1967) Moving Semi-mature Trees. *Quarterly Journal of Forestry* 61, 1–2.

QJF (1968) The Society's Examinations, 1967: National Diploma in Arboriculture. *Quarterly Journal of Forestry* 62, 75.

QJF (1971) Chain-Saws and Vibration. *Quarterly Journal of Forestry* 65, 75–76.

QJF (1975) Arboriculture: Research and Advisory Needs. A Report prepared by the Arboricultural Research Working Party. *Quarterly Journal of Forestry* 69, 75–83.

QJF (1987) Re-titling of Top Arboricultural Qualification. *Quarterly Journal of Forestry* 81, 76.

QJF (1990) The Society's Arboricultural Examinations. *Quarterly Journal of Forestry* 84, 145.

Rattigan, D. (2013) The 1950s: An Industry Booms and the Rises of the Machines. *Tree Care Industry* 24(6), 28–34.

Read, H. (2000) *Veteran Trees: A Good Management Guide.* English Nature: Peterborough.

Rollins, J. (2001) Tree UK – The Urban and Community Forest Movement. *Tree News*, Autumn/Winter, 39–41.

Ryan, J. (2009a) Paul Akers: Profiles in Arboriculture. *Essential Arb* 30, 8–10.

Ryan, J. (2009b) Peter Bridgman: Profiles in Arboriculture. *Essential Arb* 28, 22–25.

Ryan, J. (2010) Bill Matthews: Profiles in Arboriculture. *Essential Arb* 34, 18–19.

SCAE (1978) *The Present Pattern of Employment, Education and Training in Arboriculture.* Standing Committee on Arboricultural Education: York.

Sheat, W.G. (1948) *Propagation of Trees, Shrubs and Conifers.* Macmillan and Co.: London.

Shigo, A.L. (1977) *Compartmentalization of Decay in Trees.* Agriculture Information Bulletin No. 405. United States Department of Agriculture, Forest Service: Washington, DC.

Shigo, A.L., McGinnes, E.A., Funk, D.T. and Rogers, N. (1979) *Internal Defects Associated with Pruned and Nonpruned Branch Stubs of Black Walnut.* Forest Service Research Paper NE-440. USDA Forest Service: Broomall, PA.

Shigo, A.L. (1991) *Modern Arboriculture: A Systems Approach to the Care of Trees and their Associates.* Shigo and Trees Associates: Durham, NH.

Stokes, J. and Rodger, D. (2004) *The Heritage Trees of Britain and Northern Ireland.* Constable: London.

Storey, W.E. (1963) Tree Management. The Fourth Askham Bryan Horticultural Conference, 43–55. Askham Byan College: York.

Storey, W.E. (1984) Michael Robert Cudd 1945–1984. *Arboricultural Journal* 8(4), 319–320.

Strouts, R.G. and Winter, T.G. (2000) *Diagnosis of Ill-health in Trees.* Research for Amenity Trees No. 2. Second edition. Department of the Environment, Transport and the Regions. The Stationery Office: London.

Suffolk Free Press (1949) Spare that Tree Cries Council. *Suffolk Free Press*, 27 January, 5.

Tewkesbury Register (1951) Avenue of Limes Not Dangerous. *The Tewkesbury Register*, 13 March, 6.

The Scotsman (1948) Court of Session: Interdict Action. *The Scotsman*, 20 November, 3.

The Scotsman (1949) Advertisement by Essex County Council, appointment of Tree Warden. 21 April, 8.

Thomas, G.S. (1979) *Gardens of the National Trust.* Weidenfeld and Nicolson: London.

Thurman, P.W. (1983) The Management of Urban Street Trees using Computerised Inventory Systems. *Arboricultural Journal* 7(2), 101–117.

Tree News (2012) Tree Safety: A Practical Guide. *Tree News* 21, 22–23.

Tuley, G. (1985) The Growth of Young Oak Trees in Shelters. *Forestry* 58(2), 181–195.
Usher, G.A. (1953) Tree Climbing: The Case for the Climbing Chair. *Quarterly Journal of Forestry* 47, 120–122.
Watkins, R.W. (1950) Examination in Arboriculture. *Quarterly Journal of Forestry* 44, 239.
Wells, D.V. (1967) Moving Semi-mature Trees. *Quarterly Journal of Forestry* 61, 264.
Wells, D.V. (1974) The Arboricultural Association Ten Years On 1964–1974. *Arboricultural Journal* 2(7), 217–226.
Wells Journal (1956) 'Reprieve' for Stoberry Park Timber. *The Wells Journal*, 28 December, 1.
Western Daily Press (1949) Chiltern Tree Surgeons advertisement. *The Western Daily Press*, 21 June, 5.
Western Mail (1955) Welsh Tree Services Ltd, advertisement. *Western Mail*, 25 August, 3.
Western Mail (1959) Landscapes classified advertisement. *Western Mail*, 14 November, 9.
West Sussex County Times (1949) The Tree Specialist Company advertisement. *West Sussex County Times*, 7 October, 3.
West Sussex County Times (1952) The Southern Tree Surgery Company advertisement. *West Sussex Country Times*, 10 October, 2.
West Sussex County Times (1958) Comment by the Editor: Trees. *West Sussex Country Times*, 5 December, 8.
West Sussex Gazette (1949a) Rumsey, Tree Surgeon, Littlehampton, advertisement. *The West Sussex Gazette*, 28 April, 6.
West Sussex Gazette (1949b) Classified advertisement for tree workers by Tree Specialist Co., Littlehampton. *The West Sussex Gazette*, 25 August, 10.
West Sussex Gazette (1955) Advertisement by Borough of Sutton and Cheam for Forestry and Street Tree Foreman. *The West Sussex Gazette*, 26 May, 10.
West Sussex Gazette (1957a) The Tree Preservation Company advertisement. *The West Sussex Gazette*, 8 August, 1.
West Sussex Gazette (1957b) Balcombe: Public Inquiry Considers Fate of Beech Trees. *The West Sussex Gazette*, 28 November, 2.
West Sussex Gazette (1958a) The Tree Specialist Company advertisement. *The West Sussex Gazette*, 20 February, 1.
West Sussex Gazette (1958b) Advertisement for Vickers and Co, Worthing. *The West Sussex Gazette*, 20 February, 1.
Whitehead, J. (1981) Prussik Loop Climbing. *Journal of Arboriculture* 7(5), 123–126.
Wilks, J.H. (1969) Girling vs Matthews. *Arboricultural Association Journal* 1(9), 251–252.
Wilks, J.H. (1972) *Trees of the British Isles in History and Legend*. Frederick Muller: London.
Williams, P. (1972) Instant Trees – Large or Small? *Arboricultural Association Journal* 2(2), 51–53.
Wilson, K.W. (1981) *Removal of Tree Stumps*. Arboricultural Leaflet 7. Prepared for the Department of the Environment by the Forestry Commission. HMSO: London.
Winning, A.L. (1973) Arboriculture: The Emerging Specialism. *Arboricultural Association Journal* 2(5), 148–160.
Wolftree, F. (1995) Turning Over a New Leaf. *Horticulture Week*, 5 October, 18–21.
Woodland, D. and Dallimore, D.C. (1958) Transplanting Mature Trees. *The Gardeners' Chronicle* 144, 364–365.
Yorkshire Post (1948) Advertisement by Borough of Harrogate, appointment of Forester. *Yorkshire Evening Post*, 13 January, 2.

Index

Numbers in italics denote pages with illustrations
Tree species are not listed in the index